The Life of Stars

Giora Shaviv

The Life of Stars

The Controversial Inception and Emergence of the Theory of Stellar Structure

MAGNES

Springer

Prof. Giora Shaviv
Technion-Israel Institute of Technology
Dept. Physics
Technion City
32 000 Haifa
Israel
gioras@physics.technion.ac.il

ISBN 978-3-642-02087-2 e-ISBN 978-3-642-02088-9
DOI 10.1007/978-3-642-02088-9
Springer Heidelberg Dordrecht London New York

Library of Congress Control Number: 2009935999

Cover illustration: NASA/The Hubble Heritage Team (STScl/AURA/NASA)

Cover design: eStudio Calamar S.L.

Printed on acid-free paper

Springer is part of Springer Science+Business Media (www.springer.com)

Preface

It is the stars, The stars above us, govern our conditions.

William Shakespeare, King Lear

A Few Words about What, Why and How

The structure of the stars in general, and the Sun in particular, has been the subject of extensive scientific research and debate for over a century. The discovery of quantum theory during the first half of the nineteenth century provided much of the theoretical background needed to understand the making of the stars and how they live off their energy source. Progress in the theory of stellar structure was made through extensive discussions and controversies between the giants of the fields, as well as brilliant discoveries by astronomers. In this book, we shall carefully expose the building of the theory of stellar structure and evolution, and explain how our understanding of the stars has emerged from this background of incessant debate.

About hundred years were required for astrophysics to answer the crucial questions: What is the energy source of the stars? How are the stars made? How do they evolve and eventually die? The answers to these questions have profound implications for astrophysics, physics, and biology, and the question of how we ourselves come to be here. While we already possess many of the answers, the theory of stellar structure is far from being complete, and there are many open questions, for example, concerning the mechanisms which trigger giant supernova explosions. Many internal hydrodynamic processes remain a mystery. Yet some global pictures can indeed be outlined, and this is what we shall attempt to do here.

Astrophysical systems, like the Sun, the solar system, the stars, and the galaxies are very complex, and they cannot be brought into the laboratory for extensive investigation, taken apart for examination, or perturbed to learn how they respond. Consequently, progress is far from trivial, and we shall witness much controversy and debate before a consistent picture and theory eventually emerge.

It is not unusual to hear non-scientific arguments along the following lines: the famous rabbi says 'so must it be', and then his statement is quoted as the authoritative answer and the reason. One may suppose that the origin of this kind of reasoning goes back to the Talmud, The Chapter of the Fathers, Pirkei Avot: *Rabban Gamaliel said: Provide yourself with a teacher and remove yourself from doubt ....*[1] In other words, follow the doctrine of some clever fellow. The continuation of this frequently cited phrase is: ... *and do not accustom yourself to give tithes by estimate.* So when you have difficulty estimating what is the suitable donation to the poor, ask the wise man. It does not mean that one should always, and on all matters, blindly adopt the pronouncements of some 'authority'. The quotations given in the present book are not meant to convince you that this or that great scientist believed such and such and for this reason you must also believe it. On the contrary, think for yourself and find your own reason to be convinced. Let us not forget that great scientists can make great errors. In the Eddington versus Jeans controversy about the energy source of the stars, Jeans was wrong and Eddington was right, but in the Chandrasekhar versus Eddington controversy about the structure of cooling, dying stars, Eddington was wrong and Chandrasekhar was right. And nobody in the history of astrophysics knew the stars better than Eddington!

Too frequently we witness something rather opposite, namely, a lack of proper credit. It is for scientists that we quote the following: *Rabban Yochanan ben Zakkai received the Torah from Hillel and from Shamai. He used to say: If you have learnt much Torah do not claim for yourself moral excellence, for to this end you were created.*[2]

The non-existence of a scientific answer or an explanation for a phenomenon, or indeed some controversy among scientists, are often held against science. This is a misconception. Our purpose in depicting the history of the theory of stellar structure and evolution is to show that heated debate and argument among scientists are a fundamental feature of the scientific arena. The discussions lead to sharper views and tests to validate or disprove the theory. Science progresses via discussion and argument. Scientists strive at objectivity. Yet science is not objective on the short timescale, but only in the long term. Human feelings, even hatred, play a significant role on this short timescale, but science is a long term self-correcting process.

The age of the Earth and the age of the Sun fix a timescale over which most of the important elements for life were formed. It is therefore pivotal to understand how the long life of the Sun gives rise to a long age for the Earth, which in turn provides ample time for the evolution of biology. The evolution of biology on the Earth and the age of the Sun are intimately bound together! So what determines the age of the Sun?

It is remarkable how fashion can dictate scientific thinking or bias. This is probably due to the way physicists are trained, which ignores alternative explanations raised in the past. As a consequence, the average physics student accepts the preaching and indoctrination of the day without questioning its validity. While the history of failed ideas is no substitute for what we may believe today to be the 'last word', there is much to be gained from an adequate exposition of how the 'final answer' was reached. We agree on this point with Bogdan Paczinski who said: *If less than half of your ideas are wrong, you are not trying hard enough.* Consequently, quota-

[1] Pirkei Avot, Chap. A, Mishna 16.

[2] Pirkei Avot, Chap. B, Mishna 9.

tions or excerpts, even from the greatest scientists, need not always reflect the truth, or be taken as 'God's will'.

With regard to methodology, we shall refer to papers published in the professional literature (even when the ideas were completely irrational), and refrain from quoting personal letters, notes, or rumors, as these can seldom be assured to be the final word, and they cannot be expected to commit their writers. For this reason, detailed references are given. Quotations from papers are given in italics. The problem of giving proper citations and credit is thousands of years old, as we find in the Talmud: *Says Rabi Elazar in the name of Rabi Hanina: He who repeats something said by another, in that person's name, brings salvation to the world.*[3]

Many of the contributors to the theory have been immortalized by having their names attributed to craters or mountain peaks and ridges on the moon. The letter m after the year of death of a scientist indicates that some feature on the moon has been named after him.

Some Scientific Remarks

Physical systems comprising several components are said to be bound if the separation of the components requires energy. This energy is called the binding energy. The system is stable as long as there is no state with lower energy into which it can descend. A nucleus is stable only to the extent that there is no state with lower energy into which it can decay. A nucleus is radioactive whenever there exists a lower energy state.

Stars are large bound macroscopic systems which lose energy continuously. Hence, stars gradually evolve into lower energy states. The evolution of stars is an incessant decrease in the (negative) binding energy. Periods in which the star has a particular energy source, like nuclear energy, are nothing but temporary halts in this incredible pumping of energy from the stellar gravitational field into the surrounding space. Biology develops during one of these temporary stops. The process continues until the state of lowest energy is reached. At this moment, the star stops evolving and can be pronounced dead.

Different stars reach different 'last stops'. What we describe in this book is an outline of this extraordinary life of the stars, and how it was discovered and debated.

Acknowledgments

I would like to thank many colleagues for discussions and arguments. In the search for old papers I got help from Jean-Pierre Lasota and the library at the Institut d'Astrophysique Fondamentale, and Peter Ulmschneider and Rainer Wehrse from the Department of Theoretical Astrophysics in Heidelberg. I made considerable use

[3] Talmud Bavli, Tractate Megillah, Chap. I.

of the libraries in the Department of Physics at Stanford University, at the Institute for Theoretical Physics, Heidelberg University, and at the Max Planck Institute for Medicine, Heidelberg. The library of the physics department helped me find some old papers. The library in the Hebrew University allowed me to maintain connections with various archives. The library at Toronto University was extremely helpful in finding and extracting old articles and books.

Special thanks to Bob Wagoner (Stanford) and Rainer Wehrse (Heidelberg) for many discussions, very helpful comments, and reflections about the evolution of science and its practice.

Israel Institute of Technology, Haifa *Giora Shaviv*
March 2009

Contents

Chapter 1
The Controversy about the Age of the Earth

The determination of the age of the Earth is part of the question of how the Earth was formed and how it evolved to what we see around us today. It is a question of what the basic processes were that shaped the Earth, and caused it to evolve and harbor life. The long time scale of the Earth's transformation is in proportion with the time scale for the evolution of the Sun and its energy source, and it is the time scale needed for the synthesis of the chemical elements. The synthesis of the elements and the energy of the stars are connected through nuclear fusion. So the slowness of nuclear fusion is one of the key issues here.

The ages of the Earth and the Sun are tied together. The Earth probably could not have formed before the Sun, and the Sun probably could not have formed much before the Earth. According to present day ideas, the Sun and the Earth were formed roughly together. Hence, determining the age of one puts constraints on the age of the other. Today, we can measure the age of the Earth quite accurately, and thereby impose a very strict constraint on the age of the Sun. There are many inferred constraints on the age of the Sun, but this one is the most accurate. Hence, the importance attached to the determination of the age of the Earth.

1.1 The Pre-Scientific Era

With regard to the determination of the age of the Earth, the pre-scientific era lasted up until about AD 1700, and was characterized by a biblical type of calculation, or totally speculative and unfounded estimates claiming to have some scientific basis. People were sitting in their armchairs pondering about the Universe, but collecting no data, and not even attempting to calculate the age in a sensible and consistent way. Examples abound. Probably the most famous of all was the estimate by James Ussher (1581–1656), who was the Archbishop of Armagh in Ireland. In 1640, on the basis of the Bible, he 'calculated' that the Earth was exactly (in 1640) 5 644 years

G. Shaviv, *The Life of Stars*, DOI 10.1007/978-3-642-02088-9_1,

old![1] In this respect, it is interesting to learn the attitude of the greatest Jewish rabbi Moshe Ben Maimon (1135–1204), known as Maimonides, who claimed well within the pre-scientific era that it was wrong to read Genesis literally. Maimonides argued that one has to understand the Bible in a way that is compatible with the findings of science. Indeed, in his writings, Maimonides said that, if science and the Bible were in conflict, it was either because science was not understood or because the Bible had been misinterpreted. Maimonides reasoned that, if science proved a point, then the finding should be accepted, and the Holy Scriptures should be interpreted accordingly.

If you simply read the Bible literally without any sophisticated interpretation, and you count the years and days since the creation of Adam, you get some 5700 or more years. So when did the Universe start? On a particular day? Does it make sense? The story of the Six Days of Genesis, which caused people so many headaches in trying to understand science vis-a-vis the Bible, is confined to 31 sentences. That is all there is to go on. Can this be compared with present day data? Following Maimonides, it should not. We have to take the biblical description as a poem, or as an allegory, with its unsurpassed implications for humankind, and not as some scientific article.

It gradually became clear, even in those days, that the biblically based age of the Earth was too short a time to explain the present status of the Earth. So to overcome this short and implausible age, Catastrophism was invented. Catastrophism says that the Earth was created by a sequence of violent events which could accelerate its formation, and not via slow evolution (what the astronomers call secular evolution).

Catastrophism was first proposed by Baron George Cuvier (1769–1832m),[2] a French comparative anatomist by profession. His studies in comparative anatomy allowed him to draw conclusions about one part of an organism from other parts of the same organism. Cuvier extended the classification scheme of Linnaeus (1707–1778),[3] by grouping related classes into phyla. The important work in this connection is his extension of the classification system to fossils, which he recognized correctly as the remains of animals now extinct. For this reason, he is frequently declared to be the father of paleontology. Cuvier believed that *animals have certain fixed and natural characters*, and therefore rejected both the theory of evolution and Lamarck's (1744–1829m)[4] theory of inheritance of acquired characteristics. He proposed that life was created anew after periodic advances and retreats of the sea, what we would call today mass extinctions of life. Cuvier believed that the age of the Earth was indeed 6 000 years and that only catastrophic events changed its structure. Cuvier's Catastrophism conforms to the biblical thinking about the age of the Earth. According to Cuvier's catastrophe model, the changes seen within fossilized

---

[1] In fact, it was created at 8 pm sharp on 23 October −4004!

[2] Cuvier, G., *Tableau Elémentaire de l'Histoire Naturelle des Animaux*, Baudouin, Paris, 1798.

[3] Linnaeus, C., *Systema naturae per regna tria naturae, secundum classes, ordines, genera, species, cum characteribus, differentiis, synonymis, locis, Tomus I.* Editio decima, reformata, Holmiae, 1758.

[4] Lamarck, J.-B., *Histoire naturelle des animaux sans vertèbres*, vol. 3, *Radiaires, vers, crustacés, insectes*, Verdiere, Paris, 1816.

bones were a result of a previous catastrophic change, when an entire former and less developed species was wiped out in order to give rise to a new species. But no reason was given as to why the newly created creatures were more advanced than those that went extinct in the catastrophes, or why there is evolution towards more advanced forms of life.

1.2 Charles Lyell and Jean Fourier

Charles Lyell (1797–1875m)[5] was a Scottish geologist who wrote a masterful treatise called *Principles of Geology: Being an Attempt to Explain the Former Changes of the Earth's Surface by Reference to Causes Now in Operation* (see Fig. 1.1). The

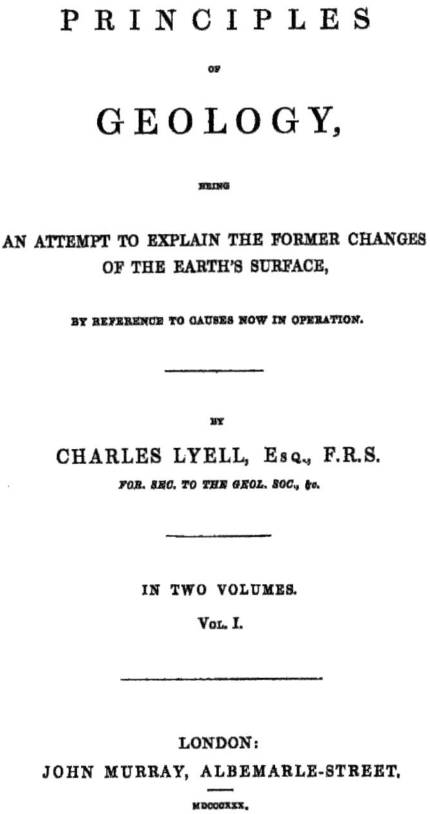

PRINCIPLES

OF

GEOLOGY,

BEING

AN ATTEMPT TO EXPLAIN THE FORMER CHANGES
OF THE EARTH'S SURFACE,

BY REFERENCE TO CAUSES NOW IN OPERATION.

BY

CHARLES LYELL, Esq., F.R.S.

FOR. SEC. TO THE GEOL. SOC., &c.

IN TWO VOLUMES.

VOL. I.

LONDON:
JOHN MURRAY, ALBEMARLE-STREET.

MDCCCXXX.

Fig. 1.1 Lyell's masterpiece

[5] Lyell, C., *Principles of Geology: Being an Attempt to Explain the Former Changes of the Earth's Surface by Reference to Causes Now in Operation*, John Murray, London, 1830.

reason for the very long name was that it contained Lyell's basic message, namely, that the Earth has evolved gradually over a long time. The monograph, published in 1833, was very popular and in such high demand that the ensuing 40 years saw 11 editions come out of the printing shop. The impact of Lyell's book was so profound that the notion of uniformitarianism, i.e., the idea that the Earth developed gradually, won general acceptance by the scientific establishment. However, the book itself did not contain any explicit calculation of the age of rocks or the Earth, but presented a huge number of geological phenomena from receding water falls to wind erosion, mountain formation, and much more, all of which indicated a gradual change in the texture of the Earth, in stark contrast to a theory of sudden (catastrophic) change.

The idea that the Earth formed hot and has been cooling ever since its formation was not new to Lyell. A few years earlier, in 1824 Jean Fourier (1768–1830m) had published[6] a memoir entitled *On the Temperature of the Terrestrial Globe and Planetary Spaces*, in which he calculated the cooling time of the Earth. The initial assumption was that the Earth formed at a high temperature, so high that it was molten, and that it had been gradually cooling since then. As the discoverer in 1807 of what is known today as the heat equation, i.e., the equation which describes the flow of heat from hot to cold objects, Fourier was the first to be able to carry out such a calculation. Among the first problems he took up was the cooling of the Earth. As a matter of fact, Fourier invented a new type of mathematical tool, known today as the Fourier series, which was such a breakthrough that it took the great scientists of the French Academy 13 years to digest before accepting it for publication!

Fourier was a very special character among physicists and mathematicians, with a particularly broad horizon of interests. Napoleon made him the governor of Lower Egypt, and after Napoleon's retreat from Egypt he was made the Prefect of the Isère, a region in the east of France. However, he did not always follow Napoleon, and when he realized the troubles Napoleon had brought upon the people of France, he turned his back on him, and paid dearly for it. Fourier knew how to combine the two extremes: politics and mathematics. Few were later willing or able to follow suit.

It was during his stay in Grenoble as the Prefect of the Isère that Fourier made his most important mathematical discovery on the theory of heat propagation in solids. The emphasis is on the word 'solid', in contrast to liquids or gases. Heat transport in liquids and gases is mainly via mass transfer, a phenomenon that cannot occur in solids. The questions were: What is heat and what is it that moves in solids? In those days the caloric theory of heat prevailed. Caloric was supposed to be a fluid, which allegedly penetrated through all matter and carried heat. It is quite astonishing how Fourier was able to reach his results, working in complete scientific isolation in the Prefect's building, where he had his apartments, and how he conducted experiments there to confirm his theory. Moreover, it is equally surprising that Fourier was able to derive the correct heat equation knowing only about the imaginary caloric. In fact, he succeeded in deriving the heat equation using the wrong model for heat propagation.

[6] Fourier, J., Ann. de Chimie et Phys. Tome XXVii p. 136, October 1824.

THÉORIE

DU

MOUVEMENT DE LA CHALEUR

DANS LES CORPS SOLIDES.

Mémoires de l'Académie Royale des Sciences de l'Institut de France, années 1821 et 1822, t. V, p. 153 à 246; 1826. Imprimerie Royale.

Fig. 1.2 Fourier's breakthrough manuscript. Note the wording of the title: 'movement', not propagation or transport

He worked for three years, and in 1804 completed his memoir entitled *On the Movement of Heat in Solid Bodies* (see Fig. 1.2). The memoir was read on 21 December 1807 at the Paris Institute before a committee consisting of some of the most prestigious mathematicians of the day, such as Lagrange (1736–1813m), who had also supervised his PhD thesis, Laplace (1749–1827m), Monge (1746–1818m), and Lacroix (1765–1843m). While the work was treated with great respect, it was also criticized. There was a basic problem which the committee had some difficulties in accepting: can a function with a discontinuous slope, like a square wave, be expressed as an infinite sum of functions which all have continuous slopes?[7] This is not a trivial question. Fourier could not convince the committee that his work was correct. Even the greatest mathematicians of the time had difficulties in understanding him, so profound and revolutionary was the work! The problem the great mathematicians faced was not a simple one, however, and a detailed discussion would carry us too far away from our subject.

Another lesser problem was Biot's objection to the derivation of the heat equation. Fourier did not cite, and for a good reason, Biot's (1774–1862m) flawed work of 1804, but this was not a strictly scientific matter!

The problem of the heat equation became a hot topic, and the Paris Institute announced in 1811 a competition to solve the problem of heat transfer in solid bodies. Two entries were submitted to the competition, one of them was Fourier's work from 1807. Fourier's submission also included research on the cooling of infinite solids and terrestrial radiant heat. A committee composed of Lagrange, Laplace, Malus (1778–1812), and Legendre (1752–1833m) was set to determine the winner and de-

[7] In mathematical terms, can a function which is discontinuous or has discontinuous derivatives be represented by an infinite series of functions which are continuous and have only continuous derivatives?

cided to award the prize to Fourier. However, ... *the derivation is not sufficiently rigorous*, declared the committee, who subsequently prevented publication.

Fourier was elected to the French Academy of Sciences in 1817. Five years later, the secretary Delambre (1749–1822m) of the Academy passed away, and Fourier, Arago (1786–1853m), and Biot competed for his job. Fourier won. It should, however, be noted that Delambre, a famous mathematician, known for his application of mathematics to astronomy, decided before his death to publish Fourier's memoir. And so, after 13 years, this seminal work was finally accepted for publication.[8] Yet it was still years before it won proper recognition. Among the most outspoken against Fourier were Laplace, Biot, and Arago. Here is yet another stunning example of how long it could sometimes take for work that revolutionized engineering and physics to be accepted by the scientific community.

Fourier's research on the cooling of the Earth was extremely profound, in particular when we recall how early on it was carried out. Fourier discovered that gases in the atmosphere might increase the surface temperature of the Earth in the same way as 'human industry', and these were the early days of the Industrial Revolution! This was the effect that would centuries later be called the greenhouse effect. Once Fourier had written down the heat equation, he was able to establish the concept of an energy balance for the Earth and planets, and this before the laws of thermodynamics had been formulated, or even before the conservation of the total energy was known!

In establishing the energy balance of a planet, Fourier discovered that planets reflect part of the solar light (what we call albedo today), as well as losing energy by infrared radiation (which Fourier called *chaleur obscure* or mysterious heat), with a rate that increased with temperature, although he did not know the exact law. Therefore, he concluded correctly, there must be some temperature at which a balance is reached between energy gains and losses.[9] He realized that the atmosphere shifts the balance point to higher temperatures due to absorption of radiation. He also recognized that the Earth's atmosphere is transparent to solar visible light and, most importantly, that the Earth's internal heat source does not contribute much to the Earth's global energy balance. The physical evidence he provided to support this was that such heat is not felt over most of the surface of the Earth. However, he incorrectly believed that there is a significant contribution of radiation from interplanetary space.

It is amazing in retrospect that Fourier guessed the existence of *chaleur obscure* or 'dark heat' emitted by the Earth, which was essentially the infrared radiation dis-

[8] Fourier, J.B.J., *Théorie Analytique de la Chaleur*, Didot, Paris, 1922.

[9] On the one hand the Earth is heated by the Sun, and on the other the Earth loses energy by radiating into space. The radiation into space depends on the temperature in such a way that, when the temperature rises, the emission of energy into space increases. If this were not the case, i.e., if the cooling of the Earth by radiation into space did not increase with temperature, the temperature on the Earth would increase indefinitely due to solar heating. So Fourier realized that the fact that the temperature of the Earth is stable means that the cooling (by means of radiation) increases with temperature sufficiently fast to stabilize the temperature of our planet.

covered by Herschel (1738–1822m).[10] Fourier understood that the rate of infrared radiation emission increases with temperature, but the exact form of this dependence, known today as the Stefan–Boltzmann law (fourth-power law), was only discovered 50 years later. After all these dramatic discoveries about the energy balance of the Earth, correct scientific facts even today, Fourier did not publish his calculated age for the cooling of the Earth, although he had all the ingredients necessary for such a calculation. The reason was given at the end of his pioneering paper:

> *We do not know how much the interior of the Earth has lost of its original heat; one can only state that at the surface, the excess of heat due to this sole cause has become insensible; the thermometric state of the globe does not vary anymore except with extreme slowness; and if one could conceive that below a depth of a few leagues one replaced the interior masses with a frozen body [...] it would take a great number of centuries before one could observe any appreciable change in the temperature of the surface. The mathematical theory of heat furnishes many other consequences of this type whose certitude is independent of all hypotheses of the interior state of the Earth.*

In summary, Fourier realized that despite the known increase in temperature with depth, it is practically impossible to calculate the age of the Earth from the hypothesis that the Earth cools.

Are there any observable signs that the Earth cools?[11] In 1785, James Hutton (1726–1797m), considered by many as the father of geology, published his *Theory of the Earth*.[12] In this book he demonstrates that Hadrian's Wall, built by the Romans, did not show any detectable changes after 1500 years. He thus suspected that the Earth was much older than 6000 years. We should mention that Pierre-Simon Laplace,[13] a mathematician, physicist, and astronomer, and a dominant figure in French science of his time, had already demonstrated by reference to astronomical observations on the circumference of the Earth carried out by Hipparchus,[14] that there had been no noticeable contraction of the Earth over the previous two thousand years. Lyell concluded in his *Principles of Geology* that the state of the Earth remains unchanged, and hypothesized that the reason may be some unknown heat source in the molten lava beneath the crust of the Earth, which keeps the Earth hot.

Remarkably, both Fourier and Lyell were right. Lyell was right in hypothesizing that the Earth's internal energy source is the cause for the perpetual changes of the Earth's surface, and Fourier was right in stating that the effect of the heat source inside the Earth on the surface temperature is minimal. Lyell meant the heat source which changes geological features like mountains, while Fourier meant the heat

---

[10] Herschel, W., RSPS **1**, 20 (1800). Herschel referred to the infrared radiation as calorific rays.

[11] Normal bodies expand upon heating and contract upon cooling. So the logic was that if the Earth cools one should be able to discover a contraction in the size of the Earth.

[12] Hutton, J., *Theory of the Earth*, Royal Society Edinburgh, Edinburgh, 1788.

[13] Laplace did not get a crater on the Moon, but a cape on the northeast edge of Sinus Iridum (the Bay of Rainbows), called Promontorium Laplace. See Sect. 5.39.

[14] The first measurements of the radius of the Earth were carried by Erastothenes (276–194 BC) and the result was a few percent off the present most accurate value. More accurate results were obtained by Hipparchus (190–120 BC). For this reason, Laplace based his argument on Hipparchus' results.

balance of the Earth. Part of the problem, as we know today, is that the Earth is far from being uniform, and lava, for example, comes out only in a few places like volcanos, while heat is not felt over most of the surface of the Earth. A second point concerns the time scale. Lyell's time scale for geological changes was much longer than Fourier's time scale for heat flow and climate change.

On the basis of fossils, Lyell estimated (in a different publication) that the age of the Earth was about 240 000 years.[15] We know today that this is a minimal age, as the fossils were formed rather late in the evolution of the Earth. So the time that elapsed between the formation of the Earth and the time when the first fossil formed is not accounted for.

While Fourier's heat equation became a pillar of science, the controversy about Fourier's assumptions and results on the cooling of the Earth did not subside for many years. Alexander von Humboldt (1769–1859m), a German who combined botany and zoology with writing about art and sociology, vacillated between the assumption of a fluid core and the assumption of a solid core.[16] If the matter in the core is in a liquid state, cooling by currents is faster than cooling by conduction through solid rocks. The English chemist Humphry Davy (1778–1829m)[17] and the French physicist André Marie Ampere (1775–1836)[18] suggested that the heat of the Earth was due to chemical reactions in the core of the Earth. The French mathematician Simeon Denis Poisson (1781–1840m) hypothesized that, as the Earth cooled by radiation into space, solidification would have started from the surface inward. Hence, the inner part would still be in the liquid state. As heat is carried relatively well by means of currents in the liquid phase, the temperature would not rise to extreme values in the core. Heat transfer by currents is so much more effective than cooling by heat propagation through solids that the temperature gradient needed (in liquids) could be much smaller. As for the source of the heat, Poisson hypothesized that it was due to a passage of the Solar System through hotter stellar regions in the galaxy at sometime in the past.

We shall now direct our attention to the age of the Sun. It is plausible to assume that this is greater than the age of the Earth. And if we find that the age of the Sun is actually less then the age of the Earth, then it is likely that we have stumbled upon a contradiction.

---

[15] It is worth mentioning that Leonardo da Vinci (1452–1519) already suspected from the fossils he had observed that the Earth was much older than what the Bible described.

[16] Humboldt, A, von, *Kosmos. A General Survey of the Physical Phenomena of the Universe*, Hipolyte Bailliere, London, 1845. It is in this book that the Island Universe hypothesis was first suggested.

[17] Davy, H., Phil. Trans., Roy. Soc. London **105**, 214 (1815). The 1805 lectures for the general audience, edited by Siegfried & Dott, Univ. Wisconsin Press, 1980.

[18] Ampere, A.M., *Théorie des phénomènes électro-dynamique*, Paris, 1826.

1.3 Energy Conservation: Helmholtz and Mayer

Hermann von Helmholtz (1821–1894m) was the first physicist to provide an apparently bona fide physical solution to the problem of the Sun's energy source. Helmholtz was in a position to do so because he discovered the law of conservation of energy, and the problem of extracting energy from the gravitational energy of the Sun required just such a conservation law.

The idea of a conservation law and the existence of conserved quantities was not new. As early as 1668, John Wallis (1616–1703) had suggested that momentum might be conserved. Gottfried Liebniz (1646–1716m) suggested the law of conservation of mechanical energy (potential plus kinetic energy).[19] Antoine Lavoisier (1743–1794m), who is often referred to as the father of modern chemistry, was the first to formulate clearly and unambiguously the law of mass conservation. So the idea that certain quantities might be conserved was not new. However, combining such different entities as heat, mechanical work, and radiation, for example, into one conservation law, was most definitely new. All forms of energy are conserved together, even if the energy transforms from one form to another. These were the days before Rudolf Clausius (1822–1888m)[20] proved in 1850 that heat is associated with the kinetic motion of molecules,[21] and many years after Rumford (1753–1814m) had shown in 1798 that the 'supposed carrier of heat, the caloric' could not be conserved. We recall how in those days physicists and chemists invented 'imponderable fluids' to solve problems: the caloric for heat, the phlogiston for burning, and the ether for light. The first of these was overthrown by Rumford, the second by Lavoisier, and the third by the Michelson–Morley experiment and Albert Einstein (1879–1955m) (in the special theory of relativity).

But despite Rumford's demonstrations, there were scientists who kept thinking in terms of the caloric theory, most notably Carnot (1796–1832m),[22] who discovered the Carnot cycle and laid the grounds for the second law of thermodynamics,[23] although the first law (the conservation of energy) was not yet known. One can, however, 'translate' Carnot's arguments into present day thermodynamics, if one

---

[19] See, Iltis, C., Isis **62**, 21 (1971).

[20] A summary of Clausius' ideas can be found in Clausius, R., *Die mechanische Wärmetheorie*, Vieweg, 1867.

[21] Clausius realized that heat is nothing but the random (casual) motion of the molecules. Consider a gas in a container at rest. If we could make a huge microscope and observe the gas molecules, we would see them moving in all directions in a completely chaotic way. The random speed of the molecules means that they have kinetic energy of motion. According to Clausius, the mean kinetic energy of the molecules is proportional to the temperature. As the container is at rest, the average velocity of all the molecules at any given time is zero, but the kinetic energy, which depends on the square of the velocity, does not vanish. Further, the long-time average of the velocity of any molecule vanishes. When the temperature rises, so does the mean kinetic energy of the molecules in the gas.

[22] Carnot, S., Annal. Sci. de l'Ecole Normale Supérieure, Ser. 2, 1872.

[23] The first law of thermodynamics states the conservation of energy. The second law of thermodynamics states that it is impossible to convert heat into mechanical energy with 100% efficiency, but that it is possible to convert mechanical energy into heat with 100% efficiency.

replaces 'caloric' by 'entropy'. Then one has the statement that the entropy increases or does not change in any process. (The concept of entropy as we know it today is due to Clausius, in 1865.)

As a matter of fact, Helmholtz was not the first to discover the law of energy conservation. The discoverer was in fact Robert Mayer (1814–1878), a German surgeon on a Dutch vessel sailing in the tropics. It was during therapeutic removal of blood that he recognized that the venous blood of the Europeans was pale red and looked like oxygenated arterial blood. It was a known phenomenon, but never before explained. Mayer supposed that the oxygen was needed to power the muscles and keep the body warm. However, the hot weather of the tropics requires a lower metabolic rate to maintain the body temperature, and hence needs less oxygen than in a colder climate. He concluded therefore that the human body needs less oxygen in tropical than in temperate zones to maintain the body at a constant temperature. Moreover, he suggested that the heat produced during muscular effort must also be derived from the chemical energy stored in food. The input of energy in the food and the output of 'force' must balance. (In those days, what we call today energy was called force.) From this, he drew the surprising conclusion that motion and heat are *different manifestations of one and the same energy*. So they *must permute, and transform into one another*. This conclusion, coming as it did out of the blue, was not easily accepted by the scientific establishment.

Upon his return to Germany, and having set up a private practice, Mayer summarized his ideas in a paper and sent them to Johann Christian Poggendorff (1796–1877), the editor of the Annalen der Physik und Chemie. In this paper, he postulated *the conservation of force*, which today we call the law of conservation of energy. However, probably owing to Mayer's lack of advanced training in physics, it contained some fundamental errors, and the paper was rejected by Poggendorff. Mayer did not give up. He began to study physics, and debated the issue with the Tübingen physics professor Johann Gottlieb Nüremberg, who naturally rejected his claim. In exchange, Mayer got some ideas about how to prove his point experimentally. Subsequently, he not only demonstrated how mechanical energy converts into heat, but also measured the conversion factor[24] of the transformation, namely, the mechanical equivalent of heat. The result of his investigations was published in 1842 in Justus von Liebig's Annalen der Chemie und Pharmacie.[25] Liebig (1803–1883m), one of the greatest chemists of the 19th century, recognized the importance of Mayer's discovery. Three years later, Mayer published the book: *The Organic Movement in Connection with the Metabolism*, in which the numerical value of the conversion factor was given, a value which deviates by only about 10% from today's value.

[24] Different units are used to measure different energies like heat, mechanical energy, electricity, and so on, because the various forms of energy were discovered independently, and without any apparent connection. Hence, one needs to find the relation between the different units, which is called the conversion factor. Once the total energy is conserved, the relation between the units becomes important.

[25] Mayer, R.J., *Bemerkungen ueber die Kraefte der unbelebten Natur*, Annalen der Chemie und Pharmacie **43**, 233 (1842).

One of the most versatile scientists who ever lived, Hermann von Helmholtz was born in Potsdam, Germany, son of a high-school teacher. Helmholtz was attracted to physics from an early age, but his family did not have the financial means to let him study physics. Instead, his father persuaded him to take up medicine, since the education of physicians was supported by the state, provided they served for several years as doctors in the Prussian army. Helmholtz attended the Institute for Medicine and Surgery in Berlin from 1838 to 1842, and served as an army surgeon from 1843 to 1848. But, his soul was always in research. Even in the army barracks, he set up a small laboratory for research in physiology and physics, the subjects he loved.

On 23 July 1847, Helmholtz presented a paper on the conservation of energy at a meeting of the Berlin Physical Society. It was a talk at a reasonably high level of mathematical sophistication, intended to convince physicists that energy is conserved in any closed[26] physical system (strictly speaking, he spoke about physical processes). The paper was submitted to Poggendorff, and was rejected as being too long and too mathematical for his readers. So Helmholtz published the results in a pamphlet that soon became recognized as one of the most important papers in physics. Poggendorff had the dubious honor of rejecting the first two independent papers containing the discovery of energy conservation.

Helmholtz's bold and ground-breaking paper, written when he was only twenty-six years old, was his first, and most fundamental statement of the principle of conservation of energy. It came at a critical moment in the history of science, when scientists and philosophers were arguing about whether conservation of energy was a truly universal principle. In his 1847 book, Helmholtz showed convincingly that the conservation of energy is indeed universally valid. Note that the full detail of the energy conservation law was published by Helmholtz in the book *Über die Erhaltung der Kraft* (On the Conservation of Force) in 1847,[27] and not in a refereed journal. It is interesting that one of the most important laws in physics was discovered by two trained physicians, not physicists.[28]

Helmholtz showed that the assumption that work cannot be produced from nothing leads to the conservation of kinetic energy. He then applied this principle to a variety of different situations. He demonstrated that, in various situations where energy appeared to be lost, it was in fact converted into heat (which is just another form of energy). This happens in collisions, expanding gases, muscle contraction, and other situations. The book looked at a broad range of applications including electrostatics, galvanic phenomena, and electrodynamics. Today we may say that Helmholtz proved the non-existence of perpetuum mobile of the first kind. A perpetuum mobile of the first kind is a machine which produces more energy than it uses, thus violating the law of conservation of energy. A perpetuum mobile of the

---

[26] A closed physical system is one which does not have any interaction with the outside, so that nothing like energy, momentum, or mass is exchanged between the 'closed system' and 'the rest of the world'.

[27] Helmholtz, H. von, *Über die Erhaltung der Kraft*, Leipzig, Engelmann, 1847.

[28] *On the Conservation of Force, Introduction to a Series of Lectures Delivered at Karlsruhe in the Winter of 1862–1863 by Herman von Helmholtz*, The Harvard Classics, 1909–14, translated by E. Atkinson.

second kind is a machine which operates forever by converting its waste heat back into mechanical work. Such a machine does not violate the law of conservation of energy (which is the first law of thermodynamics), but does violate the second law of thermodynamics, discovered earlier by Carnot.

During the same years that Mayer and Helmholtz were active in Germany, British scientists were also busy understanding the connection between heat and mechanical energy. The dominant player was James Prescott Joule (1818–1889m). In 1845,[29] he discovered what is known today as Joule's law, namely, the connection between an electric current and the heat it produces. The discovery was presented to the Royal Society, but was not highly regarded, probably because in those days (and until 1854) Joule managed the brewery he inherited from his parents and did not have the auspices of a university or research institute. Two years later he found the numerical equivalent between the electric current and the heat generated, and in 1845 he did away with the electric current and experimented with the conversion of mechanical energy to heat and vice versa. In this way he discovered what is known today as Joule's constant, namely, the conversion factor between mechanical energy and heat.[30]

Objectors to the concept of irreversibility, which Joule's experiments implied, raised the thermoelectric effect, the direct conversion of temperature differences to electric power by building a voltage difference, as a counterexample. Reversibility means that one type of energy can be converted to another type and back to the original form. Irreversibility means that, if mechanical energy, for example, is converted into heat, the heat cannot be converted back into mechanical energy without losses. This was exactly what Carnot had discovered. The thermoelectric effect is reversible, while Joule heating, the conversion of mechanical energy to heat, is not. The current flowing between the two edges of a metal object held at a temperature difference (Thomson heat) is indeed reversible, and so is the heat released or absorbed when a current flows. Reversibility holds in the sense that, when the current is reversed, the effect remains the same, but changes sign. But the processes of heat conduction in wires and heat dissipation in an electrical resistance are not reversible. Also in 1847, one of Joule's presentations at the British Association in Oxford was attended by distinguished figures like George Gabriel Stokes, Michael Faraday, and William Thomson. Stokes was inclined to believe Joule, Faraday was struck by it, although he had some doubts, and Thomson, who would later be known as Lord Kelvin, was intrigued but skeptical.

Although the initial attitude of Thomson was that Joule's results demanded a theoretical explanation, he gradually began to feel that Joule might be right. In his 1848 paper, in which he established the existence of an absolute temperature, known today as the Kelvin scale, Thomson nevertheless wrote: *The conversion of heat (or caloric) into mechanical effect is probably impossible, certainly undiscovered.* But in a footnote, he raised his first doubts about the caloric theory, referring to Joule's *very remarkable discoveries.* The use of the term 'caloric' may explain the difficulty

---

[29] Joule, J.P., Phil. Mag. **27**, 205 (1845).

[30] Joule, J.P., *On The Mechanical Equivalent of Heat*, Abstract of papers communicated to the Roy. Soc. London **5**, 839 (1843–1850).

Kelvin had with the question: what happens to the caloric upon conversion of heat into mechanical energy? Surprisingly, Thomson did not send Joule a copy of his paper, but when Joule eventually read it, he wrote to Thomson, claiming that his studies had demonstrated the conversion of heat into work, and that he was planning further experiments. Thomson replied, revealing that he was planning his own experiments and hoping for a reconciliation of their two views. Although Thomson conducted no new experiments, over the next two years, he became increasingly dissatisfied with Carnot's caloric theory, and more convinced of Joule's claims. In his 1851 paper,[31] Thomson was willing to go no further than a compromise, and declared: *The whole theory of the motive power of heat is founded on [...] two [...] propositions, due respectively to Joule, and to Carnot and Clausius.*

The correspondence between Joule and Thomson gave birth to a fruitful collaboration. Joule conducted the experiments and Thomson analyzed the results and suggested further experiments. The collaboration extended from 1852 to 1856 and culminated in the discovery of the Joule–Thomson effect. The collaboration with the highly esteemed Thomson helped to bring a general acceptance of Joule's work and the kinetic theory of gases.

Before we leave the subject of energy conservation, it is appropriate to mention Colding (1815–1888), a Danish engineer who put forward the idea that energy is not lost, but merely transformed into another form. He even carried out experiments to prove his thesis.[32] Apparently, his papers were known to Helmholtz, but his contribution was not recognized by the scientific community, probably because he considered the forces of nature as spiritual and immaterial entities, something physicists understandably dislike.

Let us also mention William R. Grove (1811–1896), who wrote about himself (in 1867) that: *I believe myself to have been the first who introduced this subject as a generalized system.*[33] However, Grove, who hated high-brow mathematics, did not properly define what he meant by 'force' to distinguish it from 'energy'. Consequently, it was regarded as popular rather than rigorous science, and hence not worthy of reference.

1.4 The Source of Solar Energy

The first genuine attempt to solve the problem of the Sun's energy source is due to Helmholtz. On 7 February 1854, in a popular address delivered in Konigsberg on

---

[31] Thomson, W., *On the Dynamical Theory of Heat, with numerical results deduced from Mr Joule's equivalent of a thermal unit and M. Regnault's observations of steam*, Trans. Roy. Soc. Edinburgh, March 1851 and Phil. Mag. IV, 1852.

[32] Colding published seven papers between 1843 and 1860, all on the same subject. See also, Dahl, P.F., *Ludvig Colding and the Conservation of Energy Principle*, The Sources of Science, No. 4, 1972.

[33] Grove, W.R., *The Correlation of Physical Forces*, 1st edn., Longmans, Green, London, 1846.

the occasion of a commemoration of Kant,[34] he suggested that the Sun was initially made of small pieces of rock, or even dust-like particles, that were spread out in space. Helmholtz was influenced by the 'nebular hypothesis' advanced by Kant and Laplace. It was thus appropriate that a new theory about the energy source of the Sun should be presented in honor of Kant.

The nebular theory, which was very popular at the time, assumed that the planets were formed from merging dust and gas in rotation around the Sun. Laplace was very much influenced by the shape of the ring nebulas (see Fig. 1.3). These are indeed nebulas in the shape of a ring, and were called planetary nebulas because of the hypothesis advocated by Laplace that they are the progenitors of planetary systems.[35] According to Helmholtz, the chunks of matter or gaseous meteors fell in toward what is now the Sun's position, releasing their huge gravitational energy upon colliding with the mass already present at the center, to form a very hot molten sphere of matter. The basic idea is therefore that matter which fell onto the Sun converted gravitational energy into heat, and subsequently released the heat into space in the form of radiation.

Helmholtz, as the discoverer of the conservation of energy, was the first to be able to realize that the accretion of mass (rocks) by the forming Sun would convert potential energy into kinetic energy, and subsequently into heat and radiation. The general process has two phases. In the first phase, the Sun heats up under the continuous rain of meteorites. In the second phase, the hot Sun cools by radiation from the surface. Assuming a heat capacity similar to that of water, this yields a temperature which, when divided by the rate at which the Sun loses energy by radiation, allows one to calculate an age. The available supply divided by the rate of spending gives the time for which the supply will last, i.e., the age. Since the heat capacity was not known, the calculation involved some uncertainty. The number Helmholtz got was a few million years. However, Helmholtz spoke only of the first phase.

The mass of meteors that should fall on the Sun in a year to supply the energy radiated away is about 6×10^{25} g or about 1% of the mass of the Earth per year. It is easy to realize that the accretion of such a mass by the Sun would affect the orbit of the Earth in a noticeable way[36] and change the length of the year by about a second per year. Such an effect would have been easily detectable even in those days. Clearly, if the meteors came from within the Earth orbit, no such change would be expected, and if they came from within the radius of Mercury, even Mercury would

[34] Von Helmholtz, H., *On the Interaction of Natural Forces*, Königsberg, 7 February 1854, Phil. Mag. **11** (series 4), 489 (1856).

[35] Today we know that the so-called planetary nebulas consist of the mass ejected by a dying star (see Chap. 7).

[36] The orbit of the Earth around the Sun is determined by the attractive force of the Sun. According to Newton's universal law of gravity, the gravitational attraction of the Sun is proportional to the mass of the Sun. If that mass increases, so therefore will the attraction of the Sun. The gravitational force is given by the product of the mass of the Sun by the mass of the Earth divided by the squared distance between the center of the Earth and the center of the Sun. So if the Sun changes its radius but not its mass, there is no change in the attraction of the Earth by the Sun. If, however, the mass falling on the Sun comes from outside the orbit of the Earth around the Sun, then the change in the mass of the Sun irrespective of its radius is important for the length of the year.

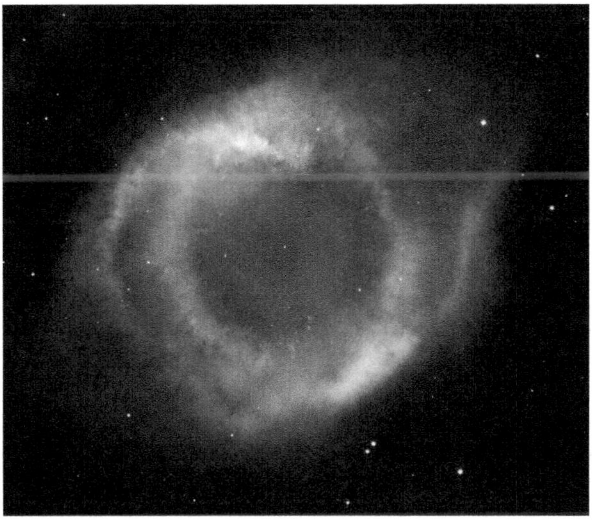

Fig. 1.3 The Helix planetary nebula. The distance of the nebula is about 650 lyrs, making it one of the nearest planetary nebulas. The diameter is about 5.1 lyrs. Credit: The Hubble Space Telescope

not feel the change in the gravitational force. But in the latter case the available energy is much smaller. These consequences of the hypothesis, which could confirm or contradict the thesis, could have been, but were not, discussed by Helmholtz. In principle, it was possible even then to see that the Sun could not be in the first phase.

A simple calculation illustrates right away that the source of the energy of the Sun must be something new and unique. Assume that the Sun generates its energy via chemical reactions, say the Sun is composed of coal. Then the total solar mass of coal would suffice for a few thousand years if the Sun always shone at the same power. Assume that the energy per molecule is 10 erg, which is an overestimate. There are $N_A M_\odot/\mu$ molecules in the Sun, where $N_A = 6.023 \times 10^{23}$ is the Avogadro number, $M_\odot = 2 \times 10^{33}$ g is the mass of the Sun, and μ is the molecular weight which we take as 30 (and that is a big overestimate).[37] The total luminosity of the Sun is $L_\odot = 3.8 \times 10^{33}$ erg/s. Hence, the predicted life of the Sun would be about 5 000 years if it were powered by chemical energy. This age estimate is even shorter than the rejected biblical time.

If falling objects convert potential energy into heat, how come people did not realize long before Helmholtz that water falling in a waterfall heats up when it hits the ground? The reason is very simple, the conversion factor from mechanical energy into heat is very small. For this reason, Helmholtz's creativity was needed. The water in a waterfall 100 meters high heats up by about 0.002°C, and this would be no easy matter to detect. All that Helmholtz's idea required was the conservation of energy, and it did not need any (numerical and unknown) conversion factors at all.

[37] The molecular weight of the sun is close to 2, as will become clear later.

Let us end with a somewhat upsetting note. In 1848, Mayer learnt about Joule's papers and wrote to the French Académie des Sciences to assert his priority. His letter was published in Les Comptes rendus de l'Académie des Sciences,[38] and Joule was quick to react. Thomson's close relationship with Joule allowed him to be dragged into the controversy. The two of them planned that Joule would admit Mayer's priority for the idea of the mechanical equivalent, but claim that experimental verification rested with Joule. Some of the greatest names in British science were drafted in to help Joule, like Rankine and Maxwell. But it did not help. On 18 May 1850, Mayer attempted suicide, and we can only guess that the controversy did not help his mental state.[39]

Several years later, in 1862, John Tyndall (1820–1893m),[40] who inherited the position of the great Faraday and was a successful scientist in his own right, continued Faraday's tradition of popular public talks and argued in a lecture entitled 'On Force' at the Royal Institution that Mayer was to be credited with conceiving and measuring the mechanical equivalent of heat. Thomson and his followers lost their temper and started an ugly campaign in the pages of the Philosophical Magazine. Historical and scientific justice prevailed, however, and Mayer's priority is now recognized. For more on the Mayer–Joule controversy, see Lloyd 1970.[41]

In 1905, the physicist Carl Barus (1856–1935) summarized for Science magazine[42] the major scientific achievements of the 19th century. The law of conservation of energy was not mentioned at all.

1.5 Charles Darwin

By the time Darwin (1809–1882m) published his Earth-shattering conjecture about the *Origin of the Species by Means of Natural Selection* in 1859[43] (see Fig. 1.4), the proponents of the biblical age of the Earth disappeared, only to resurface again in the second half of the 19th century, in the form of the creationist postulate for life and the panspermia hypothesis.[44] Is the Genesis–geology debate of any relevance today? One might have hoped that it would have become a thing of the past. But

[38] Mayer, R., Comptes rendus **27**, 385 (1848).

[39] An analogous story repeated itself in 1906, when Boltzmann, depressed by attacks on his work, though he was right and would eventually be victorious, took his own life.

[40] In 1861 John Tyndall discovered the role of water vapors and CO_2 as greenhouse gases in the Earth's atmosphere [Phil. Mag. **22**, 169, 173 (1861)].

[41] Lloyd, J.T., Notes and Records of the Royal Society of London **25**, 211 (1970).

[42] Barus, C., *The Progress of Physics in the Nineteenth Century*, Science **22**, 385 (1905).

[43] Darwin, C., *On the Origin of Species by Natural Selection*, Appelton, 1859. The fifth edition was published in 1872.

[44] Panspermia is the hypothesis that microorganisms which came to the Earth from outer space are responsible for originating life on Earth, and possibly in other parts of the universe, where suitable atmospheric conditions exist. The word comes from the Greek 'panspermi', meaning a mixture of all seeds.

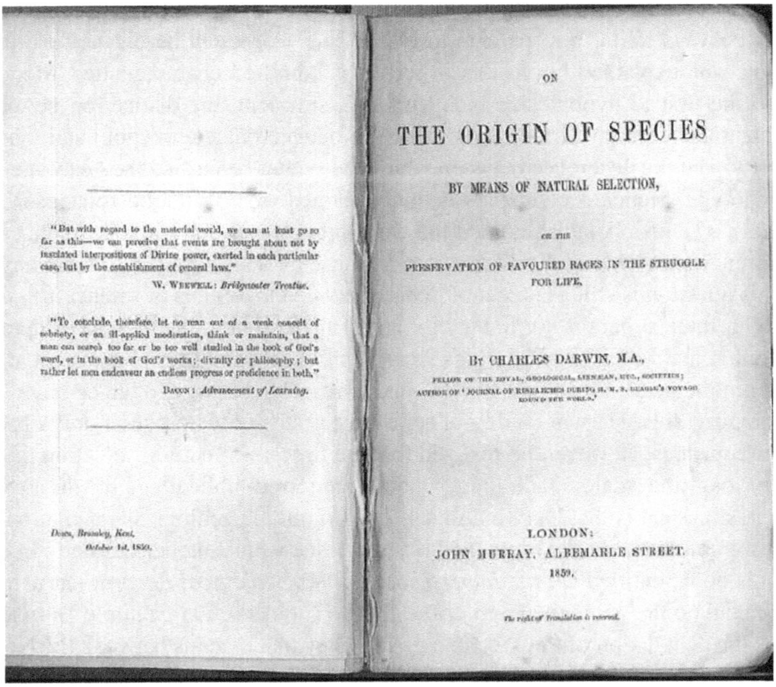
"But with regard to the material world, we can at least go so far as this—we can perceive that events are brought about not by insulated interpositions of Divine power, exerted in each particular case, but by the establishment of general laws."

W. WHEWELL: *Bridgewater Treatise.*

"To conclude, therefore, let no man out of a weak conceit of sobriety, or an ill-applied moderation, think or maintain, that a man can search too far or be too well studied in the book of God's word, or in the book of God's works; divinity or philosophy; but rather let men endeavour an endless progress or proficience in both."

BACON: *Advancement of Learning.*

ON

THE ORIGIN OF SPECIES

BY MEANS OF NATURAL SELECTION,

OR THE

PRESERVATION OF FAVOURED RACES IN THE STRUGGLE FOR LIFE.

BY CHARLES DARWIN, M.A.,

LONDON:

JOHN MURRAY, ALBEMARLE STREET.

1859.

Fig. 1.4 The front page of the revolutionary book by Darwin

alas, it is still alive. It is not rare today to see explanations of scientific evidence based on religious beliefs (recall Maimonides' approach).

While the conjecture fermented in Darwin's mind for a long time (he published the book when he was about 50 years old), it finally flowered after his famous trip on H.M.S. Beagle (1831–1836). During this trip, in which he collected data, Darwin studied Lyell's book. Darwin's conjecture is very strongly connected to Lyell's claim that the Earth is very old. Moreover, Darwin applied the same principle that Lyell had used in geology to biology, namely, the principle of gradual evolution.

The principle of natural selection, or the survival of the fittest, is an extremely simple principle, although it can manifest itself in many different forms, as detailed by Darwin himself in his book. But this is not the subject here. Suffice it to raise the following question: If God created each of the species we see today and all the species that existed on the Earth in the past, then why did He allow the extinction of so many of them?

Another surprising feature is that Darwin was able to provide a quite rigorous morphological description of the evolution of life without any knowledge of DNA. Note also that the name of the book was the *Origin of Species* and not the *Origin of Life*. The existence of genes was first suggested in the 1860s by Gregor Mendel

(1822–1884m),[45] who studied inheritance in pea plants and hypothesized a factor that conveys traits from parents to offspring.[46] Although he did not use the term 'gene', he explained his results in terms of inherited characteristics. Mendel was also the first to hypothesize independent assortment, the distinction between dominant and recessive traits, the distinction between a heterozygote and a homozygote,[47] and the difference between what would later be termed the genotype and the phenotype. Mendel's concept was finally named when Wilhelm Johannsen (1857–1927), a Danish botanist, coined the word 'gene' in 1909.[48] It appears that Darwin was unaware of Mendel's discoveries, because he does not refer to them anywhere. (This question is still a subject of debate between historians of science.)

Our interest here is not in the biological theory, but in the fact that it determines a time scale for evolution. In those days, astronomy and astrophysics could not set the time scale for evolution, and one had to resort to biology for an estimate. On the biological side, Darwin could not set a time scale for his process, only observe its consequences. However, he realized that the process of natural selection required a very long time scale, much longer than the inferred biblical age for the Earth. The interesting fact is that Darwin himself carried out the geological calculation to find the minimum age of the Earth, and it is this issue which interests us here. In Chap. 9 of his book, entitled *On the Imperfection of the Geological Record*, Darwin set out to calculate the geological age of the Earth. The particular example Darwin chose was the denudation of the Weald, a great valley that extends between the North and South Downs in the southern part of England. The data were taken from Ramsay, but Darwin had to guess the critical number: the rate of denudation. Darwin made the very rough estimate that the sea erodes into the 500 foot cliff at a rate of 1 inch per century, whence he found that it would take about 306 662 400 years to create the present day valley.

While Darwin had good reasons for the estimated rate of erosion, it was not a measured number. Another heavy assumption in this calculation was the constancy of the process over such a long period. Put differently, an average constant rate over a long period was assumed. We know today that the surface of the Earth is relatively young compared to the age of the Earth. The surface of the Earth changes continually, and the estimate Darwin made on the basis of the rate of erosion was therefore minimal, and related to the relatively short time scale of continental motion, mountain formation, and valley erosion. Darwin must have been aware of Helmholtz's calculation of the age of the Sun, but apparently decided to ignore the contradiction that the Sun was younger than his estimate for the age of the Earth, although the contradiction was in a direction that helped his case. It may be that he chose to ignore it because the age of the Sun was estimated in continental Europe. But, when the Scot Thomson pointed out the discrepancy, he could no longer overlook it.

[45] An Austrian monk, famous for his experimental work on heredity.

[46] Mendel, G., *Experiment on Plant Hybrids*. In Stern, C., Sherwood, E.R. (Eds.) *The Origin of Genetics*, Freeman, San Francisco, p. 1.

[47] Organisms that have different or the same series of genes, respectively, at a locus in homologous chromosomes.

[48] Johannsen, W., *The Genotype Conception of Heredity*, The Amer. Naturalist **45**, 129 (1911).

Today we know that errors during DNA replication, for whatever reason, e.g., cosmic radiation, may lead to a gene duplication that deviates from the original. Although the two sequences may remain the same, or be only slightly altered, they are typically regarded as separate genes. This became known only after Darwin's death.[49] In 1901, Hugo De Vries (1848–1935m), a Dutch plant physiologist, presented his conjecture that mutation can be induced by periods of stress in the environment, leading to the nearly instantaneous production of new species.[50] In 1912 Victor Hess (1883–1964m) discovered the cosmic rays.[51] Hess flew balloons to high altitude and discovered that the ionization current decreases up to about 800 m altitude, and then increases. The comparison between day and night indicated that the source of the radiation is not the Sun, so he hypothesized the existence of cosmic rays. Hess won the 1936 Nobel prize (jointly with Carl Anderson) for this discovery. In 1920 Herman Muller (1890–1967) showed that an intense X-ray flux can induce mutation and it soon became clear that the infinitely more energetic cosmic rays are an important factor in inducing mutations in living organisms.

Without wishing to belittle Darwin's colossal achievement, it should be mentioned that the concept of biological evolution was supported in Classical times by the Greek and Roman atomists, notably Lucretius. With the rise to dominance of Christianity came the belief in the biblical story of creation according to Genesis, along with the doctrine that God had directly 'created kinds' of organisms that were immutable. Other ideas surfaced, and in the 17th century the English word 'evolution' (from the Latin word 'evolutio', meaning to unroll like a scroll) began to be used to refer to an orderly sequence of events, particularly ones in which the outcome is somehow contained within the sequence from the start. However, it was Darwin who converted the idea into a theory by supplying all the phenomenological data.

Natural history developed considerably in the 18th century, aiming to investigate and catalogue the wonders of God's works. Discoveries showing the extinction of species were explained by Catastrophism, the belief that animals and plants were periodically annihilated as a result of natural catastrophes and replaced by new species created ex nihilo (out of nothing). Is this hypothesis simpler than gradual evolution? Countering this possibility, James Hutton's uniformitarian conjecture of 1785 envisioned gradual development over aeons of time.

By 1796, Charles Darwin's grandfather, Erasmus Darwin (1731–1802), who was a naturalist, had put forward ideas of common descent with organisms 'acquiring new parts' in response to stimuli, then passing these changes on to their offspring, and in 1802 he hinted at natural selection. In 1809 Jean-Baptiste Lamarck (1744–1829m) developed an analogous conjecture, with 'needed' traits being acquired, then passed on. These theories of transmutation were developed by radicals in Bri-

---

[49] In biology, mutation is a sudden, random change in a gene, the structural unit of inheritance in living organisms. Changes within single genes, called point mutations, are actually chemical changes in the structure of the constituent DNA.

[50] De Vries, H. *Die Muthationstheorie*, Veit & Co, Leipzig, 1901.

[51] Hess, V., Phys. Zeit. **13**, 1084 (1912).

tain like Robert Edmund Grant (1793–1874),[52] a physician who became a marine biologist. He was a radical free-thinker, opposed to the doctrine of the church, and saw no divine intervention in the natural world. He tried to promote the idea that the fossil record showed evidence of animals progressing from lower forms of life to higher forms.

Even for the non-specialist, Darwin's book was quite readable (as it still is), and it attracted widespread interest. Although the ideas presented in the book are now supported by overwhelming scientific evidence and are widely accepted by scientists, they are still highly controversial, particularly among non-scientists who perceive the idea of evolution as contradicting the literal interpretations of various religious texts. Various ideas have been developed to reconcile the scientific findings and the simple understanding of sanctified writings. However, these interesting developments are outside our scope here.

The greatness and ingenuity of Darwin's phenomenological investigation is that it led to the hypothesis of evolution without providing a mechanism, without any knowledge of DNA, genes, and mutation. It took science the time it needed to establish the phenomenological findings on biochemical grounds and identify responsible mechanisms. This great hypothesis preceded the evidence for the operation of a mechanism which leads to the observed evolution.[53]

Wallace (1823–1913m),[54] who claimed to be a coinventor of the theory of evolution, presented his theory of evolution to the Royal Society simultaneously with Darwin. However, Wallace is generally disregarded because he coupled human evolution with spiritual forces. He did not believe in the long time required for evolution. Wallace claimed that man had escaped the influence of the laws of natural selection.[55] And this is how:[56]

> *By his superior intellect* and *by his superior sympathetic and moral feeling, he becomes fitted for social state [...] as there is undoubtedly an advance – on the whole steady and a permanent one – both in the influence on public opinion of a high morality, and in the general desire for intellectual elevation; and as I cannot impute this in any way to 'survival of the fittest', I am forced to conclude that it is due to the inherent progressive power of those glorious qualities which raised us so immeasurably above our fellow animal.*

---

[52] Grant, R.E. *Animal Kingdom.* In: Todd R.B., (Ed.), *The Encyclopedia of Anatomy and Physiology*, Vol. 1, London, Sherwood, Gilbert & Piper, p. 107.

[53] The first significant discovery of Neanderthals was made in August 1856, three years before Darwin published his theory. A partial skeleton was found at the Feldhofer Cave in the Neander Valley, near Dusseldorf in Germany. This was the find that gave the species its name. The image of the Neanderthal man as a savage barbarian was depicted in 1908 by Marcellin Boule. Recent DNA research shows that the Neanderthals and ourselves are unrelated, and did not interbreed, even though the two races lived together for some time. Neanderthals lived in Europe and western Asia from 300 000 years ago until the last of them disappeared on the Iberian peninsula about 28 000 years ago. The prevailing theory is that modern humans arose in Africa less than 200 000 years ago, and appeared in great numbers in Europe from about 40 000 years ago.

[54] Wallace, A.R., *On the law which has regulated the introduction of new species.* Annals & Mag. Natural History **16**, 184 (1855).

[55] Wallace, A.R., *The Origin of Human Races and the Antiquity of Man Deduced from the Theory of 'Natural Selection'*, J. Anthropological Soc. London **2**, clviii (1864).

[56] Wallace, A.R., *Contributions to the Theory of Natural Selection*, Macmillan and Co, 1871.

The separation between the evolution of the human race and the evolution of the rest of biology is probably the reason why he and his theory are ignored by the scientific community.

Darwin's theory brought together biology and astronomy, an act which infuriated Kelvin: *How can biology be mixed with physics?* Years later Eddington, one of the giants of 20th century astrophysics, in his search for the source of the Sun's energy, reversed the argument and claim that:[57] *Biological, geological, physical and astronomical arguments all lead to the conclusion that this age is much too low and that the time-scale given by the contraction hypothesis must somehow be extended.* Note the order: biology first!

1.6 Devout Criticism of Darwin

Criticism of Darwin's evolution theory came from two fronts: religious and scientific. Soon after Darwin published his book in 1859, the storm raged. A notorious public debate took place during the meeting of the British Association for the Advancement of Science, held on 30 June 1860 on the occasion of the inauguration of Oxford's new cathedral of science. This public debate is known today as the Great Oxford Debate of 1860. Representatives of the Church and scientists debated the subject of evolution, and the event is often viewed by scientists (very probably wrongly) as symbolizing the defeat of theological views of creation. However, there are few eyewitness accounts of the debate, and available accounts were mostly written by rather biased scientists. The debate was widely publicised in the daily press and various cartoons filled the media, carrying the participants overnight into the hall of fame of the history of science.

It is in this public debate that Bishop Wilberforce ridiculed Thomas Henry Huxley (1825–1895m),[58] asking him whether *it was through his grandfather or his grandmother that he claimed descent from a monkey*. The legend says that Huxley muttered to Sir Benjamin Brodie (1817–1880), a chemist, president of the Royal Chemical Society at the time of the debate: *The Lord has delivered him unto my hand*, and replied to the bishop:

> *If the question is put to me whether I would rather have a miserable ape for a grandfather or a man highly endowed by nature and possessed of great means of influence and yet employs these faculties and that influence for the mere purpose of introducing ridicule into a grave scientific discussion, I unhesitatingly affirm my preference for the ape.*

---

[57] Eddington,S.A., *The Internal Constitution of the Stars*, 1926, p. 290.

[58] As the nickname 'Darwin's bulldog' would suggest, Huxley was an outspoken defender and advocate of Darwin's theory of evolution by natural selection. Huxley was trained and worked as a physician, but never held a research position. His firm support for Darwin, though with some critique, earned him the name of one of the smallest craters on the Moon, about 4 km in diameter.

Huxley's views about the evolution of man were exposed in his book[59] and lectures.[60]

Wilberforce, the Bishop of Oxford, was an outspoken adversary of the theory of evolution, and he appealed to Richard Owen (1804–1892), a well-known anatomist who disagreed with Darwin.[61] Wilberforce, a great orator and preacher, drew a large audience, and attacked Darwin to the cheering shouts of students. Huxley, the considered scientist, was no match for Wilberforce when it came to charisma, and he essentially lost the audience to Wilberforce. Seeing that, Joseph Hooker (a long-time friend of Darwin and his botanical mentor), asked to speak, attacking Wilberforce viciously. And as is often the case, both sides claimed victory. It is said that Huxley, recognizing the power of oratory, decided to perfect his speech for the next time.

Meanwhile, Darwin's nerves could not stand it, and he was treated during the debate at Dr. Lane's Hydropathic Clinic. How pivotal that session of the British Association for the Advancement of Science was in terms of shifting the weight of popular and scientific opinion to an evolutionary viewpoint is as unclear as what was actually said. Equally uncertain is the damage it did to the clerical cause against Darwinism. But the stakes were high for both sides. In any case, it seems unlikely that the debate was as spectacular as is traditionally suggested, because contemporary accounts by journalists of the day did not mention any particularly notable quotes. Furthermore, contemporary accounts suggest that it was not Huxley, but Sir Joseph Hooker who defended Darwinism most vocally at the meeting.

Barely a month later (July 1860), Samuel Wilberforce published a review of Darwin's book,[62] in which he attacked Darwin's evolution theory with considerable ferocity. Here, we quote only the reference to the required age of the Earth:

> *The Lyellian hypothesis, itself not free from some of Mr. Darwin's faults, stands eminently in need for its own support of some such new scheme of physical life as that propounded here. Yet, no man has been more distinct and more logical in the denial of the transmutation of species than Sir C. Lyell, and that not in the infancy of his scientific life, but in its full vigor and maturity.*

Wilberforce's campaign against the non-orthodox won him the special gratitude of the Low Church party, to the point that he felt strong enough to dictate lines of thought. He thus took active part in forcing Bishop J.W. Colenso of South Africa to resign for writing an article in which some non-orthodox views were expressed.[63]

---

[59] Huxley, T.H., *Evidence as to Man's Place in Nature*, Williams & Norgate, London, 1863.

[60] Huxley, T.H., *Lecture on the Elements of Comparative Anatomy*, John Churchill & Sons, London, 1864.

[61] Owen's statements on evolution were contradictory. In later years he alternately denied its validity, admitted ignorance on the matter, or claimed to have come up with the idea himself almost ten years before. Owen invented the term 'dinosauria', which means 'terrible lizards'. See also Camardi, G., *Richard Owen, Morphology and Evolution*, J. History Biology **34**, 481 (2001).

[62] Wilberforce, S., *On the Origin of Species by C. Darwin*, The Quarterly Review **108**, 225 (1860).

[63] Wilberforce was killed on 19 July 1873, when he fell from his horse near Dorking, Surrey. Huxley commented that Wilberforce's brains had at last come into contact with reality, and the result had been fatal.

An interesting and typical case is St. George Jackson Mivart (1827–1900). In 1871,[64] twelve years after the publication of Darwin's *Origin of Species*, and the very same year in which Darwin published[65] *The Descent of Man and Selection in Relation to Sex*, describing how man had evolved, Mivart published his essay entitled *On the Genesis of the Species*. Mivart was a well-known biologist and a Catholic convert, whom Darwin apparently appreciated. Mivart did not flatly reject the Darwinian thesis. On the contrary, he supported 'evolution', but tried to argue that science and religion should be separated. By maintaining the creationist theory of the origin of the human soul, he attempted to reconcile his evolutionism with the Catholic faith.

In spite of being a scientist, Mivart brought quotations as evidence, writing as follows:

> *It must be borne in mind that for a considerable time after the last of these writers (St. Augustine and St. Thomas Aquinas) no one has disputed the generally received view as to the small age of the world or at least of the kinds of animal or plants inhabiting it. It becomes therefore, more striking if views formed under such a condition of opinion are found to harmonize with modern ideas regarding 'creation' and organic life.*

In short, Mivart brought quotes as a substitute for scientific facts. Worse than that, Mivart quoted Darwin by shortening sentences and omitting words in a way that led Darwin to say:[66] *Though he means to be honourable, he is so bigoted that he cannot act fairly*.

Five years later, Mivart was rewarded. It appears that Mivart's attack on Darwin so pleased the Catholic church and Pope Pius IX that it was decided, in a very unusual move, to confer upon him a Doctor of Philosophy in 1876.[67] But Mivart continued to vacillate, and expressed views which eventually did not please either 'sides' in the controversy, and consequently was effectively excommunicated by Darwin's supporters and opponents.

1.7 Scientific Criticism of Darwin

John Phillips (1800–1874m) was a well-known English geologist. A year after the publication of *On the Origin of Species*, Phillips, then the president of the Geological Society of London, delivered a lecture at Cambridge, England. In the lecture, he attacked Darwin's weak point, the calculation of the denudation of the Weald, and claimed that Darwin had made many errors of arithmetic. Phillips estimated the age of the Earth at 100 million years. Other evidence that Phillips presented in his lecture[68] had even more far-reaching ramifications than the age estimates, but

[64] Mivart, S.G., *On the Genesis of Species*, Appelton, NY, 1871.

[65] Darwin, C., *The Descent of Man and Selection in Relation to Sex*, John Murray, London, 1871.

[66] Browne, J., *Charles Darwin: The Power of Place*, Cape, London (2002) p. 329.

[67] The Catholic Encyclopedia, 1913, Item Mivart, G.J.

[68] Reprinted in book form as: Phillips, J., *Life on the Earth: Its Origin and Succession*, Cambridge University Press, 1860.

go beyond our scope here. However, Darwin would have been happy with this age estimate.

As an aside, Sir George Darwin (1845–1912) was Charles Darwin's fifth son.[69] As a famous astronomer and mathematician who worked intensively on the theory of the Earth's tides, Darwin hypothesized that the Moon had separated from the Earth, in what is called the fission theory. He proposed that the Earth was rotating so fast that a large piece, the Moon, had broken off. Since then, via the mutual action of tides, the separation between the Earth and the Moon had increased continuously. In 1905, George Darwin[70] claimed that:

> *If at every moment since the birth of the Moon tidal friction had always been at work in such a way as to produce the greatest possible effect, then we should find that sixty million years would be consumed in the portion of evolutionary history. The true period must be much greater, and it does not seem extravagant to suppose that 500 to 1000 million years might have elapsed since the birth of the Moon.*

In this way, the long time scale dictated by astronomy vindicated his father's estimate. The fission theory is out of favor today. The Earth could never have rotated fast enough to throw a moon into an orbit, and the escaping moon would have been torn apart while moving within the Roche limit.[71]

1.8 First Attempt to Quantify the Meteor Theory

As early as 1860, Waterston[72] (1811–1883) tried to use Joule's new result to calculate the temperature of the Sun. His first attempt was to calculate the temperature that would be reached by a meteor hitting the surface of the Sun. Assuming the meteor and the Sun to behave like water, he found the fantastic temperature of 55 million degrees.[73] He argued that, if the Sun were made of iron, the temperature would have been nine times higher. Waterston simply converted the kinetic energy of the falling meteor into heat using Joule's conversion factor. He went on to calculate the mass of the meteor that would supply the present power emitted by the Sun, and reasoned that if such a meteor did not fall into the Sun, then the latter would

---

[69] George Darwin studied law and was admitted to the bar, but on account of ill-health, abandoned his profession to undertake some scientific work for Kelvin (his father's most ferocious critique), which among other things included the reduction of a huge collection of Indian tide observations, the aim being to understand the problem of the rigidity of the Earth.

[70] Address to the British Association in South Africa, On Cosmical Evolution, The Observatory, October 1905.

[71] The Roche limit is the orbital distance at which a small (theoretical) liquid planet will begin to be torn apart tidally by the body it is orbiting. The planet is held together by its own gravity. Tidal forces arise from the fact that the side of the planet close to the big star is attracted more strongly then the far side, and the difference between these two unequal attractions is equivalent to a tearing force on the small planet.

[72] Waterston, J., Fall of meteor into the Sun, MNRAS **20**, 198 (1860).

[73] The correct value using his data would in fact be 16 million degrees.

cool by the amount the meteor would have supplied. This strange argument led him to conclude that the Sun cools by 4.59°C per year. Had he pursued this reasoning, he would have deduced that the lifetime of the Sun was 22 million years, a conclusion the old Darwin would have appreciated. Interestingly, Waterston calculated the effect that mass accreted by the Sun would have on the length of the year on Earth, and rejected the idea that objects as massive as the Earth could fall on the Sun every year, because he calculated that such a mass added to the Sun would change the length of the year by 130 seconds per year. Such a modification would certainly be noticeable to astronomers.

1.9 Kelvin

Lord Kelvin played a special role in the battle against Darwin's theory of evolution (1824–1907m).[74] It is difficult to comprehend the enormous influence Kelvin had on science in the second half of the 19th century and the first part of the 20th century. William Thomson (later Lord Kelvin) can be credited with the formulation of the second law of thermodynamics[75] and many other discoveries in physics. However, Thomson was knighted, not because of his achievements in theoretical physics, but thanks to his inventions in practical physics.

As an experimental physicist, he was interested in the transmission of electrical signals in cables. His work on the theory of the electric cable started in 1855, when the idea of an Atlantic cable was first mooted. The mathematical analysis of the behavior of electrical transmission lines is based on the works of James Clerk Maxwell (1831–1879m), Lord Kelvin, and Oliver Heaviside (1850–1925m). In 1855 William Thomson formulated the governing equation of the submarine electric cable. This equation, known today as the telegraph equation, correctly predicted the poor performance of the 1858 transatlantic submarine telegraph cable. The signals were so weak that ordinary receiving methods were useless. Thomson's solution was the invention of the mirror galvanometer.[76] This extremely simple instrument solved the problem. More than seven hundred messages had been received and the problems presented by the new transatlantic cable were apparently solved, when suddenly the signals stopped coming. The cable had broken and was beyond repair. *We must build a new and better cable*, touted Thomson. He busied himself with the plans, arranging to have a cable ship, the Great Eastern, carry the whole length of the required cable. He equipped the ship for the maneuvers needed to lay the cable. Two

---

[74] A crater (Thomson) and a mountain ridge Promontorium Kelvin were named after Kelvin on the Moon.

[75] Thomson, W., *On the Universal Tendency in Nature to the Dissipation of Mechanical Energy*, Phil. Mag. **4**, 256 (1852).

[76] The idea is simple and brilliant. A small mirror is attached to the axis of the galvanometer. The mirror is illuminated by a source of light. A tiny current will cause a very small movement in the mirror, but when the reflected light by the mirror is cast on a distant scale, the tiny movement is enhanced and easily detected.

attempts were made before a successful line was laid in 1866. As electrical engineer for the expedition and the man most responsible for its success, William Thomson was knighted by Queen Victoria, and became Lord Kelvin. However, as a scientist, Kelvin's most important contribution was the laws of thermodynamics. The resulting technological inventions and scientific discoveries propelled Kelvin to a unique position in science in the second half of the 19th century. As a result, he sometimes acted as though he were top dog in the world of science.

Kelvin carried out two independent calculations: one was the age of the Earth and the other was the age of the Sun. The idea that the Sun and Earth should have the same age had not yet been proposed at that time. The calculation of the age of the Sun was carried out in 1862.[77] First, Kelvin examined the idea that the heat generated in the solar atmosphere by falling meteors might be sufficient to generate the entire power of the Sun (an idea sometimes later attributed to Robert Mayer, although no mention of him could be found in Kelvin's paper). Kelvin compared this idea with the assumption that the Sun is an incandescent liquid mass, losing the initial heat generated by past falling meteors. Kelvin rejected the first possibility (which was Helmholtz's first version) by noting that the mass accreted by the Sun in 2 000–3 000 years would be sufficient to change its mass in such a way as to affect the length of the year. The amount of mass per year needed to generate the solar luminosity should be 1/47 the mass of the Earth, which is 1/15 000 000 of the mass of the Sun, and for this to be possible, one has to assume that the total mass of meteors in the plane of the planets is of the order of 1/5 000 of the mass of the Sun. However, such a mass would affect the motion of the planets to a measurable extent, giving rise to deviations from Kepler's law, a phenomenon which had not been observed.

We recall that, as early as 1840,[78] Urbain LeVerrier (1811–1877m) had calculated the advance of the perihelion of Mercury due to perturbation from other planets.[79] While there was a small discrepancy, later to be explained by Einstein, it was

---

[77] Thomson, W. (Lord Kelvin), *On the Age of the Sun's Heat*, Macmillan's Magazine **5**, 288 (1862). Transaction of the Royal Society of Edinburgh, April 1864. *On the Dynamical Theory of Heat, with numerical results deduced from Mr Joule's equivalent of a thermal unit*, Phil. Mag., 1853; *On the secular cooling of the Earth*, Phil. Mag., 1863; *On the Reduction of Observations of Underground Temperature*, Trans. Roy. Soc., Edinburgh, 1860.

[78] LeVerrier, U.J.J., *Sur les variations séculaires des éléments elliptiques des sept planètes principales: Mercure, Vénus, la terre, Mars, Jupiter, Saturn et Uranus*, J. Math. Pures Appl. **4**, 220 (1840). Ann. Obs. Imp. Paris, 1859.

[79] The perihelion is the closest point to the Sun in the orbit of a planet. According to Newton's law of gravity and Kepler's laws of planetary motion, the planets move around the Sun in ellipses that are fixed in space. However, astronomers found that the ellipse corresponding to the orbit of Mercury is not fixed in space, but rotates by about 575 arc seconds per century, i.e., the perihelion moves in space. This means that some other bodies perturb the motion of Mercury, and LeVerrier assumed that it was the other planets in the Solar System. He thus calculated that the effect of all the planets, and in particular Jupiter, amounts to exactly this rotation minus 44 arc seconds per century. The Newtonian theory of gravity failed to explain the latter extremely small effect, and it was left for Einstein to explain it in his new theory of gravity some 60 years later. Here we note the high accuracy with which the motions of the planets were already known by the mid-18th century, providing very strong constraints on both the theory and the mass of the Sun. From the

too small to explain the resistance due to a swarm of meteors, such as was needed to explain the energy of the Sun. To avoid disturbance to the motion of Mercury, the meteors had to be very close to the Sun, thereby significantly reducing the energy released whenever they might fall onto the Sun.

Kelvin was therefore forced to assume that the Sun was heated at the time of its formation, and that since then it had been radiating its energy. In other words, he concluded that the Sun was cooling. Kelvin quoted the total energy the Sun releases (6×10^{30} times the heat needed to raise the temperature of one pound of water by 1°C) from Herschel and Pouillet. He then assumed that the Sun is made from similar material to the Earth (and here he relied on the spectroscopic observations of Kirchoff and Bunsen, who discovered in the Sun many elements found on the Earth). Next, Kelvin assumed that the specific heat of the solar material resembles that of water, dividing the mass of the Sun times the specific heat by the energy output. In this way he derived the cooling time of the Sun. As the Sun cools, it must contract, behaving like every material on the Earth, expanding upon heating and contracting upon cooling. Kelvin found that, at the present rate of cooling, the Sun should contract by about 1/120 000 of its diameter per degree centigrade that it cools. This implies that, if solar material behaves like water, than the Sun must contract by 1% every 860 years, a change that would be noticed by astronomers. Since a contraction at such a rate had not been observed, Kelvin assumed arbitrarily that the heat capacity of the material making up the Sun must be 10 000 times greater than the heat capacity of water!

Kelvin continued his line of thought, realizing that, upon the contraction of the Sun, different parts of the Sun must do work which he could not calculate, because he did not know the density inside the Sun. If the density is constant inside the Sun, as Helmholtz assumed, then the energy of the Sun should suffice for 20 000 years. This number may increase if the density increases towards the center of the Sun. Next, Kelvin argued:

> *It is in the highest degree improbable that mechanical energy can in any case increase in a body contracting in virtue of cooling. It is certain that it really does diminish very notably in every case hitherto experimented on. It must be supposed, therefore, that the Sun always radiates away in heat something more than the Joule equivalent of the work done on its contracting mass, by mutual gravitation of its parts. Hence, in contracting by one tenth per cent of its diameter, or three tenth per cent in its bulk, the Sun must give out something either more, or not greatly less, than 20 000 years' heat.*

By this argument, in which the Sun contracts by 1% every 20 000 years, Kelvin solved the problem of the unobserved contraction of the Sun, to his apparent satisfaction.

Kelvin ends this discussion by sneering at Darwin's calculation of the 'denudation of the Weald'. One strong storm, so he claimed, would have eroded the cliff 1 000 times more than the rate of 1 inch per century assumed by Darwin. This was surely a wild guess by Kelvin. The number was not known to him. It is puzzling that

---

perturbations to the orbit of Uranus, LeVerrier calculated the position of the then unknown planet causing the perturbations. The discovery of Neptune, the first planet to be discovered in modern times, is considered to be one of the crowning achievements of science.

Kelvin blamed Darwin for assuming an unreasonable rate of erosion, at least, unreasonable according to Kelvin, while at the same time assuming an imaginary and completely arbitrary value for the heat capacity of the Sun, just to fit his presupposed result.

After showing that the Sun was in fact cooling, Kelvin compared the energy production per square foot on the surface of the Sun with the energy produced in the furnace of a locomotive, and concluded that they are similar. Since he assumed that the Sun was a liquid, he did not calculate any heat transfer inside the Sun, assuming the heat to be carried effectively by currents.

Finally, came the question of the origin of the total amount of solar heat. Kelvin concluded that the theory of meteoritic showers must be rejected in favor of the theory making the hypothesis that all solar heat is generated by past massive meteor showers. Here he returned to the hypothesis put forward by Helmholtz. Lastly, he concluded that the age of the Sun cannot be less than 10 million years if the density is constant, and may even be 100 million years if the density increases inward, but that it is definitely not as long as 300 million years, as suggested by Darwin.

Kelvin concluded his article with words that would be appropriate even today, though with a twist in the reasoning:

> *As for the future, we may say, with equal certainty, that inhabitants of the Earth cannot continue to enjoy the light and heat essential to their life for many million years longer unless sources now unknown to us are prepared in the great storehouse of creation.*

Kelvin's calculation was wrong on many counts! The most fundamental error was pointed out by Eddington, who showed years later that the effective specific heat of stars is negative, that is, stars lose energy by radiation and heat up, rather than cool![80]

The calculation of the age of the Earth was (and still is) more complicated then the calculation of the contraction of the Sun, and significantly more uncertain. Kelvin knew about Fourier's work, and even declared it to be *a poem*, but he dismissed Fourier's conclusion, namely, that the then available data did not allow the calculation of the cooling rate.

The first question to answer when attempting to calculate the age of the Earth concerns the initial state of the Earth. Kelvin assumed that the Earth was molten and had had some initial temperature. Next, one has to make some assumption about how well the various layers in the Earth conduct heat. Then one needs to know the dependence of temperature on depth, that is, over what depth the temperature increases by one degree. Kelvin took yearly averaged temperatures measured in mines. Finally, one has to assume the heat capacity of the Earth's material, and its heat conductivity. The age obtained was a few tens of thousand of years, with a large error due to inaccurate data.

[80] Eddington showed correctly that a star that loses its energy by radiation will contract, and in this way extracts energy from the gravitational field. This energy in turn heats it up. So half the energy released by the gravitational field is radiated out and half goes to heat the star. The reason is that the contracting star increases the gravitational pull by the compression and, to balance this extra pull, the temperature must rise so as to enhance the outward pressure of the gas which balances the gravity. In this way the apparently paradoxical result of negative specific heat is created.

As we know today, Kelvin's calculation was wrong because he ignored the heat source inside the Earth, which Fourier suspected to exist. It is astounding that Kelvin the scientist was not bothered by his own finding of a discrepancy between the ages of the Sun and the Earth.

1.10 The Darwin–Kelvin Controversy

The vocal arguments that the widely esteemed, but also highly self-esteemed Kelvin had with the community of geologists would occupy many pages, but they are not the subject here. Kelvin was wrong and the geologists who refused to learn physics from him were right. Adding to his irritation was the infiltration of biology into his subject, personified by Charles Darwin, and promoted by the *Origin of Species* and several more books shortly afterwards.

As mentioned above, in his 1862 paper on the age of the Sun, Kelvin mocked Darwin's attempt to estimate geological ages. Five years later, Kelvin's friend Fleeming Jenkin wrote a long review of the *Origin of Species*, dismissing Darwin's foray into quantitative geology as a calculation of the kind engineers refer to as *guess at the half and multiply by two*. But to no avail. By that time Darwin's book had already gained popularity and had gone through several editions. In the discussion part of his book, Darwin still required a long time for evolution to take place, but admitted defeat on this count. Moreover, the discussion about the age estimate based on the erosion of the Weald valley was removed from subsequent editions. Kelvin's scientific bullying had won the day.

Even by the mid-1860s, Kelvin's arguments about the age of the Earth were practically ignored by geologists, who were at that time not such quantitative scientists as they are today. It disturbed him that his *rock-solid thermodynamic arguments* were rejected on the grounds that physics and geology should not be mixed. Physicists think correctly that all systems must obey the laws of thermodynamics, because they are so general. In 1867, when the British Association for the Advancement of Science met in Dundee, Kelvin argued with Andrew Ramsay, who had provided Darwin with his geological data. Ramsay was of the opinion that geological ages are very long, even billions of years. Kelvin objected that the Sun, being a finite body, could not possibly shine for so long. Ramsay responded that this point of physics had nothing to do with him: *I am as incapable of estimating and understanding the reasons which you physicists have for limiting geological times as you are incapable of understanding the geological reasons for our unlimited estimates.* Kelvin responded by saying that physics can be explained to anyone who is really willing to listen and understand. As far as Kelvin was concerned, he was not telling geologists how to conduct their science, only that their theories could not disregard the universal laws of thermodynamics. Kelvin wanted the geologists to listen to him, but at the same time, he refused to listen to them.

Very little had changed in the data since Fourier arrived at his conclusion about the inaccuracy of calculation of the age of the Earth, but the arrogant and self-assured Kelvin ignored what the very careful mathematician Fourier had written fifty years earlier, and proceeded with his calculation of the cooling of the Earth. Darwin could console himself that Kelvin argued not only with him, but with many others on various topics. He had argued at the beginning with Joule, but then had been convinced and collaborated with him. He had argued with the entire community of geologists. He had argued with Tyndall about magnetism, but had been found to be wrong. Tyndall fought Kelvin on many fronts. It is interesting to mention here that in 1874 Tyndall[81] attacked Kelvin's version of the Sun's energy source. He calculated à la Kelvin the amount of energy that would be obtained if all the planets fell into the Sun (including their rotation energy) and found an age of 45 586 years, which he rejected as implausible. He then cited Helmholtz's original idea of the contraction of the Laplace original nebula, and got a temperature of 28 million degrees.[82] On this basis, he also calculated that it would take the sun 17 million years to cool to the temperature of the Earth.

Kelvin was a very colorful figure, and occupies a special position in the history of science due to his extreme self-confidence, his outspoken nature, and his many fantastic quotes, which have subsequently entered the folklore of physics. Let us mention just two here: in 1895 he declared that *heavier-than-air flying machines are impossible* (Australian Institute of Physics), to be followed in 1896 by the statement: *I have not the smallest molecule of faith in aerial navigation other than ballooning [...]. I would not care to be a member of the Aeronautical Society.* An interesting assertion by Kelvin, which is relevant to our story here, was made in a speech at an assembly of physicists at the British Association for the Advancement of Science in 1900: *There is nothing new to be discovered in physics now. All that remains is more and more precise measurements.*

However, the most famous statement must surely be the following. On 27 April 1900, Lord Kelvin gave a lecture to the Royal Institution of Great Britain. The title of the lecture was *Nineteenth-Century Clouds over the Dynamical Theory of Heat and Light*. In his characteristic way, Kelvin admitted that the *beauty and clearness of theory* was overshadowed by *two clouds*. He was talking about the null result of the Michelson–Morley experiment,[83] and the problems of black body radiation.[84] In

[81] Tyndall, J., *Heat as a Mode of Motion*, Appleton & Co, 1873, pp. 488–9.

[82] Today we know that this number is wrong by a factor of 2.

[83] The experiment was carried out in 1887 by A.A. Michelson (1852–1931m) and E.W. Morley (1838–1923m). The aim was to discover the ether in which light was assumed to travel, by observing the motion of the Earth around the Sun, this being assumed to create an 'ether wind'. The experiment gave a null result. In 1907, Michelson was awarded the Nobel Prize for Physics for *his optical precision instruments and the spectroscopic and metrological investigations carried out with their aid*, and became the first American to win the Nobel Prize. It should be stated that neither Michelson nor Morley considered that the experiment disproved the ether hypothesis, although others did. There is not a single word about the experiment and relativity in his Nobel address. Morley worked with D. Miller on attempts to prove the existence of the ether after his work with Michelson.

[84] In physics the term 'black body' refers to a perfect absorber of light, i.e., one that absorbs all light that falls on it. At the same time, the black body is a perfect emitter. Of all bodies at a given temperature, the black body is the best emitter. Calculation of the radiation emitted by a black

fact, he could not have chosen the 'clouds' better. The null result in the Michelson–Morley experiment led to the theory of relativity, and the failure of the classical theory to explain the behavior of black body radiation led to the discovery of quantum theory. These two theories constitute the pillars of the scientific and philosophical revolution of 20th century physics. Newton's theory of motion had served physics well into the 19th century, and Kelvin was confident that these two problems would soon be explained and cleared up within the realm of what we call today classical physics. Kelvin proposed his own solutions to the *two clouds* based on a classical point of view. Kelvin argued that light is a 'vibration' that can be treated by Newton's laws of motion. A necessary, but not sufficient condition for this explanation to be valid is that the light or 'vibration' should propagate through some sort of physical medium called the ether. However, the Michelson–Morley experiment had dealt a death blow to this explanation. What is more, these two theories, which Kelvin did not expect to emerge, are fundamental to stellar evolution and the fate of the stars.

Did Kelvin notice the discrepancies that his calculations led to? We do not know, and in any case it is not mentioned in his writings. But Kelvin was one of the greatest physicists, so how can we explain his thinking and behavior? In fact, Kelvin was a very religious person. He could not live with Darwin's evolution of life and human beings. While Darwin did not address the question of the origin of life, Kelvin did. In August 1871, he addressed the British Association for the Advancement of Science meeting held in Edinburgh and exposed his hypothesis about the origin of life. First he discussed the hypothesis that *under meteorological conditions very different from the present, dead matter may have run together or crystallized or fermented into 'germs of life' or 'organic cells' or protoplasm*, and claimed that science brought evidence against this hypothesis, although he did not provide any manifestation of such evidence. Kelvin claimed that *dead matter must be under the influence of life matter to become alive*, and then suggested searching for *spontaneous generation of life*.

Kelvin went on to ask how life originated on Earth, and stated that:

> *If a probable solution, consistent with the ordinary course of nature, can be found, we must not invoke an abnormal act of Creative Power.*

Kelvin then continued with the example of how wind carries the seeds of vegetation onto the cooled lava of a volcano, and from there soon reached the idea of panspermia.[85] The essence of the theory à la Kelvin was that:

---

body on the basis of classical physics gave an infinite rate of emitted radiation, and hence posed a problem.

[85] Panspermia is a theory that attempts to explain how life propagates in the cosmos from one habitable location to the other. The fundamental assumption of the theory is that seeds or germs which contain all the ingredients of life are able to propagate in space. The theory offers no solution as to how, if at all, life was created. Notable believers in the panspermia hypothesis have been Anaxagoras (500–428 BCE), Hermann von Helmholtz, Svante Arrhenius (1859–1927m), who was the first to formulate the theory in a scientific way [Arrhenius, S., Die Umschau **7**, 481 (1903); Scientific American **196**, 196 (1907)], and Fred Hoyle (1915–2001). In different versions of the theory, the carriers of the elements of life are transported by light pressure (Arrhenius), unmanned

Because we all confidently believe that there are at present, and have been from time immemorial, many worlds of life besides our own, we must regard it as probable in the highest degree that there are countless seed-bearing meteoric stones moving about through space.

Thus, Kelvin claimed that:

The hypothesis that life originated on this Earth through moss-grown fragments from the ruins of another world may seem wild and visionary; all I maintain is that it is not unscientific.

The amazing thing is that Kelvin realized that his idea required evolution, and indeed stated that:

All creatures now living on Earth have proceeded by orderly evolution from some such origin.

Kelvin stated his effective belief in evolution, and just after, quoted from *The Origin of Species* by Darwin. However, he omitted two sentences as he explained:

I have omitted two sentences [...] describing briefly the hypothesis of the 'origin of species by natural selection', because I have always felt that this hypothesis does not contain the true theory of evolution, if evolution there has been, in biology.

And here comes the crux of the matter:

I feel profoundly convinced that the argument of design has been greatly too much lost sight of in recent zoological speculations.

This explains, at least on the face of it, why Kelvin was not bothered by the inconsistencies in his arguments.

Darwin died on 19 April 1882, aged 73, not knowing whether he was right about the age of the Earth. Kelvin died on 17 December 1907, not before radioactivity was discovered and had become a tool for geological dating, knowing that he was wrong on many issues, including the age of the Earth, the emergence of new physics, and the existence of the ether, to list but a few. Lord Kelvin's obituary in the Times of London was 13 columns long and included a discussion, more than a column long, about the age of the Earth and the Sun, and how Kelvin had fought the geologists.

Michael Faraday (1791–1867m), James Clerk Maxwell, and Lord Kelvin were leading scientists in the second half of the 19th century, and all made colossal contributions to science. Yet their names are hardly known outside narrow scientific circles, in dramatic contrast to the name of Darwin. Faraday, Maxwell, and Kelvin were scientists, and very religious. In particular, the first two completely separated their religious beliefs from their scientific lives, and are not known to have expressed opinions on the controversy over the age of the Sun. Maxwell is known to have been interested in astronomy, since he solved the problem of Saturn's rings.[86]

---

spaceships (Crick's directed panspermia), meteorites (ballistic panspermia), or comets (Hoyle and Wickramasinghe's modern panspermia).

[86] Maxwell proved that they are composed of small rocks. *On the Stability of the Motion of Saturn's Rings*, an essay which obtained the Adam's Prize for the year 1856, in the University of Cambridge. An abstract of the essay can be found in MNRAS **19**, 297 (1859).

Darwin on the other hand was not an atheist. He described himself as an agnostic, and it is likely that he retained a belief in some kind of personal God, although not a deity that interferes continuously in the evolutionary process, or in human affairs.

So what is the reason for Darwin's fame and the relative obscurity of such great scientists? It seems probable that it was because Darwin threatened to bring science into the domain of religion, unlike Kelvin, Faraday, or Maxwell, even without discussing the origin of life. This explanation goes along with the following observation. Many people have heard the names Newton and Einstein, and even though they may not understand what these great men actually achieved, they trust and believe their theories. Einstein has even become an adored universal emblem. Darwin on the other hand created a disturbing animosity in about half the population. The reason appears to be the apparent clash between religious dogma, or literal interpretation of canonized writings, and science. To most people, the relativity of space and time is so far removed from such considerations that it could not rock their basic faith. But at the same time, a negator of Darwin would be unlikely to obtain any kind of university position!

Was Kelvin wrong? His calculations were correct, but his great failure was to not recognize that something was missing from his assumptions, and hence that there was a fundamental flaw. This was his great fiasco. The problem, so it appears, was his loss of critical wherewithal when it came to matters that touched upon holy writings. He did not have the hindsight needed to spot that something fundamental was missing, and this was the heating of the Earth by radioactivity. But was this accidental? Probably not. And so he said:

> *I cannot admit that, with regard to the origin of life, science neither affirms nor denies Creative Power. Science positively affirms Creative Power. It is not in dead matter that we live and move and have our being, but in the creating and directing Power which science compels us to accept as an article of belief.*

As an example of what the discrepancy might have stimulated, it is interesting to consider the comment by T.C. Chamberlain, a leading geologist, in one of his addresses in 1899:

> *Is present knowledge relevant to the behavior of matter under such extraordinary conditions as obtain in the interior of the Sun sufficiently exhaustive to warrant the assertion that no unrecognized sources of heat reside there? What the internal composition of the atoms may be is as yet an open question. Is it not improbable that they are complex organizations and the seats of enormous energies? [...] No cautious chemist would probably venture to assert that the component atomecules, to use a convenient phrase, may not have the energies of rotation, revolution, position, and be otherwise, comparable to those of a planetary system. Nor would he probably be prepared to affirm or deny that the extraordinary conditions which reside in the center of the Sun may not set free a portion of this energy.*

In short, Chamberlain claimed that the discrepancy called for an examination of all assumptions, and he was right.

1.11 The Birth of the Theory of Stellar Structure

The only models of stars and the Sun at this time (the beginning of the second half of the 19th century) were models of liquid stars. The logic was simple. The mass and the radius of the Sun are known,[87] so it is simple to calculate the mean specific density of the Sun. The value is 1.342 g/cm^3. The Sun is on the average a bit denser than water. We do not know on Earth any substance with such a high specific density which is in a gaseous state under standard conditions. A density of 1 g/cm^3 on Earth corresponds to liquids or solids. The specific density of air, for example, is a thousand times smaller. Hence, it was natural to assume that the Sun and the stars were liquids. As water is practically incompressible. The assumption about liquid stars is equivalent to the assumption that stars are incompressible.

Many theorems were proven about such stars, and researchers discussed problems like their stability and the possibility of fission when rotating fast or cooling, as discussed by Kelvin and Helmholtz. On the other hand, and contrary to a widely expressed view, Kelvin and Helmholtz could not discuss gravitational energy derived from compression of the Sun, as the Sun in their view was incompressible. The configuration of rotating liquid masses acted upon by self-gravity was a topic investigated by some of the greatest mathematicians, like Colin Maclaurin (1698–1746m) (who gave his name to the Maclaurin ellipsoids), Carl Jacobi (1804–1851m) (whose name is associated with the triaxial objects now called Jacobian ellipsoids), Jules Henri Poincaré (1854–1912m) (who found that these objects can transform into pear-shaped objects), and George Darwin (Charles' son). Ritter turned these fascinating mathematical models into an exotic topic of no relevance to real stars by considering gaseous stars. The first to consider gaseous stars was Zöllner (1834–1882m) in 1871.[88] However, Zöllner assumed that the temperature was constant throughout the star. This assumption is problematic in gaseous models, because it leads to an infinite stellar mass, which is an unacceptable solution.

August Ritter (1826–1908m) was a professor of mechanics at the polytechnic university of Aachen. During the period 1878–1883,[89] he published a series of 18 papers in Wiedemann's Annalen in which he made two basic assumptions. The first was that the stars, including the Sun, are gaseous. While the papers were crucial to the theory of stellar structure, they did not attract the attention they deserved from the scientific community, and in 1898, the editor, George Ellery Hale,[90] of the newly formed Astrophysical Journal, decided that Ritter's series of papers was so important that he initiated the publication of the sixteenth paper in the Astrophysical

---

[87] The mass of the Sun, denoted by $M_\odot$, is 1.895×10^{33} g, and the radius $R_\odot$ is 6.96×10^{10} cm.

[88] Zöllner, E. *Über die stabilität kosmischer Masses*, Leipzig, 1871.

[89] There is a lunar crater named after Karl Ritter (1779–1859) and August Ritter.

[90] The Astrophysical Journal was established in 1895 under the editorship of George Ellery Hale (1868–1938m) and James Edward Keeler (1857–1900m). Edwin Brant Frost (1866–1935m) served first as an assistant editor and then as editor 1902–1935. The crater on the moon commemorates two Hales, George Hale and William Hale (1797–1870m).

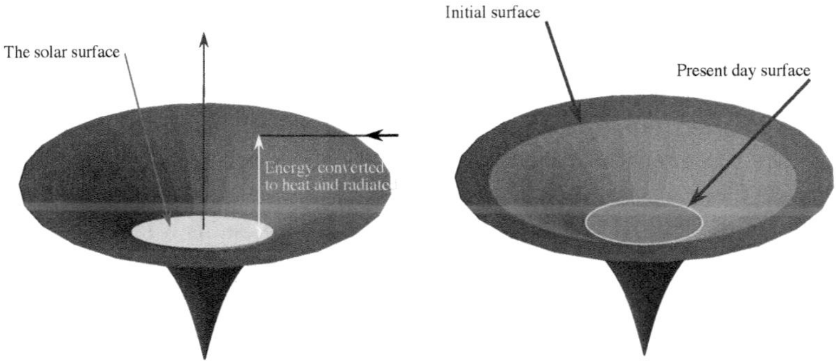

Fig. 1.5 *Left*: The Helmholtz–Kelvin picture. The Sun is inside a gravitational potential well with a fixed radius. A meteor coming from infinity has positive kinetic energy. When the meteor hits the Sun, the original potential energy is converted into kinetic energy during the fall of the meteor onto the potential well, and converted into heat when the meteor finally comes to rest. This heat is the energy released by the falling meteor. *Right*: The Ritter picture. The Sun is inside a gravitational potential well. The initial Sun had a much larger radius and bigger surface. As it contracts, it sinks more deeply into the potential well and releases the difference in energy. The Sun changes its radius in time and releases gravitational energy

Journal.[91] It is inspiring to see how, in his twelfth paper,[92] Ritter applied the newly discovered law of the emission of radiation by a black body, discovered by Jozef Stefan (1835–1893m) in 1879,[93] to relate the emissivity of two stars. Ritter found that the stars emit energy according to the fourth power of the surface temperature and the surface area! Ritter discovered one of the most fundamental equations in stellar structure, namely, that the luminosity L of a star is given by $L = 4\pi R^2 \sigma T_e^4$, where σ is the Stefan–Boltzmann constant, R the radius of the star, and T_e the surface temperature of the star. Moreover, Ritter succeeded in obtaining a relation connecting the central temperature and the surface temperature. This was a major breakthrough,

[91] This paper, Astrophysical Journal **8**, 293 (1898), includes a special preface by the editor, a very unusual move. It also contains a complete reference to the entire series of papers.

[92] Ritter, A., Wiedemann Annalen **14**, 610 (1881).

[93] The Irish physicist John Tyndall (the same Tyndall who challenged Kelvin) measured the ratio of the radiation emission from a platinum wire at 1200°C to the same at 525°C, obtaining the value 11.7 [Tyndall, J., *Heat Considered as a Mode of Motion*, Longman, Green, Longman, Roberts and Green, London (1865) Chap. 12]. Stefan [Stefan, J., Sber. Math. Naturw. Classe K. Akad. Wiss Wien **79**, 391 (1879)] found in 1879 that the ratio of 1200+273 to 525+273 raised to the fourth power is 11.6, and stated the law with just one data point! A few years later, Ludwig Boltzmann (1844–1906m) derived the law theoretically [Boltzmann, L., Ann. Phys., Lpz. **21**, 21 & 291 (1884)]. Two coincidences worked in Stefan's favour: first the emission of platinum at this temperature is far from that of a black body, and secondly, the accurate ratio is 18.6 and not 11.6. The law says that the intensity of the radiation emitted by a perfect body is proportional to the fourth power of the temperature. One can assume with great accuracy that the stars behave like black bodies, i.e., any radiation that falls on them is fully absorbed. Although Stefan made at least two errors in his empirical fitting of the data, he was proven right by Boltzmann's theoretical derivation. For details, see Dougal, R.C., Phys. Educ. **14**, 234 (1979).

since the temperature at the center was found to be tens of millions of degrees K, not a few thousand degrees, as derived by Kelvin.

Ritter was the first to convert what is unjustifiably called the Kelvin–Helmholtz hypothesis into the form it has today, namely, that the Sun has a fixed mass and is gaseous. Upon contraction/compression, the Sun sinks deeper into the gravitational potential well, heating up and radiating away only half of the energy extracted from the gravitational field. While the temperature at the surface is a few thousand degrees, it is higher inside, and Ritter assumed correctly that at such a temperature the matter must be in a gaseous form, vindicating a posteriori his assumption about the state of the matter in the Sun. Moreover, this gas is assumed to obey the ideal gas law,[94] the simplest known gas law. It should be stated that the gases in the Earth atmosphere obey this law to very good accuracy, and the higher the temperature, the better the description fits the behavior of gases, i.e., the more sense the assumption made.

Without any knowledge of an energy source or how energy is transported from the core to the surface, Ritter had to make a second fundamental assumption, namely that the run of temperature and density throughout the Sun is adiabatic. Usually 'adiabatic' means that no heat enters or leaves a particular element of the gas. Here, the meaning is slightly different. Take an element at a given point of the star. Clearly, it is subject to the force of gravity and pressure, and when these forces balance each other, the element is in equilibrium (at rest). Now move this element to another location in the star. As the pressure in the new location is different, the element may contract, for example, this raising its pressure and consequently its temperature, until it is at the same pressure as the new location. If the displacement of the element from one location to the other is carried out without adding/removing heat to/from the element, then we define the displacement as adiabatic. If the temperature lapse in the star is equal to the temperature inside an adiabatically moving blob, then we say that the temperature lapse is adiabatic. The first to discover the adiabatic lapse was Kelvin, when he analyzed the variation of the temperature in the atmosphere of the Earth in 1861.[95] So what Ritter did was simply to assume the same temperature lapse for the Sun. (But note that Ritter actually cited Mohn and Guldberg 1878.[96] Did Ritter ignore British papers?) Once he had made this assumption, he could solve for the run of the temperature, pressure, and density throughout the entire Sun.

As soon as Ritter had the solution for the density throughout the Sun he could carry out the first correct calculation of the gravitational energy release by a contracting Sun, without any additional assumptions concerning the heat capacity, initial

[94] An ideal gas is one described by the simplest model for a gas, based on the assumption that individual molecules in the gas do not interact with one another, and only collide with the container walls. In such a gas, the pressure is given by the density times the temperature times the gas constant divided by the molecular weight of the gas. Ritter cited the Boyle–Mariotte law, which states that the pressure times the volume depends only on the temperature. Although the law was first discovered by Robert Boyle in 1662, and later by Edme Mariotte in 1676, Ritter called it the Mariotte law.

[95] Sir W. Thomson (Lord Kelvin), Mathematical and Physical Papers **3**, 255 (1911), Cambridge.

[96] Mohn, H., Guldberg, C.M., *Uber die Temperaturänderung in verticaler Richtung in der Atmosphäre*, Ztschr. d. Österr. Ges. f. Meteorologie, Nr. 8 (1878).

temperature, initial state of the Sun, and so on. Moreover, Ritter dispensed with the meteor shower, with the cooling of the Sun, and all the problems of meteor observations, i.e., locating and evaluating the maximum mass of meteors that can be hidden without disturbing the orbit of Mercury, and so on. A simple derivation led Ritter to the fundamental formula that the gravitational energy E_{grav} is given by $E_{\mathrm{grav}} = \alpha GM^2/R$, where α is a numerical constant with a value of order unity, G is the universal constant of gravitation, and M is the mass of the star. Thus, Ritter got the correct clean relation between the rate of contraction of the radius and the rate of energy loss from the surface. He was unable to calculate what the rate of energy output of the Sun should be. This was done fifty years later by Eddington. But given the power output of the Sun, he calculated the rate of change of the Sun's radius. Assuming that the Sun radiates at a constant rate all the time, Ritter found that the radius of the Sun must have been 215 times its present radius 5 509 864 years ago. As the radius of the Earth's orbit around the Sun is 215 times the radius of the present day Sun, this result meant that the maximum age of the Earth was about 5 million years. Ritter also checked what would have happened when he relaxed the assumption of constant power output by the Sun, but found only small changes in the results.

It is amusing to note that, as Ritter correctly stated, when the radius of the Sun was 215 times its present value, the Sun as seen from the Earth would have occupied half the sky at midday, i.e., the Sun filled the entire sky! Ritter ended his calculation of the age of the Sun by writing that he could only give a maximum age of 4 million years, since the *original assumption must still be regarded as hypotheses imperfectly satisfied*. In his last paper, Ritter discussed stellar evolution (see later). Darwin's supporters were not happy with the solar age found by Ritter. In this respect he did not solve the problem of the age of the Sun, but brought it to a crisis.

It is astonishing to note that, 25 years after Kelvin had discovered the idea of the adiabatic temperature run in the atmosphere, and after Ritter had completed his series of papers on gaseous stars, Kelvin solved the problem of the equilibrium of a gas under its own gravitation alone, and independently derived most of Ritter's results.[97] The 1887 paper promised a follow-up paper which appeared more than 20 years later, posthumously,[98] and contained a nice summary of the problem prior to the publication of the monumental work by Emden. It is the opinion of the present author that the time scale for gravitational contraction should be called the Ritter–Kelvin–Helmholtz time scale, if not the Ritter time scale. For why should the name of the last to discover it be better than the name of the first?

While Ritter was busy in Europe, Lane was similarly occupied in the US. Jonathan Homer Lane (1819–1880m) was an American mathematician who served in the US coast survey, and from 1869 till 1880 was associated with the bureau of weights and measures. He devoted considerable attention to astronomy, and was often sent under the auspices of the coast survey to observe solar eclipses in various places.

---

[97] Thomson, W., Phil. Mag. **22**, 287 (1887). The paper makes reference to Lane, who published in the USA (see later), but not to Ritter who published in Germany.

[98] Thomson, W. (Kelvin), *The Problem of a Spherical Gaseous Nebula*, Collected Papers **5**, 254 (1908).

As part of an attempt to determine the surface temperature of the Sun, Lane[99] wrote down for the first time the set of equations describing a gaseous sphere in hydrostatic equilibrium. Lane is frequently credited with constructing the first physical model of the solar interior in particular, and the stellar interior is general. (He did not assume an adiabatic lapse, but used instead a more general assumption to be discussed later.) Lane supposed that stars are in an equilibrium between the gravitational force which attempts to pull inward and the gas pressure differences which attempt to push outward. The state of the star is then determined by the balance between these two forces. Interestingly, although his model predicted the central temperature of the Sun reasonably well (when compared to the known present day value), his predicted surface temperature of 30 000 K was way off the mark. This is because Lane's work was carried without the aid of Stefan's radiation law, of which he was unaware. Instead he relied on the earlier work of Dulong and Petit and of Hopkins on the rate of radiant energy from heated surfaces.

Not knowing how the energy is generated and transferred through a star forced astrophysicists to make alternative assumptions. A complete theory which circumvents the unknown energy source, the polytropic theory,[100] was invented and widely used to build stellar models. Of course, several simplifying assumptions had to be made, but a good idea about the possible structure of the star could be obtained.

In 1902, back in Europe, Robert Emden (1862–1940m) published a paper on the structure of the Sun, and in 1907 he published the monograph *Gaskugeln* (gaseous spheres).[101] Emden knew about Ritter's and Lane's papers and extended their results. Emden's book ended the era of stellar models without energy transfer. The most famous equation governing the hydrostatic equilibrium of stars is called the Lane–Emden equation,[102] and it served as the starting point for all theoretical work on stars until the early 1950s, when new methods and computers were introduced, and above all, when the stellar energy source became known.

Although this was an amazing achievement, a full understanding of the complex processes was still not to hand. The polytropic description, although informative, still did not address the question of why stars radiate, or what the stellar energy source might be. The first question was answered with the advent of the quantum theory of black body radiation by Max Planck (1858–1947m) in 1900. After the introduction of the black body spectrum, it was determined that a star was essentially radiating according to the rules of a black body, with the notable exception of the

---

[99] Lane, J.H., Am. J. Sci. 2nd ser. **50**, 57 (1869).

[100] In general, the condition of hydrostatic equilibrium controls the run of the pressure as a function of the density. Since the energy source was not known, it was impossible to determine the run of the temperature. Hence, some assumption had to be made about the temperature run. If one considers a general thermodynamic cycle, like heating with constant effective specific heat, and if one takes it that the same assumption holds in stars, then it can be shown that in this case the pressure goes as $P = k\rho^n$, where n is called the polytropic index constant. If one now includes the ideal gas equation $P = (R_{\rm gas}/\mu)\rho T$, one obtains the temperature run. Ritter's adiabatic lapse is accurately described with $n = 3/2$. Zöllner's constant temperature is given by $n = 1$.

[101] Emden, R., Gaskugeln, Teubner, Leipzig 1907.

[102] Presumably, historical justice would claim that the proper name should be the Lane–Emden–Ritter equation.

Fraunhofer absorption lines. Once the assumption that a star could 'live' off its gravitational energy had been rejected, a new assumption was made, namely, that stars have some sort of unknown internal energy source.

With this in mind, Karl Schwarzschild (1873–1916m) began his work around 1906[103] on how radiation is transferred by the stellar atmosphere into space. The stellar photosphere is defined as the layer in the atmosphere of the star from which radiation escapes into space. In this way, the first accurate estimate of the abundances of the elements in stars could be carried out. Before Schwarzschild, only the existence of the element could be stated with confidence. In this seminal paper Schwarzschild also discovered what is known as the Schwarzschild convection criterion. This is the condition that determines when the atmosphere carries the energy flux by radiation and when it carries it by convection, that is, by mass motions. Despite this ground-breaking work, utterly fundamental to stellar theory, Schwarzschild is probably most famous for the discovery of the black hole solution to Einstein's equations of general relativity (see later).

1.12 Was Solar Contraction Observed?

Do stars have fixed radii and is their luminosity constant in time? When we observe the stars, we soon find that they scintillate, that is, their brightness varies on a short time scale. These fast scintillations are due to the motion of hot air in the Earth's atmosphere. The scintillations are fast, even reaching a rate of a few tens of scintillations per second. However, if one removes these fast atmospheric variations, one finds that stars are very stable, and shine every night with the same luminosity. At least, this is the general rule. Stars that erupt or explode, or brighten suddenly in a fantastic way and then disappear, are not included in the discussion here. They will be discussed later.

In the later part of the 18th century, about 6 variable stars were discovered. By variable stars, we mean stars that change their luminosity in a well-defined, periodic way. Edward Pigott (1753–1825) wondered whether there were any other variable stars that had not yet been discovered? To answer this question, he and his cousin John Goodricke, started to observe the sky systematically. In 1783 Goodricke discovered the light variations of Algol,[104] and attributed them correctly(!) to eclipses by an unobserved companion. In other words, it was an external reason that had nothing to do with the structure of the star. Only in 1890 did Vogel[105] prove Goodricke to be right, when he discovered indications of the companion in the spectrum of Algol (known as a spectroscopic binary). Three years after the discovery of Al-

[103] Schwarzschild, K., Nachr. Kön. Gesellsch. d. Wiss., Göttingen, No. 1, 1906.

[104] Goodricke, J., *A Series of Observations on, and the Discovery of, the Period of the Variation of the Light of the Bright Star in the Head of Medusa, Called Algol*, Phil. Tran. Roy. Soc. London **73**, 474 (1783).

[105] Vogel, H.C., AN **123**, 289 (1890).

gol's variations, Goodricke discovered[106] that the brightness of the star δ Cephei changes by a factor of about 2.5 in intensity over a period of 5 days, 8 hours and 37 minutes.[107] The star had a peculiar asymmetric light curve, quite distinct from that of Algol, which is highly symmetric. Later, when many more such objects had been discovered, stars in this group were called the classical Cepheids. Goodricke realized that there was a difference between δ Cephei and Algol, but could not offer any explanation.

Almost a century later, in 1879, Ritter suggested that the source of the light variation in δ Cepheid type stars was a periodic change in the radius. For Ritter, who was an engineer, the pulsation of a gaseous sphere seemed natural. Incompressible stars could not pulsate, so pulsation was good evidence to support his basic assumption that the stars were in fact gaseous. Yet the idea had to wait almost 40 years for Eddington to provide a mechanism, and most importantly for us here, derive the most fundamental result for gaseous stars, namely, the period–density law. This law states that the period of pulsation times the square root of the mean density of the star is constant ($P\sqrt{\rho} = \text{constant}$). Thus, if the variable stars derive their energy from contraction, the mean density must increase and the period decrease as a function of time. So in 1917,[108] Eddington examined 126 years of measurements of the period of δ Cephei, and found that the maximum change was 0.106 ± 0.011 seconds per year! The contraction theory would demand a period change of 40 seconds per year to account for the luminosity of δ Cephei. And so Eddington found that the contraction theory was in contradiction with observation.

The earliest documented search for changes in the size of the Sun is probably due to von Lindenau,[109] even before the idea was born that contraction might supply the solar energy. Von Lindenau used a transit instrument (an instrument which measures when a celestial object crosses the meridian) to follow the Sun for over two years. The results indicated a periodic change in the diameter of the Sun, in agreement with similar observations carried out in Greenwich about 50 years earlier. On the basis of these results, Lindenau concluded that the Sun is an ellipsoid and hence must be rotating along its longer axis. The radius of the equator was smaller than polar radius by about 1/280 to 1/140. The results and the conclusions were criticized right away by Bessel,[110] who claimed that the error resulted from periodic changes in the instruments.

---

[106] Goodricke, J.B., *A Series of Observations on, and a Discovery of, the Period of the Variation of the Light of the Star Marked δ by Bayer, Near the Head of Cepheus*, Phil. Tran. Roy. Soc. London **76**, 448 (1786).

[107] Actually, on 10 September 1784, Piggott discovered the variability of η Aquilae, preceding Goodricke's discovery of δ Cephei by a week. The two stars show similar light variations. Had Pigott published his results first, this kind of variable star would have been known as the η Aquilae type.

[108] Eddington, A.S., The Observatory **40**, 290 (1917).

[109] Von Lindenau, F.B.A., in Zach, *Monatliche Correspondenz*, June 1809.

[110] Bessel, F.F.W. Zach, *Monatliche Correspondenz*, July 1809.

Piazzi (1746–1826m)[111] and Bianchi (1791–1866)[112] repeated the von Lindenau measurements and got the result that the Sun is indeed ellipsoidal, but that the polar radius is smaller than the equatorial radius by about 1/249. These contradictory results apparently convinced astronomers that the Sun was round.

Secchi was mainly interested in sunspots and prominences (eruptions from sunspots).[113] In 1871, he approached the problem of the dimensions of the Sun[114] and discovered a correlation between the number of sunspots and the diameter of the Sun, namely that the diameter was maximal when the number of sunspots was maximal. This result was confirmed by observations at the Palermo Observatory.[115]

The results by Secchi, though based on a relatively small number of observations (187 in total and during one year), attracted attention because they came after the publication of Helmholtz's and Kelvin's hypothesis. So Auwers[116] thoroughly examined the observational evidence due to Secchi and his predecessors, and concluded that fluctuations in the solar diameter cause these observational errors, and that there was no foundation to the claim that the solar diameter changed with time.

A year later, Newcomb and Holden[117] reached the conclusion that solar variability with a period of several days or longer can be excluded, but that short-time variability, on a scale of hours, could not be ruled out. This new result spurred Auwers into action once more,[118] and he decided to reduce all the data using an equation which allowed periodic variations as well as a gradual secular change in the radius. This time the data reduction indicated that the variations in the number of Sun spots were indeed correlated with changes in the solar radius. But Auwers was unhappy with the results, and continued to accumulate data from various observers. He discovered that some of the results were periodic, while some showed abrupt changes. Moreover, the results of Dunkin showed a secular contraction of 0.006 arc seconds per year, while the results of Downing showed an expansion by 0.01 arc second per year, which corresponds to a change of 6×10^{-6} in the radius per year. Auwers' careful conclusion, after examining 26 000 observations, was that the Sun does not show any long-period variation. All results can be attributed to the variations in the temperature of the instrument.

---

[111] Piazzi, G., Specola Astronomica di Palermo, LIV, VI. This is a case in which Piazzi Smyth (1819–1900) and his godfather Giuseppe were immortalized on the Moon for astronomical discoveries.

[112] Bianchi, G., AN, No. 213, **9**, 365 (1831).

[113] Secchi, A., MNRAS **32**, 226 (1872).

[114] Secchi, A., Atti del'Accademia del Lincei, Jan. 1872.

[115] Hilfiker, J., *Ueber die bestimmung der constante der sonnenparallaxe MIT besonderer berucksichtigung der oppositionsbeobachtungen*, Bern, Buchdruckerei B.F. Haller, 1878.

[116] Auwers, A., *On the Alleged Variability of the Sun's Diameter*, MNRAS **34**, 22 (1873). This paper is a summary by Lynn of the original one published in Berlin: *Ueber eine angebliche Verän derlichkeit des Sonnendurchmessers*, Monatsberichte of the Royal Academy of Sciences at Berlin, May 1873.

[117] Newcomb, S., & Holden, E.S., *On the Possible Periodic Changes in the Sun's Apparent Diameter*, Am. J. Sci. Art., Oct. 1874.

[118] Auwers, A., *Neue Untersuchungen über den Durchmesser der Sonner*, I, II, Sitzungsberichte of the Berlin Academy, Dec. 1886 and June 1887.

The question as to whether the solar variations were real or not was taken up by Poor, who carried out a very thorough analysis of all possible ways to measure changes in the Sun,[119] concluding that:

> *The exact shape of the Sun is not known. The generally accepted idea that the Sun is a sphere is at least open to question. Practically every series of measures heretofore made show departure from a spherical form; but these departures are extremely minute.*

The most recent result on the variation of the radius of the Sun with time is due to Parkinson et al. 1980,[120] who used the transit of Mercury (the passage of Mercury in front of the Sun, when it is seen as a full dark circle on the background of the Sun) and eclipse data to conclude that there has been no detectable change in the solar radius since 1700. A similar conclusion was reached by Shapiro[121] in 1980.

The deviations of the Sun from a sphere, if they exist, should be of paramount importance to the question of the precession of the perihelion of Mercury. If the Sun is not a perfect sphere, then the gravitational force of the Sun acting on the planet is not only proportional to the inverse of the distance squared, but contains an additional component that behaves as the inverse of the distance cubed, and this additional force may be the cause of the precession.

The search for solar oblateness resumed in 1967 when Dicke and Goldenberg[122] discovered a solar oblateness which was sufficient to explain the precession of the perihelion of mercury, implying a major correction to Einstein's general theory of relativity. Several researchers[123] looked for confirmation, but could only set an upper limit of $1 : 10^{-6}$, i.e., the Sun is an extremely perfect sphere. During this extensive work to find the shape of the Sun, it was discovered that the Sun oscillates with many periodicities. These oscillations developed into a new field called helioseismology, which today provides extremely valuable and interesting information about the interior of the Sun.

If the Sun extracts energy from meteors, we should witness meteors falling onto it. Moreover, it is natural to expect to see meteors in the vicinity of the Sun, even if they do not actually fall onto it. Indeed, in 1879, Penrose,[124] in the wake of an unusual eclipse and observed corona, reported meteors moving through the solar corona (see Fig. 1.6). However, the accounts of dark objects falling onto the surface of the Sun were very scant. Denning examined these accounts in 1914,[125] and dismissed them all. In summary, there was no evidence whatsoever that any matter falls onto the Sun.

---

[119] Poor, C.L. *An Investigation of the Figure of the Sun and of Possible Variations in its Size and Shape*, Annal. NY Acad. Sci. **18**, 385 (1908).

[120] Parkinson, J.H., Morrison, L.V., & Stephenson, F.R., Nature **288**, 548 (22 Dec. 1980).

[121] Shapiro, I.I., Science **208**, 51 (1980).

[122] Dicke, R.H. & Goldenberg, M., PRL **18**, 313 (1967).

[123] Hill, H.A., Clayton, P.D., Patz, D.L., Healy, A.W., Stebbins, R.T., Oleson, J.R., Zanoni, C.A., PRL **33**, 1497 (1973).

[124] Penrose, F.C., The Observatory **2**, 302 (1879).

[125] Denning, W.F., The Observatory **37**, 417 (10 Oct. 1914).

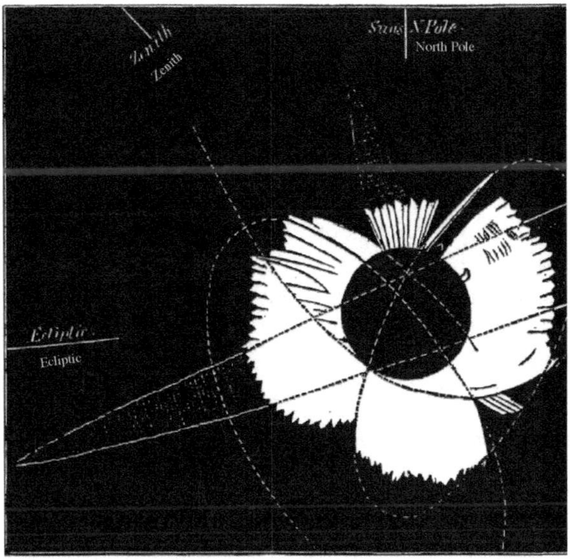

Fig. 1.6 The image of the Sun during an eclipse, as reported by Penrose 1879. The solar corona appeared to be traversed by meteors

Siemens (1823–1883) was a German inventor who worked on thermal and electrical energies. In 1881,[126] he hypothesized that solar energy is conserved. Bothered about the one-way energy dissipation, he suggested that the Sun might conserve its heat by circulating its fuel in space:

> *The elements dissociated in the intense heat of the glowing orb*[127] *rush into the cooler regions of space, and recombine to stream again towards the Sun, where the process is renewed.*

The hypothesis was a daring one and evoked a great deal of discussion, to which Siemens responded in a book.[128]

[126] Siemens, C., *On the Conservation of the Solar Energy*, Proc. Roy. Soc. London **33**, 389 (1881).

[127] In astrology, 'orb' means a radius of up to 10° around the celestial object. It seems surprising that Siemens used astrological terms, but apparently being both an inventor and an industrialist are no guarantee against maintaining certain beliefs. Indeed, it is not at all clear what Siemens meant in this statement.

[128] Siemens, C., *On the Conservation of Solar Energy: A Collection of Papers and Discussions*, Macmillan and Co., 1883.

1.13 Is the Sun Really Liquid?

A very interesting and apparently little noticed article was published in 1898 by See[129] from Montgomery City, Missouri, USA, which is hardly renowned for scientific research. See revisited Helmholtz's hypothesis in the light of Ritter's theory, and examined the Sun. He concluded that the Sun is gaseous and heats up because of the gravitational contraction. When the Sun reaches a sufficiently high density, the gases will liquify, so argued See, stop contracting, and start cooling because it is impossible to compress liquids.

Despite quite convincing arguments and calculations, See's claim either went unnoticed or was rejected, as can be seen from Halm's[130] general exposition on the Sun. Halm supported the general view that the Sun is liquid and cools, rejecting See's hypothesis by means of the following argument:

> *It seems impossible to imagine that the Sun, or in fact, any gaseous star at the Sun's temperature, can ever be liquified by increase of pressure if the temperature increases at the same time. Thus Dr. See's argument that the gaseous state of the Sun at present is to be considered a proof of its being still on the ascending branch of the temperature curve appears to be untenable. This theory, coming into conflict as it does, with one of the fundamental laws of nature, leads to no result which can be adduced against the generally adopted opinion that our Sun, although gaseous, has already passed the point of culmination, and belongs to what may properly be called the class of cooling stars.*

This is clearly a circular argument. It effectively says that, because the Sun is cooling, it must be cooling. However, See had had an interesting idea: the gaseous Sun contracts and heats until it liquifies and stops contracting. From this point on the Sun can only cool. For several years Halm was considered wrong and See right, but later it became clear that both were wrong.

---

[129] See, T.J.J., A.N. No. 3540 **148**, 177 (1898).

[130] Halm, J., *Contributions to the Theory of the Sun*, Annals of the Edinburgh Observatory, **1**, 71 (1902).

Chapter 2
What Stellar Classification Tells Us

2.1 Secchi. The First Steps

When Bunsen and Kirchoff observed the stars spectroscopically, they opened up a new field of observation, and stellar spectroscopy soon became a routine. What the astronomers discovered was a bewildering variety of stellar spectra, to the point of confusion and disarray. There was an urgent need for some order, for once order is established, theories about the evolution of stars, as well as their energy source, can be conceived and checked against observation. It was stellar classification that revealed to astrophysicists where the elements are synthesized and how stars evolve. However, it was not an easy ride.

These were the days of Darwin, Kelvin, and the debate that opposed the theory of evolution and the church, while Galileo's exploits had not yet been forgotten. Yet the church needed the stars. As a matter of fact, the Vatican Observatory is one of the oldest in the world. A major problem with the Julian calendar arose around 1500 AD. The Julian calendar, introduced by Julius Caesar in about 46 BC, was not sufficiently accurate, and accumulated about 10 days of deviation from the solar year over 15 centuries of use.[1] Pope Gregory XIII appealed to the Jesuit mathematicians and astronomers of the Roman College to solve the problem and fix the calendar. Using the Vatican's Tower of the Winds, which housed the Meridian

[1] The Julian calendar was devised to reproduce the tropical year, the time it takes the Earth to go around the Sun, which is 365.242 190419 days long. The Julian calendar is based on 365 days divided into 12 months plus a leap day added to February every fourth year (provided one wants the day and the hour to be of fixed duration). Hence, the average Julian year is 365.25 days long. The small difference between the actual length and the average accumulates, and there is a need to shorten the average Julian year. This is corrected by skipping a leap year every 400 years. The difference is now $0.00781 \sim 0.01$ day per year. In 400 years there are 100 leap days and if one is omitted, it reduces the difference by 0.01 leaving 0.002190 of a day to be corrected. Since the adoption of the Gregorian calendar varied from country to country, astronomers use the Julian day number. For example, 1 January 2006 is Julian day 2 453 750.

G. Shaviv, *The Life of Stars*, DOI 10.1007/978-3-642-02088-9_2,

Hall,[2] these astronomers were able to propose the required correction to the Pope and generate the modern Gregorian calendar (completed 1582). They essentially adopted the proposal by the Italian physician Aloyius Lillus (circa 1510–1578). This was the moment the Pope recognized the power of astronomical research, and from this time on the church began to support it.

In the middle of the nineteenth century, our story encounters the Jesuit[3] Father Angelo Secchi (1818–1878m), who in another sense can also be called the father of stellar classification. This ground-breaking contribution by a priest induced Pope Leo XIII to establish the Vatican Observatory (Specola Vaticana) in 1891.

So what did Father Secchi do to deserve this honor? What he noticed was that some stars have many absorption lines in their visible spectra, while others have relatively few. Between 1863 and 1867, Secchi carried out a remarkable study of the spectra of some 4 000 stars,[4] using a visual spectroscope[5] on the telescope of the Roman College Observatory. He then sorted the stars into five groups, based on the number of absorption lines he could detect by eye. The use of photography for spectroscopy had not yet been invented. With these limitations Secchi defined five different classes of stars (see Table 2.1). A close examination of the classification reveals that:

- the Type V stars are very different from all other types, as they show emission lines and not absorption lines,
- there are two classes of red stars.

No physical explanation was given for the different classes, and neither was there an explanation for the two classes of red stars. Yet, following his scientific instincts, Secchi felt that the two groups of stars were different. It would take 40 years to clarify this observation, and to understand that the stars have a range of temperatures that correspond to the various lines seen in the spectra.

As Secchi was a scientist and a priest, it is of interest to quote some of his writings. For example, in 1856, he wrote:

> *It is with sweet sentiment that man thinks of these worlds without number, where each star is a sun which, as minister of the divine bounty, distributes life and goodness to the other innumerable beings, blessed by the hand of the Omnipotent.*

Did he intend to imply that there might be life around other stars? In 1858, Secchi observed the planet Mars and saw thin lines crossing the surface. He was the one who coined the term 'canali' to describe them. In the same period, Giovanni Vir-

---

[2] The Meridian Hall was a room with a hole in the wall and a straight line on the floor. When the Sun crossed the meridian, it lit the hole, which cast a beam of light on the floor. The time the Sun crossed the meridian, i.e., noon, was thus determined.

[3] The Jesuit order, which also specializes in scholarship, runs the Vatican Observatory.

[4] Secchi, P.A., Catalogo delle stelle di cui si e determinato lo spettro luminoso all', Osservatorio del Collegio romano, Parigi, Per Gauthier-Villars, 1867, 32 pages, Compt. Rend. **63**, 626 (1866).

[5] The spectroscope, or prism, was attached to a telescope. Since there were no means for recording the spectra, such as photography, Secchi made his observations by eye.

Table 2.1 Father Secchi's stellar classification of 1866

Class	Properties	Prototypes	Color
Type I	Strong hydrogen lines	Sirius, Vega	White–blue
Type II	Numerous metallic lines (Na, Ca, Fe), weak hydrogen lines	Sun, Capella, Arcturus	Yellow–orange
Type III	Bands of lines which get darker towards the blue (TiO_2), and metallic lines as in Type II above	Betelgeuse, Antares	Red
Type IV	Bands that shade in the other direction. Faint stars, few visible to naked eye		Deep red
Type V	Bright emission lines, either in conjunction with, or instead of, absorption lines		

ginio Schiaparelli (1835–1910m) published detailed maps of the surface of Mars.[6] Imagine a priest being inspired by a fantastic story about dying life on a neighboring planet.

2.2 Huggins and Lockyer. Scientific Astrophysical Spectroscopy

Two key figures in stellar classification were Huggins and Lockyer, working in England during the latter part of the 19th century. They can be characterized as astronomers who combined spectroscopic experiments in the laboratory with detailed examinations of a relatively small number of stars, the aim being to discover the composition and physical conditions of those stars, rather then to explore a large number of stars and classify them into groups. However, before the contribution of these great astronomers can be discussed and understood, we should mention that all the identifications were carried out by comparisons. There were several attempts to standardize observations, for example, by Kirchoff, who published a spatial scale and a list of lines, but the general state of spectroscopy was really something of a mess. In 1868, Angstrom[7] (1814–1874m) suggested an absolute scale of wavelengths with a unit length of 10^{-10} meters. Later this unit was called the angstrom.

William Huggins (1824–1910m) was an amateur astronomer who built a private observatory in 1856, and devoted his time to spectroscopy. After his marriage in 1875 to Margaret (1848–1915) they published jointly some of the earliest spectra of astronomical objects. In 1864, William Huggins succeeded in matching some of the dark Fraunhofer lines in the spectra of several stars with terrestrial substances,

---

6 Schiaparelli, G.V., *La Planete Mars*. See also Schiaparelli, G.V., *The Distribution of Land and Water on Mars*, PASP **5**, No. 31, 169 (1893).

7 Angstrom, A.J., *Recherches sur le spectre solaire: le spectre normal du soleil*, Uppsale, 1868, p. 1.

demonstrating that stars are made of the same earthborne elements, rather than some kind of exotic substance. Huggins and Miller[8] tried to find an explanation for the fact that different stars have different colors, rejecting all the explanations proposed by Sestini,[9] and suggesting that the difference in composition of the atmosphere gives rise to the different colors. Recall that these were the days when Kelvin's hypothesis about the cooling liquid Sun still prevailed. So the authors assumed that all stars had liquid interiors, which emitted the same light. The core was assumed to be covered by an atmosphere, whose composition determined which part of the light would be absorbed and which would go through unimpeded, thereby creating the color of the star.

Huggins and Miller concluded that the differences between the stars were very small, and yet that these small differences were sufficient to give rise to the variation in color. They went on to argue that:

> *We may infer that the stars, while differing the one from the other in the kinds of matter of which they consist, are all constructed upon the same plan as our Sun, and are composed of matter identical, at least in part, with material of our system.*

This seems to be the first scientifically checked conclusion that elemental composition might be uniform throughout the universe. They also claimed that at least some of the laws of physics prevailing on Earth were valid in the stars, but they did not provide a proof. They went on to hypothesize the existence of solar systems like ours around similar stars. Their conclusion about possible life on other planets was of course stretching their scientific logic and evidence a bit too far.[10]

A correct stellar classification should be carried out without any prejudice or theory of stellar evolution. Lockyer was apparently an adamant follower of Kelvin and Helmholtz, although no reference was made to them in any of his many papers on the subject. According to Lockyer:

> *New stars, whether seen in connection with nebula or not, are produced by the clash of meteor swarms, the bright lines seen being low temperature lines of elements, the spectra of which are most brilliant at a low stage of heat.*

Lockyer published this theory for the first time in 1877,[11] and tried to explain all phenomena on the basis of this theory. He was not generally successful in his attempts. Consider, for example, the phenomenon of nova. A nova is a star that erupts suddenly, increasing in brightness by a prodigious amount, whereafter the light decays over a period of several months. Such a phenomenon disturbed Lockyer, as it did not fit in with the theory. He made attempts to resolve it,[12] but to no avail, and the arguments did not convince his contemporaries.

[8] Huggins, W., & Miller, W.A., Phil. Trans. Roy. Soc. London **154**, 423 (1864).

[9] Benedict Sestini (1816–1890) was a Jesuit astronomer and mathematician who published a *Catalogue of Star Colors*, Memoirs of the Roman College (1845–1847).

[10] Note that Huggins and Lockyer could identify the existence of the same elements on the Earth and on the stars, but could not determine the relative abundances. There was still no theory that predicted how spectral lines form in a stellar atmosphere.

[11] Lockyer, N.J., Nature **16**, 413 (1877).

[12] Lockyer, N.J., Phil. Trans. Roy. Soc. London **182**, 397 (1891).

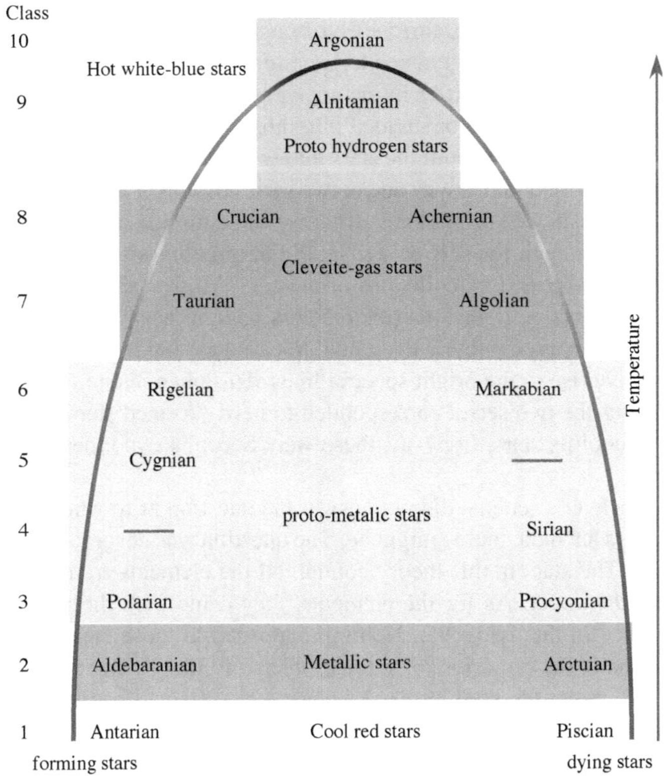

Fig. 2.1 Lockyer's chemical classification of stars (1899). Names refer to typical stars belonging to the relevant class. For example, Algolian implies a spectrum similar to the one found for Algol

Lockyer devised a stellar classification system around the idea that there are two sequences of stars: those that are heating up and those that are cooling down (see Fig. 2.1).[13] According to his meteoritic theory, the stars form cool on the left side of the diagram. They then heat up, and during the gradual heating process, they move through classes 1 to 10. When the stars reach a maximum temperature, they begin to cool off, dropping back down through the classes. As the temperature rises, the spectrum changes. At the beginning of a star's evolution, when its temperature is low, it is red, and appears to be made of metals. When the star reaches its maximum temperature, it exhibits mostly hydrogen, and when it cools down, it eventually disappears as a dead star. All stars experience the same evolution. Lockyer used the term 'proto' to indicate vapors (in contrast to liquids).

[13] Lockyer, N.J., Phil. Trans. **184**, 724 (1893), and a paper entitled *On the Chemical Classification of the Stars*, read before the Royal Society on 4 May 1899 [Proc. Roy. Soc. **65**, 186 (1899)].

But how did Lockyer decide which stars were increasing in temperature and which were decreasing? As a trained spectroscopist, Lockyer noticed the following: stars with the same color (and hence temperature) and consequently belonging to the same class, appeared to differ in the appearance of the hydrogen lines. Stars in the same spectral class could be divided into those with wide, medium, and narrow lines. Hence, Lockyer had to split the stars into two groups according to the shape of the hydrogen lines, and in this way he got two series of stars. Lockyer also noted that one series of stars showed bright lines (what we call emission lines today), while the other did not show such lines. How was he to interpret this situation? According to Lockyer, the stars formed by collisions of meteors. Most of the impacts would not be head-on, but grazing collisions (the meteors passing near one another and rubbing together). This type of collision would release gas, and it was this gas that was supposed to give rise to the bright spectral lines. This phenomenon helped Lockyer decide which of the two series corresponded to newly formed stars and which corresponded to cooling stars. In a way, these were Secchi's two types of stars, but in another form.

Clearly, Lockyer's scheme did not answer the question as to where the elements came from, or what their source might be. The question was never raised by Lockyer (or Huggins). The stars in this theory contain all the elements from the beginning, heat up, and then cool. As for the elements, they came with the birth of the star and were buried in the dying star. Nothing happened to those elements during the entire evolution of the star. Moreover, the problem of binary stars raised by Huggins and Miller was not addressed at all. A binary star system is a pair of stars which revolve around a mutual center of gravity. About 2/3 of all stars are binaries, so the phenomenon is rather widespread. Logic would say that the stars in a binary system were formed at the same time.[14] If so, the spectral class of the pair should be identical, whereas observations show that this is not the case in most binaries. Thus, instead of using the state of the observed binary stars as evidence against his picture, Lockyer argued that it was impossible for the two stars to have formed at the same time. And here lay an unresolved problem. Some twenty years were needed to solve the puzzle.

Helium, the second most abundant element in the universe was discovered in the Sun before it was found on Earth. Pierre-Jules César Janssen (1824–1907), a French astronomer, noticed a yellow line in the Sun's spectrum while studying a total solar eclipse in 1868.[15] Lockyer realized that this line, with a wavelength of 5 874.9 Å,

[14] If the stars were not formed at the same time, one star must capture the other. But the capturing of a star is very complicated, because the binary state means that the stars are bound together, and hence that their binding energy is negative, whereas two free stars would have positive energy. Consequently, for capture to take place, special mechanisms would be required to remove the positive initial energy and leave the system with negative energy. In summary, unless there is a third star around, or some kind of mechanism which dissipates the extra energy, it is difficult to work out a scenario for capture to take place.

[15] Janssen, M., Astronomical Register **7**, 107, 131 (1869). However, these reports on the eclipse do not contain a word about any new element. He discussed only the bright lines he saw from the protuberances on the surface of the Sun. The discovery was announced in Compt. Rend. **67**, 838 (1868).

could not be produced by any element known at that time. Since this yellow line was close to the famous sodium D lines, it was called the D_3 line.[16] Lockyer drafted in the well-known chemist Frankland (1825–1899)[17] to help with the identification of the mysterious line. The paper described how they mixed different gases but could not reproduce the D_3 line. Lockyer hypothesized, therefore, that a new element on the Sun was responsible for this mysterious yellow emission. The unknown element was named helium by Lockyer. Imagine, a single unidentified spectral line was observed and a new element discovered! Moreover, it would turn out to be the second most abundant element in the Universe. It is important to note that it was discovered during a solar eclipse, when the Sun was covered. As late as 1896, Lockyer[18] reached the conclusion that the D_3 line does not form as a part of the spectrum emerging from the solar corona.

Lockyer's biographers[19] claimed that the name helium was coined by Lockyer. Frankland, on the other hand, was more hesitant, as there were quite a number of claims concerning new elements. In later publications on the Sun, Lockyer used the name helium extensively, while in other publications,[20] he used the name cleveite (see later). Lockyer tried the same technique several more times,[21] but luck did not strike twice, and no new lines were discovered.

Lockyer's discovery of a new element was accepted with skepticism. Shuster, for example,[22] wrote:

> *If Mr. Lockyer is right we must look forward to finding some trace of helium, or calcium or hydrogen in the discharge taken from iron poles. When this is done, and not till then, will this theory be considered as proved.*

But one does not find traces of helium in such a discharge, and Lockyer's chemistry (not the evolutionary sequence of the stars) was right after all.

2.3 Is There a Universal Abundance?

In 1880, Plummer[23] suggested that there was an effective universal abundance of elements, pointing out that, out of 16 elements discovered in meteorites, 14 had

---

[16] If you put a grain of salt in the flame of a gas range, you will see immediately bright yellow light. This is the famous sodium D line. The line is actually a double line, but cannot be seen as such with the naked eye. It gives rise to the yellow color of sodium lamps used to light streets.

[17] Frankland, E., & Lockyer, N., Proc. Roy. Soc. **17**, 288, 453 (1869); ibid. **18**, 79 (1869).

[18] Lockyer, J.N., Phil. Trans. Roy. Soc. London, Ser. A **187**, 551 (1896).

[19] Lockyer, T.M., & Lockyer, W.L., *Life and Work of Sir Norman Lockyer*, Macmillan, London 1928, p. 42.

[20] Lockyer, J.N.,Proc. Roy. Soc., London **61**, 148 (1897).

[21] For example, Lockyer,J.N., *On the Unknown Lines Observed in the Spectra of Certain Minerals*, Proc. Roy. Soc. London **60**, 133 (1896–1897).

[22] Shuster, A., *On the Chemical Constitution of the Stars. And Additional Remarks*, Proc. Roy. Soc. London **61**, 209 (1897).

[23] Plummer, J.I., Obs. **3**, 581 (1880).

been seen in the Sun, while the other two were trace elements. Consequently, either meteorites fall into the Sun and make it up, or the Sun ejects them in its frequent eruptions. This universality in the abundance of the elements was thus used to support Helmholtz's idea.

In 1882, Hildebrand used a spectroscope to examine the uranium mineral cleveite,[24] and discovered spectral lines of a mysterious unidentified gas, which was called cleveite gas after the mineral in which it was found. The hunt for helium on the Earth ended in 1895, when Ramsay conducted an experiment with cleveite. He exposed the cleveite to mineral acids and collected the gases thereby produced. He then sent a sample of these gases to two scientists, Lockyer and Sir William Crookes, who were able to identify the helium within them. Two Swedish chemists, Abraham Langlet[25] and Cleve, independently identified helium in cleveite at about the same time as Ramsay.

How come helium was discovered during an eclipse and not in any previous observations of the Sun? During an eclipse, the Moon covers the Sun, but not the corona of the Sun. The apparent diameter of the Moon is just equal to the apparent diameter of the Sun as viewed from the Earth (sometimes a bit less and sometimes a bit more depending on how close the Earth is to the Moon during the eclipse). The temperature of the surface of the Sun is about 5 800 K, while the temperature of the corona is 2 000 000 K. The tenuous corona is a million times less bright than the dense Sun, so it can only be observed during an eclipse.[26] At the relatively low temperature of the Sun's surface, helium lines are not excited (owing to the properties of helium), and hence no helium is seen on the Sun in regular observations. At the surface temperatures of the hottest stars, about 30 000 K, many helium lines appear. At the high temperature of the corona, most of the helium lines are no longer seen (the temperature already being too high), except for the strong D_3 yellow line. It is a pure coincidence that the temperature of the corona leaves one line of helium, while no lines of helium are seen in the solar photosphere.

2.4 Harvard and Potsdam

Before the turn of the 19th century, the centers for stellar classification research moved to Potsdam and Harvard, and the leading figures were Herman Carl Vogel (1841–1907m), who was the director of the Potsdam Observatory from 1882 until his death in 1907, and Edward Charles Pickering (1846–1919m), who was the director of the Harvard College Observatory from 1877 until his death. Pickering and Vogel independently discovered the first spectroscopic binary stars. The irony

[24] Named after Per Teodor Cleve (1840–1905), who was a Swedish chemist and geologist.

[25] Langlet, A., Fresenius J. Anal. Chem. **36**, 79 (1897).

[26] According to Stefan's law, the hot corona should have been $(2\times 10^6/5800)^4$ times brighter than the Sun, because it is so much hotter. But the corona is far from being a black body, because it is so tenuous. One can observe stars through the corona, as was done in an experiment to verify the general theory of relativity.

was that Pickering discovered lines of ionized helium (helium atoms which have lost one electron) in the hot star Zeta Puppis in 1896,[27] and identified it incorrectly as a special form of hydrogen. Later these lines were found in other hot emission line stars and Wolf–Rayet stars.[28] Pickering was convinced that the lines were due to hydrogen under unknown temperature and pressure conditions.[29] Lockyer, who also misidentified the lines, called the spectrum of ionized helium proto-hydrogen (see his stellar classification scheme).

With two influential and charismatic leaders, no wonder stellar classification became such a competitive arena between the old and the new worlds. The Harvard College Observatory was founded in 1839 and was one of the first observatories in the New World. The Potsdam observatory, not far from Berlin, was established in 1874 and quickly became one of the most important centers for astrophysical research. A notable event occurred at the Potsdam observatory in 1881, when Michelson attempted his first reliable experiment to detect the Earth's motion with respect to the ether, in the cellar under the eastern dome. His persistent lack of success in detecting any motion in this and later experiments in America led eventually to the overthrow of the ether theory by Einstein, and set the scene for the special theory of relativity.

2.5 Vogel. The Helium Stars

The first catalogue of stellar spectra was published by Vogel in 1874.[30] The catalogue also contained a classification of spectra. The latter was based on the same mysterious element discovered by Lockyer in the Sun, the element that the physicists and chemists refused to recognize for many years.

In 1895, Vogel upgraded his classification of stellar spectra.[31] In this year, Ramsay confirmed that cleveite gas was indeed helium. Ramsay identified the strong D_3 line Lockyer had seen. Shortly afterwards, Runge and Paschen[32] provided a complete list of spectral lines for cleveite gas, and this allowed a secure identification of the stellar gas with the terrestrial gas.

Vogel himself explained that:

---

[27] Pickering, E.C., ApJ. **4**, 369 (1896). The Brackett line of ionized helium, which has a series limit at 364.4 nm, is called the Pickering line, and is observed in helium stars. This line is seen in O type stars.

[28] Wolf–Rayet stars are hot stars with a high rate of mass loss, and surface temperatures in the range 25 000–50 000 K. The mass loss is essentially due to a strong fast-moving wind that blows continuously out from the star.

[29] The irony is that what were then called the 'additional hydrogen lines' or the Pickering series could be fitted to the Balmer formula, provided half integral quantum numbers were allowed.

[30] Vogel, H.C., A.N. **84**, 113 (1874).

[31] Vogel, H.C., Ap. J. **2**, 333 (1895).

[32] Runge, C. & Paschen, F., Ap. J. **3**, 4 (1896). This paper is a reissue of Sitz. d. K. Akad. d. W. Berlin, July 1895, pp. 639, 759.

> *A rational system of classification is conceivable only on the basis that the different spectra of the stars are indications of different stages of development. In my opinion, it is to be regretted that, in the comprehensive spectroscopic Durchmusterung [survey] of stars [...] to faint stars, which Pickering has undertaken [...], the stars are classified without reference to any general considerations but are merely divided into sixteen classes.*

In other words, Vogel wanted a theory-biased classification, and criticized Pickering for avoiding such a scheme. Vogel complained that his original scheme, suggested over twenty years earlier,[33] had been 'proved' by observations, though it was not clear how a classification can be proven right or wrong. And of course, Vogel maintained that his scheme showed continuous transition between the classes. As stars with bright lines were supposed to be the first stage of development à la Vogel (and Lockyer), they had to belong to the first class. Vogel's spectral classification contained only three classes: white stars, yellow stars, and red stars,[34] and not ten as in Lockyer's classification.

In a meeting of the Berlin Academy on 8 February 1894,[35] Vogel reported on the peculiar double spectrum of β Lyrae and suggested that the motion of the spectral lines might be caused by the motion of two or more bodies. This meant then that there were at least two stars revolving around their center of gravity. So Vogel inferred that *the two stars should have the same composition but differ with respect to density and state of incandescence*. In this way, Vogel reached the correct conclusion that the different conditions on the two stars give rise to two different spectra, in spite of the fact that their composition is the same. While Vogel stressed in his papers that his classification supported the theory of the evolution of stars, his papers never specified exactly what theory of stellar evolution that might be.

2.6 The Henry Draper Project

One of the problems of stellar spectroscopy at the time was that all observations were carried out by eye. There were no technical means to register observations. The breakthrough came in 1872 when Henry Draper (1837–1882m) made the first photograph of a stellar spectrum. The honor of being the first star to have its spectrum photographed went to Vega. This trait ran in the family, because his father John William Draper, made the first photograph of the Moon in 1840, on what were known at the time as Daguerre plates (named after the inventor), while his niece Antonia Maury shocked the establishment with her work on stellar classification (see Sect. 2.7).

---

33 Vogel, H.C., A.N. No. 2000, **84**, 113 (1874).

34 The original classification which appeared in A.N. No. 2000, contained no explanation of the theory of stellar development that Vogel claimed his classification agreed with. I could not find any such theoretical explanation in Vogel's later papers. I can only guess that the combined influence of Helmholtz, Kelvin, and Lockyer was sufficiently strong to affect Vogel's perception of stellar evolution from hot to cold stars.

35 Vogel, H.C., Sitzugsberichte der k. Akad. zu Berlin, 1895, p. 947.

There were still problems with the kind of film used, and efforts were required to increase its sensitivity. However, this was a huge step forward. Draper was also among those to carry out spectral classification of stars, and his original scheme is an expansion on Secchi's four-class classification. Draper used capital letters (A,B,C,...,P) running alphabetically, followed by numerical subcategories (A1, A2,...). It should be mentioned that Draper was a physician who practised medicine, and was even the dean of the medical faculty of the New York City University, where he was a professor of physiology and chemistry. The astronomical community, however, appreciated his work as an amateur. After his death, his wife established the Henry Draper Memorial Fund at Harvard Observatory, supporting the extensive work on the Henry Draper catalogue of stellar spectra. Today astronomers joke by asking for the 'telephone number' of an object, when they need the Draper catalogue number, e.g., HD 12389, so deeply rooted this catalogue has become in astronomers' night life!

2.7 Oh Be A Fine Girl Kiss Me

Edward Charles Pickering was a leading physicist and astronomer who, having come from a prominent New England family, attained a full professorship at MIT at the age of 22, before moving on to Harvard in 1877 to become the director of the Harvard College Observatory at the age of 30. Pickering, quite justifiably, decided to classify a large number of stars without reference to any theory of evolution of the stars. But the job was colossal and well beyond the power of a single man. So Pickering hired assistants, all female, who became known as Pickering's women, to help him with this work. The most prominent names are Willimina Flemming (1857–1911m) (who was a teacher, converted due to circumstances to Pickering's housemaid), Annie J. Cannon (1863–1941m), Antonia Maury (1866–1952m), and Henrietta Swan Leavitt (1868–1921m), who excelled in their work and rose to eminence for the admirable work they did. Rumor had it that the reason for hiring women was the low salary they were paid at the time, about half to a third that of men. What Pickering could not guess, however, was the standard of excellence that would be achieved by these women.

Annie J. Cannon studied physics and astronomy and was hired by Pickering in 1896. In spite of her ardent and important work, it was only in 1938, two years after her retirement, that she got a regular Harvard appointment as William C. Bond Astronomer. The American Astronomical Society established the Annie J. Cannon Award in Astronomy in 1934, while Annie was still alive. The Cannon Award is distributed annually to a woman resident of North America, who is within five years of receipt of a Ph.D., for distinguished contributions to astronomy or for similar contributions in related sciences, which have an immediate application to astronomy.

Antonia Caetana de Palva Pereira Maury was the granddaughter of J.W. Draper and the niece of Henry Draper. Due to disagreements with Pickering about her proposed changes in the classification and their meaning, she left the Harvard College

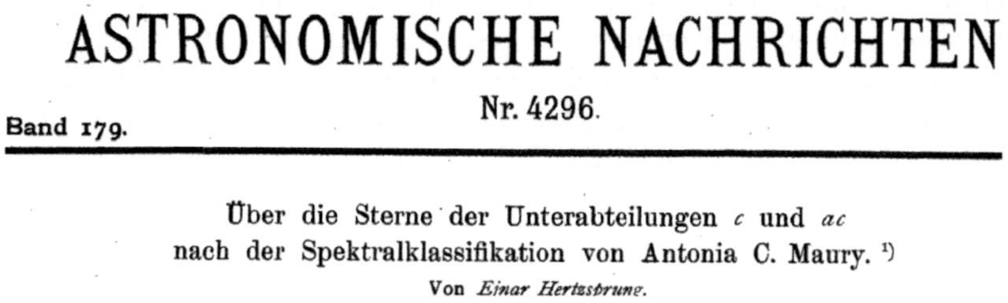
ASTRONOMISCHE NACHRICHTEN

Band 179. Nr. 4296.

Über die Sterne der Unterabteilungen *c* und *ac* nach der Spektralklassifikation von Antonia C. Maury. ¹)

Von *Einar Hertzsprung.*

Fig. 2.2 Hertzsprung's paper, explaining the importance of Maury's unique classification

Observatory to teach in New York. However, in a seminal paper in which he actually discovered what is known today as the Hertzsprung–Russell diagram,[36] Hertzsprung referred to Maury's classification, and gave it a fundamental meaning. The title of the paper contains Maury's name. Maury returned to HCO when Harlow Shapley became the director in 1920, and remained active for many years. In 1943 she was awarded the Annie J. Cannon Award in Astronomy by the American Astronomical Society.

The original Henry Draper Catalogue classification system runs alphabetically, but the Harvard group decided to change the classification to OBAFGKM, so that, out of 22 classes, only 7 were left. The ensuing difficulties in remembering this strange combination gave rise to many mnemonics, the most famous being: Oh Be A Fine Girl Kiss Me.[37]

The anecdote about how the alphabetical order changed to become a famous acronym in astrophysics is blended with many stories that are not always faithful to the events. We prefer to follow the author's own account, that is, the version due to Cannon.[38] To begin with, the letters A to Q were assigned to stellar spectra. This classification was purely empirical, based wholly on external appearances, without any intention of expressing differences of temperature, stages of evolution, or any other physical parameter. The first classification of 10 351 stars was carried out by Mrs Flemming.[39] Miss Antonia C. Maury[40] then discovered small peculiarities in the classification, and made detailed studies of wavelengths and line intensities. As a result, she formed 22 groups of spectra, using Roman numerals instead of letters. Differences in the width of the lines were designated by *a, b*, and *c* to express medium, wide and narrow lines. It is this extra classification that was the center of the

[36] Hertzsprung, A., Uber die Sterne der Unterabteilungen c und ac nach der Spektralklassifikation von Antonia C. Maury, AN 4296, **179**, 373 (1909).

[37] The new order B,A,F,G,K,M appears already in the paper: Pickering, E.C., Ap. J. **6**, 349 (1897), and it is stated that it *indicates divisions in a continuous sequence*, but without mention of a temperature or any other continuous parameter. Pickering, E.C., Ap. J. **7**, 139 (1898).

[38] Cannon, A.J., The Henry Draper Memorial, J. Roy. Astr. Soc. Canada **9**, 203 (1915).

[39] Volume XXVII Part I Harvard Annals.

[40] Annals of the Harvard College Observatory **28**, 1 (1897).

Table 2.2 The Henry Draper Catalogue stellar classification. (The temperatures do not appear in the original catalogue)

Class	Color	Spectral features	T [K]
O	Blue	Strong lines of ionized helium, ionized metals, weak hydrogen lines	40 000
B	Blue	Neutral helium lines, hydrogen lines stronger	25 000
A	White	Strong hydrogen lines, ionized calcium	9 500
F	White	Strong ionized calcium lines, neutral metals	7 200
G	Yellow	Numerous strong ionized calcium lines, strong neutral metal lines	5 800
K	Orange	Numerous strong lines of neutral metals	4 900
M	Red	Numerous strong lines of neutral metals, strong molecular bands	3 600

controversy between Maury and Pickering, and which resulted in Antonia Maury leaving the HCO.

Recall that Secchi had also observed differences in the width of the lines, and decided to separate them, while Lockyer had based his entire theory on these small differences. So it was not a new phenomenon. And yet it did not have any explanation. At the same time, Pickering had qualms about the extra a, b, and c. In 1897, Miss Cannon undertook the classification of the bright southern stars.[41] Cannon noticed that the appearance of some of the letters, such as C, D, and E, were not confirmed by later and better photographs. Similarly, class H was found to be identical with class K when better spectra were obtained. Consequently, these letters were dropped from the sequence. In 1891, Pickering wrote: *The principal question now outstanding is to determine what substance or substances cause the characteristic lines in the spectra of stars of the Orion type.* The Orion stars are a group of very bright stars found in the Orion constellation. The reason for Pickering's problems were the lines of the mysterious cleveite gas seen in these stars, the very lines used by Vogel for his classification.

Before Ramsay identified the cleviete gas as helium (1895), Vogel[42] identified the lines of terrestrial helium with those of the Orion stars, and called them cleveite gas stars. After the identification by Ramsay and the acceptance of helium as a genuine element, the preponderance of such stars in the Orion constellation and the detection of helium in these stars led to them be called helium stars. As it had been clearly demonstrated by the Harvard classification that these spectra precede the spectrum of Sirius (as could be inferred from the hydrogen lines), the letter B, which was

[41] Harvard Annals, Vol. 38 part II.

[42] Vogel, H.C., *On the Occurrence in Stellar Spectra of the Lines of Cleveite Gas, and on the Classification of Stars of the First Spectral Type*, Ap. J. **2**, 333 (1895).

assigned to the Orion stars, clearly had to be placed before the letter A (which is the spectrum of Sirius), or otherwise all the stars previously labeled A and B had to be swapped. Since several thousand stars had already been classified and published, a change in the order of the letters was the only practical course. The remaining original classes were B, A, F, G, K, and M, and they represented the sequence of gradual changes in line properties from one class to another, as far as it was then established.

Further classifications of helium stars discovered that the letter B could not stand for all of them, with their varied line intensities and differences in the number of lines present in their spectra. Cannon therefore decided to divide each class into 10 subclasses, like B1, B2, etc. Again, even this fine division could not overcome the problem of the variations in the widths of the lines, so the groups *a*, *b*, and *c* remained. With the fine division, Cannon found that, even by dividing class O into 10 subclasses, she could find a connection between classes O and B. To put it simply, Cannon found stars (for example 29 Canis Majoris) with a spectra that was just between O and B0. So once again the natural order of the alphabet had to be broken, and O was then placed before B in the stellar sequence. This is how, in Cannon's own words: *The sequence O,B,A,F,G,K,M formed a continuous sequence.* Note that 'continuous' meant that the change in the line ratio and strength was continuous, while the term 'temperature' was not mentioned.[43]

However, even this major step forward, classifying the stars by a physical quantity, turned out to be insufficient to describe the wealth of phenomena exhibited by the stars. Additional sorting was therefore invented, into the so-called luminosity classes. The fundamental underlying question was: can the stars be described by a single physical variable, say the temperature of the surface, or are further physical parameters needed to describe the star in a unique way? It seemed that Maury's classification pointed in that direction. The scientific instincts of Cannon and Maury, leading to the final classification sequence, laid the ground for Hertzsprung's and Russell's discoveries.

2.8 Anjar Hertzsprung. First Correlations

Anjar Hertzsprung (1873–1967m) was a Danish astronomer who trained as a chemical engineer, and was an expert in photochemistry. This may explain why his great discovery was first published in Zeitschrift fur Wissenschaftlishe Photographie and not in a known astronomical journal. After gaining experience as a chemist, he became an independent astronomer, and in 1902 was invited to Göttingen to work with Karl Schwarzschild, and later followed Schwarzschild to Potsdam in 1909, where he became the director after Vogel's death. It was during these years that he carried out the work that brought him fame, in the form of the Hertszprung–Russell diagram, which has since become the single most important tool in understanding the theory

[43] Annals of the Astronomical Observatory of Harvard College, Vol. 91, The Henry draper Catalogue, by A.J. Cannon and E.C. Pickering, 1918.

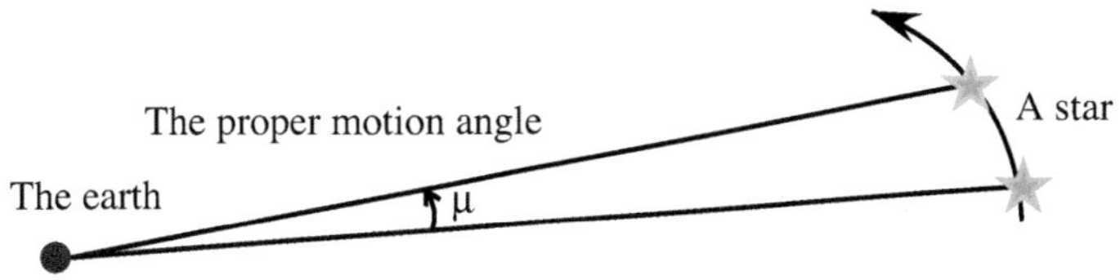

Fig. 2.3 The proper motion μ of a star is the angle the star appears to move through in one year

of stellar evolution. Hertzsprung published about 200 papers, all as sole author. The bulk of the papers were published in observatory publications like Astronomische Nachrichten, which was at that time the bulletin of the Potsdam observatory, or BAN of Leiden, and not in traditional refereed journals. As a rule, his papers were seldom more than three pages long.

In the first paper, Hertzsprung[44] discussed the implications of the spectral classification. He first noticed the refinements to the classification introduced by Miss Maury. As Hertzsprung mentions, Maury guessed that the stars belonging to her classes *a* and *b*, in contrast to class *c*, form a *collateral series of evolution, that is to say, not all stars have the same spectral development*, and he set out to determine whether this was true. It was not the first time that such a possibility had been mentioned. The fundamental and crucial question Hertzsprung posed, and tried to answer, was this: if we brought all stars to the same distance, would we see differences between stars of the same spectral class? The observed stars are at different distances. We observe the bright stars at large distances and the fainter ones only when they are at smaller distances. Does this fact change our perception of the classification? To answer this question, Hertzsprung had to find the distances to the stars. He did so by using their proper motion.

The proper motion of a star is its apparent velocity across the sky expressed as the angle crossed per year (see Fig. 2.3). When the star is very far away, the proper motion is usually a very small angle and cannot be measured. If the star is close, one can expect a high proper motion. The so called 'fixed' stars are not really fixed in the galaxy. They move with speeds of tens of kilometers per second. But as the distances are so large, the constellations appear to us as fixed. Furthermore, with this approach, only the component of the velocity perpendicular to the line of sight is measured, and not the true velocity in space. One can measure the velocity of a star towards or away from the Earth by means of the Doppler effect, provided the velocity is large enough. In any case, Hertszprung had at his disposal only the transverse component of the velocity in the form of proper motion.

Altogether, Hertzsprung had 308 stars with good data. In analyzing the data he discovered that, while the stars of class A all have about the same brightness, the stars of classes G and M each split into two groups, one very bright and one faint. Hertzsprung did not represent the data graphically, but presented it in the form of a table. He hypothesized that the bright red stars form a second collateral evolu-

[44] Hertzsprung, A., Zeitschrift fur Wissenschaftlishe Photographie **3**, 429 (1905); ibid. **5**, 89 (1907).

Fig. 2.4 The Pleiades open star cluster. This cluster is about 425 light years away. Credit NASA

tionary sequence, and gave two lists of stars that form the two parallel sequences of evolution. According to his hypothesis, one sequence has sharp lines, while the other does not.[45] The idea was reminiscent of Lockyer's theory, yet Lockyer was not mentioned in this paper. Hertzsprung ended his paper by stating that: *This result confirms Maury's assumption that the c stars are something unique*. Indeed, this was a colossal discovery. This was the giant branch of stars.

One should point out that two years after Hertzsprung left Göttingen, Hans Rosenberg published the diagram for the Pleiades cluster[46] (it was sent for publication June, 1910) the way we are used to seeing it today, that is, with the log luminosity depicted on the y axis and spectral type along the x axis (see Fig. 2.5). The figure included 41 stars altogether, but only the main series, the liquid cooling stars, were clearly visible. Only 5 stars with spectral type later than A5 were seen. Most of the stars were B class. Rosenberg was the first to draw a diagram of a cluster of stars. The advantage was that, in a stellar cluster, all stars are at the same distance from us, and hence there is no problem of distance determination. We know today that the stars in a cluster of stars were formed at the same time from the same initial cloud of gas, and hence have the same age and composition. This unique property makes the stellar cluster the ideal object for such investigations. Unfortunately, however, Rosenberg's contribution has hardly been recognized.[47]

[45] Hertzsprung's first series was referred to as the main series, because it contained the liquid cooling stars. Eddington would later change the name to 'main sequence'.

[46] Rosenberg, H., A.N. **186**, 71 (1911).

[47] Rosenberg notes that the idea of observing a star cluster was due to Schwarzschild. He also mentions the special classification by Miss Maury.

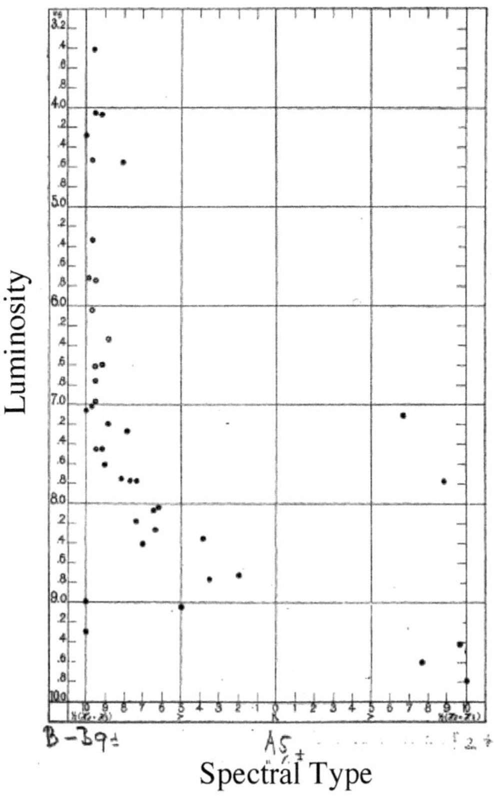

Fig. 2.5 The Rosenberg (1911) diagram for the Pleiades star cluster. The ordinate is the absolute magnitude, i.e., the logarithm of the brightness divided by some standard brightness

While Hertzsprung corresponded with Pickering, communicating all his results, Pickering seems to have chosen to ignore Hertzsprung. But when Karl Schwarzschild visited Harvard for a conference in 1910, he advertised Hertszprung's results and nobody, including Russell and Pickering, could ignore them any longer.

At the same time, Hertzsprung[48] was working on the Hyades star cluster. The Hyades is a relatively small group of stars located at a distance of about 150 light-years from us, which is considered a short distance in the galaxy. The unique feature of a star cluster is that all the stars are to a very good approximation at the same location in space, and hence at the same distance from the Earth. Consequently, the problem of bringing all the stars to the same distance in order to compare their brightness does not exist, and one can compare the stars directly. Hertzsprung did not calculate the temperature of the stars although he had the data for doing it, but instead calculated the wavelength at which the stellar light intensity was maximal

[48] Hertzsprung, A., Potsdam Publ. No **63**, 26 (1911).

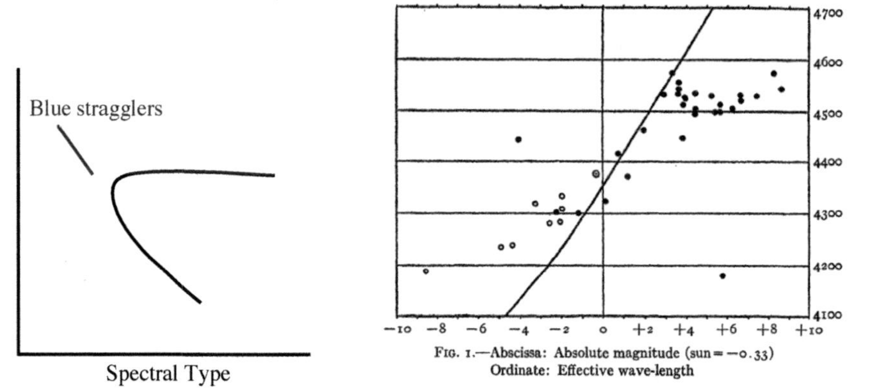

Fig. 2.6 *Left*: The blue stragglers discovered by Hertzsprung are the blue continuation of the main series towards the bluer stars. Hence the name blue. The word ‘straggler’ implies that these stars somehow wandered to this location in the diagram. *Right*: The first HR diagram produced by Hertszprung. The diagram is of the Hyades star cluster. The effective wavelength is related to the surface temperature

from the measured colors of the stars. In this way he plotted a diagram in which the abscissa represented the brightness of the star, increasing from right to left, and the ordinate represented the temperature, increasing from top to bottom. So strange were the coordinates and the diagram that astronomers did not recognize it, let alone understand and appreciate it. So when Russell drew the diagram in 1914 in the form we know it today, it was called the Russell diagram. Nevertheless, several years later astronomers realized that Hertzsprung had indeed drawn what we call to day the Hertzsprung–Russell diagram, or the HR diagram for short, and the name of Hertzsprung was added to the title.

Hertzsprung’s findings were nothing but a vindication of Lockyer’s evolutionary theory, although he refrained in his papers from expressing such ideas. However, in 1925 he discovered a phenomenon which he claimed did not agree with Lockyer’s hypothesis, nor with any other hypothesis advanced at the time. Hertzsprung found that the diagram of the Hyades contained a group of stars that were situated at the continuation of the main group of stars, but with a gap between them and the rest of the stars (see Fig. 2.6). There appeared to be no continuity between this group of stars and the rest. Today these stars are called the ‘blue stragglers’, stars that somehow wandered to this location, and their true nature is still a mystery.

Hertzsprung probably suspected that the concentration of the stars along the horseshoe might not be the evolutionary track of the stars, as Lockyer had hypothesized, but the location of stars with different masses. This may sound a small difference, or even purely semantic, but it had major implications. So he decided to check the mass dependence of the diagram. In 1915,[49] he used the 60-inch telescope on Mount Wilson to observe another star cluster. Once again the plots were in strange units, but

[49] Hertzsprung, E., Ap. J. **101**, 1 (1915).

when presented in terms of our present day system, we discover that he managed to check the effect of the mass, only to find out that he could not discern such an effect. The constancy of the brightness for absolute magnitudes +3 to +8 remained (see Fig. 2.6). Hence, it did not appear to be a mass effect. Hertzsprung's explanation for brightnesses above +3 (due to the unique astronomical notation, this actually means fainter stars!), which corresponds to a temperature of 3 400 K for a black body the size of the Sun, was that relatively dark solid matter forms on the surface of the star, blocking the light. This was, of course, completely wrong, and there was not a shred of evidence to point in this direction. If the dark matter absorbed the light, it would soon heat up, rather than stay cool.

2.9 The 1910 Referendum: Science by Popular Vote?

Any classification of continuous properties by a small number of classes poses a problem: when is the change sufficient to warrant a new class? There were therefore astronomers who classified the stars into 3 or 4 classes, and those who preferred to use a larger number of classes. Next surfaced the problem of what principles should guide the classification: should it express a priori some assumed evolution of stars, or should it be independent of any theory? The use of different classifications duly gave rise to problems and confusion, and by 1904 some two dozen stellar classifications had appeared in the literature. In 1904, Frost[50] asked:

> *Is it not time that a beginning be made by the organization of an international committee to consider the question of a new classification of stellar spectra, representative of the observable facts of the first decade of the twentieth century?*

Eventually, it was agreed by leading spectroscopists to try to resolve the classification question in the 1910 meeting of the International Solar Union meeting in Pasadena. As summarized by F. Schlesinger,[51] the leading contenders were:

- The Draper Classification developed by Harvard,
- Miss Maury's classification, which also originated in Harvard, and
- Vogel's classification devised in Potsdam.

Schlesinger mentioned the classification systems of Lockyer and McClean as used in important research projects, but not as leading classifications.

A series of five questions was composed and sent to leading spectroscopists. These were:

1. It will be noticed that, at the meeting reported above, there seemed to be a practically unanimous opinion that the Draper Classification is the most useful that has thus far been proposed. Do you concur with this opinion? If not, what system do you prefer?

---

[50] Frost, E.B., Ap. J. **20**, 342 (1904).

[51] Schlesinger, F., Ap. J. **33**, 260 (1911). Schlesinger was the secretary of the Classification of Stellar Spectra of the International Union for Cooperation in Solar Research.

2. In any case, what objections to the Draper Classification have come to your notice and what modifications do you suggest?
3. Do you think it would be wise for this committee to recommend at this time or in the near future any system of classification for universal adoption? If not, what additional observations or other work do you deem necessary before such recommendations should be made? Would you be willing to take part in this work?
4. Do you think it is desirable to include in the classification some symbol that would indicate the width of the lines, as was done by Miss Maury in Annals of the Harvard Observatory, Vol. 28?
5. What other criteria for classification would you suggest?

Present day pollsters will tell you right away that questionnaires are formulated in a biased way. This one was no exception. You can find all the replies in the above report by Schlesinger. It is interesting to note that half of the committee were Americans, and eight were Germans, while later, Alfred Fowler, a former student of Lockyer, was added. Lockyer was not on the committee and Vogel had passed away three years earlier.[52] The structure of the committee may be interpreted as an American bias, but it may also be viewed as the rise to dominance of the new world in the field of spectroscopy.

The respondents were unanimously in favor of the Draper Classification, suggesting a few changes here and there, none of which were accepted. In a few cases the idea of mixing stellar evolution into the classification scheme was suggested, and again (correctly) rejected. Some of the comments by the respondents are interesting. Cannon noted that the Draper Classification is based only on wavelengths between 388.9 and 492.2 nm, which is less than the visible range, and in this way many of the stars showing many lines at longer wavelengths were not properly classified. Hertzsprung mentioned the Maury sub-classification as valuable. He also suggested adding a new dimension to the classification, namely, the brightness of the star. As will be seen, this was exactly what he did. The astronomers (and theoreticians) accepted the additional classification only much later. Sometimes it takes the scientific community a long time to accept new ideas. Maury, such a superb observer, preferred a system based on (speculative) evolutionary concepts (something that should not be done), while Russell was strongly against feeding any theoretical conside-

[52] The people asked and reported by Schlesinger were: Adams, W.S., Mount Wilson, USA, Albrecht, S., Cordova, Argentina, Campbell, W.W., Lick Observatory, USA, Cannon, A.J., Harvard College Observatory, USA, Cortie, A.L., Stonyhurst, England, Curtis, H.D., Lick Observatory, USA, Curtis, R.H., Ann Arbor, USA, Ludendorff, H., and Eberhard, G., Potsdam, Germany, Flemming, W.P., Harvard College Observatory, USA, Frost, E.B., Yerkes Observatory, USA, Hamy, M., Paris, France, Hartmann, J., Gottingen, Germany, Hertzsprung, E., Potsdam, Germany, Hough, S.S., Cape of Good Hope, South Africa, Kustner, F., Bonn, Germany, Lord, H.C., Emerson McMillin Observatory, USA, Lunt, J., Cape of Good Hope, South Africa, Maury, A.C., Hastings-on-Hudson, N.Y., USA, Parkhurst, J.A., Yerkes Observatory, USA, Pickering, E.C., Harvard College Observatory, USA, Plaskett, J.S., Ottawa, Canada, Russell, H.N., Princeton University Observatory, USA, Scheiner, J., Potsdam, Germany, Schlesinger, F., Allegheny Observatory, USA, Schwarzschild, K., Potsdam, Germany, Sidgreaves, W., Stonyhurst College, England, Slipher, V.M., Lowel Observatory, USA, Wilsing, J., Potsdam, Germany.

rations into the classification. Moreover, Russell claimed that the bizarre choice of letters prevented anyone from thinking of the classification as based on a theory of evolution.

In reply to questions (4) and (5), Cannon asserted that, although the peculiar spectra fitting Maury's spectral types *c* and *ac* were rare, they should be investigated. No recommendation for special classification was given, however. Flemming's replies resembled Cannon's. Maury's reply to the fourth question is interesting. She relied on Hertzsprung's papers,[53] and recounted that these stars *led him to the conclusion that [they were] bodies at great distance and of super-normal light energy*. There was thus a mutual dependence on each other's research results. She then mentioned a list of stars prepared by Cannon, which she classified as *c*. These stars showed enhanced silicon lines and formed collateral series, and as Cannon noted, the series ended towards Secchi's Type III. This was an indirect statement that Maury's observation was correct.

Maury did not answer question (5) explicitly. In fact, most respondents ignored question (5), while some even preferred pictures. Some suggested using chemical elements as a means of classification instead. It is surprising that Hertzsprung made no reply at all to the question, and did not mention Maury's classification, even though her classification was the starting point for his discovery! Pickering provided a short reply to (4), and none to (5). Russell just suggested replacing Maury's *a*, *b*, and *c* for the width of the lines by Greek letters. There was not a word about the importance of this additional classification. Schlesinger, the secretary of the committee, wrote: *I regard this matter of specifying the width of lines as being of minor importance as compared with other questions that the committee is considering.*

Schwarzschild admitted that the Draper classification was the best, although he was against a recommendation by the committee to adopt it as the unique system for all purposes. His scientific instincts induced him to draw attention to Maury's classification and Hertzsprung's results, and he raised the possibility that there might be more than two variables that determine the spectra (and structure) of stars. Schwarzschild speculated that there might be different abundances in different stars, and that this might show up in the spectra. Slipher, a leading astronomer, simply replied that: *It is important to investigate the width of the line issue*. No more than that. It is clear from the replies that some of the respondents were confused by Maury's *a*, *b*, and *c* classes. The tacit question was: do they run in parallel with the regular classification or not? Russell cited Hertzsprung when he discussed the effect of the brightness on the classification.

We have gone to great lengths here to report the views of this group of leading astronomers, because it is surprising to say the least how such a critical point as the meaning of Maury's classification was not properly appreciated by so many accomplished scientists, even after Hertzsprung had demonstrated its great importance. The doorway to understanding stellar evolution was standing ajar, and few if any saw and appreciated the fact.

[53] AN **179**, 373 (1909); Zeit. fur Wissenschaftliche Photographie **3**, 429 (1905); and **5**, 86 (1907).

Russell asked that the use of 'early' and 'late', terms now so frequently used in describing spectra, be discontinued in favor of 'white' and 'red'. For many years astronomers called the spectral classes A and B 'early type' meaning that these were the first stars, while the spectral types K and M were called 'late type', meaning that these were the old stars. It is unclear when the qualifiers 'early' and 'late' were invented and started to hint, quite incorrectly, at a supposed evolution of stars.

We stress that even in 1912 Pickering still believed that the spectral line he discovered in 1896 in the spectra of Zeta Puppis was hydrogen, even after Alfred Fowler,[54] Lockyer's student, had shown that these lines could be produced in a laboratory by a mixture of hydrogen and helium.

It was only in 1922 that the Draper classification, which was generally accepted by the International Solar Union in 1910, was finally adopted by the recently formed International Astronomical Union.[55]

2.10 Warning Signs

In 1910, while working on a completely different problem, the systematic motions of stars, Jakob Halm (1866–1944) from the Royal Observatory in the Cape of Good Hope discovered[56] a connection between the velocity of stars and their location in the HR diagram. His first question concerned the motion of the Sun with respect to the stars in the galaxy. The Sun is not fixed in space, but moves with a speed of 24.7 km/s with respect to the stars in the galaxy. Halm realized that, when one has a group of stars with different masses in the galaxy, the average speed is inversely proportional to the square root of the mass of the star, i.e., $v \propto 1/\sqrt{M}$.

Consider a collection of particles with different masses, and assume the particles are in equilibrium. This means that the particles exchange energy between them as they collide with one another. The thermodynamic principle in this case tells us that the energies of the particles will be the same, but not the velocities or the momenta. As a consequence, when one has a mixture of gases in the atmosphere, one has molecules of different masses that behave exactly like the stars in the galaxy, moving in such a way that the kinetic energy (the mass times the velocity squared) is constant. Hence, the average speed of the molecule/star is inversely proportional to the square root of the mass. Numerically, the atomic weight of a hydrogen molecule is 2, while that of oxygen 32. Accordingly, the hydrogen molecule has an average

[54] Fowler, A., MNRAS **73**, 62 (1912).

[55] The International Astronomical Union (IAU) was founded in 1919. Its mission is to promote and safeguard the science of astronomy in all its aspects through international cooperation. Its individual members are professional astronomers all over the World, at the Ph.D. level or beyond, and active in professional research and education in astronomy. However, the IAU also maintains friendly relations with organizations that include amateur astronomers in their membership. National Members are generally those with a significant level of professional astronomy. The IAU is composed of 8 993 Individual Members and 62 National Members worldwide (according to statistics in February 2006).

[56] Halm, J. MNRAS **71**, 610 (1911).

Table 2.3 The connection between average velocity and spectral type according to Halm in 1911

Spectral type	Average speed [km/yr]	Number of stars
B-B9	6.0	64
A-A5	11.2	18
F-F8	14.5	17
G-G5	12.6	26
K-K5	15.4	55
M	19.3	6

speed $\sqrt{32/2} = 4$ times greater than the average speed of the oxygen molecule. So Halm found that stars belonging to different spectral classes have different average speeds (see Table 2.3) and masses.

Halm went on to compare the brightness of stars, and found that the Orion type stars were on the average 2.29 times brighter than stars of class A, while stars of class A-K were 5.25 times fainter than stars of class A. Halm reached the very important conclusion that: *The intrinsic brightness and mass are in direct relationship.* This landmark conclusion, which could have drastically shortened the path to the meaning of the Hertzsprung–Russell diagram, was overlooked by everyone in the astrophysics community, save Eddington.

2.11 Henry Norris Russell

Henri Norris Russell (1877–1957m) was the leading American astrophysicist of his day, and an expert in spectroscopy. He was known to physicists through the Russell–Saunder coupling in atomic physics (showing how to calculate the properties of a collection of electrons) and to astrophysicists through the Hertzsprung–Russell diagram. He was deeply interested in stellar evolution and many of his scientific papers dwelt on related problems. Three of the leading American astrophysicists were Russell's doctoral students; Harlow Shapley (1885–1972m), Donald Menzel (1901–1976m) and Lyman Spitzer (1914–1997).[57]

In 1913, Russell addressed the meeting of the Royal Astronomical Society[58] and described the ongoing and still unpublished research in Princeton, explaining how he found his diagram. He took the brightness (luminosity) of each star and plotted it as a function of Pickering's and Miss Cannon's spectral determinations (see Fig.2.7). In this way he discovered that the stars populated certain restricted regions in the plane of brightness versus spectral type. A star of a given spectral class cannot have an arbitrary brightness, and its brightness is actually fixed by the spectral type.

[57] The Spitzer Space Infra Red Telescope carries the largest infrared telescope in space and is one of NASA's Great Observatories.

[58] Russell, H.N., The Observatory **36**, 324. Also, Nature **93**, 227 (1914).

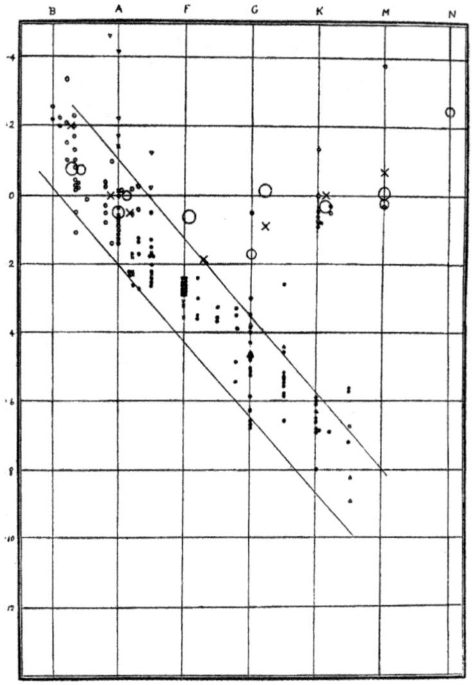

Fig. 2.7 The first spectral class–luminosity diagram drawn by Russell in 1913

Russell found that all faint stars were very red and all blue stars were very bright, but that not all red stars were faint. He noted that the red stars classified as M class stars could be divided into very bright red stars and faint red stars. However, there were no faint blue stars. The phenomenon of two groups of stars was also exhibited in other spectral classes, but as the color became bluer, the difference in brightness decreased, until the two groups merged at class B. Russell's description is drawn schematically in Fig. 2.8. After showing the diagram for stars whose distance had been measured, and whose intrinsic brightness could thus be calculated, Russell repeated Hertzsprung's trick of observing stellar clusters, four in number, for which there was no need to bring all stars to the same distance and for which the brightness comparison could therefore be carried out without any additional correction or calculation.

DeVorkin[59] claims that Russell learnt about Hertzsprung's discoveries from Schwarzschild during a meeting in Harvard in August 1910. A year later Russell wrote to Pickering suggesting follow-up of Hertzsprung's work. Although Russell could already have produced his diagram in 1910, according to the historian DeVor-

[59] DeVorkin, D.H., *The Origins of the Hertzsprung–Russell Diagram.* In: *In Memory of Henry Norris Russell*, Davis-Philip, DeVorkin (Eds.), Dudley Observatory Report 13, Proceedings of IAU Symposium 80 (1977).

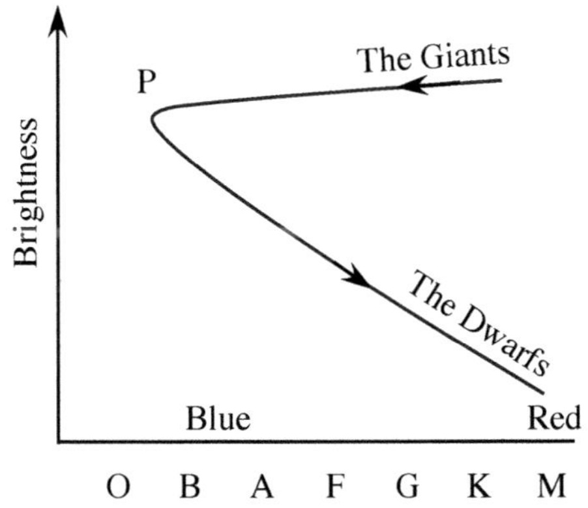

Fig. 2.8 The schematic diagram drawn by Russell in 1913 during the talk given at the Royal Astronomical Society, London. The stellar evolution is marked with *arrows*

kin, he did not do so because he was worried about the meaning of the great luminosity differences between the giants and the dwarfs, a terminology, so it appears, that Russell himself invented. As a matter of fact, Hertzsprung wrote to Pickering as early as 15 March 1906, describing his recent discoveries based on Maury's findings. There is no evidence that Pickering transferred this information to anyone, including Russell. It appears that Pickering was not happy with these findings, and did not attribute any significance to them.

In 1933, Russell[60] credited Hertzsprung as the inventor of the term 'giant'. However, the first time I have found Hertzsprung using this term was in the obituary he wrote on K. Schwarzschild in 1917.[61] It seems likely that the terms 'dwarf' and 'giant' were in fact invented by Russell in 1907, when he lectured in Princeton. Russell used this terminology in his address before the Royal Astronomical Society in 1913, and already at that point attributed its invention to Hertzsprung.

So how did Russell explain the two groups of red stars? Could it be that the brighter stars were more massive? In those days it was impossible to determine the mass of a single star. But the masses of binary stars could be determined by observing their orbits, and using Newton's and Kepler's laws for the motion of objects around their mutual center of gravity. In this way Russell could find the masses of several stars and show that the brighter stars are not always more massive than the faint ones, thus confusing the issue.

It was in his 1913 lecture that Russell used the terms 'giant' and 'dwarf' in public for the first time, to describe the branch of bright stars and the branch of fainter stars. 'Giant' meant extremely bright, while 'dwarf' meant faint. The original terms did not relate to the physical dimensions. Russell concluded that the differences in brightness were not due to differences in mass, but rather to differences in mean

[60] Russell, H.N., JRASC **27**, 375 (1933).

[61] Hertzsprung, A., *Karl Schwarzschild*, Ap. J. **45**, 285 (1917).

density, so that the bright stars had a much larger radius and hence volume, and were therefore brighter for this reason. It is not very convincing logic. The luminosity may be fixed, and as the star expands and increases it radiating area, the radiation per unit area can go down, so that the total luminosity and energy production in the star do not change. As it turned out, Russell guessed correctly, but for the wrong reasons.

Finally, Russell explained how the new findings did not support the then-accepted theory of evolution of the stars, but instead supported Lockyer's theory. The gaseous stars contract, releasing gravitational energy and heating up. The stars move up along Lockyer's left-hand series, as marked by the arrow, which corresponded to Russell's newly discovered giant branch, until they reach such high compression that they liquify. This happens close to the top of the curve (point P in Fig. 2.8). From this point on, the stars cool off and descend along the second series (which according to Russell is the location where most stars are observed). This sounded like a victory for stellar evolution à la Kelvin and Lockyer.

2.12 The Discussion on the Diagram

A year after the lecture in London, Russell sent his results to be published in Nature, the journal Lockyer edited. However, Eddington, who listened to Russell's lecture in London, had already published his criticism,[62] even before the official publication. The conventional hypothesis at the time was that the young stars are those of classes B and A, and for this reason they were called early type. The old stars, according to the conventional hypothesis, were the M stars, and for this reason they were called late type. Eddington based his criticism of the Russell hypothesis on the observational findings that dwarf M stars, which according to Russell (Lockyer and Kelvin) were the oldest of all stars, have on the average spatial velocities exceeding those of the giants. The adopted evolutionary hypothesis would have it that the stars were born with high speed and decelerated with time. It made no physical sense to assume that the stars moved faster as time went by, which was the implication of Russell's evolution theory. Eddington started his career in 1906 as chief assistant to the Astronomer Royal, and his main duty was work on stellar motions. This explains his direct knowledge of stellar velocities.

Russell admitted[63] that the objections to the *conventional picture that the stars of class B are effectively the youngest and those of class M the oldest are so serious that it appears surprising to the writer that this hypothesis is not oftener called in question.* Russell then presented three possibilities:

- The process of star formation has ended. No more new stars are formed.
- Stars undergo the initial contraction very fast, whence it cannot be observed.
- Contracting stars exist and we see them. If this is true, then the giant stars are those undergoing contraction.

---

[62] Eddington, A.S., The Observatory **36**, 467 (1913).

[63] The Observatory **37**, 165 (1914).

Russell then contended that the first two possibilities appeared to him less probable, and consequently he did not discuss them any further. He explained that, according to the Ritter–Lane theory,[64] the more massive the gaseous star is, the higher is the temperature it will reach before starting to turn back. Thus the stars of class B should be the most massive of all stars. Indeed, Ludendorff[65] investigated the masses of binary stars and found that the average mass of class B stars was 4.27 times the solar mass, while the average mass of a star in classes A and F was 1.4 times the solar mass.

But according to the theory, as the massive B stars cool, they should pass through classes A and F, and hence massive cold stars should be found in these classes as well. On the other hand, the most massive stars found in classes A and F were 2.19 solar masses, well below the average for class B. Russell went on to explain this fact by stating that it seemed that stars of class B lose more than 2/3 of their mass before cooling to classes A and F. This was not exactly what Kelvin and Lockyer had had in mind. Moreover, in this case, signs of mass outflow should have been observed in the spectra of the stars, but this was not the case.

As for the objection raised by Eddington (that the velocity of class B is lower than the velocity of classes K and M), Russell brought other pieces of data indicating that the brightest stars belonging to classes F and G have small velocities (and hence could be those stars that were on their way to becoming hot class B stars before starting to cool down). Nobody was convinced by the argument.

2.13 Summary for 1915

The Manchester meeting of the British Association, Section A, 9 September 1915 held a discussion about the spectral classification of stars and the order of stellar evolution.[66] The discussion is interesting because it provides a glimpse into the thinking of various scientists who participated in the discussion. Alfred Fowler (1868–1940m), a well-known spectroscopist, summarized the state of stellar classification. In Fig. 2.9, Fowler showed the relation between the stellar classifications, and he concluded by presenting the generally adopted theory, which was a variation of Lockyer's theory, and the two rival theories, the original Lockyer theory and Russell's new theory.

The generally accepted theory was based on the sequence from gaseous nebulas to red stars. About 99% of all stars, so estimated Fowler, were readily placed at some point on the series. In short, the right-hand column in Fig. 2.9 was interpreted as the evolution of stars. Stars were born in gases and died as red stars. The classification

[64] Note that Russell called the theory of stellar contraction the Ritter–Lane theory, rather than the Kelvin–Helmholtz theory.

[65] Ludendorff, H., *On the Masses of Spectroscopic Binary Stars*, A.N. **4520**, Band 189, 8 (1911).

[66] The Observatory **38**, 379 (Oct. 1915).

Secchi.	Draper.	Colour.	
—	P.		Nebulæ.
[V.]	O.		Wolf-Rayet.
[O.]	B.	White.	Helium stars.
I.	A.		Sirian.
I.-II.	F.		Procyonian.
II.	G. K.	Yellow.	Solar.
III.	M.	Red.	Antarian.
IV.	N.		Piscian (19 Piscium).

Fig. 2.9 Comparison between the classification schemes, according to Fowler in 1915

included the Wolf–Rayet phase as the second stage in the evolution.[67]. Stars form from a nebula, first become very hot stars like the Wolf–Rayet stars, and then cool down to die as red stars, confirming Kelvin and Helmholtz's hypothesis. The vertical column therefore represents stars with decreasing temperatures, from 10000°C for the B stars to 3000°C for the M stars. Actually, claimed Fowler, the best evidence to support this interpretation was Lockyer's laboratory work on how the spectra of various elements change with temperature. The conclusion was therefore that the temperature is the sole factor changing along the series. The chemical composition was identical in all stars.

While the majority of the scientific community supported this hypothesis, Fowler mentioned two who opposed it strongly: Lockyer and Russell. The difference between Lockyer and Russell was minute: Lockyer assumed that M stars were in the early phase of evolution and in the late phase of evolution, while Russell split the M stars into giants, which are young, and dwarfs, which are old. Thus Lockyer ignored the physical size of the stars.

In summary, there were two main hypotheses:

1. The Kelvin–Helmholtz hypothesis: There is a continuous progression from nebulas to red stars, in the order indicated by the Draper sequence, viz., O, B, A, F, G, K, M.
2. The Lockyer–Russell hypothesis: The history of the star begins with a cool nebulous mass, which first condenses into a red star of type M, continues through the yellow to the white stage with increasing temperature, and subsequently, with falling temperature, passes back through the yellow to the red stage, i.e., the order of the evolution, so far as it had been specified, was M, K, G, F, A, B, A, F, G, K, M.

Frank Dyson(1868–1939m) was the ninth Astronomer Royal, and is best known for directing the observations of the Sun and nearby stars in the famous 1919 solar

[67] Wolf–Rayet stars are hot stars with many spectral lines in emission, unlike most stars. Note that hot gases also produce emission lines. Hence, the 'normal' location for Wolf–Rayet stars, according to this classification, would have been between the nebulas and the hot B stars

eclipse.[68] These observations aimed to confirm general relativity. Dyson argued that the evolution was not monotonic in the temperature, but rather in the density. Stars evolved by increasing their density continuously, due to contraction. The contraction continued until a mean density of about 0.1 g/cm^3 was reached, at which point the gaseous star liquified under the enormous pressure and stopped behaving like a sphere of gas.

Father Cortie suggested that the Secchi classification went from simple to physically complex. Cortie therefore supposed that this physical complexity carried some physical meaning, which he hoped to preserve. But what about the nebulas? These were still undetermined, claimed Father Cortie.

Rutherford, who had heard Fowler, asserted that *astronomers may be proceeding too much on the assumption that the evolution proceeds only in one direction. [...] I see no reason why the evolution should always proceed in the direction of condensation.*

Lindemann argued that:

> *If the radioactive evidence for a great age of the Earth is to be trusted, there must be some other unknown source of heat in the Sun and stars. In this case, a revision of astronomical theories based upon Lane's and Ritter's work would be necessary.*

He then made the prediction:

> *If indeed gravitational energy is the sole energy source of stars, then there should be many 'dark stars', which are 'red dwarf stars', which continue to cool.*

Thirty five years later, Mestel essentially confirmed Lindemann's hypothesis.

Eddington, who was the first to react to Fowler's summary, claimed that no physicist would believe that the stars depended on a single parameter, and yet this appeared to be the case. He admitted that he was not an 'out-and-out' supporter of Russell's theory, because it played havoc with a great deal of what seemed orderly and intelligible. He mentioned the problem of the star velocities. As a matter of fact, right after Fowler's summary in The Observatory, there appeared an article entitled *The Relation between the Velocities of Stars and their Brightness* by Eddington,[69] in which he claimed that: *The feebly luminous stars move with much higher average speeds than the bright stars.* Eddington ended by suspecting that the cause for the difference between stars was the mass, so that the bright stars were more massive than the light ones. While the reason was obscure at the time, it posed a serious problem to all evolution theories presented up to then. As will be evident, Eddington had had the right hunch.

In 1917, Adams and Joy[70] carried out extensive research, in which they measured the luminosities and distances of 500 stars. It was the biggest effort so far, and they used the largest telescope in the world at the time, the 100 inch telescope on Mount Wilson, inaugurated in November 1917! The results of Adams and Joy provided complete approval of the giant dwarf branches of stars. But why was this approval so

[68] The expedition was led by Eddington, a fact that added significantly to Eddington's reputation.

[69] Eddington, A.S., Obs. **38**, 392 (1915).

[70] Adams, W.S., & Joy, A.H., Ap. J. **46**, 46 (1917).

important? The number of bright M stars in a given unit of space is very small. The bright M stars are rare. So in order to collect a large number of them and establish the result, one had to look far away. But distant stars hardly move, so one needs a big telescope to detect their motion in the sky.

Stars live for many years and we observe any particular star at a single moment in time. It is conceivable that the composition of stars might change in time, and in this way foul up the meaning of the classification, and with it our interpretation. This question interested Chapman,[71] who explored the way the different elements behave in the gravitational field of a gaseous star, if undisturbed for a long time.[72] Since metals are heavier than hydrogen, his calculations showed that (with time) the heavy element would sink and the light hydrogen would float, so that we should see only hydrogen on the surface of stars. And yet we see metals as well as hydrogen. Hence, concluded Chapman, there must be some agent which prevents the metals from sinking into the star. The explanation was the existence of convective motions, which continuously brought the heavy elements to the surface and prevented the star from settling into Chapman's equilibrium.

This result is extremely important, because it explains how it comes about that we see heavy elements on the surface of stars. If it were not for the convective currents, all stars would expose surfaces to us with only hydrogen in them, and we would not have known about the existence of other elements in stars, let alone been able to determine the relative cosmic abundance of the elements. Chapman treated the stars as gaseous. In this way he could actually carry out the calculation. He added a note in proof after hearing Jean's Bakerian lecture to the effect that his results were probably wrong for the dense dwarf stars, which are not gaseous. The question of mixing haunted the theory of stellar structure and evolution for many years, and continues to beleaguer the theory even today, as will become evident from recent observations of stellar explosions.

The Henry Draper spectral classification remained untouched for many years and continued to play a dominant role in the theory of stellar evolution. In 1935, Russell, Payne, and Menzel[73] reached the conclusion that the classification brought with it a host of problems, and that new principles and physical prerequisites were therefore needed. However, so much was invested in the Draper classification that it is actually impossible to replace it. In the words of the authors:

> *From its first days this system served only to place the spectra in convenient pigeonholes, from which those worthy of special study could be withdrawn, and redistributed with labels, such as Miss Maury's a, b, and c.*

They listed some 13 different problems. However, these changes had to wait.

[71] Chapman, S., MNRAS **77**, 539 & 540 (1917).

[72] Chapman assumed a steady state, and did not calculate how long it would take for the steady state to be established. Further, he assumed the matter to consist of unionized atoms, an assumption which later turned out to be wrong. Finally, Chapman did not provide the time scale for the sinking of the heavy elements.

[73] Russell, H.N., Payne Gaposchkin, C.H., & Menzel, D.H., Ap. J. **81**, 107 (1935). The first in Princeton and the other two in Harvard, where the Draper system was devised.

Chapter 3
The Dawn of a New Era

3.1 The Ultimate Answer to the Earth Age Problem: Radioactivity

The later years of the 19th century and the beginning of the 20th saw dramatic and rapid progress in our understanding of atomic structure. The discovery of radioactivity played a pivotal role in the unraveling of the atomic microworld. And radioactivity is important to our subject for several reasons:

- it provides the heat source of the Earth,
- it allows accurate dating of rocks,
- it opens the door to meteorite dating and determination of the age of the Solar System,
- it indicated for the first time that there is an end to the periodic table,
- it allowed Rutherford to discover the structure of an atom in which most of the mass is concentrated in the nucleus.

3.2 The Complicated Structure of the Earth

In 1897, a discovery by the Irish geologist Richard Dixon Oldham (1858–1936)[1] provided early clues about the nature of the Earth's interior. By analysing the propagation of seismic waves, he found that these waves move through the interior of the Earth in a non-isotropic manner, i.e., the speed of propagation varies with direction.

There are many types of seismic wave, and Oldham discovered the two most important: the P and S waves. Primary or P waves are pressure waves (the P can also stand for pressure), while the secondary or S waves are shear waves (the S can also stand for shear). An S wave can propagate only through media like solids

[1] Oldham, R.D., *Report on the Great Earthquake of June 12th, 1897*, Office of the Geology survey, Paul, Trench & Trübner & Co.

G. Shaviv, *The Life of Stars*, DOI 10.1007/978-3-642-02088-9_3,

that can transmit shear forces. On the other hand, P waves can propagate through any medium which can be compressed, namely, solids, liquids, and gases. The two waves have different speeds of propagation. The speed of the primary wave is the greater (hence the name primary, since it arrives first). The speed of both waves varies with density, a phenomenon which can be used to 'measure' the run of the density with depth. Thus the speed of the P waves changes from about 6.4 km/s in the outer crust of the Earth to almost twice that speed in the core.

By 1906, Oldham's investigations had progressed to the clear identification of the core of the Earth.[2] The core appeared as a shadow, or a fata morgana, for the P waves. The bending of the waves around the core did not allow one to 'see' the core,[3] providing evidence that the Earth is composed of different layers. In 1909, Andrija Mohorovicic (1857–1936m)[4] analyzed an earthquake in Croatia and discovered a sharp boundary between the crust and the mantle of the Earth. The boundary is known today as the Mohorovicic discontinuity, or the Moho for short. In 1914, Beno Gutenberg (1889–1960)[5] estimated the diameter of the core to be about 4 375 miles (7 000 km), a figure that still stands today.[6]

The recognition that the structure of the Earth is very complex implied that geological methods based on erosion in the continental plates would at best provide an age estimate for the continent, but not an age for the Earth. An independent method was needed, free from all the geological evolution of the Earth.

3.3 The Invention of the Cathode Ray Tube

In 1855, Heinrich Geissler (1815–1879m), a glass blower from Bonn, Germany, invented a new vacuum pump using mercury. The advantage of using mercury is that it has a low gas vapor pressure. The new invention allowed physicists to reach sufficiently high vacuums to allow gas discharges to take place continuously. Geissler had already used such tubes to get different colors when he filled them with different gases and applied a high voltage to the electrodes.

The first to apply a magnet to the glowing gas was Julius Plucker (1801–1886). He discovered that the diffuse light could be concentrated by the magnet. The effect did not depend on the type of gas used. In 1869, Wilhelm Hittorf (1824–1914), who was Plucker's student, placed a solid object inside the tube and discovered that it cast a shadow on the walls of the tubes. He noticed that only the cathode cast this

[2] Oldham, R.D., J. Geol. Soc. London **62**, 456 (1906).

[3] In 1926, Harold Jeffreys discovered a similar and more pronounced phenomenon in S waves, which indicated that the core is liquid.

[4] Mohorovicic, A., Das Deben, Jb. met. Obs. Zagreb **9**, 63 (1909).

[5] Gutenberg, G.Z. Geophys. **14**, 1217 (1913).

[6] In tribute to this discovery, the boundary between the mantle and the core is called the Gutenberg discontinuity. If it were a crater on the Moon or Mars, there would be hope of observing the discontinuity directly.

shadow. As a matter of fact, he discovered the cathode rays, but did not identify the phenomenon as such.

Many improvements in cathode ray tubes were introduced by William Crookes (1832–1919m). Like his predecessors, he played with the shape of the tube and the shape of the cathodes (e.g., concave and convex ones), he changed the pressure of the gas by improving the vacuum, and towards the end of the 1870s, he asserted that the cathode rays were negatively charged particles. By placing a paddle wheel composed of mica inside the tube, he was able to observe how the radiation turns the wheel. He then concluded that the radiation is composed of 'some particles'. The particles, so he suggested, were molecules that hit the cathode, and as a consequence captured a negative charge.

Lockyer[7] examined the spectra in the Geissler tube and compared them with stellar spectra. Moreover, he succeeded in showing how the resulting spectral line varies with the pressure in the tube.

When a large voltage is set up between two metal plates, there is a tendency for the charged plate to discharge by a short spark, like lightning. This is called an electrical discharge. The lower the pressure, the sooner the discharge takes place. At sufficiently low pressures, the discharge generates cathode rays. Cathode rays are invisible, but various effects caused by them disclose their existence.

3.4 Dispersing Kelvin's Clouds. A New Horizon

Wilhelm Wien (1864–1928) was the son of a landlord, destined to follow in his father's footsteps. However, an economic crisis forced him to change course, and study mathematics and natural sciences. Wien was a student of Helmholtz, and prepared his PhD thesis under his guidance. In 1900 he succeeded Röntgen in Wurzburg, and in 1902 he was invited to succeed Boltzmann in Leipzig, but refused. In 1893, Wien[8] discovered that the wavelength of the light emitted by a hot body changes with temperature, a law which was later called Wien's law of displacement.[9] The radiation emitted from a hot object is a continuum. It vanishes for short and long wavelengths, and in between has a maximum (see Fig. 3.1). Wien's law relates the temperature T of the emitting body and the wavelength of the maximum λ_m, in a very simple way, viz., $T\lambda_m = \text{constant}$.

In 1894, while discussing the thermodynamic properties of radiation in space, Wien defined the ideal body, which he called a black body, as the body which absorbs all radiation incident upon it. In 1896, he published his phenomenological formula for the law of displacement, and discovered a short time later that his for-

[7] Lockyer, J.N., *Recent Research in Solar Chemistry*, 1875. See also, Lockyer, J.N., Proc. Roy. Soc. London **61**, 148 (1897).

[8] Wien, W., Annal. der Phys. **58**, 662 (1896).

[9] Wien's law gives the wavelength at which the intensity of the black body radiation reaches its peak. The law states that $T\lambda_{peak} = 0.23$, where the temperature T is given in degrees kelvin and the wavelength λ_{peak} of the peak intensity in centimeters.

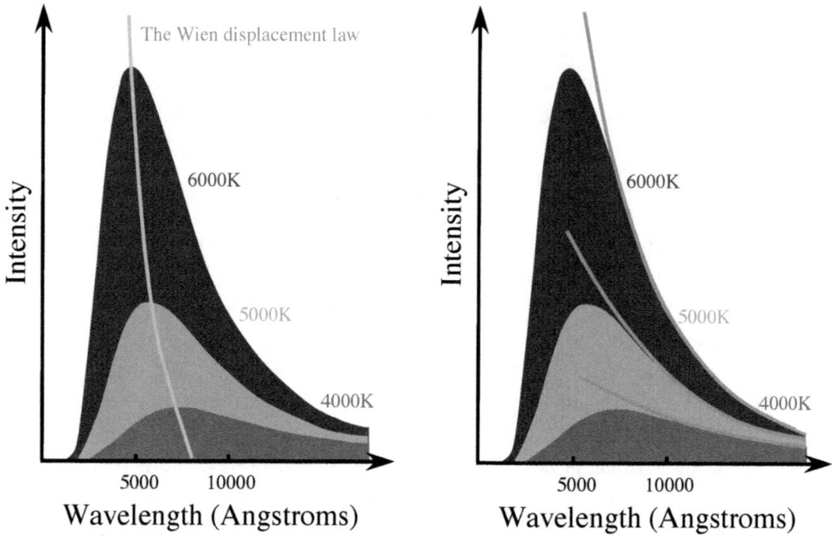

Fig. 3.1 *Left*: The black body radiator, the perfect emitter. The emitted radiation has a maximum. In contrast to the predictions of classical physics, the maximum depends on the temperature. Wien's displacement law states that the higher the temperature, the shorter the wavelength. *Right*: Radiation from a black body and the Rayleigh–Jeans law. The *green curves* are the Rayleigh–Jeans classical results for long wavelengths. Agreement breaks down before the maximum is reached

mula worked only for short wavelengths, failing badly at longer wavelengths. It was for this discovery, which could not be explained by the theory, that he was awarded the Nobel Prize in 1911.[10] This discovery was Kelvin's famous first cloud, when he described the state of physics in the late 19th century.

In physics, a black body is a perfect absorber, that is, it absorbs all the radiation falling onto it. The classical black body is a cavity (see Fig. 3.2). As can be seen from the figure, a ray which penetrates the cavity is partly absorbed and partly reflected by the walls, depending on the properties of the material. However, because the opening is very small relative to the size of the cavity, the incoming ray is finally absorbed by the walls of the cavity before it has a chance to escape.

On the other hand, the walls of the cavity are kept at a fixed temperature and emit radiation. This radiation also has difficulty leaving the cavity, and is reflected/absorbed many times before a small part of it can escape. So what happens in the cavity? The walls emit radiation, and the radiation interacts back with the walls, and as a consequence the radiation in the cavity is in equilibrium with the walls. If the walls absorb too much radiation they will heat up and vice versa. The radiation 'feels' the walls, and the walls emit radiation which is characteristic of their tempe-

[10] In his Nobel address, Wien explained that the displacement law for which he had won the prize had already been elucidated by Planck in 1900. However, the theoretical explanation by Planck was only awarded the prize eighteen years after the discovery, in 1918. Wien also won the prize eighteen years after his discovery.

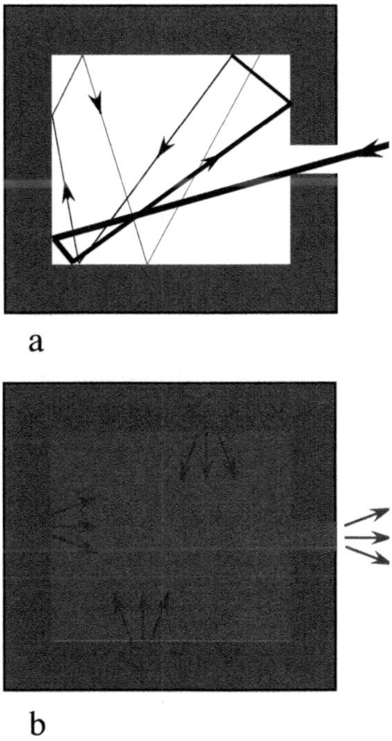

Fig. 3.2 The cavity as a black body. (**a**) A ray entering the cavity is absorbed after many reflections and absorptions. The width of the line reflects the intensity of the ray after each reflection/absorption. (**b**) The thermal emission by a cavity

rature. A small part of the radiation escapes the cavity, and this is what is called the thermal emission of a black body. It is difficult to overestimate the important role the notion of black body plays in physics and astrophysics.

The nature of the cathode rays turned into an international controversy. Most British scientists followed Crookes, while the Germans believed they were disturbances in the ether. When Wien moved to Aix-la-Chapelle in 1896 to succeed Lenard, he took over Lenard's laboratory and started to investigate the nature of cathode rays. He soon confirmed Perrin's earlier discovery that the fast-moving cathode rays were composed of negatively charged particles. He then measured the charge-to-mass ratio and essentially discovered that cathode rays are beams of electrons.[11] However, he was scooped by J.J. Thomson in Cambridge, who had already identified the electron earlier that year using another method.[12] This discovery was important because it was the first time an object smaller than the atom, in fact a constituent of the atom,

[11] Wien, W., Annal. der Phys. **301**, 440 (1898).

[12] Thomson, J.J., Phil. Mag. **46**, 528 (1898).

had been discovered. Weichert[13] and Kaufmann[14] carried out similar experiments and got the same results, but they did not have the courage, or the confidence, to reach the conclusions Thomson had proclaimed.

Wien was not discouraged by being scooped, and in 1898, he investigated the canal rays discovered by Goldstein, discovering that they were composed of positively charged particles. He managed to measure the mass-to-charge ratio of the particle and found that it was much greater than the mass-to-charge ratio of the electron. Unfortunately, he was unable to measure the charge and the mass separately. However, this effective discovery of the proton was not recognized by the scientific community. It was only in 1913, after J.J. Thomson had refined Wien's experimental method, which became the basis for mass spectroscopy,[15] and after Rutherford's work in 1919, that Wien's discovery of the proton was confirmed but not credited. Wien's consolation for being scooped twice was that he won the 1911 Nobel Prize. The citation mentioned only Wien's contributions to the study of black body radiation.

The first measurement of the charge of the electron was carried out by Townsend in 1897.[16] Townsend found that, as he separated weak acids into hydrogen and the acid radical by means of electrolysis, the hydrogen bubbled through the solution and formed a cloud of charged particles. By assuming that all charges were identical he could estimate the basic unit of charge. The experiment was perfected by Thomson[17] and Wilson[18] in 1903.

Against this background, Millikan (1868–1953m) started his long effort to obtain an accurate value for the elementary charge. He used a somewhat different technique. He sprayed oil drops into a vessel and watched them fall. The friction between the falling drops and the air charged the drops. Millikan then applied an electric field to halt the fall of the drops. Once a droplet was hanging in the air, he could easily determine its charge from the value of the voltage needed to halt the fall. This famous experiment, known today as the oil drop experiment,[19] won him the Nobel Prize.

In the fall of 1895, Wilhelm Conrad Röntgen (1845–1923m), who was a professor of physics and the director of the Physical Institute of the University of Wurzburg, became interested in the nature and properties of cathode rays. He obtained a special tube built by Lenard, and repeated some of Lenard's experiments.

---

[13] Weichert, E., Ann. Phys. **544**, 61 (1897); Weid. Ann. **69**, 739 (1898). In the first paper Weichert invented the term 'elektron'.

[14] Kaufmann, W., Wied. Ann. **61**, 544 (1897).

[15] Mass spectroscopy is a method based on separation of the masses of particles in a beam. In light spectroscopy, the beam is split into different wavelengths. The components are the wavelengths. In mass spectroscopy the beam of particles is separated into particles with different charge-to-mass ratios.

[16] Townsend, J.S., Proc. Camb. Phil. Soc. **9**, 244 (1897).

[17] Thomson, J.J., Phil. Mag. **46**, 528 (1898).

[18] Wilson, H.A., Phil. Mag. **5**, 429 (1903).

[19] Millikan, R.A., Phys. Rev. **2**, 109 (1913).

Röntgen was very industrious and worked long hours, frequently even sleeping in the laboratory. He was interested in the nature of the radiation inside the cathode ray tube, and in particular the effect it had on air. For this reason, he replaced the regular screen with a very thin piece of aluminium foil. Three elements combined to make an exceptional discovery which catapulted Röntgen to fame: the fact that he was interested in the effect on air and hence used aluminium foil with a particularly high voltage to get strong radiation, the fact that he worked late hours and the laboratory was dark, and the fact that he was an extraordinary experimentalist whose attention and observation nothing could escape.

The perception that he was on the verge of a great discovery occurred to him one evening when he was working late in the laboratory. The details are important, because research with cathode ray tubes was being carried out in many laboratories, but only Röntgen made the discovery, because he carried out the experiment in a dark laboratory. Starting the experiment, he noticed a green light emanating from the screen. Röntgen realized that the green light could not have been produced by the cathode rays, since it was well known that they could not pass through the walls of the tube. Visible light could not be the source either, since during the experiment the tube was covered with a shield that was opaque to light. So Röntgen hypothesized correctly that he must have been producing some unknown type of radiation.

As soon as Röntgen had convinced himself of his discovery of X rays, he sent copies of his manuscript to the leading scientists of the day, including the French mathematician Henri Poincaré. When Poincaré got the manuscript he was excited, and three weeks later, announced it at the meeting of the French Academy of Science.[20] Poincaré reported Röntgen's new discovery, and wondered to what extent luminescent material[21] could emit X rays.[22] In 1852, George Stokes discovered the Stokes law for fluorescence, which says that the emitted wavelength is always longer than the absorbed wavelength. So what Poincaré wondered was whether the opposite might be possible, that is, could X rays be emitted by fluorescent materials? Today we know from Planck's theory that the longer the wavelength of a photon, which is a particle of light, the lower is its energy. Hence, Stokes law states that the emitted light has a smaller energy than the absorbed light (see Fig. 3.3). Poincaré made the

[20] Röntgen, W., Invited talk at the Wurzburg meeting of the Physico-Medical Soc. **9**, 132 (1895). A summary of the talk can be found in Röntgen, W.C., Nature **53**, 274 (1896).

[21] Luminescence is a phenomenon in which a material absorbs light of one wavelength (color), and emits light with another wavelength (color). Fluorescence is the phenomenon where a material is illuminated by, say, visible light, absorbs the light, and re-emits (non-identical) light later. The absorbed light causes one of the electrons to jump to a higher state. The electron does not stay in the higher state, and after a while 'wants' to return to the original state. However, the way back may be through other, intermediate states, and not necessarily directly to the original level. Fluorescence is a form of luminescence which is mostly found as an optical phenomenon in cold bodies, in which a molecule absorbs a high-energy photon, and re-emits it as a lower energy (longer wavelength) photon. The energy difference between the absorbed and emitted photons ends up in molecular vibrations (heat). Usually, the absorbed photon is in the ultraviolet, and the emitted light (luminescence) is in the visible range, but this depends on the absorbance curve and Stokes shift of the particular fluorophore. Fluorescence is named after the mineral fluorite (calcium fluoride), which exhibits this phenomenon.

[22] Poincaré, H., Revue Générale des Sci. Pure et Appliq. **7**, 52 (1896).

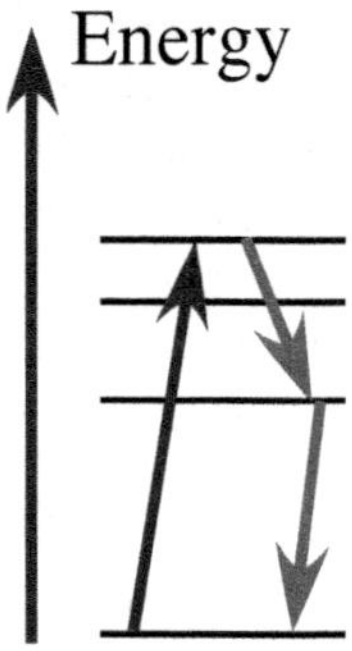

Fig. 3.3 An electron absorbs a photon and jumps to a higher energy level (*blue arrow*). The electron can return via an intermediate state, and hence emit photons with longer wavelength (*red arrows*)

opposite conjecture, asking whether the absorption of light could trigger the emission of more energetic light. It was only in 1912 that Max von Laue (1879–1960m) succeeded in showing quite definitely that X rays were electromagnetic radiation similar to light, but much more energetic.[23] Röntgen had the unique honor of being the first to be awarded the Nobel Prize in 1901, when the prize had just been established.

Henri Becquerel (1852–1908m) heard about Poincaré's conjecture and got excited. His father, Edmund, was a physicist and an expert on fluorescence. Among the materials he had studied were several uranium compounds. Edmund discovered peculiarities in the emitted light of uranium compounds, and Henri continued to study these compounds with the hope of clarifying the nature of the peculiar fluorescence light. He would place his samples in the Sun and later observe the emitted light (in the dark). At first, Becquerel thought he had discovered that sunlight induced his uranium crystals to produce something like fluorescence, blackening the photographic plates through the paper they were wrapped in. But fortunately, February 1896 was a grey winter in Paris, with heavy overcast skies and no Sun to illuminate his crystals, so he put them away in a drawer. To his surprise, when he developed the plates a month later, they were nevertheless blackened. Invisible radiation from his crystals had caused this. At first, Becquerel discovered that the three uranium compounds he was investigating were all fluorescent, emitting invisible light. But later he found that there exist uranium compounds that do not display any fluorescence and yet nevertheless emit the penetrating radiation.[24] He appeared to have confirmed Poincaré's conjecture with the first compounds, but then found that the penetrating radiation was emitted without any external excitation! The radiation was emitted even when the sample was left alone.

We know today that uranium ore is radioactive and emits rays – initially called Becquerel rays – which easily passed through the wrapping paper and caused a chemical process in the plates, as if the latter had been exposed to visible light. Becquerel managed to show that the radiation was made up of different parts with

[23] Friedrich, W., Knipping, W.P., & Laue, M., Sitz. Bayrische Acad. der Wissen. **303**, 322 (1912).

[24] Becquerel, H., Comp. Rend. **122**, 420 (1896).

different penetration powers. The three kinds of radiation were not identical, but had different energies.

A crucial part of the glorious work on radioactivity was carried out by the Polish born Marie (1867–1934) and French Pierre (1859–1906m) Curie, who became one of the most famous couples in the history of science.[25] The couple worked intensively on radioactive ores and identified the pure element thorium as well as the sequence of radioactive elements which follow its decay. In 1898, they distilled[26] the radioactive ore pitchblende[27] to discover the new element polonium and to separate radium. After various chemical distillations and treatments with various acids, they succeeded in isolating a new element which led them to state that:

> *We believe therefore that the substance which we have removed from pitchblende contains a metal not yet reported, close to bismuth in its analytical properties. If the existence of this new metal is confirmed, we propose to call it polonium from the name of the country of origin of one of us.*

The Curies noted also that their attempt to confirm the new element spectroscopically had not succeeded, because U, Th, and Ta have very complicated spectra, and this prevented the identification of a new element. Only after a definite spectrum of radium had been obtained by Eugene Demarcay[28] in 1898 could Soddy report[29] to the British Chemical Society (as late as 1905) that radium had a well-defined spectrum, of the kind an element was expected to posses. Chemists were reluctant to accept the new element until such well-defined spectroscopic results had been assured. The isolation of radium by the Curies, and the ability to accumulate significant amounts of it,[30] were decisive steps in research, because radium is a very strong source of radiation and hence allows new experiments requiring such sources. A few years later, the Curies themselves coined the term 'radioactivity'.

In 1897, J.J. Thomson (1856–1940m) discovered the free electron in cathode ray tubes. He managed to measure the charge and the mass separately, something

---

[25] Pierre and Marie shared (half) the 1903 Nobel Prize (with Henry Becquerel) for the study of 'spontaneous radiation' as the Nobel citation states, and the 1903 Davy medal. The definition of spontaneous radiation reflects the confusion and surprise in the scientific community: *Unexpectedly there are elements which emit suddenly intrinsic radiation.* Marie Curie also got the 1911 Nobel Prize for Chemistry, for the discovery of new unstable elements. (Some chemists did not approve of this, because they considered that the evidence for a new unstable element was not sufficiently strong. Chemists did not like the idea of *an element which decays into another element.*) This time she got the prize alone, although the discoveries were made with her husband. Pierre was killed in an accident in 1909. Their daughter Irene and her husband, Frederic Joliot, shared the Nobel Prize in chemistry 1935. The element curium $^{247}Cm_{96}$ is named after the Curie couple.

[26] Curie, M.P. & Curie, P., Comptes Rendus **127**, 175 (1898).

[27] Pitchblende contains uranium and thorium. Uranium was first isolated by Eugene M. Péligot in 1841, and thorium was discovered by Breezeless in 1828.

[28] The discoverer of europium in 1896. The relevant paper is: Demarcay, E., *Sur le Spectre d'une Substance Radio-Active*, Comp. Rend. (26 December 1898).

[29] Soddy, F., Nature **69**, 297 (1904).

[30] The Curies accumulated many honors, among them a unit of radioactive decay. Thus 1 becquerel (Bq) is one decay per second, while 1 curie (Ci) is 3.7×10^{10} Bq. The ratio of the units does not reflect the relative importance of their discoveries.

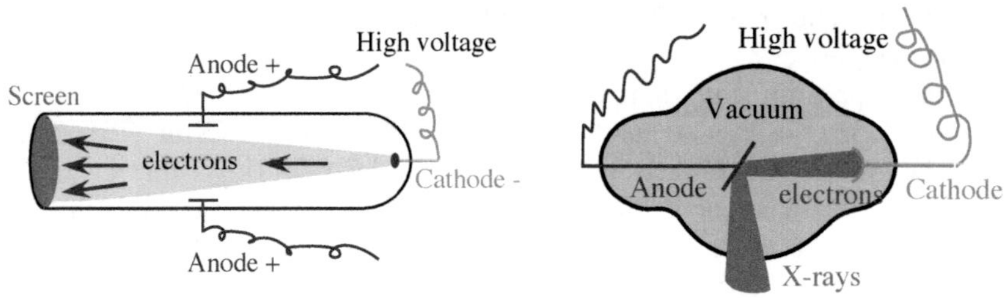

Fig. 3.4 *Left*: The basic structure of the cathode ray. A high voltage is applied between the cathode and the anode. As a consequence, negative electrons are emitted from the cathode and move to the positive anode. The electrons go on to hit a screen coated with a material that releases light when struck. *Right*: The basic structure of the X-ray tube. A very high voltage is applied between the cathode and the anode. As a consequence, negative electrons are emitted from the cathode and move to the positive anode. The electrons hit the metallic anode, and as a consequence the metal emits X rays

that Wien had not succeeded in doing. The atom, that fundamental building block of matter hypothesized by Dalton almost a century earlier,[31] is actually made up of smaller particles and has an internal structure. Atoms are neutral, and if they contain negative electrons, then clearly, there should also be an equal amount of positive charge. But how are the two opposite charges arranged? Are they at rest or moving?

Ernest Rutherford (1871–1937m), originally from New Zealand, but who then went to Montreal, Canada, was a student of J.J. Thomson in England. He started his long and illustrious investigation of radioactivity in 1899, embarking on a journey that completely overturned many disciplines of science. His first goal was to explore the nature of radioactive radiation, and his first move was to dispose of the clumsy photographic plate and use the much more accurate electrometer. Next, he used a magnetic field to show that one component of the radiation, which he called β rays, was strongly affected by the field, while the α part remained unaffected, essentially confirming Becquerel's and the Curies' claims. Rutherford's first guess, that natural radioactive rays were similar to artificially generated X rays, turned out to be wrong. Six years later, Rutherford began to investigate the nature of the α rays, and discovered that they are in fact nuclei of helium, the element Lockyer had discovered in the Sun.[32]

In 1900, Becquerel[33] proved that β rays are electrons and Villard (1869–1934) discovered[34] a third kind of radiation emitted by radioactive matter, viz., γ rays, while carrying out chemical research on uranium. (This was a natural name, since Rutherford had used α and β to denote the previously discovered rays.)

[31] Dalton, J., *A New System of Chemical Philosophy*, Philosophical Library NY, 1808.

[32] Rutherford's own summary of his discoveries in this period can be found in his Bakerian lecture, Proc. Roy. Soc. London **73**, 493 (1904).

[33] Becquerel, H., Compt. Rend. **130**, 809 (1900).

[34] Villard, P., Compt. Rend. **130**, 1010, 1178 (1900).

Table 3.1 Types of radioactive radiation

Name	Nature	Salient properties
α	Helium nucleus	Positive charge, $e/m = 1/2$
β	Electron and positrons	Negative and positive charges, respectively, $e/m = \pm 1/1840$.
γ	Electromagnetic radiation	Neutral

In 1900, in a completely different and apparently unrelated field of research, the radiation emitted by a black body, Planck (1858–1947) formulated[35] the idea that light is quantized, and succeeded in explaining the radiation emitted by a black body, Wien's ideal absorber. Two theories about the nature of light existed at the time. One said that light was waves in some strange, mysterious, and elusive matter called the ether, while the other considered that light was made of particles.

Until the experiments by Thomas Young (1773–1829m) and Jean Fresnel (1788–1827, Promontorium Fresnel on the Moon) at the beginning of the 19th century, physicists were inclined to believe in the particle theory of light, as advocated by Newton. Although the experiments by Young and Fresnel had tilted the balance towards the wave theory, one difficulty remained with it: what does the light wave propagate in? Of course, particles propagate in vacuum without any problem, but what about these waves? To answer this question, Young[36] had revived an idea due to Huygens (1629–1695)[37] according to which a 'luminiferous ether' must pervade all material bodies. Then light could propagate in the ether as waves propagate in the sea. In this way, a medium was invented through which light could propagate.

This was a unique medium. On the one hand, it was not disturbed by matter moving through it, and on the other hand, it penetrated other bodies like water. Of course, various questions came up concerning the details of this hypothesis. Does the ether accumulate near large bodies like the Earth or the Sun? Is it still, or does it move? Each possibility has consequences that physicists tried to sort out. Planck himself, like most physicists, believed in the existence of an ether, and even had a hypothesis of his own concerning its properties (see later). It is against this background that we have to appreciate the Planck's bold assumption, viz., that light is made of particles whose energy is proportional to the frequency. This was 'all' that Planck dared to assume, but with it, he was able to derive the observed distribution of the radiation emitted by a black body.

The discovery can be considered as the official birth of quantum theory. It was essential to Bohr, for his model of the atom, and to Einstein, in his explanation of the photoelectric effect, for which each of them got the Nobel Prize. During the coming twenty five years, the discoveries of Planck and Bohr, together with the

[35] Planck's idea developed gradually. The peak was in two communications to the Berlin Academy on 19 October 1900 and 14 December 1900. See also, Annal. der Phys. **4**, 533 (1901).

[36] Young, T., Phil. Trans. Roy. Soc. **94**, 1 (1804).

[37] There is a Huygens crater on Mars, as well as a space mission that carries his name.

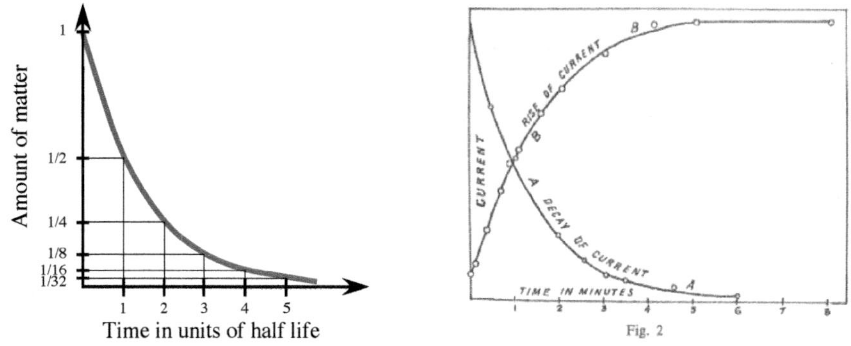

Fig. 3.5 *Left*: The exponential law. Note that the time is given in units of the half-life. Hence, after every half-life interval, half of the amount that existed at the beginning of the time interval is left. This law is typical to all phenomena where the rate of decay of some quantity is proportional to the quantity itself. For example, the rate of cooling of a cup of tea is proportional to the difference between the temperature of the tea and the outside air temperature. *Right*: Rutherford's original discovery that radioactive isotopes decay according to an exponential law with a typical half-life that varies from one isotope to another. In Rutherford's own words: *Curve A shows the relation existing between the current through the gas and the time. The current, just before the flow of air is stopped, is taken as unity. It will be observed that the current through the gas diminishes exponentially with time. The current through the gas is proportional to the intensity of the radiation emitted by the radioactive. Consequently, the intensity of the radiation emitted by the radioactive particles falls off in a geometrical progression with the time, namely, exponentially*

growing understanding of radioactivity, would merge into the quantum theory, and give birth to nuclear physics. Kelvin's first cloud turned out to be one of the two colossal breakthroughs of 20th century physics, namely, quantum theory.

In parallel, Rutherford discovered that the activity of each radioactive element decays exponentially, and he defined the concept of half-life. The exponential law has the unique property that after every lapse of time equal to the half-life, just half of the original amount of material remains (see Fig. 3.5).[38] The decay rate, as was found by Rutherford, does not depend on the amount of radioactive matter, nor on the conditions in the laboratory. It is amazing how nothing could be found to affect the decay or stop it. This is important. Radioactivity goes on irrespective of the outside conditions, so that radioactive decay should go at the same rate inside the Sun and inside the Earth, even if the pressures are extremely high.[39] The 'spontaneous decay' by emission of a particle caused severe logical problems. What determines the decay? What determines which atom should emit radiation and when? These problems, which appeared to contradict the fundamental notion of cause and effect

[38] The original discovery of the exponential law was in Rutherford, E., Phil. Mag. **49**, 161 (1900). Rutherford did not use the term exponential, but referred to geometrical progression. See Fig. 3.5 for the original graph given by Rutherford.

[39] Only at the astronomical pressures and densities found in special stars called white dwarfs can negative β radioactivity be stopped. See later.

(since the cause was not observed), obsessed the physical world for many years, and were only partially solved by George Gamow almost thirty years later.

Also in 1900, two Gymnasium (of Wolfen-Büttel, Germany) physics teachers, Julius Elster and Hans Geitel[40] investigated why charged bodies discharge when surrounded by air in closed vessels, and discovered that the Earth's atmosphere and soil contain radioactive elements. The radioactive rays, emitted by the radioactive elements, create charges in the atmosphere which cause the air to conduct electricity and discharge the charged body. Everything they touched was found to be polluted with radioactivity.

And in the same eventful year of 1900, Rutherford[41] investigated the radioactive substance emitted from thorium to discover what he referred to as the emanation, that is, a new radioactive gas. At about the same time, the Curies[42] found similar effects when they investigated the emission from radium. In the Curies' case the emitted gas remained radioactive for a month.[43]

In 1901, the young student Rayleigh[44] and Crookes put forward the hypothesis that the α rays are positively charged. Crookes was known to have 'strange' ideas about the discovery of 'meta elements' or radiant matter, or the 'fourth state of matter',[45] but he had important scientific discoveries to his credit. These included the discovery that α radiation causes scintillation when it hits a crystal of zinc sulfide, a phenomenon that soon became the main tool in radioactive research. Crookes' scientific record convinced Rutherford to look carefully into his claims. So Rutherford obtained the strongest magnet available at that time and eventually showed in 1902,[46] following several unsuccessful trials, that the α rays were indeed massive positively charged particles.

---

[40] Elster, J., & Geitel, H., Phys. Zeit. **2**, 116 (1900); ibid. **2**, 560 (1901).

[41] Rutherford, E., Phil. Mag. **49**, 1 (1900).

[42] Curie, P., & Curie, M., Compt. Rend. **129**, 714 (1899).

[43] There are quite a few citations in the literature (over 200) claiming that F.E. Dorn, who was the first to show that Becquerel rays are deflected by an electrostatic field [Dorn, E., Physik. Zeit. 337 (1900)], also discovered that radium releases not only radioactive elements, but in addition a radioactive gas [Dorn, E., Abhl. Naturf. Ges. Halle **22**, 155 (1900)]. However, as was demonstrated by Marshall and Marshall [Marshall, J., & Marshall, V., Bull. Hist. Chem. **28**, 76 (2003)], this almost invariable assertion that Dorn was the discoverer is actually false. The 'cut-and-paste' of references without reading them appears to be an old phenomenon. The original paper by Dorn, which is so often cited without consultation, appeared in a disregarded and remote publication of the Nature company of Halle, and gave the credit to Rutherford. Furthermore, the nature of the emanation was not studied by Dorn, although it was by Rutherford. The two historians travelled to Halle in search of the original paper, and indeed found the original German version. The article Kleinert, A. von, Die Naturforschende Gesellschaft zu Halle. Acta Historica Leopoldina **36**, 247 (2000), gives a brief history of the Naturforschende Gesellschaft zu Halle, which was founded in 1779. Its last documented meeting was in 1920, and it was last listed in the address directory of Halle in 1935.

[44] His original name was John Strutt.

[45] Crookes, W., Proc. Roy. Soc. London **30**, 469 (1880).

[46] Rutherford, E., Phil. Mag. **5**, 177 (1903).

In 1902, Rutherford and Frederick Soddy[47] (1877–1956m) concluded that radioactive elements transmute spontaneously from one form into another. One element breaks down into another, lighter element, releasing α, β, or γ radiation in the process. Natural al-chemistry had been discovered! At long last it looked as though the alchemists' dream had been realized! Somehow the emitted radiation changed the nature of the atom. Rutherford and Soddy did not understand the energy source, and stated that: *In the case of the three naturally occurring radioactive elements, however, it is obvious*, so they claimed that: *there must be a continuous replacement of the dissipated energy, and no satisfactory explanation has yet been put forward.* They also demonstrated that a particular radioactive element decays into another element at a distinctive rate. Each element has its own distinct half-life.

It became evident that some radioactive nuclei had such a short half-life that they disintegrated in the laboratory. However, other radioactive elements like uranium, thorium, and radium have long half-lives, as long as a billion years. The conclusion from the fact that these elements still exist on the Earth is that their decay time, though not known accurately, is of the order of the age of the Earth or longer. Otherwise they would have decayed away. This fact led Rutherford to suggest using them to measure the age of the Earth by determining the relative proportions of radioactive materials in geological samples. Radioactive dating was born in 1905 in a lecture Rutherford gave at Harvard University, where he suggested using the uranium/helium ratio to date rocks. But note that, in reality, radioactive elements heavier than lead decay into another radioactive element. Thus, the mother nucleus and the daughter nucleus are both radioactive. The series of radioactive elements ends only when the daughter is one of the isotopes of lead.

Let us turn our attention to another discovery of prime importance. In discussing the consequences of the disintegration theory, Rutherford and Soddy pointed out that any stable substance produced during the transformation of the radio-elements should be present in a certain quantity in the radioactive minerals, where the processes of transformation have been taking place for a long time. This suggestion was first put forward in 1902:

> *In the light of these results and the view that has already been put forward of the nature of radioactivity, the speculation naturally arises whether the presence of helium in minerals and its invariable association with uranium and thorium, may not be connected with their radioactivity [...]. It is therefore to be expected that if any of the unknown ultimate products of the changes of a radioactive element are gaseous, they would be found occluded, possibly in considerable quantities, in the natural minerals containing that element. This lends support to the suggestion, already put forward, that possibly helium is an ultimate product of the disintegration of one of the radioactive elements, since it is only found in radioactive minerals.*

At that time it was not yet known that the α particles are in fact helium nuclei. So on the one hand they knew about the radiation, and on the other hand they discovered helium inside radioactive ores, but the connection was yet to be made.

---

[47] Rutherford, E., & Soddy, F., Phi. Mag. **4**, 370 (1902); Phil. Mag. **5**, 106 (1903); Phil. Mag. **4**, 453, 579, 582 (1902); Phil. Mag. **5**, 453 (1903); Rutherford, letter in Nature **69**, 20 August 1903.

Fig. 3.6 Nagaoka's Saturnian model. All electrons move in the same plane, like the rings of Saturn

During the years 1902–1904 Kelvin and J.J. Thomson[48] put together the first model of the atom. Clearly, it had to be neutral. Electrons as atomic building blocks were discovered before, while the nature of the positive charge was still unknown. So a logically possible model was a continuously distributed positive charge with the negative small spheres, the electrons, dispersed within it like raisins in a cake (the pudding model). In this view, the entire volume of the atom was supposed to be full of particles and matter (see Fig. 3.9).

In the same issue of the journal, just a few pages away, Nagaoka (1865–1940m) from Japan hypothesized[49] a completely different model, something that *resembles the Solar System*, that is, a massive positive charge concentrated in the center of the atom, with the light electrons forming a ring around the nucleus rather like the rings of Saturn. This was called the Saturnian model (see Fig. 3.6).[50] Nagaoka assumed that the electrons were attracted by the large central mass and repelled one another. However, he did not examine the stability of such a system. The basic problem was that atoms emitted radiation in lines and not continuously. Hence, a successful model of the atom had to explain how the spectral lines form, and what determines the specific wavelengths of the spectral lines.

Nagaoka adopted a special and unconventional approach:

> *Instead of seeking to find a system whose modes of vibration are brought into complete harmony with the regularity observed in the spectral lines, inasmuch as the empirical formulae are still a matter of dispute, I propose to discuss a system whose small oscillations accord qualitatively with the regularity observed in the spectra of different elements.*

This was a fundamentally new, and very unorthodox approach. Nagaoka was happy to explain trends and general behavior, giving up, at least for the time being, accurate predictions of observed data. Indeed, the vibrations of such a ring of electrons are very complicated, and there are many vibrational modes, none of which could accurately reproduce the one that was actually observed. Moreover, Nagaoka shied away from discussing the most serious problem, namely, radiation losses.

It has long been known that the an accelerating charge, like the electron, radiates electromagnetic radiation, and hence suffers energy losses. If the electrons move in circles, they are continuously accelerated, and hence must radiate energy away, eventually losing it all and collapsing into the nucleus. Kelvin and Thomson

48 Kelvin & Thomson, J.J., Phil. Mag. Ser. 6, 7, **39**, 237 (1904).

49 Nagaoka, H., Phil. Mag. Ser. 6, 7, **39**, 445 (1904).

50 In 1859, Maxwell [Maxwell, J.C., MNRAS **19**, 297 (1859)] had shown that the rings of Saturn had to be composed of very small particles which exerted a negligible gravitational force on Saturn. This essay won the Adams' prize for the year 1856, and the paper cited is just a summary.

avoided the problem by assuming that the electrons were at rest in the pudding. Nagaoka simply did not discuss the difficulty. And yet this model of the atom granted Nagaoka a name on a lunar crater.

A crucial discovery for our story here was made in 1903, when Pierre Curie and Albert Laborde[51] showed that radium was a self-heating substance, and was always hotter than the surrounding air. It seemed probable from the beginning that stopping the fast-moving particles would cause a conversion of kinetic energy to heat. However, nobody realized that the amount of heat would be sufficient to heat the radium!

If the amount of radium is small, the α particles escape from it, but if the amount of radium is sufficiently large, the α particles cannot escape, and are absorbed by the radium itself, depositing their energy in it. But since the amount of radium is large, heating it requires more particles. The amount of energy emitted per particle was so high as to appear to many as ‘creation of energy from nothing’. It was the first indication that the energy emitted during the radioactive decay is very high.

This important fact was confirmed by the work of Rutherford and Barnes in 1903,[52] who showed that three quarters of the heating effect in radium was not directly due to the radium but to its product, the radium emanation discovered previously (the gas released by radium and still unidentified at this time), and that each of the different elements produced in the radium decay released heat in proportion to the energy of the α particles expelled from it. These experiments exposed the enormous energy, compared with the mass of matter involved, which was emitted during the transformation of the radium emanation. It could readily be calculated that one kilogram of the radium emanation and its products would emit energy at the rate of 14 000 horsepower, and during its life would give off energy corresponding to about 80 000 horsepower for one day (the numerical example is from Rutherford and Barnes). Nobody had dreamt at that time of anything remotely resembling what we call nuclear energy today, whence the overall surprise.

It was thus clear that the heating effect of radium was mainly a secondary phenomenon resulting from bombardment by its own α particles. It was also evident that all the radioactive substances must emit heat in proportion to the number and energy of the α particles expelled per second.

Geologists quickly realized that the discovery of radioactivity was going to revolutionize geological dating. The standard models assumed that the Earth and Sun were created simultaneously at some time in the past and had been steadily cooling since that time. Radioactivity provided a process that generated energy. The astronomer George Darwin[53] and geologist John Joly (1857–1933) were the first to suggest in 1903 that natural radioactivity might partially account for the heating

[51] Curie, P., & Laborde, A., Compt. Rend. **136**, 673 (1904).

[52] Rutherford, E., & Barnes, H.T., *Heating Effects of the Radium Emanation*, Nature **68**, 622 (1903); Nature **69**, 126 (1903); Phil. Mag. Ser. 6, Vii, 202 (1904); *The Heating Effect of the Gamma Rays from Radium*, Nature **71**, 151 (1904); *Heating Effect of the Gamma Rays from Radium*, Phil. Mag. Ser. 6, ix 621, (1905).

[53] Darwin, G., *Radioactivity and the Age of the Sun*, Nature **68**, 496 (1903).

of the Earth.[54] In 1908,[55] Joly discussed the role of uranium radioactive decay in heating the Earth. In later years, Joly worked on how radioactivity affects the formation of minerals in the Earth, and explained the halos[56] found around inclusions in micas as due to radioactivity, with the implication that radioactivity is abundant in the Earth.

Two very interesting and far-reaching comments by Rutherford and Soddy[57] appeared in the above paper. One was this:

> *If elements heavier than uranium exist it is probable that they will be radioactive. The extreme delicacy of radioactivity as a means of chemical analysis would enable such elements to be recognized even if present in infinitesimal quantity.*

The two researchers realized that there is an end to the periodic table. Elements heavier than uranium, if formed, are bound to be unstable and decay. The second remark was this:

> *The maintenance of solar energy [...] no longer presents any fundamental difficulty if the internal energy of the component elements is considered to be available, i.e., if processes of sub-atomic change are going on.*

This was the first time that the term 'sub-atomic energy' had been used. It was used extensively until it was replaced by 'nuclear energy'. The idea that radium might be the source of the Earth's heat was pronounced by Rutherford in 1905.[58]

Among the first to appreciate the huge amount of energy released in radioactive decay was James Jeans (1877–1946m). Jeans came out with an extremely imaginative idea to explain the phenomenon of radioactivity and the associated energy.[59] Jeans remarked that external conditions like temperature and pressure have no effect on the disintegration of the nucleus. The nucleus disintegrates without any outside intervention. So, concluded Jeans, it must be an internal process. But what could it be? Jeans adopted the idea of Osborne Reynolds (1842–1912) that the source of the instability was *some agitation of the 'grains' of which the ether is constituted*. The velocities of these 'grains' must be very high, and when one of them collides with another one, it gives rise to a restructuring of the ether. Jeans argued that a process of this kind would be unaffected by temperature and pressure. It seemed probable, hypothesized Jeans, that:

> *[...] the restructuring would consist of the combination and mutual annihilation of two ether strains of opposite kinds, in the coalescence of a positive and negative ion, and would therefore result in a disappearance of a certain amount of mass.*

---

[54] Joly, J., *Radium and the Geological Age of the Earth*, Nature **68**, 526 (1903); also *Radioactivity and Geology: An Account of the Influence of Radioactive Energy on Terrestrial History*, Archibald Constable & Co, London (1909).

[55] The British Association for the Advancement of Science, Dublin 1908. The contribution of John Joly to geology earned him a name on a crater on Mars, but not on the Moon, and for this reason the m is missing.

[56] Joly, J., Phil. Trans. Roy. Soc. London A **217**, 51 (1918).

[57] Rutherford, E., & Soddy, F., Phil. Mag. & J. Sc. 6, **5**, 576–591 (1903).

[58] Rutherford, E., *Radium. The Cause of the Earth's Heat*, Harpers Mag. (1905) p. 390.

[59] Jeans, J., *A Suggested Explanation of Radioactivity*, Nature **70**, 101 (1904).

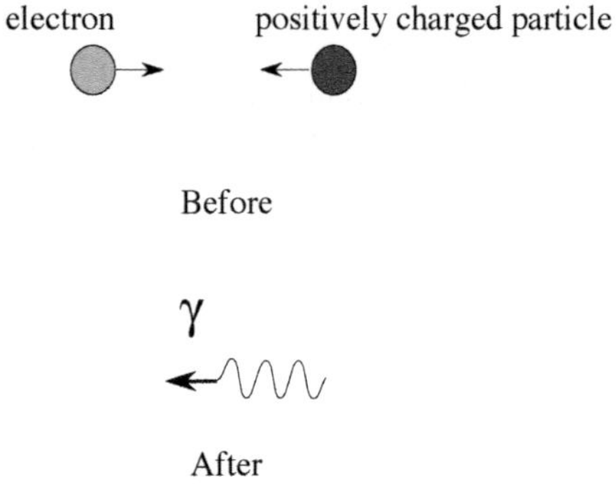

Fig. 3.7 The conversion of mass into radiation according to Jeans. A positive proton annihilates a negative electron, yielding a *single* photon

According to Jeans, neither mass nor material energy is conserved (separately). The process of radioactivity was thus an increase of material energy at the expense of the destruction of a certain amount of matter. This was the first time the idea of matter conversion into energy, or matter annihilation, appeared in the literature. Jeans elaborated on what he had called the mechanism of radiation[60] a few years earlier. Recall that this happened just one year before Einstein came up with the $E = mc^2$ result, and with the special theory of relativity, which finally did away with the idea of the ether! So if the Jeans hypothesis is purged of the ether reference, we are left with the idea that positive and negative charges may annihilate each other and convert their mass into energy. This very short paper in Nature is considered as the birth of the idea of mass annihilation. At that time only the massive positively charged proton and the light negatively charged electron were known, and it would take close on thirty years (Hughes and Jauncey[61]) to realize that this process could not take place the way Jeans had suggested, because it did not satisfy the well-established conservation laws of physics (to be discussed later).

Jeans' idea did not attract much attention at the time, but over a decade later became a central issue in one of the biggest controversies in astrophysics. The nature of the positive charge was not yet known. On the other hand, once the β rays had been identified with electrons, it was clear that the atom left behind by the emitted electron had somehow to get rid of its positive charge. The simplest option was that suddenly, for some as yet unknown reason, an electron annihilates a positive charge in the atom. As the number of positive charges decreased by one, the extra electron had to leave the atom.

Other ideas were thrown up. Crookes,[62] for example, suggested that the energy was taken from the surrounding air. An air molecule somehow collided with the

[60] Jeans, J.H., Phil. Mag. **2**, 421 (1901).

[61] Hughes, A.L., & Jauncey, G.E., Phys. Rev. Let. **45**, 217 (1934).

[62] Crookes, W., Compt. Rend. **128**, 176 (1899).

radioactive element, and endowed it with energy that was stored until it managed to burst out in a rush of released energy. Kelvin, who was now 83 years old, also tried to explain radioactivity.[63] Kelvin argued that the α particles were atoms or molecules of radium which had lost an unspecified number of negative charges. Kelvin could not accept that the decay was spontaneous, and completely discredited the idea that the atom could store such huge amounts of energy. Like Crookes, he believed that the energy must be supplied from the outside. However, all the ideas failed, including the idea that, when the atoms were produced, they did not have the same strength and hence become unstable at different times,[64] as if the energy stored in the atoms was primordial, existing from the time the atoms were formed. This was not too far from what was hypothesized over 50 years later.

In 1904, Rutherford[65] took up his own idea and worked out a detailed description of how the heat of the Earth and Sun could be accounted for by radioactive decay. Late in 1904, Rutherford took the first step toward realizing his idea of radiometric dating, which he had proposed two years earlier, by suggesting that the α particles released by radioactive decay could be trapped in a rocky material as helium atoms. At the time, Rutherford was only guessing the relationship between α particles and helium atoms, a connection he would prove four years later, but not before Ramsay and Collie[66] had identified the spectrum of radium emanation as helium.

The first application of Rutherford's method by a geologist was carried out in 1905 by Bertram Boltwood (1870–1927). Boltwood, then at Yale, heard Rutherford's lecture when he visited Yale in 1904, and got excited. He suggested that, because we know the half-life of uranium decay into lead, then by measuring the amount of lead in a lump of uranium ore, it might be possible to determine the age of the rock in which the ore was found. Since the original amount of lead was not known (and might depend on the process in which the lead and uranium were synthesized) only a maximum age could be found in this way. On the other hand, Rutherford suggested using the uranium-to-helium ratio. Since the helium is a result of radioactive decay, then clearly it starts forming only after the formation of uranium. However, if the uranium were liquid, or even solidified into a porous rock, helium could escape, and hence the age so measured would be a minimal age. At the same time, Strutt used the radium-to-helium ratio to date various old rocks, and got ages of 400 to 2 000 million years.[67] In the same year, Rayleigh confirmed Rutherford's suggestion that the occurrence of radioactivity was one of the 'missing' factors in Kelvin's calculations of the cooling of the Earth. They forgot about Fourier, who had written about it long before.

---

[63] Radioactivity: (Sound recording) a talk by Lord Kelvin, 1905, American Inst. Phys. Center for History of Physics.

[64] J.J. Thomson, Rep. Brit. Assoc. Avn. Sci. 3 (1909).

[65] Rutherford, E., *Heating Effect of the Radium Emanation*, Trans. of Aus. Assoc. for Adv. of Sci. 87–91 (1904).

[66] Ramsay, W., & Collie, J.N., Proc. Roy. Soc. London **73**, 470 (1904).

[67] Strutt, R.J., Proc. Roy. Soc. London A **76**, 88 (1905). The paper was communicated by Lord Rayleigh's father.

The year 1905 is well known in science for the publication of Einstein's theory of special relativity.[68] The theory did away with the ether hypothesis and resolved the conundrum of the Michelson–Morley experiment. Kelvin's second cloud had been dispersed. The basic assumption of the theory is that the speed of light is constant in all systems, and observers moving with constant speed relative to each other must see the same physical laws. It sounds simple, but in reality the impact on all laws of physics was extremely profound, but with one exception, namely the laws of electrodynamics as formulated by Maxwell. Unbelievably, since Maxwell had written his laws some forty years earlier, they nevertheless satisfied all the requirements of the new theory.

The events in the investigation of radioactivity remained for a while unaffected by Einstein's new theory, as if there were no connection between the two. It took 15 years for the impact of $E = mc^2$ to be felt on the theory of the stellar energy source. This was rather strange, because in his 1905 paper,[69] Einstein suggested using radioactivity to check experimentally the theoretical result that $E = mc^2$. In his own words:

> *It is not impossible that with bodies whose energy-content is variable to a high degree (e.g., with radium salts) the theory may be successfully put to the test.*

For many years, nobody really paid attention to Einstein's remark.

In 1905, Chamberlin,[70] with the help of Moulton, started to construct his theory of the formation of the Earth. The basic idea was that eruptions on the surface of the Sun were amplified by a visiting star. As a result of the additional pull of the star, two long arms of matter were formed, extending to large distances, so that the Sun was converted into a spiral nebula. The matter in the spiral arms cooled and condensed, and formed a series of blobs, or planetesimals, which eventually became the planets. Jeans investigated this idea in 1916, and found that the passage of a star sufficiently close to the Sun would give rise to huge tides that would lead to spiral arms, whence there was no need for eruptions on the Sun.[71] Over the same period, George Darwin considered the possibility that the Moon was pulled out of the Earth via the same mechanism. The age of the Earth could then be significantly smaller than the age of the Sun.

---

[68] Einstein published 5 seminal papers in 1905, concerning the theory of special relativity, the photoelectric effect, and Brownian motion. All these papers were revolutionary. Some call this the annus mirabilis.

[69] Einstein, A., *Does the inertia of a body depend on its energy-content?*, Annalen der Phys. **18**, 639 (1905).

[70] Chamberlin, T.C., *Fundamental Problems of Geology*, yearbook no. 8, Carnegie Institution, Moulton, F.R., *Introduction to Astronomy*, New York (1906) p. 463; Chamberlin, T.C., *The Origin of the Earth*, Chicago Univ. Press, 1916. A summary of the theory can be found in the Royal Astronomical Society of Canada, November 1916, where, upon the request of the editor, the article from Scientia, 1914, was reprinted.

[71] We know today that young forming stars, which are in the state appropriate to the forming Sun, do show very strong eruptions from the surface, but this has nothing to do with the formation of the Solar System.

Still in 1905, Jeans published a paper in the Philosophical Magazine which showed the impossibility of the ether reaching thermal equilibrium with matter. This meant that either heat flowed from the radiation to the ether, or vice versa. Of course, Planck had announced his formula for black body radiation five years earlier, now known as Planck's radiation formula, but Jeans was strongly opposed to Planck's results. Today Jeans' paper can be seen as a mathematical 'proof' that classical physics breaks down. We should also note that Jeans' paper was written after the Michelson–Morley experiment had disproved the existence of the ether, and in the same year that Einstein published the special theory of relativity, which removed the need for the ether, not to mention his explanation for the photoelectric effect, for which Planck's idea was essential.

Confusion over the nature of radioactivity rumbled on. It became difficult to keep track of so many newly discovered elements in the radioactive series. On 9 August 1906 Kelvin wrote to the London Times arguing against the idea, by then widely accepted, that radioactive decay involved the transmutation of one element into another. His line of thinking was as much semantic as physical. He proposed that heavier elements were compounds like molecules, composed of lighter elements, which split into their various components when they disintegrated. In other words, radium was a compound of helium and other lighter elements, and not a true element in its own right. Nowadays, knowing that atomic nuclei are built from protons and neutrons, we say that the different elements are all combinations of the same ingredients. One might almost suggest that Kelvin was reaching in this direction, but since no one at that time had any clear idea of what atoms were made of, the debate had no real substance.

It must be stated that the use of the daily newspapers as a scientific medium was an exception. Other physicists wrote in to disagree, while many chemists agreed with Kelvin. The Times, in an unusual move for the press, pitched in with an editorial(!) asking for Kelvin's views to be taken seriously, on account of his great reputation and experience. However, at the same time Reade (1832–1909)[72] a well-known correspondent of Darwin, expressed some of the relief geologists now felt when he wrote: *The bugbear of a narrow physical limit to geological time being got rid of, we are free to move in our own field of science.*

Notwithstanding, young scientists have little respect for seniority. Frederick Soddy wrote to the newspaper on 31 August putting the case against Kelvin, and concluding that:

> *It would be a pity if the public were misled into supposing that those who have not worked with radio-active bodies are as entitled to as weighty an opinion as those who have. Atomic disintegration is based on experimental evidence, which even its most hostile opponents are unable to shake or explain in any other way.*

[72] Reade, T.M., *Radium and the Radial Shrinkage of the Earth*, Geological Magazine Series 5, **3**, 79 (1906). In 1920 Reade was posthumously awarded (with three other deceased geologists) the Liverpool Geological Society Silver Medal for his contributions to the determination of the geological ages.

Summarizing these inconclusive exchanges a few weeks later in Nature,[73] Soddy's veiled tribute to Kelvin came close to condescension:

> *Whatever opinion may be formed of the merits of the controversy, all must unite on admiration for the boldness with which Lord Kelvin initiated his campaign, and the intellectual keenness with which he conducted, almost single-handed, what appeared to many from the first almost a forlorn hope against the transmutational and evolutionary doctrines framed to account for the properties of radium. The weight of years and the almost unanimous opinion of his younger colleagues against him have not deterred him from leading a lost cause, if not to a victorious ending, at least to one from which no one will grudge him the honors of war.*

Great scientists can be obstinate.

Even Ernest Rutherford, the great pioneer of radioactivity and atomic theory, who had written to his mother years ago of his admiration for Kelvin, could not help but think of the aging natural philosopher as a child. They met at a scientific party at Terling, Lord Rayleigh's estate. Rutherford described the proceedings in a letter to his wife:[74]

> *Lord Kelvin has talked radium most of the day, and I admire his confidence in talking about a subject of which he has taken the trouble to learn so little. I showed him and the ladies some experiments this evening, and he was tremendously delighted and has gone to bed happy with a few small phosphorescent things I gave him.*

In 1907, just before Kelvin's death, Boltwood[75] calculated the ages of 26 samples of rocks containing uranium, using the method suggested by Rutherford (U/Pb). He estimated that the youngest rock was 410 million years old, while the oldest was about 2.2 billion years old. It is interesting to note that, though Kelvin had formulated the second law of thermodynamics fifty years earlier, and no doubt this was a colossal achievement in physics, he was not awarded the Nobel Prize, which was established in 1901.[76]

3.5 Rutherford's 1907 Address

In 1907, Rutherford was invited to address the Royal Astronomical Society of Canada,[77] and spoke on 'Some Cosmic Aspects of Radioactivity'. He described how Elster and Geitel had discovered that the environment is highly polluted with natural radioactivity. He explained that he took their discovery seriously, and had repeated

---

[73] Soddy, F., Nature **75**, 35 (1906).

[74] Joseph Press, *Degrees Kelvin: A Tale of Genius*, Invention and Tragedy (2004).

[75] Boltwood, B., Amer. J. Sci. Series 4, **23**, 77 (1907).

[76] His competitors were Röntgen (1901), Lorentz and Zeeman (1902), Becquerel, Pierre Curie and Marie Curie (1903), Rayleigh (1904), Lenard (1905), J.J. Thomson (1906), and Michelson (1907). Only Lorentz was a theoretician, and he got the prize for the theory of electrons and light propagation.

[77] Rutherford, E., J. Roy. Ast. Soc. Canada, May 1907.

their experiment with McLennan, finding that the falling rain and snow were radioactive and showing that most of the Earth's atmospheric radioactivity was due to radium emanation.

Repeating the measurements in other places produced similar results. So Rutherford concluded that:

> *We must bear in mind that all of us are continuously inhaling the radium and thorium emanations and their products and ionized air. In addition, we are continuously undergoing a type of mild X-ray treatment, for the β and γ rays from the Earth and atmosphere continuously pass into and through our bodies.*

Rutherford warned that this radiation might have physiological effects. He then continued to evaluate the amount of radioactive matter in the Earth's crust, citing Boltwood, who had detected the existence of radium in a deep-seated spring, and McLennan and Burton who had shown that the petroleum from the oil wells in Ontario contained radium.

Rutherford described how radioactive material releases heat:

> *A pound of radium in the course of a year will emit as much heat as that resulting from the combustion of 100 pound of good coal.*

He noticed that the lifetime of radium is about 20 000 yrs. However, radium is the product of uranium decay, the half-life of which he still did not know accurately at the time of the talk, but estimated to be about 1 billion years. This meant there was a long term supply of radium in the Earth. Rutherford quoted a calculation he had carried out in 1902, according to which 1.7×10^{-13} g/cm^3 of radium spread in the soil would suffice to maintain the temperature run observed in deep mines. He then turned to the results of Strutt (who later became Lord Rayleigh 1842–1919m),[78] who had measured the amounts of radium in the Earth's crust.

Rayleigh measured many rocks and found that most contain radium in quantities that vary between 9.56×10^{-12} g/cm^3 (granite from Rhodesia) to 0.613×10^{-12} g/cm^3 (Basalt, Ovifak, Disco Island, Greenland). The amount Rutherford calculated as needed to maintain the heat of the Earth was more than 10 times smaller. Hence, Rayleigh suggested that the distribution of radioactive matter is not uniform throughout the Earth, but confined to a thin surface shell of the Earth (about 45 km thick). Rutherford and Rayleigh did not know about sedimentation and fractionation during the formation of the Earth, but knew from geologists that the Earth is not uniform. It is interesting to note that Rayleigh predicted volcanic activity on the Moon, where the temperature gradient, according to his calculations, was expected to be 8 times bigger then in the Earth. However, no such activity is observed. He overlooked the fact that there is no water on the Moon, an essential ingredient for volcanic activity.

As he continued his talk, Rutherford turned to the old problem of measuring the age of the Earth. He suggested the following method to measure the age. Helium

---

[78] He got the Nobel Prize in 1904 *for his investigations of the densities of the most important gases and for his discovery of argon in connection with these studies*. Also in 1904, Ramsay got the Nobel Prize for Chemistry, for the discovery of all noble gases.

is released during the radioactive decay of radium. The helium is stored in the rock when the rock solidifies. If one measured the amount of helium stored in the rock, one could find the time at which the rock had solidified. The minimal ages Rutherford got in this way were 500–1 000 million years. The age of the Earth had therefore to be longer.

Rutherford's conclusions are also interesting because he discussed the Sun's energy source. He observed that the Sun contains a lot of helium (which he knew to be released by radioactive decay, and had been observed by Lockyer to exist on the Sun). Hence, Rutherford concluded that the Sun was made of radioactive elements, and that helium was the product. As addition evidence, he cited Lord Rayleigh, who had found that many meteors contain as much helium as the Earth's crust. He showed that, if the Sun derives its heat from radioactive decay, he could extend the age of the Sun way beyond the uncomfortably short Kelvin–Helmholtz–Ritter time.

But how did the radioactive elements get into the Sun? Rutherford quoted spectroscopic observations which indicated that the composition of the outer parts of the Sun resembled those of the Earth. So Rutherford speculated that, at the enormous temperature of the Sun, it was possible that ordinary matter might become radioactive. If this were the case, then somehow energy was being driven into the radioactive elements. Rutherford was happy with this idea, but he avoided the obvious question it raised: what was the origin of the energy to be stored in the radioactive matter? In short, the suggestion seems to solve the age problem, but it creates an energy problem.

Rutherford turned the Earth age problem upside down. The age was not determined by the cooling of the Earth, but from radioactive measurements of rocks. The age calculation became simpler and more reliable. The Earth was heated by radioactive elements, and it remained only to determine how much radioactive material there was inside the Earth. He found that, if he assumed that radioactive elements were spread uniformly through the Earth, then there was too much radioactive material. But by then it was clear that the Earth was stratified and not uniform, so it was safe to assume that the radioactive elements were concentrated mainly in the crust.

In 1908, Rutherford was awarded the Nobel Prize for Chemistry *for his investigations into the disintegration of the elements, and the chemistry of radioactive substances*. In his Nobel address he confessed that he had no idea whether the radioactivity was due to an internal mechanism or an external excitation, and added:

> *In all probability, the α particle was a helium atom which carried two unit charges. According to this view, every radioactive substance which emits α particles must give rise to helium. This suggestion offered at once an explanation of the fact observed by Debierne that actinium, as well as radium, produced helium. It was pointed out that the presence of a double charge of helium atom was not altogether improbable for reasons to be given later.*

In the same year, Gray and Ramsay[79] isolated the radium emanation and named it niton. The name radon ($^{222}Ra_{86}$), as a product of radium, was given in 1923. Radon is colorless and odorless, does not react chemically, and is the densest known

[79] Gray, R.W., & Ramsay, W., Proc. Roy. Soc. London, Ser. A **84**, 536 (1911).

gas.[80] This is very important because radon is released by natural radioactivity and hence settles in basements and deep caves. Radon is a noble gas which decays into polonium and loses its chemical nobility through the decay. The half-life of the longest lived isotope is 3.82 days.

3.6 The Atom Is Mostly Empty

One of the most important experiments in physics took place in 1909 when Hans Geiger (1882–1945m)[81] and the undergraduate student Ernest Marsden (1889–1979), under the direction of Rutherford, sent α particles towards a very thin foil of gold, and discovered that the majority of them passed through the foil without hitting anything (see Fig. 3.8). Only a tiny number of particles were scattered back (towards the source) after hitting the nucleus of a gold atom.

The results of the experiment were analyzed by Rutherford,[82] and led to several far-reaching conclusions. The first was that most of the atom is empty (see Fig. 3.9)! The nucleus occupies only the 10^{-15}th part of the volume of the atom. Earlier experiments had given information about the size of the atom (a radius 10^5 times bigger than the newly found radius of the nucleus). Thus, the orbits of the outermost electrons (which define the radius of the atom) are far away from the nucleus. The second conclusion was that the force acting between the positive charge in the nucleus and the charged scattered projectiles obeys the Coulomb law. In short, Rutherford discovered that the atom resembles the Solar System, i.e., there is a very massive object at the center, while the light particles (the electrons being analogous to planets) move in orbits around this central object.[83]

---

[80] The density of radon is 9.73 kg/m^3 at 0°C. For example, the density of hydrogen, the lightest gas is 0.08988 kg/m^3.

[81] A year earlier, Geiger had invented the Geiger counter (with Walther Müller, then his student) to measure radioactive radiation, and consequently Rutherford invited him to England. After carrying out the famous experiment, Geiger returned to Germany and became a member of the Uranverein (Uranium Club) in Nazi Germany, the group of German physicists who, during World War II, tried to create the German atomic bomb. His loyalty to the Nazi Party led him to betray his Jewish colleagues, many of whom had helped him in his research before he became a member of the Nazi Party. Very few physicists were members of the Nazi Party. Notorious exceptions were Pascual Jordan and Philipp Lenard.

[82] Rutherford, E., Phil. Mag. Ser. 6, **21**, 669 (1911). The experimental results are described by Geiger and Marsden in Proc. Roy. Soc. Ser. A **82**, 495 (1909), and later in Phil. Mag. Ser. 6, **25**, number 148 (April 1913).

[83] Rutherford was lucky on two counts. The first was that the results for scattering are only identical in the classical and quantum theories in the case of the Coulomb force. The Planck constant, the hallmark of quantum theory, does not appear in what has become known as Rutherford's formula. Of course, quantum theory did not exist when Rutherford analyzed the experiment. The quantum mechanical result was first found by Mott, N.F., Proc. Roy. Soc. London A **118**, 654 (1929), and by Gordon, W., Zeit. fur Phys. **48**, 11 (1928). The second piece of good fortune has to do with the fact that the atom is mostly empty and hence, even though the thin foil of gold contained many atomic layers, no two atoms overlapped, so that Rutherford could justify considering a single scatterer.

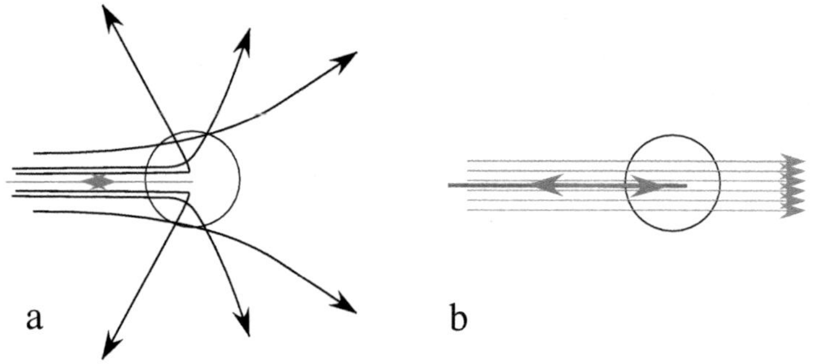

Fig. 3.8 The famous experiment by Geiger and Marsden, sending α particles onto a thin gold foil. The experiment revolutionized our model of the atom. The result was that the majority of the α particles moved straight through, without interruption, while a small fraction of them were deflected through an unrestricted range of different angles. The Thomson pudding model predicts that very few α particles will pass straight through in this way, and that the majority will be deflected, as shown in (**a**). In the experiment it was found that (i) most particles went straight through, and (ii) those that were deflected behaved as though they had been acted upon by a Coulomb force, or a gravitational repulsive force. The experiment provided the radius of the nucleus directly

In many respects this picture is one of the most fundamental in physics.[84] We note that at this time Wien's discovery of the proton had not yet been approved. The nucleus was found to contain the positive charge, but its breakdown was not clear. Note also that Rutherford got his Nobel Prize before he analyzed the α scattering experiment and made this fantastic contribution, for which he definitely deserved the Nobel in physics. The fact that the atom is mostly empty and the mass is essentially concentrated in a very small nucleus was to become extremely important for the question of the source of stellar energy. The smallness of the nucleus is the reason for the large energy.[85]

[84] In the case of the Solar System, the fact that the masses of the planets are very small relative to that of the Sun allowed Newton (1643–1727m) and Kepler (1751–1630m) to find the laws of planetary motion. If the masses of the planets had been of the same order as the mass of the Sun, the interactions between the planets would have been intricate enough to obscure the law of gravity, and complicate the motion immensely. In the case of the atom, Bohr (1913) was able to devise his atomic theory by neglecting the interactions between the electrons, and consider only the interaction between the electrons and the nucleus. Once the fundamental and dominant interaction had been found, corrections due to the interaction between the planets (electrons) could be taken into account, and even used to discover new planets.

[85] According to the uncertainty principle, the uncertainty in the momentum times the uncertainty in the location cannot be smaller than the Planck constant. Hence, if the particle is constrained to move in a small volume, its momentum must be very high, and hence its energy must also be very high. If the particles in the nucleus are restricted for whatever reason to move in the small nucleus, their energy must be much greater than the energy of the electrons which are free to move in the large atom. This explains why chemical energy, which is due to changes in the electronic structure, is of the order of a few electron volts, while nuclear changes are of the order of millions of electron volts. Chemical energy, which comes from relatively large structure, like atoms or

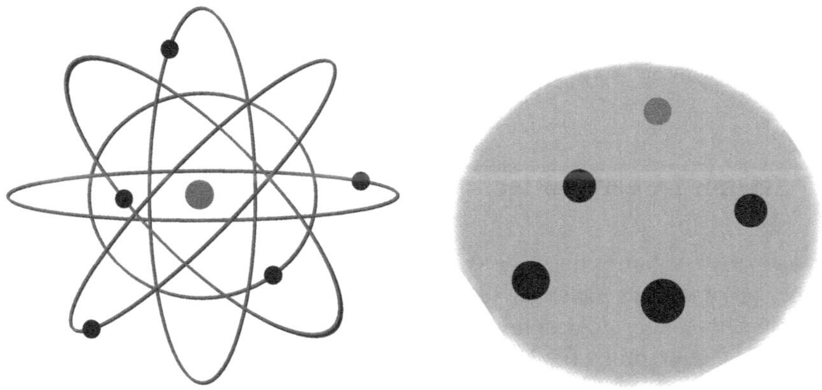

Fig. 3.9 *Left*: The difference between the new Rutherford model of the atom and the old Thomson–Kelvin model. The *red sphere* represents the protons in the nucleus, and the *blue spheres* are electrons. In the Rutherford model, the electrons move around the massive nucleus. *Right*: The Thomson model for the atom. The positive charge is continuous and fills the entire volume of the atom. The electrons are embedded in the positive charge like raisins in a cake. The electrons do not move, and their locations are fixed

In 1909, Francis Aston (1877–1945m) joined Thomson and they worked on improving Thomson's positive ray apparatus. Using the improved mass spectrometer, it took Thomson two years[86] to confirm Wien's discovery of the proton, although the final word was left for Rutherford (in 1919). Thomson wrote:

> *In 1898, however, Wien, by the use of very powerful magnetic fields, deflected these rays and showed that some of them were positively charged; by measuring the electric and magnetic deflections he proved that the masses were more than a thousand times the mass of a particle in the cathode ray. The composition of these positive rays is more complex than that of the cathode rays, for whereas the particles in the cathode rays are all of the same kind, there are in the positive rays many different kinds of particles.*

Indeed, different ions came out, but the one with the smallest mass was the proton. Thomson separated the ions according to their charge and mass. His major conclusion was that:

> *All results point to the conclusion that the occurrence and magnitude of the multiple charge are connected with the mass of the atom rather than with its valence or chemical properties.*

He was close, but not quite home. Once the radioactive elements could be distilled, so as to obtain significant amounts of them, the α particles were used as projectiles and fired against various targets. The first case was the α scattering experiment. Next, in 1919, Rutherford[87] bombarded nitrogen with α particles, and identified

molecules, cannot supply the solar energy, while nuclear energy, which comes from such a small volume, can.

[86] Thomson, J.J., Proc. Roy. Soc. A **89**, 1 (1913).

[87] Rutherford, E., Phil. Mag. A 6, **37**, 571 (1919).

the nucleus of hydrogen as the emitted particle. This was the first ever artificial nuclear reaction with element transmutation. A year later,[88] he accepted the hydrogen nucleus as an elementary particle, calling it the proton.

3.7 Science by Committee

Around 1910, Arthur Holmes (1890–1965)[89] began his estimates and measurements of the age of various rocks. He used the U/Pb and the U/He methods to produce the first calibrated geological time scale for the various geological periods, whence the past evolution of the Earth was no longer considered as one long period, but a succession of different ones. Two years later, before obtaining his PhD, he suggested the first radioactivity-based time scale. The first estimate of Earth's age was 4 billion years.[90]

In spite of the crudeness of the basic data known at the time (half-lives) and the unavoidable assumption that the initial uranium did not contain any lead (so that his ages were maximal[91]), his initial results are very close to the accepted values today. Around 1930, Holmes[92] suggested a mechanism that could explain Alfred Wegener's theory of continental drift: convection currents in the hot liquid Earth. Currents of heat and thermal expansion in the Earth's mantle, he suggested, could force the continents toward or away from one another, creating new ocean floor and building mountain ranges. [The theory was later expanded by Harry Hess (1906–1969m).[93]]

Holmes was a widely respected geologist by then, but he was a few years too late to support Wegener (1880–1930m), and about 30 years too early to have hard data to back up his theory. He warned that his ideas were *purely speculative* and could *have no scientific value until they acquire support from independent evidence*. Yet he had come very close to describing the modern view of the Earth's tectonic plates and the dynamics between them. Among other things, continental drift explains why the age of the rocks on the surface of the Earth can be significantly younger than the age of the Earth. The idea that the surface of the Earth changes continuously was a revolution in itself. If the radioactive elements are confined to the outer layers and

---

88 Rutherford, E., Bakerian lecture on *Nuclear Constitution of Atoms*, 3 January 1920, Proc. Roy. Soc. A **97**, 374 (1920).

89 Holmes, A., Proc. Roy. Soc. London A **85**, 248 (1911).

90 Holmes, A., *The Age of the Earth*, Harper Brothers, NY (1913).

91 More importantly, the experimental results that had accumulated by 1915 showed that different ores came with different mixes of isotopes and atomic weights, indicating that the amount of lead was not constant.

92 Holmes, A., Trans. Geol. Soc. Glasgow **18** III, 559 (1931) (1928–29 published 1931).

93 The two distinguished scientists Harry Hammond Hess and Victor Franz Hess are unrelated, but share the same mountain on the Moon.

are not spread uniformly in the Earth, then continental drift and circulation must be limited to the layers with radioactive elements.[94]

At that time, many geologists felt that the new discoveries made radiometric dating so complicated as to be worthless. Boltwood gave up and shifted his interest to other problems. But Holmes felt that there was plenty of room for improvements in the radiometric techniques, and he pushed forward. His work was generally ignored until the 1920s, though in 1917,[95] Joseph Barrell, a professor of geology at Yale, redrew geological history as it was understood at the time, to conform with Holmes's radiometric dating. Barrell's research determined that the layers of strata had not all been laid down at the same rate, so that current rates of geological change could not be used to provide an accurate timeline for the history of the Earth. But Barrel was the exception, as the majority of geologist's were not yet convinced of the superiority of radiometric techniques.

Holmes's persistence finally began to pay off in 1921, when the speakers at the yearly meeting of the British Association for the Advancement of Science came to a rough consensus that the Earth was a few billion years old, and that radiometric dating was reliable. The consensus did not create a mass migration towards radiometric dating, since the die-hard geologists resisted.[96] They never cared for attempts by physicists to intrude into their domain, and had successfully ignored them, so great was the legacy left by the Kelvin campaign against their findings. The growing weight of evidence finally tilted the balance in 1926, when the National Research Council of the US National Academy of Sciences decided to resolve the question of the age of the Earth by appointing a committee to investigate the data and methods. Doing science with committees was never a good idea. But fortunately, Holmes, who was one of the few experts in radiometric dating techniques, was a committee member, and in fact it was he who wrote most of the final report.

The report concluded that radioactive dating was the only reliable tool for pinning down geological time scales. Questions of bias were deflected by the great and exacting detail of the report. It described the methods used, the care with which measurements were made, and their errors and limitations.

---

[94] Continental drift refers to the movement of the Earth's continents relative to each other. The hypothesis that continents 'drift' was developed by Alfred Wegener in 1912. However, only with the development of the theory of plate tectonics in the 1960s could sufficient geological evidence be found, and the causes of their movement be explained. Wegener, A., *Die Entstehung der Kontinente*, Peterm. Mitt.: 185, 253, 305 (1912).

[95] Barrell, J., & Huntington, E., *The Evolution of the Earth and Its Inhabitants*, Yale University Press, New Haven (1922). Series of Lectures Delivered Before the Yale Chapter of the Sigma Xi During the Academic Year 1916–1917.

[96] According to Planck, an important scientific innovation rarely makes its way by gradually winning over and converting its opponents. What does happen is that the opponents gradually die out.

Fig. 3.10 The solar corona. Very hot gas extends many solar radii away from the Sun. It can only be seen during eclipses, when the much brighter Sun is hidden. The brightness of the solar corona is about 10^{-6} times that of the Sun, so it cannot be observed under normal conditions. NASA, Marshall Space Flight Center, `http://solarscience.msfc.nasa.gov/corona.shtml`

3.8 New Elements or Misled by the Stars

During the total solar eclipse of 7 August 1869, William Harkness (1837–1903) and Charles Young (1834–1908) discovered an emission line of feeble intensity in the green part of the spectrum of the corona. Young identified the line as the 531.68 nm iron line, No. 1474 in Kirchoff's catalogue of iron lines. However, iron has a huge number of lines and it was impossible to understand how only one line showed in the spectrum. As the identification of iron was not confirmed, the unavoidable conclusion was that it must be a new element, duly named coronium.

The suspicion that something was bogus about coronium had already been expressed by Agnes Mary Clerke (1842–1907m).[97] Agnes Clerke was a noted historian of astronomy in the second half of the nineteenth century,[98] to the point that a lunar crater was named after her. She had excellent relations with all the well-known astronomers of the day, and in particular with Huggins. And so wrote Clerke:

> *The behavior of coronium in the Sun is highly anomalous. It shows no signs of being subject to gravitational pressure, or of participating in solar atmospheric motion [...] it does not belong to the chromospheric spectrum as such, although inevitably seen projected upon it. [...] this lightest of gases – as it is assumed to be – comports itself in the chromosphere precisely after the manner of metallic vapour. [...] outside the Sun, coronium has not been convincingly identified.*

[97] Clerke, A.M., Obs. **21**, 325 (1898).

[98] Brück, M.T., Irish Astr. J. **24**, 193 (1997).

Nebulium, first observed by Huggins in 1864,[99] four years before Lockyer's discovery of helium, is another example of an unidentified spectral line which was suspected of being a chemical element. Huggins observed two greenish lines in the spectra of the nebula NGC 6543 in the Dorado constellation which he could not identify with any lines emitted by an element in the laboratory. The mystery persisted for 64 years before it was finally resolved by Bowen (see later). In retrospect, all these 'discoveries' of new elements in the cosmos should have stopped as soon as Henry Moseley (1887–1915m) discovered the connection between atomic number and X-ray spectra, because all possible locations for chemical elements were occupied.

The two misidentifications, which required so many years to unravel, arose from a lack of knowledge as to what spectra should look like when atoms have lost one or more electrons. It took time for astronomers to realize that, under stellar conditions (mostly high temperatures), atoms may be stripped of one or more electrons, and appear as completely new elements.

In his 1904 paper about the ether, Mendeleev[100] cited the discovery of coronium by Harkness and Young, as well as the observation by Nashini, Anderlini and Salvadori (1893) of coronium in volcanic gases, and the fact that coronium lines were observed far away from the Sun, where hydrogen lines are no longer seen, as evidence *that coronium should have less density and atomic weight than hydrogen.* In 1919, Cady and Elsey[101] developed special spectroscopic equipment in an attempt to resolve the question as to whether certain lines that were frequently detected in the spectrum of helium derived from natural gas were due to coronium. They suggested that certain observed lines did indeed belong to coronium.

The mysterious coronium lines resisted identification until 1939, when they were identified as lines of highly ionized iron by Walter Grotrian (1890–1954m)[102] and Bengt Edlén (1906–1993).[103] At the very high temperatures of the corona (about 2 million degrees), the collisions between atoms are sufficiently powerful to tear several electrons from them. Since such high temperatures were not available on the Earth, the spectral lines of the highly stripped atom were not known. The theory needed to calculate the position of the spectral lines did not yet exist.

In an attempt to explain the observed lines of the nebula, which are not lines of known elements measured in the laboratory, John Nicholson (1881–1955), a mathematician, resorted[104] to:

> *[...] the atom of nebulium, a hypothetical element predicted by the theory, when electrically neutral, contains four electrons, each with a charge −e− rotating uniformly at equal distances in a circle round a positive nucleus whose charge is* 4e. *If one of the electrons*

---

[99] Huggins, W., Phil. Trans. **154**, 437 (1864).

[100] Mendeleev, D., *An Attempt Towards a Chemical Conception of the Ether*, Trans. Kamensky, Longmans, Green and Co., London (1904). Mendeleev was haunted by the ether assumption, and spent many years looking for it.

[101] Cady, H.P., & Elsey, M.H., Science **18**, 71 (1919).

[102] Grotrian, W., Naturwiss. **27**, 214 (1939).

[103] Edlén, B., Zeit. f. Astrophys. **22**, 30 (1942).

[104] Nicholson, J.W., MNRAS **72**, 49, 139, 677, & 729 (1911).

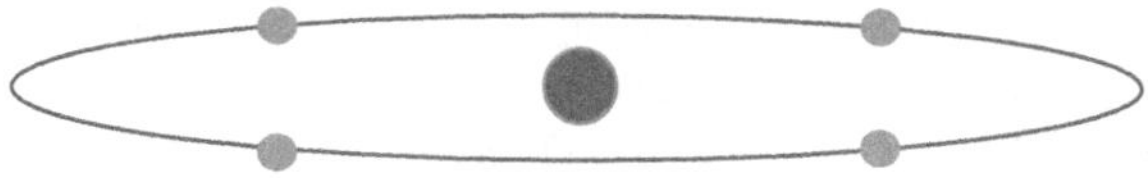

Fig. 3.11 Nicholson's ring model of nebulium. The electrons form a planar ring and move in fixed locations in the ring

is missing, the other three can take up equidistant positions and rotate in a new orbit, the system then consisting of an atom of nebulium with a single positive charge. In a similar manner, the atom may take up more electrons and acquire a negative charge.

So Nicholson's idea[105] was that all electrons are in the ground state, but that the energy of the ground state changes with the number of electrons. The energy difference between the two ground states is the energy needed for ionization of the atom. At the same time, the ring of electrons can vibrate, and the differences in energy between the different vibrations are emitted as photons of fixed wavelengths, that is, as spectral lines. As the system can vibrate in a multitude of ways, Nicholson chose to discuss only those vibrations which were perpendicular to the plane of the ring of electrons. In a way, Nicholson based his model on Thomson's[106] and Nagaoka's.[107]

The idea of a ring of electrons was devised so as to minimize the radiation losses by the accelerating electron. However, the radiation losses did not vanish. According to Larmor,[108] if the vector sum of all the central accelerations vanishes, as is the case with a symmetric arrangement of electrons, there is no radiation loss. This is known as the Larmor condition. This is the reason why the model assumed that all electrons moved in the same plane.

Unlike previous researchers, Nicholson calculated the frequencies of the vibrations such a ring would possess. The idea of electron rings was quite popular, and several researchers carried out extensive calculations with this model.[109] The complete calculation cannot be carried out without a model for the electron, and hence three different models with three different results were examined by Nicholson. All in all, the calculation was very complicated and lengthy, and ended with a complicated formula for the frequencies, a formula which did not accurately reproduce the observed spectra. The operative word here is 'accurately'.

---

[105] An atom with four electrons is the chemical element beryllium, which was already known at the time (discovered in 1798 by Vauquelin), but Nicholson did not refer to this element, or a nearby element in the periodic table which had lost an electron. If nebulium really existed in nebulas, then its abundance would have been many orders of magnitude greater than the known abundance of beryllium. Coronium was assumed by Nicholson to have 5 electrons and he named it protofluorine.

[106] Thomson, J.J., Phil. Mag. (March 1904) p. 237.

[107] Nagaoka, H., Phil. Mag. **7**, 445 (1904).

[108] Larmor, J., Phil. Mag. **44**, 503 (1897).

[109] See Schott, G.A., Phil. Mag. (1906) p. 21; ibid. (February 1907) p. 210; ibid. (January 1908) p. 180; ibid. (April 1908).

Of course, the calculation did not include any of the assumptions of quantum theory. Moreover, the classical theory did not predict the relative strengths of the various transitions between the oscillation modes. In a lecture given to the Royal Astronomical Society,[110] in the same year that Bohr published his quantum theory of the hydrogen atom, Nicholson raised his objections to the implementation of Planck's quantum theory to the atom:

If the atoms are Planck's resonators, it is evident that on the present view they have a fundamental periodicity whose frequency is that of their chief vibration. They should accordingly give out energy in multiples of this frequency. Can a physical process be imagined which would satisfy this necessity?

As a matter of fact, the wavelengths of the spectral lines, like those of hydrogen,[111] do not come in simple harmonics and factors of two. Nicholson ended his talk by stating:

This theory allows an interpretation of nearly the whole known spectrum of the corona, and simultaneously the possibility of such an interpretation is one of the strongest indications of the essential truth of much of the modern physical theory.

Nicholson compared the theoretical spectra with the observed lines of nebulium[112] as given by Wright,[113] and inferred that the atomic weight of nebulium had to be $M_{\mathrm{neb}}/M_{\mathrm{hyd}} = 1.31$ *with a possible error of unity in the last figure*. The error in Nicholson's fit amounted to 3.6 angstroms and he claimed, wrongly, that if the wavelength of the line was about 4 400 angstroms, this difference in wavelengths amounted to a relative error of 0.000 8, which by non-spectroscopic considerations was very small. But spectroscopy is extremely accurate, and a deviation of 3.6 angstroms is quite large. Yet Nicholson complained:

It is unfortunate that the wavelengths in the coronal spectrum cannot be found with similar accuracy, so as to allow a more precise deduction of the atomic weights of the substance concerned.

In other words, he expected his theory to be more accurate than the observations. Nicholson accompanied his theory with some more general philosophical remarks like:[114]

The possibility of astrophysics as an arbiter of the destinies of ultimate physical theories is of course clear.

---

[110] Nicholson, J.W., Obs. **36**, 103 (1913).

[111] The spectral lines of hydrogen are given by the formula $\lambda^{-1} = R_y\left(m^{-2} - n^{-2}\right)$, where $n \geq m$ takes the values $1, 2, 3, \ldots$. The series with $m = 1$ is known as the Lyman (1874–1954m) series, $m = 2$ the Balmer (1825–1898m) series, $m = 3$ the Paschen (1865–1947m) series, $m = 4$ the Brackett (1896–1988m) series, $m = 5$ the Pfund (1878–1949) series, and $m = 6$ the Humphreys (1898–1986) series. R_y is the Rydberg constant.

[112] Nicholson, J.W., MNRAS **72**, 49 (1911).

[113] Wright, W.H., Ap. J. **16**, 53 (1902). Note that Wright did not call the lines nebulium lines, but instead used the term 'nebular lines'.

[114] Nicholson, J.W., Obs. **36**, 103 (1913).

He was philosophically right, but physically wrong.

Three years later, Nicholson's conclusion was substantiated by Buisson, Fabry and Bourget[115] who used a newly devised interferometer[116] (an instrument which compares waves) on the Orion nebula, and claimed in the discussion part of the paper that:

This result shows that the unknown gas which emits the double ultra-violet lines has an atomic weight higher than that of hydrogen. The ratio of the two atomic weights is 2.74. A figure in the neighborhood of 3 is therefore the probable value of the atomic weight of this gas.

We remark that the authors did not give any reference to Nicholson, though their result supported his.

In 1918, five years after Bohr published his model of the atom (see below), Nicholson returned to the problem of the atomic weight of the elements in nebulas,[117] and got the same result again. Clearly, this result for the atomic weight of nebulium only enhanced the mystery of the nature of the element. It was a clear case for applying Ockham's razor. What is the preferred theory, a classically well established theory with a very complicated model that inaccurately predicts the experimental results, or a new theory with a bold and highly unorthodox assumption, but which predicts the experimental results to high accuracy? Science showed that the latter was the correct solution.

Bowen[118] was inspired by Russell's conclusion that:

It is now practically certain that the spectral lines must be due not to atoms of unknown kinds but to atoms of known kinds shining under unfamiliar conditions.

By the words 'unfamiliar conditions', Russell was referring to the very low density. As Bowen explained, when the electron jumps to a high energy level, it stays there for some time before it jumps to a lower level. But atoms are never alone in the universe. Any nearby atoms will collide with the atom at a frequency which depends on the density (and the temperature). When the density is high, the resulting frequent collisions may hit the atom with the electron in the high level, and as a consequence the electron can be kicked out of the level before it has time to jump back down to the lower energy level. Thus, at a high density, certain lines may not be observed. When the density is low the electron may stay long enough in the high energy level before it jumps back. The spectral lines emitted by atoms placed in low and high density conditions are therefore expected to be different.

The low densities in space may be of the order of 1 000 atoms per cubic centimeter or less, and the time between collisions as long as 10^4 to 10^7 seconds.[119]

[115] Buisson, H., Fabry, Ch., & Bourget, H., Ap. J. **40**, 241 (1914).

[116] This is one of the first applications of the then newly devised interferometer, known today as the Fabry–Perot interferometer used extensively today. It was invented in 1901 [Fabry, Ch.,& Perot, A., Ap. J. **13**, 265 (1901).

[117] Nicholson, J.W., MNRAS **78**, 349 (1918).

[118] Bowen, I.S., Ap. J. LXVII, 1 (1928). See also PASP **39**, 295 (1927).

[119] Note that one year is 3.1×10^7 seconds, whence the rate of collisions is one collision every few months.

Such low densities cannot yet be created in the best vacuum constructed on Earth. In spectroscopy, such lines are said to be forbidden, for the following reason. The probability of transition from the high to the low level may be very low, so that the electron has to stay a long time in the high level before it jumps to the low one. On Earth, where the density is high, this means that there is a fair chance that collisions between atoms will remove the electron from the high level long before it has the chance to jump to the lower level. Hence, these lines are not observed on the Earth. But in the very tenuous gas in space, where atomic collisions are much less frequent, these spectral lines appear as very strong lines.

So in 1928, without mentioning the element nebulium by name, Bowen explained and calculated how the mysterious line arose from doubly and singly ionized oxygen, as well as singly ionized nitrogen. As a matter of fact, Bowen identified all but two or three of the 'nebulium' lines with a high accuracy (fractions of angstroms). Only then did it become clear that atoms can lose their electrons in the environment of the nebula, with its extremely low density (and temperatures of a few thousand degrees). The long-standing mystery of possible elements which exist on stars or nebula but not on the Earth had almost come to an end. Interestingly, as early as 1920, in his Bakerian lecture[120] entitled *Nuclear Constitution of Atoms*, Rutherford had discussed the bothering problem of nebulium and summarized it by saying:

> *It is not easy at the moment to see how the new atoms from oxygen or nitrogen can be connected with the nebular material.*

It should, however, be remarked that Bowen's hypothesis did not go without criticism.[121] Even correct solutions are sometimes difficult to digest.

Bowen's solution was heralded with enthusiasm by leading astrophysicists like Eddington,[122] Fowler,[123] and Russell,[124] who immediately recognized how the extreme astrophysical conditions could give rise to such spectral lines. This was the beginning of the end of 'elements which exist in stars but not on Earth'.

3.9 The First Atomic Quantum Theory

The Rutherford model of the atom created some new unsolved theoretical problems. It is known from classical electrodynamics that an accelerating charge loses energy by radiation. Now an electron which moves in circles is accelerating, and so should radiate and thereby lose energy. However, it does not, because if it did lose energy, it would fall quickly into the nucleus. In the same way, Einstein discovered in his

---

[120] Rutherford, E., Proc. Roy. Soc. A **97**, 374 (1920).

[121] Bartlett, J.H., Phys. Rev. **34**, 1247 (1929).

[122] Eddington, A.S., MNRAS **88**, 134 (1927).

[123] Fowler, A., Nature **120**, 582, 617 (1927).

[124] Russell, H.N., Phys. Rev. **31**, 27 (1928).

general theory of relativity, that the Earth, which moves around the Sun in an elliptical orbit and hence accelerates, should radiate gravitational waves and lose energy. Indeed, the phenomenon was discovered years later: it causes a close binary system of stars to move closer towards each other, in such a way that they will eventually coalesce.[125]

This problem did not exist in the Kelvin–Thomson model, because in this model the particles are at rest (but there are many other problems, the first of which is that it does not agree with observations). Jeans noted in 1915[126] that, if the electric force acting between two charged particles is as assumed by Rutherford, inversely proportional to the square to the distance, the charges cannot approach too close to one another, otherwise the force will tend to infinity. Consequently, Jeans concluded that there cannot be point charges as advocated by Rutherford. The nucleus of positive charge must be finite, though small relative to the size of the atom. Rutherford got the size of the nucleus from the experimental result. He might already have realized then that having the positive charge enclosed in such a small volume would involve huge energies and very strong forces.

The formation of the spectral lines, which was one of the fundamental tools for identifying chemical elements on the Earth and in the cosmos, eluded explanation until 1913, when Niels Bohr[127] attacked the problem of the structure of the atom, and invented the first model of the atom based on the Planck hypothesis and the Rutherford picture. Bohr assumed that the electron must move in special (quantized) orbits, and that the spectral lines are formed when the electrons jump from one orbit to a lower one.

More precisely, Bohr's hypothesis was that all quantities which are conserved in classical physics, such as the total energy, are quantized, and can only assume certain discrete values. Although the law of gravity and the Coulomb force behave in the same way, planets can move at any distance from the Sun while electrons are restricted in their orbits. Since the energy of the orbits is fixed and not every orbit is possible, only fixed amounts of energy, the energies corresponding to the lines, are emitted. This almost necessary assumption explained why hydrogen atoms always emit the same lines whether they are on the Earth, on the Sun, or on a distant star. If electrons could move anywhere without restriction, there would not have been fixed wavelength spectral lines, and there would have been no quantitative astrophysics.

This was a huge victory, because Bohr explained the entire spectrum of hydrogen. This spectrum is characterized by several series of spectral lines, named after their discoverers. In 1888, the Swedish physicist Johannes Rydberg (1854–1919m) discovered a simple formula which described all known spectral series of hydrogen

---

[125] Joseph Taylor discovered that the binary system denoted PSR 1913+16, composed of two neutron stars which revolve around their center of gravity, loses energy at the rate predicted by the general theory of relativity. Nobel Prize 1993, Weisberg, J.M., & Taylor, J.H., *Binary Radio Pulsars*, ASP Conference Series, Vol. 328, conference held 11–17 January 2004.

[126] Jeans, J.H., *The Mathematical Theory of Electricity and Magnetism*, Cambridge Press, 3rd edn. (1915) p. 168.

[127] Bohr, N., Phil. Mag., Ser. 6, **26**, 1 (1913).

to extreme accuracy, but could not offer any physical explanation. It was Bohr who derived the formula theoretically and exactly.

While the Bohr atom was a fantastic success, it left many questions unanswered. For example, why do the electrons not radiate when they move in the quantized orbits around the nucleus, radiating only when they jump from one orbit to another? Where is the electron, and what happens to the electron, when it jumps between the levels? Why does classical physics not work on the atomic scale? In many respects, while Bohr did indeed apply the idea of quantization to atomic systems, the picture of the electron was still classical.

At the same time, Thomson did not give up his attempts to build a model for the atom, and claims that no one had proven the electron to be spherical symmetric, and devised a second model of the atom in which it was rigid.[128]

3.10 Kelvin and the Age of the Earth. An Epilog

How fast the wheel can turn. In 1894, approaching 70 years of age, Kelvin had every reason to feel confident in himself, as multiple attempts to determine the age of the Earth seemed to show that it was at most 100 million years. The geologists could only suggest (correctly) that Kelvin did not have all the facts, while they still believed that the Earth was significantly older. However, once radioactivity was discovered, the rate of dramatic discoveries accelerated, and it was not long before the wheel had indeed turned. The evidence for a billion year old Earth could no longer be discredited.

Kelvin was apparently never convinced by the new discoveries, although most other physicists were. One can find in Rutherford's memoirs the following amusing story about the confrontation between Rutherford and Kelvin. In 1904, Rutherford was about to give a speech on radioactivity in which he disagreed with Kelvin's estimates of the age of the Earth. When he realized that Kelvin was in the audience:[129]

> *I realized I was in for trouble at the last part of the speech [...]. Then a sudden inspiration came and I said Lord Kelvin had limited the age of the Earth, provided no new source of heat was discovered. That prophetic utterance refers to what we are now considering tonight, radium! Behold! The old boy beamed upon me.*

Rutherford concluded this speech before the Royal Society with a dramatic statement of the new order of things:

> *The discovery of the radio-active elements, in which their disintegration liberates enormous amounts of energy, thus increases the possible limit of the duration of life on this planet, and allows the time claimed by the geologist and biologist for the process of evolution.*

Kelvin never published any acknowledgment that radioactivity was supplying heat to the Earth's crust, and that as a consequence his calculations of the age of the

[128] Thomson, J.J., Phil. Mag. Ser. 6, 21, **125**, 669 (1913).

[129] Burchfield, J.D., *Lord Kelvin and The Age of the Earth*. University of Chicago Press, Chicago (1990).

Earth were not accurate. Indeed, in 1906 and 1907, he even published several letters and papers denying that radium could be a source of heat within the Earth or the Sun. But J.J. Thomson wrote in his memoirs that Kelvin admitted in private that his theories had been overthrown.[130]

Darwin died without knowing that he was right about the age of the Earth. Kelvin died knowing that he was wrong on many counts, including the clouds he had seen hanging over physics, and the age of the Earth. The ashes of Rutherford, who had suggested the radioactive heat source in the Earth, were buried in Westminster Abbey, just west of Sir Isaac Newton's tomb, and beside the tomb of Lord Kelvin, whose theory he had destroyed. Upon his death, Darwin's family arranged for him to be buried in St. Mary's churchyard in the village of Downe. However, William Spottiswoode, the President of the Royal Society wrote to the Dean of Westminster Abbey, requesting that Darwin be buried in its prestigious cemetery. Despite Darwin's controversial work and the fact that he was a self-professed agnostic, the Dean of Westminster responded positively to Spottiswoode's request. Once Darwin's family had agreed to the interment, his body was sent to the Abbey for a service and burial, and in this way all the heroes of this story found eternal rest under the same roof.

3.11 What We Know Today about the Age of the Earth

The complexity of the Earth's structure (continental drift, non-spherically symmetric heating) shifted research on the age of the Earth and the Solar System to meteorite dating, where the chemical and physical processes are simpler, and since in any case such objects constitute the primordial material of the Solar System.

The traditional assumption about the original composition of the Earth is that it resembled the carbonaceous chondrites, which are the most primitive meteorites. However, the Earth's present composition may be quite different from the composition of the meteorites we observe today to fall upon it, because of various chemical processes that the Earth has undergone. Clearly, volatile elements would have been depleted during the period of hot formation. The large metallic core indicates that the Earth as a whole is a reduced body,[131] although at least the crust and the outer shells of the mantle are oxidized.

The best age estimate for the Solar System is based on four very old lead ores, which evolved from the lead isotopic mix at the time the Solar System was formed, as recorded in the Canyon Diablo iron meteorite.[132] The age obtained is 4.54 ± 0.02

[130] Burchfield, ibid., p. 56, note 52.

[131] Reduction is the opposite of oxidation. Oxidation involves losing an electron, while reduction involves gaining one.

[132] The giant diamond-containing Canyon Diablo meteorite is unique in a number of scientifically important aspects. When it fell, about 49 000 years ago, it formed the Arizona Barringer meteorite crater, about 1 220 meters in diameter.

Table 3.2 The various radioactive elements used to determine the age of meteorites and rocks

Parent	Daughter	Half-life [billions of years]
Uranium ^{238}U	Lead ^{206}Pb	4.47
Uranium ^{235}U	Lead ^{207}Pb	0.704
Thorium ^{232}Th	Lead ^{208}Pb	14.0
Potassium ^{40}K	Argon ^{40}A	1.25
Rubidium ^{87}Rb	Strontium ^{87}Sr	48.8
Samarium ^{147}Sm	Neodymium ^{143}Nd	106
Rhenium ^{187}Re	Osmium ^{187}Os	43.0
Lutetium ^{176}Lu	Hafnium ^{176}Hf	35.9

Table 3.3 Summary of the heat flux from the Earth's interior

Vertical temperature gradient	Heat generation
10–80°C/km	$0–8 \times 10^{-6}$ watt/m^3

billion years.[133] The first signs of life are about 3.8 billion years old, and these were primitive single cell organisms without a nucleus. Fossil evidence indicates that about 600 million years later, a very short time relative to the age of the Earth, there were already many forms of primitive life.

3.12 Heat Flow

The heat flow out of the Earth is now a tool for measuring the amount of radioactive material. The rate at which the mantle loses heat appears to have been overestimated, while the available energy source in the interior has been underestimated.

Estimating the total heat flow is not simple, and the numbers given in Table 3.3 are approximate. The heat flow is far from uniform, there are plumes and volcanos, as well as hydrothermic activity through which the heat comes out. It is not clear to what extent the Earth may be slowly contracting, thus giving up gravitational energy. Most models of the crust assume that the mantle did not undergo accretional differentiation, and hence that it retains the primordial amounts of the radioactive elements. However, it seems that this assumption may be too much of a simplification.

Currently, the Earth interior is cooling by a combination of thermal conduction through the surface and advection by slabs of cold material in the interior (i.e., mass motion). The heat generated in the interior is transferred to the surface in a way

[133] Patterson, C.C., *The isotopic composition of meteoritic, basaltic and oceanic leads, and the age of the Earth*, Proc. Conf. Nuc. Proc. In: *Geologic Settings* (1953) p. 36; Cosmochimica Acta **10**, 230 (1956).

Table 3.4 The breakdown of the heat flux according to Pollack, H.N., Hurter, S., & Johnson, J.R., *The new global heat flow compilation*, Univ. Michigan (1991)

Region	Mean heat flow [milliwatt/m^2]
Oceans	101±2.2
Continents	65±1.6
Globally	87±2.0

that is too complicated to be discussed here in any detail. We may just summarize by saying that the solar flux is 1 360 watts per square meter (known as the solar constant), while the flux from the Earth is about ten thousand times smaller.

Chapter 4
Towards a Complete Theory of Stellar Structure

4.1 Eddington. The First Stellar Model with Radiative Transfer

The major theoretical breakthrough in the theory of stellar structure came in two seminal papers by Eddington (1882–1944m).[1] The radiative transfer of energy through the entire star was modeled for the first time, and the state of matter under the extreme conditions expected inside a star was analyzed. Eddington is known to the general public as the person who brought news to the English speaking world of the discoveries of a German scientist by the name of Einstein, during the first World War. The discoveries of the two theories of relativity captivated Eddington's imagination, and he became one of their staunchest advocates. However, Eddington is mainly known to the scientific community for his groundbreaking discoveries in the theory of stellar structure.

4.2 Stars Are Gaseous

Just after the beginning of the twentieth century, Eddington believed in Lockyer's hypothesis of gaseous giants and liquid dwarfs, and his first assumption was that the stars were gaseous. He therefore limited the scope of his modeling to the giants. The assumption not only simplified the calculations, but also avoided the problem that the equations of liquids and very dense gases were simply not known at the time.

---

[1] Eddington, A.S. MNRAS **77**, 16 (1916) and MNRAS **78**, 28 (1917). The paper Ap. J. **48**, 205 (1918) is a summary of the first two papers. Eddington realized that only a few American astrophysicists read his papers, which were published in Europe, and hence found it appropriate to provide a summary in the Astrophysical Journal, published in the USA. However, he explained that *after two years of experience it has become possible to make some simplifications in the treatment, and I hope that the following explanation of the theory will be more easily understood.*

G. Shaviv, *The Life of Stars*, DOI 10.1007/978-3-642-02088-9_4,

4.3 Radiation Pressure Plays a Crucial Role in Stars

According to the theory of relativity, $E = mc^2$. This result can be interpreted as follows: a given amount of energy in the form of radiation has an equivalent mass of E/c^2. The radiation in the star represents energy and hence is equivalent to mass. As this 'mass' moves with the speed of light c, it has momentum exactly like any other moving mass, given by $p = E/c$. This momentum acts on the layers of the star and represents pressure, the radiation pressure.[2] This radiation pressure, claimed Eddington, adds to the gas pressure in balancing the gravity of the star. If the radiation helps to balance the star against gravity, which pulls the layers of matter inward, then you can expect a connection between the mass which provides the inward pull and the luminosity/radiation which provides the outward push. This is the basic physical reason for the connection between the mass of the star and its luminosity, a relation called the mass–luminosity law.

The logic spelled out here may look very simple and straightforward, and yet it took about 8 years to clarify! The fact that radiation has momentum and hence exerts pressure (when absorbed or scattered) follows from Maxwell's equations,[3] and hence the new formula $E = mc^2$ is not really required if we only discuss the pressure exerted by the radiation. Some even say that it is not surprising that Einstein derived his famous equation, because he assumed the Maxwell equations in his derivation. However, in 1935 Einstein[4] derived $E = mc^2$ without relying on the Maxwell equations, proving its universal validity once more, if that was needed. In summary, Eddington applied the equivalence of mass and energy and its consequences to stars, where the density of radiation is high, with dramatic consequences. But the acceptance of the idea that radiation pressure is important in gaseous stars was difficult to digest. This was a long uphill battle.

The radiation pressure was predicted by Maxwell as a result of the existence of stresses in the ether. Poynting[5] arrived at the idea of radiation pressure from

[2] The pressure of an ideal gas, sometimes also called a perfect gas, is produced by the molecules hitting the walls of the container. The molecules have momentum, which is transferred to the wall when they hit it. The molecules bounce back as a result of the collision with the wall. Hence, the change in the momentum of the molecule or the momentum delivered to the wall, is twice the momentum of the molecules (being in one direction before and in the opposite direction after the collision with the wall). The molecules of an ideal gas are so small that they do not collide with one another, but only with the wall. Similarly, a photon absorbed by an atom delivers its energy and momentum to the atom. In short, pressure is an expression for the delivery of momentum to the wall in the case of molecules and absorption of photons in the case of radiation.

[3] Maxwell published his equations, the equations of the electromagnetic fields (the unification of the magnetic and electric fields) in 1867. Maxwell's theory compares in importance to Newton's second law ($F = ma$). Yet, Newton is better known to the general public than Maxwell. By the time Eddington applied the idea of radiation pressure, Maxwell's equations were already fairly well accepted by the scientific community, but not before Maxwell's untimely death in 1879 at the age of 48. On the other hand, it took decades before people like Kelvin and Helmholtz accepted Maxwell's colossal innovations. See Darrigol, O., *Electrodynamics from Ampère to Einstein*, Oxford Univ. Press (2000).

[4] Einstein, A., Amer. Math. Soc. Bull. **41**, 223 (1935).

[5] Poynting, J.H., Proc. Phys. Soc. London **19**, 475 (1903).

consideration of the flow of momentum along the line of light propagation in the ether. Poynting[6] applied the idea of radiation pressure to the Solar System. The powerful solar radiation exerts pressure, and the ratio of this pressure to gravitation increases as the size of the body decreases. Larmor[7] reached the same conclusions on the basis of the electromagnetic wave theory of light in the ether. Once the action of radiation pressure outside the stars was discovered, it was a question of time until someone like Eddington would come along and implement the idea for the pressure inside the stars.

Nicholson[8] and Klotz[9] worked out the value of the radiation pressure when the size of the obstructing mass decreases gradually, ultimately being reduced to the scale of the wavelength of light. In this case, the effect of repelling light pressure gradually preponderates over any gravitational force to which the particle may be subjected. At that time it was believed that there is a limit to this reduction process. If the particle is too small, it is no longer capable of acting as a barrier to the advancing light wave and consequently experiences no radiation pressure. It appeared from these investigations that, for particles of molecular size (radius $= 10^{-8}$ cm), the effect of light pressure is totally evanescent. Hence, many concluded that Eddington's idea of applying the radiation pressure to the atoms and electrons in a star was wrong.

But Saha[10] claimed that Nicholson's and Klotz's conclusions contradicted *the requirements of astrophysics* to explain the tails of comets. The idea that solar radiation pressure generates cometary tails was found by Biermann[11] to be wrong, in 1951. It is the solar wind that is responsible for the tails of comets. Thus, in 1920, Saha reached the right conclusion that the radiation acts on atoms, but for the wrong astrophysical reason (which became clear 30 years later). However, he gave excellent physical arguments.

Saha carried out the following calculation. He took a 'pulse' of light. He then implemented Planck's and Einstein's ideas and wrote that the impulsive momentum is $h\nu/c$, as if the 'pulse' had a mass of $h\nu/c^2$, and calculated the recoil velocity that a molecule would get if it absorbed this 'pulse'. Note that the term 'photon' was not used. Saha's work predated the first experiments[12] in which the photon was treated as a particle, and the kinematics was solved using relativity and Planck theory. Saha, as a matter of fact, repeated Einstein's theory of the photoelectric effect, but with one difference: Einstein assumed the expression for the energy of the photon to be

---

[6] Poynting, J.H., Phil. Trans. Roy. Soc. London **202**, 525 (1904); Science **26**, 602 (1907).

[7] Larmor, J., Phil. Mag. (November 1914).

[8] Nicholson, S.B., MNRAS **74**, 425 (1914).

[9] Klotz, O., Journal of the R.A.S. of Canada **12**, 357 (1918).

[10] Saha, M.N., Ap. J. **50**, 220 (1919).

[11] Biermann, L., Zeit. f. Astr. **29**, 274 (1951); Obs. **77**, 109 (1957).

[12] Compton, A.H., Proc. Amer. Nat. Acad. Sci. **11**, 303 (1925). However, Compton carried out such experiments for several years before his work culminated in the discovery of what is now known as the Compton effect. Compton won the Nobel prize for the experiment. Saha's calculation was not recognized.

Planck's, while Saha used it to obtain the momentum. Saha's final conclusion was that:

> *Radiation pressure may exert an effect on the atoms and molecules which are out of all proportion to their actual sizes. It also shows that the radiation pressure exerts a sort of sifting action on the molecules, driving the active ones radially out along the direction of the beam. The cumulative effect of the pulses may be sufficiently great to endow the atoms with a large velocity – the velocity with which the tops of solar prominences are observed to shoot up.*

Saha thus corroborated Eddington's application of the radiative pressure to molecules and atoms.

Before Eddington implemented the radiative pressure, Lebedew (1866–1912m)[13] demonstrated the existence of radiation pressure on molecules of CO_2, methane, etc. The classical continuum theory (which did not use Planck's idea) failed to predict this effect. When there is a large difference between the wavelengths of the radiation and the size of the object, the probability of absorption is small.

A contradictory experimental result was found by Campbell.[14] Campbell conducted an experiment which showed that the atom had to be illuminated for 15 minutes before it acquired enough energy for the emission of an electron, while the emission actually takes place instantaneously. It is not clear what went wrong in this experiment, which was not confirmed by others.

The equations for the radiation in the atmosphere of stars were developed by Schwarzschild[15] in 1906. Schwarzschild was interested in the radiation emerging from the star, and saw no reason to extend his theory to stellar interiors, because stars were supposed to be liquids, and the energy transfer in liquids is completely different.

Eddington's first step was to extend Schwarzschild's theory of radiation to the interior of gaseous stars without including the radiation pressure in his equations. But in doing so he discovered that the calculated temperature at the center rose so much that radiation pressure became important. He then applied the by now old theory of Lane and Ritter to calculate the structure of the star. The inclusion of the radiation pressure gave very different results from those of Lane and Ritter. In particular, Eddington derived a relation between the mass and the mean density on the one hand, and the effective temperature, on the other hand. The victory was that the theory predicted that giant stars would have a constant luminosity, as is actually observed.

Since the dwarf stars are dense and were believed to be liquid, the contribution of the radiation pressure was expected to be negligible. But when more physics was discovered, this expectation turned out to be wrong for the more massive dwarf stars.

---

[13] Lebedew, P., Annalen der Physique **32**, 411 (1910).

[14] Campbell, N.R., Phys. Rev. **7**, 18 (1916).

[15] Schwarzschild, K., Göttingen Nachrichten, 1906, p. 41.

4.4 Basic Astrophysics: The Stars Obey Our Laws of Physics

Do the stars obey the same physics that we are familiar with here on Earth? Newton had already shown that the law of gravity was universal. But what about the interior of stars, those regions we cannot see? It was in this paper that Eddington forged the essence of theoretical astrophysics by arguing that one can use physical laws discovered and confirmed on Earth to understand the stars! In Eddington's own words: *There are some physical laws so fundamental that we need not hesitate to apply them even to the most extreme conditions.* And the laws of radiation were such. The stars obey the same laws of physics as we discover and know to exist on the Earth! The trouble was, as he discovered shortly afterwards, that many laws were not actually known at that time! And one of the most important was the way matter absorbs (and emits) radiation.[16]

4.5 Absorption of Radiation. A Key Issue

The basic physics that remained unknown to Eddington at that time was how matter absorbs radiation. This may sound a rather dull physical subject, but it turns out to be essential for stars and planets to harbor life. The way the atmosphere absorbs the radiation coming from the Sun (mainly in the ultraviolet and the near infrared) and the way it absorbs the far infrared radiation emitted by the Earth, causes the greenhouse effect, which keeps the atmosphere at a temperature about 30°C higher than the temperature a planet would have if it had no atmosphere.

Deep inside stars the radiation is that of a black body at the local temperature. Wien's law states that, as the temperature rises, the maximum of the radiation shifts to shorter and shorter wavelengths. The temperatures in the interior of stars are 10 to 40 million degrees, and at these temperatures the peak of the radiation is in the X-ray range. If one could see into the interior of the Sun, one would find mostly X rays. The Sun is the greatest X-ray machine you can think of. This radiation is deadly for life. As the radiation propagates outward, it is absorbed and re-emitted many times.[17] The envelope of the star, which serves to generate the enormous pressure

---

[16] The absorption process is the opposite of emission. All atomic physical processes can go both ways. In physics, this is called reversibility. Time reversibility means that, if you take a movie of a physical system, say a pendulum, you cannot determine whether the movie is being projected forward or backward. The processes of absorption and emission of radiation are the time reversals of one another, and knowing one of them means knowing the other. A process is time reversable if it goes equally in both directions. Processes which increase the entropy of the universe are not reversible. Microscopic processes like photon absorption and emission are reversible.

[17] A mean free path is the (average) distance a particle/photon moves between collisions or between the point of emission (birth) and the point of absorption (death). The mean free path of an X-ray photon in the Sun is less than 1 cm. Hence, after 1 cm on average, the photon is absorbed and re-emitted. This means that the gas in the Sun is so dense that one cannot see more than 1 cm away! The radius of the Sun is $R_\odot = 6.96 \times 10^{10}$ cm, and hence a photon would need about 6.96×10^{10} absorptions and emissions to get from the center to the surface. But since the photon can be emitted

needed to keep the very hot core compressed so that nuclear reactions can take place, serves also to convert the X ray emitted by the core into visible light which emerges from the surface and provides an energy source for life without destroying it. If the stars had only a core, no life nor any complex molecule could survive near the stars, at no matter what distance! In terms of energy, the mean energy of the photons at the center of the Sun is around 1 000 eV, while the mean energy of the photons leaving the Sun is 1/2 eV. The stellar envelopes convert high energy photons, very dangerous to life, into low energy photons, which life and vegetation can actually use.

4.6 Two Logical Steps Leading to the Role of Radiation Pressure

As soon as Eddington attempted to calculate the structure of the star, he realized that he did not know how matter absorbs X rays! The discovery of X rays was already 20 years old, and yet many of their properties were not yet known, although physicists had been working extensively on the problem. It is one thing to measure how X rays are absorbed in the laboratory, and a completely different thing to discover how matter behaves at a temperature of several million degrees. So Eddington reversed the argument. He assumed his theory to be correct and tried to calculate what the absorption coefficient of the matter should be at stellar temperatures. The result was that the absorption coefficient should be 6.2×10^6 cm^{-1}, which is about a million times too large on the basis of initial experiments carried out on the Earth. Eddington considered the result to be *quite absurd*, because it indicated that *the entire radiation would be absorbed in* 10^{-6} *cm*. Eddington then tried to substitute a reasonable guess for the absorption coefficient, and found that the temperature at the center of the Sun should be as low as 130 000 K, which he also considered to be unacceptable.

Eddington then used the formula $E = mc^2$ to find the equivalent weight of the radiation inside the star, and found that 1/40 of the total mass of the star should be in the form of radiation. Since this is a relatively small number, he found no difficulty with it. So what was wrong, and why did the theory not agree with observation? Eddington tried to drop different assumptions, like the assumption that the absorption does not depend on the temperature (because no one knew to what extent this was the case, and if the answer were affirmative, how the absorption would depend on the temperature), and found that the discrepancy just became worse. The most annoying result was the energy balance. The *total imprisoned radiant energy*, or in our words, the total energy generated during the giant phase, and now stored in the star to be spent by cooling, was 5.85×10^{52} ergs in this model, while the total energy

---

in all directions and not only in the forward (outward) direction, the photon moves in a random way so that the length of the way out is equivalent to $R_\odot^2$ steps, where $R_\odot$ in expressed in cm, and that makes 4.84×10^{21} steps. If we assume that the absorption and emission take no time, a photon born at the center of the Sun would need more than 5 000 years until its great, great, ... grand photon will appear on the surface of the Sun. The exact time is 3×10^7 years, identical to the Kelvin–Helmholtz–Ritter time.

generated by contraction was 1.18×10^{48} ergs, off by a factor of about 50 000. So *one naturally asks where all the radiant energy has come from.*

Of all the suggestions put forward to explain the discrepancy, only the idea that radiation exerted pressure like a normal gas pressure gave sensible results! The general impression in those days was that gases are not subject to radiation pressure because radiation can flow through a transparent gas.[18] This cannot be the case in hot gaseous stars, because if the radiation were not absorbed and re-emitted, and degraded on the way out from the core, we would have been flooded with X rays by now! It was clear that the radiation emerging from the surface of the Sun is what emerges from the core after being duly absorbed and reprocessed by the solar envelope.

So Eddington returned to the fundamental Lane–Ritter theory and substituted in the idea that gravity is balanced by the *gas pressure and the radiation pressure.* The results, when applied to the giant stars, were in fantastic agreement with observations. The coefficient of absorption reduced to 29.5, which was reasonable. In particular, the total radiant energy came out to be only 0.238 of the contraction energy, and not several thousand times bigger. The prediction of the mean density as a function of the surface temperature also turned out to be in agreement with observation. Eddington satisfied himself by showing that *the total luminosity of a gaseous star is independent of its stage of evolution and depends only on its mass,* in agreement with Russell's conclusion for the gaseous star branch. This was very close to the idea of a connection between stellar mass and stellar luminosity! Furthermore, the range of stellar masses was found to be $1/3$–$3M_\odot$, also in agreement with the masses of binary stars as known at that time.

4.7 Is the Contraction Energy Sufficient?

But what about the energy of the star? On the one hand, the radiant energy is small because the contraction energy is small. But on the other hand, it is insufficient for the energy subsequently radiated during the cooling along the dwarf series. Eddington was forced to hypothesize that the rest of the energy was probably provided by *radioactive elements formed during the gravitational contraction.* This was a physical error. The synthesis of the radioactive elements needs energy, and since the full contraction energy is not sufficient for the later cooling phase, the use of some contraction energy to form radioactive elements which will later emit energy during the dwarf phase cannot be a solution. Could the radioactive elements be synthesized before they were put inside the stars? Maybe, but this is not a solution to the energy problem of the stars, because it only shifts the problem elsewhere.

---

[18] This was an error. The gases in the air are transparent to visible radiation, but not to X rays or infrared radiation.

4.8 Some Surprising Expressions

The year is 1916, eleven years after Einstein finally did away with the ether theory by means of his special theory of relativity. Yet Eddington, the champion of relativity, wrote that:[19]

> *It is interesting to note that the greater part of the radiant energy resides in the eather [...] it consists of eather waves traveling in all directions but unable to escape except very slowly through the meshes of matter which imprison them.*

Amazing!

4.9 Can We Model the Sun?

Finally, Eddington applied his theory to the Sun, which is a dwarf star and thus had to be a liquid. Indeed, he found a discrepancy, because the calculated surface temperature came out to be 19 300 K, which was more than a factor of three bigger than the observed temperature of 5 800 K. Eddington explained the discrepancy by the fact that the Sun is in a liquid rather than gaseous state, and hence the theory was not expected to apply. Here, Eddington erred on several counts. He did not realize how inaccurate his theory was, and that such a discrepancy could easily be the consequence of the simplifications he had to implement in order to solve the problem. The most critical error Eddington made was to assume that the mean molecular weight of the Sun is 54. If, as he assumed, the Sun derives its energy from radioactive elements, then these elements have a very high molecular weight, and it was only logical to assume a molecular weight of 54. Next, if the Sun formed from meteors as Helmholtz hypothesized, then it clearly contained mostly heavy nuclei like iron and nickel, and practically no hydrogen or helium. Eddington checked what would have happened if he were to assume that the molecular weight was 18, and found no dramatic change. (Today we know that the molecular weight is about 2, but this will come later.) As for the X-ray absorption coefficient, the number derived from observations by means of his theory compared well with measurements by Bragg (1862–1942m).[20]

4.10 The Achilles Heel

As Eddington did not know what the energy source of the stars was, or whether it was at the center of the star or spread all over, he had to make some assumption, e.g.,

---

[19] There were various spellings of the word 'ether'.

[20] Bragg, W.H., *X Rays and Crystal Structure*, p. 177. William Lawrence Bragg was awarded the Nobel Prize in 1915 at the age of 25, and up to now he is the youngest ever laureate. This is also the only case of a father (Henry) and a son (Lawrence) getting the prize jointly.

that the source was distributed uniformly throughout the star or restricted to a small core. Actually, as he assumed the energy source to be radioactive, although his main results are independent of this assumption, this meant that the energy source was spread uniformly throughout the star. He checked and found that the results were not sensitive to the assumption about the unknown energy source.[21] But the doubt remained. It is this particular assumption that allowed him to solve the equations, and was for a long time the target of criticism by many, notably Jeans and Milne. There is no physical reason why the energy production would vary like the absorption coefficient, and Eddington could not provide one. The only excuse was that this assumption allowed a simple solution of the equations. Jeans and Milne argued that this was not a physically acceptable justification.

4.11 The Second Paper

Shortly after the first paper was published, Eddington had conversations with several scientists, in particular Newall, Jeans, and Lindemann, who convinced him that the assumptions about the state of the matter in stars were not appropriate. These led Eddington to publish a second paper where major new discoveries were made.

4.12 The Mean Molecular Weight

The term 'mean molecular weight' was introduced by Eddington in his second paper, but it is rather misleading because there are no molecules involved! Eddington meant 'molecule' in the sense of *the ultimate particle*, and not in the chemical sense. The correct term should be the 'mean atomic weight of the particles'. The 'mean molecular weight' is therefore the mean atomic weight of the particles expressed relative to the weight of the hydrogen atom.[22]

Consider, for instance, the atom of iron. The atomic weight of an atom of iron is 55, meaning that it is 55 times heavier than the hydrogen atom. If the iron atom has all its electrons (26 in total), then we have a single particle composed of a nucleus plus 26 electrons. The mean atomic weight is thus 55/1=55, and it is equal to the atomic weight of an iron nucleus because the electrons are extremely light compared with the nucleus. But when the atom of iron is stripped of all its electrons, the number of particles is 26 electrons plus a single nucleus, namely, 27 particles in total. The total mass is then the mass of the nucleus, 55 in this case, exactly as before, because the mass of the electron is negligible. But the mean atomic weight is

[21] Since the way the source of energy is assumed to be spread through the star does not sensitively affect the results, he assumed it to be proportional to the radiation absorption coefficient.

[22] One frequently finds that the unit is 1/16 the mass of the oxygen atom and not the mass of the hydrogen atom. For simplicity in the explanation, we use the hydrogen unit.

now $(55+1)/27 = 2.04$. If the hydrogen atom loses its electron, the mean molecular weight is $1/2 = 0.5$.

We thus find that the question regarding the molecular weight of the matter in a star is really two questions: first, what is the composition, and then, what is the state of the electrons, i.e., are they bound to the atom as on the Earth, or are they detached from the atom?

4.13 Why Is the Molecular Weight Important for a Star?

Consider an ideal gas of particles. The particles have a distribution of energies, that is, not all particles have the same energy. However, the mean energy is always $(3/2)k_BT$, where k_B is the Boltzmann constant. This law applies even if the particles have different masses. For example, suppose we have a mixture of hydrogen and helium. The mass of the helium atom is 4 times greater than the mass of the hydrogen atom, but in a mixture of hydrogen and helium the mean energy of the hydrogen atoms and the mean energy of the helium atoms is the same, just $(3/2)k_BT$. This is called the law of equipartition, discovered by Boltzmann (1844–1906m).[23] It states that all particles in a mixture, irrespective of their masses, have the same mean energy, even if they do not have the same mass. Actually, the mean energy depends only on the temperature and not at all on the mass.

Next, the pressure of a gas is given by the total energy of the particles of the gas divided by the volume they occupy. Since the mean energy of the particles in the above mixture of hydrogen and helium is the same, the pressure they exert is the same. For simplicity, suppose that the number of hydrogen and helium particles is the same. The mass is of course different. Hence, if you consider one gram of this matter, 4/5 is helium and only 1/5 is hydrogen, yet each species contributes 1/2 of the total pressure. Thus, if you take four hydrogen nuclei and combine them into one helium particle, the mass remains the same, but the pressure, which goes as the number of particles, decreases. If you have just one gram of matter and you want to get the maximum pressure, put it in the particles with the smallest mass. Consequently, the exact composition of the matter, or the molecular weight, is very important when considering the pressure exerted by the gas.

[23] Beginning in 1870, Boltzmann published a series of extremely important papers in which he laid the foundations of statistical mechanics, including the theorem on the equipartition of energy. His work came under significant attack, and was for a long time misinterpreted. He had to wait for the discoveries of atomic structure of matter and Einstein's theory of Brownian motion in 1905 to confirm his basic assumptions. His letters to Mach and Brentano disclose a perpetual struggle with the new physics on the one hand, and the skeptical scientific community on the other, a personal conflict which ended by Boltzmann taking his own life. Einstein's theory of Brownian motion, the random motion of minute particles immersed in liquid solutions, provided a confirmation of the atomistic theory which was at the base of Boltzmann's theory and statistical mechanics.

4.14 A New Phase of Matter

On Earth, we are used to rigid atoms. An atom may lose or accept one or two electrons to or from another atom during the formation of a molecule. But as a rule, the atoms hold firmly onto their electrons. It was during 1917 that Eddington started to realize that, under the high temperatures and pressures in stars, the atoms are smashed by the temperature and pressure and cannot keep their electrons, the latter being stripped from the atoms to leave the nuclei bare, without any electrons. In short, there are no more atoms! Such matter is composed of a sea of electrons, with the nuclei of the atoms carried along inside it. In physical jargon, the atoms are completely ionized.

If this were so, argued Eddington, the molecular weight should be 2. The basic idea that the high temperature would cause all atoms to disintegrate and lose their electrons was originally proposed to Eddington by Newall, Jeans, and Lindemann. Eddington told his readers that: *Jeans has convinced me that a rather extreme state of disintegration is possible and indeed seems more plausible.* Physical knowledge at the time was not sufficient to calculate the ionization from first principles.[24] As Eddington described:[25]

> *It will be remembered that the temperatures within the stars are chiefly from a million to 10 million degrees. The radiation at these temperatures consists mainly of waves a little longer than X rays, having strong ionizing power. But, since we do not know how fast recombination of the ions with the electrons takes place, it is difficult to predict what proportion of the atoms are ionized at any moment.*

Eddington claimed correctly that, when the electron is inside the atom, the entire atom acts as a single particle. But when the atom loses say N electrons, there are now $N+1$ particles (the stripped nucleus and the N electrons) which contribute equally to the pressure of the gas, so that the pressure of the gas increases by a factor of $N+1$. If the atom has weight A, the new mean weight will be $A/(N+1)$, and for the most abundant elements, this is about 2.[26] Eddington discovered that the state of the matter in stars is what has been called since the 1920s a plasma,[27] in which, under the conditions prevailing in stars, all atoms are stripped of their

---

[24] The correct physical argument of Jeans was that, at a temperature of a few million degrees, the kinetic energy of the atoms is as high as the energy which binds the electrons to the different atoms. As a consequence, in collisions between the atoms, the latter lose their electrons. In addition, the mean energy of the photons is the same as the mean energy of the atoms, and hence, in a photon–atom collision, the atoms lose their electrons.

[25] The photon absorbed by an atom releases an electron from it. We have then a free electron and an ionized (charged by one charge) atom. If the free electron collides with the ionized atom, there is a fair chance that the atom will recapture the electron. So the question was: how long does the electron stay free before recapture? There should be a dynamic balance between free electrons and ionized atoms. For those captured electrons, there are others which are set free by the radiation. However, nobody knew how to write down the proper balance equation.

[26] Take oxygen, for example. The atomic weight is 16, and it has 8 electrons. Hence, if it loses all its electrons, we find that the mean molecular weight is $16/(8+1) = 1.78$, which is close to 2.

[27] The physical state called a plasma is not related to blood plasma.

electrons. Sometimes the plasma is called the fourth state of matter (solid, liquid, gas, and plasma). Moreover, since most of the visible matter in the universe is in stars, one can state with confidence that the plasma state is the most abundant state in the cosmos, though here on Earth it exists only in electric discharge tubes. There is one reservation, however. About 90% of the matter in the universe is 'dark matter', which is detected only through its gravitational force, and the composition of this matter is still a mystery.

The new phase consists of charged positive and negative particles attracting and repelling each other. So how could Eddington assume at the same time that the plasma behaves like an ideal gas, where the particles do not exert any force on one another? If one takes a sufficiently large volume with a large number of particles, then on the average the numbers of positive and negative charges are equal and hence neutralize each other. It is only on a smaller scale that the particles exert electric forces on each other. Thus, the plasma behaves like an ideal gas in spite of the Coulomb force acting between the particles.[28]

The year is 1916 and amazingly Eddington examined the possibility that the nuclei might also be smashed by the pressure and radiation, but rejected it because it would require significantly more energy (higher temperatures) than there is in the radiation in stars. (Actually, years later it was found that, in late phases of the evolution, the radiation field becomes so intense that it breaks the nuclei. In Eddington's day, people never imagined that such extreme conditions could exist in stars. For Eddington the particles in the core of the star were at an energy of 1 keV, while the disintegration of nuclei requires energies of 1 MeV. Such high energies are found only in very late stages of stellar evolution, stages that were not known or even conceived of in Eddington's time.) The theory was not supposed to be applicable to dwarf stars, but as an academic exercise, Eddington attempted to use it and see what results he would obtain. And lo and behold, a comparison between the luminosities of the giants and the dwarfs of class M agreed with the observations of Russell and Adams! The agreement between the theory of gaseous stars applied to the dwarf stars and observations did not (yet) alert Eddington to the fact that the dwarfs might not be liquid. So compelling was Kelvin's influence! Enlightenment came about 8 years later.

[28] More accurately, if the total energy of the Coulomb interactions is small relative to the kinetic energy, then the plasma behaves like an ideal gas. In the case of the Sun, the Coulomb energy is less than 5% of the kinetic energy, and hence still negligible.

4.15 How Much Can a Star Radiate?

Eddington's theory was summarized in two mathematical expressions:

- What is known as the quartic equation, which provides the amount of radiation pressure for a given stellar mass.[29]
- The luminosity of a star with a given mass. Eddington actually discovered that stars act like Wien's black body cavities. There is plenty of radiation inside and only a tiny part of it leaks out. The expected connection between mass and luminosity, amazing as it may seem, is like a connection between the mass of the cavity and the energy leakage.

In the same paper, Eddington already discovered the mass–luminosity law, although he did not yet realise it! The law does not depend on the (as yet unknown) energy source, although in the paper he did make certain assumptions about it. The fact that one can derive such an expression without knowledge of the energy source puzzled scientists, who consequently doubted Eddington's theory. But Eddington was right. The resulting formula, which is astoundingly simple, states how much luminosity a star of a given mass can produce. The formula, which appeared for the first time in 1918, is[30]

$$L = \frac{4\pi c G}{k} M(1-\beta) , \tag{4.1}$$

where L is the luminosity of a star of mass M. Here c is the speed of light, G is the universal constant of gravity, and k is the coefficient of absorption of radiation in the star, i.e., how much radiation the stellar matter absorbs, which is a property of the matter. β is the ratio of the pressure of the gas to the pressure of the radiation.

Note that the result was found several years before atomic physics was developed, before it was known how matter absorbs radiation, and before it became known that the matter in stars is in a plasma state, in short, before almost everything needed for stellar calculations was known. It is surprising that the luminosity of the star depends only on the total mass and does not depend on the surface temperature or the total surface area of the star, or directly on the composition of the energy source! The proof in the 1918 paper is clean and makes no mention of the energy source or any assumption about what it is made of. However, since Eddington did not know the absorption coefficients, he could not turn the equation round and use it to predict the maximum luminosity a star of mass M could have.

Eddington's result has many far-reaching consequences. The total energy of the star is Mc^2. For simplicity, let us assume maximum efficiency in conversion of mass into energy. Then if the lifetime of the star is t and the luminosity is L, we have $Lt = Mc^2$. Using Eddington's formula, we find that the maximum lifetime of the

---

[29] The formula is $\beta^4/(1-\beta) = C/m^4M^2$, where m is the mass of the electron, M is the mass of the star, and β is the ratio of the gas pressure to the radiation pressure. C is a numerical constant that depends on the gravitational constant, the gas constant, and the radiation constant.

[30] Eddington, A.S., Ap. J. **48**, 205 (1918).

star is given approximately by

$$t = \frac{c}{G}\frac{k}{4\pi(1-\beta)} = 1.08 \times 10^9 \times \frac{k}{1-\beta}\ \text{years}. \tag{4.2}$$

Different stars will have different lifetimes depending on how the absorption coefficient changes and on how much the radiation contributes to the total pressure. Stars without radiation, that is, where β tends to unity, will live forever. But, stars in which there is radiation and it leaks out have a finite lifetime, and the higher the radiation pressure, the shorter the lifetime of the star.

Usually, k and β vary throughout the star. For simplicity, and because he did not have computers to help him, Eddington assumed that β and k were constant throughout the star. It then follows that the luminosity is proportional to the mass of the star. A star with a given mass cannot radiate more than (4.1) allows it to radiate, irrespective of what the energy source is! If β becomes very small, that is, if the pressure of the gas is very small relative to the radiation pressure, one obtains the maximum luminosity that a star with given mass M can have. This luminosity is known today as the Eddington limit. In this limit, the star is fully supported by the radiation. This is like lifting someone by means of a strong beam of light.[31]

Eddington himself did not use his result to obtain the maximum possible luminosity. Since the absorption coefficients and the state of the matter inside stars were very poorly known, he used the measured mass and luminosity to estimate the actual radiation absorption coefficient. By some algebraic manipulations, he derived a formula which connects β with the mass of the star. The formula is known as Eddington's quartic formula (see footnote 29), because β appears to the fourth power. What this formula implies is that, as the mass of the star increases, β decreases, meaning that the radiation pressure becomes increasingly high. However, claimed Eddington, when the mass of the star is about 20 times the mass of the Sun, the radiation pressure is so dominant that:

> *We should expect that masses in which radiation pressure counterbalances the greater part of gravitation would very readily divide under the influence of comparatively small disturbance [...] and continue to do so until the radiation pressure no longer dominates.*

Eddington predicted that there is a limit to the masses of stars, but he did not calculate its value.

At this time Eddington still believed that Lockyer's hypothesis was correct, but his result was that stars with different masses should have different luminosities, a contradiction which apparently escaped Eddington's critical eye. As for the energy source of the star, Eddington brought mass annihilation in as a possibility, but again, the energy source was not really needed for the theory which related the luminosity of the star to its mass.

---

[31] Observations have shown that during the nova eruptions, a star manages to exceed the Eddington luminosity by a factor of 10 to 100 without disintegrating. Nir Shaviv [Shaviv, N.J., Ap. J. **494**, 193 (1998); Ap. J. **549**, 1093 (1998)] explains this phenomenon by a transition to a new inhomogeneous state, not perceived by Eddington. During this phase, a strong wind blows from the star and the star loses mass at a prohibitive rate.

4.16 Variable Stars as Theory Testers

Variable stars offer a unique tool for investigating the structure of stellar envelopes. One of the methods most frequently used by physicists to investigate the structure of a system is to disturb the system and watch how it behaves and how it returns to the original unperturbed state (if indeed it does return). Variable stars provide an opportunity to apply this idea to stars. While we cannot disturb the equilibrium the stars are in, there are various natural perturbations which can do the job for us. As a consequence of the perturbation, the star either oscillates around the equilibrium or moves to another equilibrium, or becomes unstable and collapses. The frequency of the oscillations or the shift to the new equilibrium teaches us a lot about the structure of the star.

Variable stars are stars in which, for some reason, the luminosity is not constant, but changes periodically in time. Something happens in the star which causes its luminosity to oscillate. Consider a binary system composed of two stars which differ in luminosity. Furthermore, let the system be so far away from us and the components so close to each other that no telescope can see them as two stars. The technical term is to 'separate' or 'resolve'. Finally, suppose that they revolve around the center of gravity in a plane that coincides with our point of view. So when the fainter star comes between us and the brighter star, the total luminosity of the binary system decreases. When the bright star hides the faint star from us, there is another minimum in light, but a small minimum. At other times, the luminosity is constant and equal to the sum of the two luminosities. In this case the stars do not change their luminosity, but the luminosity we get from the system does change. The variations in luminosity are not intrinsic to the stars. The first confirmation of this explanation came in 1889 when Pickering[32] noticed spectral shifts in Mizar (of the Mizar–Alcor system) which could be explained by its being a binary system. A few months later, Vogel[33] noticed analogous shifts in Algol, although this time the companion was too faint to record a spectrum. This was the first observational proof that certain variable stars are binary systems.

Of all types of variable stars, the Cepheids are the most important and famous because of the special role they played in understanding the structure of the stars and in helping astronomers like Hubble and Shapley determine the distances to other galaxies and the expansion of the Universe. However, Cepheids are also very important to the theory of stellar structure, because in this case the star itself oscillates in luminosity, that is, the change is intrinsic and hence provides information about the structure of the star. As already mentioned, Goodricke hypothesized that the variability of δ Cepheid is not caused by eclipses by a companion. In an eclipsing binary system, the light variations are very symmetric, unlike the light variations

---

[32] Pickering, E.C., Philadelphia Meeting Nat. Acad. Sci. 13 November 1889, Amer. Jour. Science **39**, 46 (January 1890).

[33] Vogel, H.C., AN No. 2947, **123**, 289 (1890), December 1889.

of δ Cephei.[34] But for lack of a better idea, astronomers interpreted the δ Cepheid stars as a binary system, until Shapley and Eddington came along in 1914–6 and shattered the idea.

4.17 Leavitt's Cepheid Variable Observations

Henrietta Swan Leavitt (1868–1921m) was one of the brighter women in Pickering's group. It was Leavitt who revealed the first indication of a period–luminosity relation in Cepheids. Following the examination of hundreds of photographic plates obtained between 1893 and 1906 at the Harvard College Observatory in Peru, she produced a catalogue of 1777 variable stars in the Magellanic Clouds.[35] Among these stars, 16 appeared in a sufficient number of plates for their periods to be determined. When tabulated in order of increasing luminosity, a pattern emerged. Leavitt observed succinctly that:[36] *It is worthy of notice that in Table VI the brighter variables have the longer periods.*

In 1912, Leavitt[37] produced more data on the period–luminosity relation. She had now managed to obtain the luminosities and periods for 25 variables in the Small Magellanic Cloud. And so she wrote: *A remarkable relation between the brightness of these variables and the length of their periods will be noticed.* As a matter of fact, Leavitt had already noticed in 1908[38] that the brighter variables have the longer periods, but at that time she felt that the number of stars examined was too small to warrant the drawing of general conclusions. However, the periods of the 8 additional variables that had been determined since that time obeyed the same law.

Leavitt's original graphs of the period–luminosity relation are reproduced in Fig. 4.1.[39] In Leavitt's words (Leavitt 1912):

---

[34] Some personal data: the distance to δ Cephei is 1 340 light years and its period is 5 days 8 hours 37.5 minutes. The intrinsic luminosity is about 10 000 times the solar luminosity. The masses of the Cepheids are a few solar masses.

[35] The Small Magellanic Cloud, known as SMC, and the Large Magellanic Cloud (LMC) are two small galaxies close to the Milky Way, first observed by Ferdinand Magellan in 1519. Today we know that the SMC and LMC are two small galaxies relatively close to our Milky Way, at a distance of about 170 000 lyrs (which is less than twice the diameter of the Milky way), but at the time Miss Leavitt carried out her observations the SMC was not known to be that far away (no distance was known), and nor was it known to be an independent galaxy. Yet Miss Leavitt had the scientific instinct to assume that the stars inside the SMC could be considered as being the same distance from the Earth, whence their intrinsic luminosities could be compared. Since the distance to the SMC was not known at the time, she could discover the relative law, but she could not determine the absolute value of the luminosity.

[36] Leavitt, H.S., Ann. Harvard College Obs. **60**, 109 (1908).

[37] Leavitt, H.S., Harvard College Observatory, Circular **173**, 1 (1912). The circular is signed by Pickering with the introductory statement: *The following statement regarding the period of 25 variable stars in the Small Magellanic Cloud has been prepared by Miss Leavitt.*

[38] Leavitt, 1908, ibid.

[39] The present day form of the law is $L = aP^{1.124}$, where a is a calibration constant.

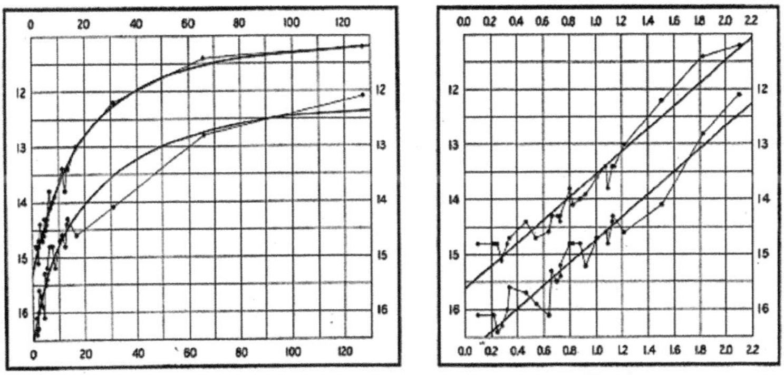

FIG. 1. FIG. 2.

Fig. 4.1 The original discovery of the period luminosity relation by Leavitt in 1912. The *abscissa* is the period in days and the *ordinate* is the astronomical magnitude (which is equivalent to the logarithm of the luminosity). Two graphs are shown in Fig. 1. The first corresponds to the minimum and the second to the maximum of the same star, whence each star provides two points. In Fig. 2, the *abscissa* is the logarithm of the period while the *ordinate* remains the logarithm of the luminosity. A straight line is obtained. This means that the luminosity is proportional to the period raised to some power. The exponent determines the slope

> *A straight line can readily be drawn among each of the two series of points corresponding to maxima and minima, thus showing that there is a simple relation between the brightness of the variables and their periods. The logarithm of the period increases by about 0.48 for each increase of one magnitude in brightness.*

Whilst Fig. 2 on the right of Fig. 4.1 does not represent a calibration of the period–luminosity relation in the modern sense (there is no zero point, since the distance to the SMC for which the relation was found was not known), these data were nevertheless used extensively by Hertzsprung and others.

Ritter had already derived a relation between the period and the mean density, and so did Eddington somewhat later. This relation is a direct consequence of the mechanical properties of the star and is derived from its mechanical balance. The period–luminosity relation is a consequence of the energy transfer in the star and should be derived from its energy balance. Hence, the two relations are independent of each other. As the absorption coefficient was still not known at that time, the complete energy equation could not have been written down explicitly, let alone solved.

4.18 Some Time Perspectives

Leavitt died in 1921 at the age of 53. However, her accumulated data were published by Luyten and Shapley in 1924 and 1930.[40] Leavitt is best known for her discovery of the period–luminosity law, which allowed Hubble and Shapley to measure distances to objects outside our own galaxy for the first time, and in this way to get modern cosmology under way. However, her work on astrographic standards, obtaining standards for the determination of the brightness of stars, is no less important, even though it may sound less glorious. Four years after her death, the Swedish mathematician Gösta Mittag-Leffler considered nominating her for the Nobel Prize for her work in formulating the relationship between the periodicity and luminosity of Cepheid variables. However, as she was already dead, she was never nominated.

Fernie[41] wrote that: *Leavitt is sometimes unjustly accused of not having appreciated the significance of her discovery.* Careful reading of Leavitt's 1912 paper reveals, to my mind, a very significant insight, as she wrote:

> *They resemble the variables found in globular clusters, diminishing slowly in brightness, remaining near minimum for the greater part of the time, and increasing very rapidly to a brief maximum.*

And she added:

> *Since the variables are probably at nearly the same distance from the Earth, their periods are apparently associated with their actual emission of light, as determined by their mass, density, and surface brightness.*

And tantalizingly, a few lines later she wrote:

> *It is to be hoped, also, that the parallaxes of some variables of this type may be measured.*

Of course, the term parallax was and remains a synonym for distance. But she did not expand further on this point. However, few discoverers of a phenomenon have ever predicted right away the full impact of their discovery. In most cases, it is only with hindsight that this impact can be appreciated in science, once all the building blocks in the construction have been laid on top of one another.

I would like to suggest after carefully reading many of her papers that, if Leavitt were guilty of any weakness, it was merely a degree of understatement and a good measure of scientific caution. Perhaps modern astrophysics has drifted away from such qualities, and from the perspective of time Leavitt's reputation has suffered as a result of the present fashion for hasty speculation.

---

[40] Leavitt, H.S., & Shapley, An. Har. **85**, 1, 157, 143 (1930). Leavitt, H.S., Luyten, W.J., Har. Cir. **261**, 1 (1924).

[41] Fernie, J.D., *The Period–Luminosity Relation: A Historical Review*, PASP **81**, 707 (1969).

4.19 A Death Blow to the Binary Hypothesis

The success of the binary hypothesis in explaining some of the famous variable stars like Algol as binary systems was so great that astronomers were inclined to suppose that all variable stars were in fact binary systems, including the Cepheids. Even the discovery of periodic oscillations in the spectral lines did not shake the belief that Cepheids were binaries. The spectral lines in the spectra of Cepheids changed their wavelength periodically, though not in a symmetric way. Wavelength changes were interpreted in terms of binary motion (Doppler shifts) and not in terms of the motion of the stellar envelope. The revolution came when Shapley[42] noticed that, if δ Cepheid had been a binary star, the radius of the orbit would have been less than the radius of the star, so that the companion would have to have been moving inside the variable star. The binary hypothesis was thus no longer tenable. Here is a typical example of how an observation can be misinterpreted by a large community, or in Shapley's words:

> *It seems a misfortune, perhaps, for the progress of research on the causes of light variation of the Cepheid type, that the oscillations of the spectral lines in nearly every case can be so readily attributed, by means of the Doppler principle, to elliptical motion in a binary system. The natural conclusion that all Cepheid variables are spectroscopic binaries has been the controlling and fundamental assumption in all the recently attempted interpretations of their light variability and the possibility of intrinsic light fluctuations of a single star has received little attention.*

Moreover, Shapley mentioned previous observations[43] which indicated that the binary hypothesis was wrong, but the required conclusion was not drawn by the community. In a footnote, Shapley described *a growing but half concealed discontent with the double star explanation of Cepheids*, and gave reference to Ludendorff[44] and Plummer.[45] Shapley ended his landmark paper with the following hypothesis: *A surface of approximately constant area progressively changes its spectral type as a result of a periodic flow and ebb of heat,* a hypothesis proposed by Schwarzschild[46] for another type of star (η Aquilae). No physical analysis of the hypothesis was given, let alone what might be the source of the oscillations, and what might determine the period. Finally, while Shapley's paper was so important in disposing forever of the binary hypothesis as a cause of variability in Cepheids, it is hard to understand why he did not make any mention at all of the period–luminosity rela-

[42] Shapley, H., Ap. J. **40**, 448 (1914).

[43] Shapley refers to the observations of ζ Geminorum by Campbell, W.W., Ap. J. **13**, 94 (1901); Russell, H.N., Ap. J. **15**, 260 (1902); and Plummer, J.I., MNRAS **73**, 661 (1913), and observations of W Sagittarii by Curtis, H.D., Lick Obs. Bull. **3**, 36 (1904), which indicated irregularities in the velocities and observation of color changes (i.e., changes of surface temperature) by Schwarzschild [Publikationen der v. Kuffnerchen Sternwarte **5**, C100, (1900)], Kohlschütter [Kohlschütter, A., A.N. **183**, 265 (1910)], Wirtz [Wirtz, C.W., A.N. **154**, 327 (1901)], and in particular Wilkens [Wilkens, A.W., A.N. **172**, 316 (1906)].

[44] Ludendorff, F.W.H., A.N. **184**, 384 (1910).

[45] Plummer, J.I., MNRAS **74**, 660 (1914).

[46] Schwarzschild, K., Publikationen der v. Kuffnerschen Sternwarte **5**, C125, (1900).

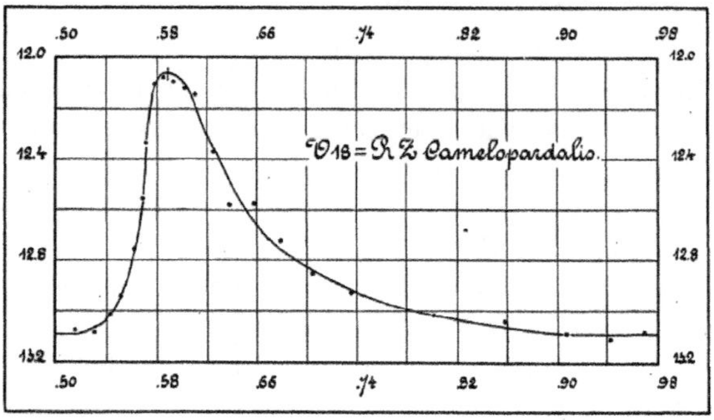

Fig. 4.2 The light curve of the Cepheid variable RZ Camelopardalis taken from Nijland, RAOU **8**, 1 (1923)

tion discovered by Leavitt in his own institution just two years earlier. Moreover, the existence of such a law could not easily be interpreted by the binary hypothesis, and hence provided support to the stellar self-oscillation theory.

But some ideas are difficult to do away with. Despite the obvious discrepancy between the predicted light curve according to the binary theory and the observed light curve, the belief in the binary theory persisted. In contrast, it must be confessed that theoretically simple pulsations, in which the heat equation is not included, yield a sinusoidal and symmetric curve that would not have agreed much better with observations. Astronomers did not want to replace a very bad theory with a poor one. Four years after Shapley's forceful paper, Perrine[47] argued that the Cepheids were binary systems, supplying observational evidence, but no real calculations. And nor did he provide any reference to Shapley's paper.

Three years later, Pannekoek[48] referred to an extremely long paper by Nijland (1868–1936m),[49] in which the asymmetric light curve (see Fig. 4.2) was explained as due to gravitational attraction by the secondary star. Pannekoek showed why this explanation did not work. As much as a decade later, Vogt[50] returned to the binary explanation and even provided a table with a list of Cepheids for which he estimated the mass of the companion.

Three decades later Hoyle and Lyttleton[51] came up with a new explanation for the $P\sqrt{\bar{\rho}} = \text{const.}$ law and the period–luminosity relation, based on the binary hypothesis. Assuming a binary system and Kepler's law, they got that the orbital period times $\sqrt{\bar{\rho}}$ is a function of the orbital separation and the radius of the star and not really a constant. Nonetheless, if one assumes that the two stars touch each other, one can get rid of these two parameters and find that the law $P\sqrt{\bar{\rho}} = const.$ only holds in touching binary stars. Even the asymmetry in the light curve has a similar explanation.

---

[47] Perrine, C.D., Ap. J. **50**, 81 (1919).

[48] Pannekoek, A., BAN **215**, 227 (1922).

[49] Nijland, A.A., RAOU **8**, 1 (1923).

[50] Vogt, H., AN **212**, 473 (1921); Ibid. **229**, 125 (1927).

[51] Hoyle, F., & Lyttleton, R.A., MNRAS **103**, 21 (1943).

4.20 The Theory of Stellar Pulsation

Eddington[52] was the first to hypothesize correctly about what goes on in Cepheids, using as a basis the idea of gaseous stars.[53] To follow his logic, we have to analyze the observations (see Fig. 4.3). The Cepheids exhibit a simultaneous change in the luminosity, surface temperature, and velocity (of the layers from which the spectral lines emerge). The unique feature of the pulsation is that, when the velocity increases from -20 km/s to $+20$ km/s, the temperature decreases by about 30%. Then as the velocity decreases quickly to -20 km/s, the temperature rises to the original value. When the velocity is positive, the star expands, and conversely, when the velocity is negative, it contracts. The star cools upon expansion and heats upon contraction. The star converts radiative energy into mechanical work by expanding. Eddington realized that the star:

> *[...] must behave as an engine in the thermodynamic sense: that is to say, it must take in heat when it is at a higher temperature than the average and give out heat at a lower temperature – just the opposite to what usually happens in natural conditions.*

Eddington found that the typical Cepheid expands by about 1 million kilometers in a day and a half, which corresponds to an acceleration of only 14 cm/s^2, while the surface gravitational acceleration of the star is about 300 cm/s^2. So the acceleration

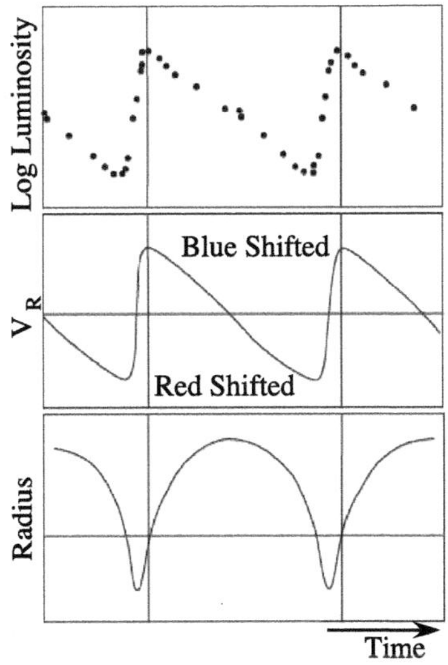

Fig. 4.3 The luminosity, velocity, and radius variations as a function of time in a typical Cepheid variable

[52] Eddington, A.S., The Obs. **40**, 290 (1917).

[53] As early as 1879, Ritter had considered the pulsation of a gaseous star. He even derived the relation between the period and the mean density of the gaseous star.

during the expansion of the envelope is in fact small compared to the gravitational force on the surface of this star. The dimensions are astronomical, but the accelerations rather negligible.

The almost complete theory of Cepheids was developed by Eddington in 1918–19.[54] with additional comments and ideas in 1926, when he published his famous book. The basic open question was: what supplies and drives the 'heat engine'? So Eddington hypothesized that the radiation absorption coefficient depends on the temperature and density, so that, when the star expands, the absorption coefficient decreases and the star becomes more transparent, releasing radiation, but when the star contracts, the absorption coefficient increases, and more radiation is absorbed. This is the famous 'kappa mechanism' (kappa from the symbol for the absorption coefficient). At that time the absorption coefficient was very poorly known, and he could not calculate explicitly the way his mechanism would work. As liquids like water cannot be compressed, it was clear that only gaseous stars could expand and contract, absorbing and releasing heat as Eddington had conceived.

Eddington's basic idea, and this is the reason why we have digressed here to discuss this theory, was that gaseous stars experience oscillation. While the cause of these oscillations was still elusive at that time, once the star oscillates it was simple to calculate the period of the oscillation. Since the star is gaseous, any perturbation must propagate in the star like acoustic waves. The speed of sound is proportional to the square root of the temperature. But the temperature in a star made of an ideal gas is related to the mass of the star via $GM/R = \alpha MT = \alpha_2 M v_s^2$, where M is the mass of the star, T the temperature, v_s the mean speed of sound, and α_1 and α_2 some numerical constants. The period of the wave or the oscillation is the time it takes the sound wave to cross the radius of the star, namely, $P = R/v_s$. If we substitute for the speed of sound and take $\rho = M/(4\pi R^3/3)$ for the mean density, then we find that

$$P\sqrt{\rho} = \text{constant}\,, \qquad (4.3)$$

which is the fundamental result for the pulsation theory of stars. We have specified in detail the way the analysis is carried out in order to stress the simple logic and physics which applies to a gaseous stellar configuration.

In the second paper, Eddington noticed that the unobserved change in the period of the Cepheids as a function of time, that is, the fact that the period was fixed, imposed a strict limit on the gravitational contraction theory as energy source. Since the period of the oscillation depends on the mean density of the star, as the star contracts, we should see a corresponding change in the period of oscillation. The period should decrease with increasing density. The null observation of a period change hammered a further nail in the coffin of gravitational contraction as an energy source.

The second part of Eddington's theory of pulsation appeared a year later.[55] In most stellar bodies, the oscillations decay due to dissipation and some viscosity inherent in most physical processes. As a result, even if the star is perturbed, the amplitude of the oscillations does not grow above a very small value which is so

[54] Eddington, A.S., MNRAS **74**, 2 (1918); MNRAS **74**, 177 (1919).

[55] Eddington, A.S., MNRAS **79**, 177 (1919).

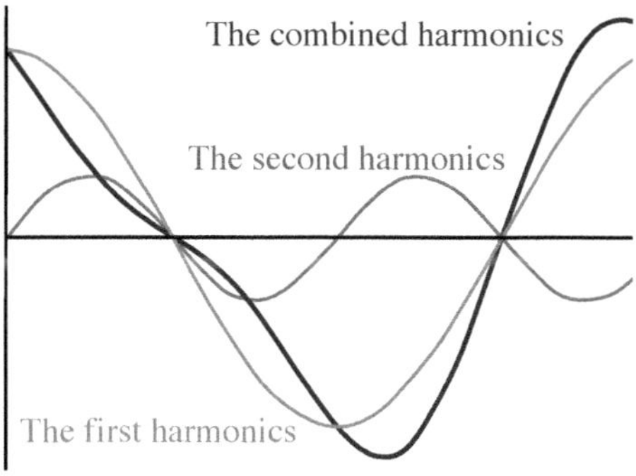

Fig. 4.4 Eddington's explanation for the non-symmetric light curve of the Cepheids. The sum of two harmonics leads to a non-symmetric light curve

small that it is unnoticeable. The Cepheids appeared to be different. A mechanism was required to feed the oscillation with energy. Two alternative reasons for the oscillations were suggested:

- During a certain stage, conditions are such that an oscillation having the appropriate period would tend to increase, so that the pulsation would start automatically.
- At a certain stage there is a sudden change in the state of the stable equilibrium and the collapse to the new state throws the star into a pulsation, which could last for a period of the order of 1 000 years. This figure is an estimate based on the fact that no change was noticeable in the oscillation during the time the Cepheids were observed.

Thus, every star with sufficient mass becomes a Cepheid variable for a brief part of its life. For the first time, self-oscillation provided an explanation for the non-symmetric velocity and light curves. When we consider, for example, the vibrations of a violin string, it can vibrate in the fundamental and higher harmonics (see Fig. 4.4). We know that the 'color' of the tone depends on how much energy there is in the fundamental harmonics relative to the higher harmonics, and it is up to the violinist to control the mix of these fundamental and higher harmonics. Similarly, when a star oscillates, it has several harmonics, and the right mix of harmonics can give rise to non-symmetric light curves.

As soon as more accurate observations had been made, it became clear that this explanation by Eddington for the shape of the light curve was not correct. The Cepheids oscillate in the fundamental mode and the shape is determined by the properties of the 'heat engine', the absorption coefficient. The variation of the absorption coefficient with temperature and density depends on the composition and the atomic energy levels. Once these had been properly taken into account (after extensive calculations), the predicted and the observed light curves agreed very well. But it took several decades to reach this agreement.[56]

The conclusions drawn were that (a) the binary hypothesis had to be ruled out, and (b) the assumption that the light variations were due to expansion and contraction of a single gaseous star led to agreement between theory and observation. The period could be determined quite accurately (to within a factor of 2) and the non-symmetric light curve could be easily reproduced. Because of the high absorption of radiation, the prediction was that the time during which the star pulsates is about 1 500 years. Finally, recall that the theory was relevant only to gaseous stars, and vindicated Eddington's assumption about gaseous stars. In the case of stars with periods longer than 3 days, Eddington found that the central density was about 1/10 the density of water, so that one could safely assume that the star was gaseous. In short, the Cepheids proved Eddington's gaseous star hypothesis.

4.21 Hertzsprung Again. A Small Digression

As soon as the period–luminosity law was discovered, Hertzsprung[57] quickly realized that the relation could be calibrated, whereupon the absolute value of the luminosity could be found, and then the intrinsic luminosity of δ Cephei stars might be determined directly from their periods. It should then be a straightforward matter to obtain their distances by comparing the measured intrinsic luminosity (derived from the observed period) and apparent luminosity. With this idea Hertzsprung and Leavitt laid the foundations for one of the most important ways to determine the distances to remote galaxies.

Actually, Hertzsprung found that the apparent luminosity of δ Cephei stars in our own galaxy, having the same period as the Cepheids observed in the Small Magellanic Cloud (see Fig. 4.5), is higher by a factor of 120. In this way he was able to determine for the first time the distance to the Small Magellanic Cloud, namely, about 30 000 light-years, a result which is off by a factor of 6. But most importantly, he devised a new method for determining distances, a method whose importance it is hard to overestimate. A few years later, in 1929, Hubble used the Cepheids to discover the expansion of the Universe.[58]

---

[56] Christy, R.F., Ap. J. **136**, 887 (1962).

[57] Hertzsprung, E., A.N. 4692 (1914).

[58] The classical method for measuring the distance to a star directly is to observe the star at intervals of 6 months. During 6 months, the Earth moves from one side of the Sun to the other, and hence the two observations are twice the Earth–Sun distance apart. Since the viewing position is different, the star is observed at different angles, and it is a matter of simple trigonometry to find the angle at which the Earth's orbit around the Sun is seen from the star – the parallax – and of course, find the distance of the star from the Sun. This method works up to a distance of about 100 light years. Stars at greater distances are so far that they are seen at the same angle all year round. Thus, the traditional annual parallax techniques are incapable of determining distances to even the closest Cepheids. For this reason, Hertzsprung had to resort to statistical (and mean secular) parallax methods. The basic idea of statistical parallax is to apply the fact that the Sun is not fixed

Fig. 4.5 The Small Magellanic Cloud. Credit NASA

4.22 A Confusing Issue. The Bizarre White Dwarfs

In science, one frequently encounters a situation where there is too much information and it is difficult to figure out what is not important and should be ignored (at least at the beginning). The sequence of discoveries we are about to report brought new information about stellar structure, but was misunderstood and caused confusion for quite some time.

Friedrich Wilhelm Bessel (1784–1846m) left the gymnasium at the age of 14 because he was fed up with Latin (not so strange really). He worked at first as an

---

in space but moves inside the galaxy. The speed of the Sun is about 13 km/s (in the direction of the Hercules constellation), so that the Sun (and the entire Solar System) moves about 2.8 times the Earth–Sun distance every year. Moreover, the measurement can extend over a few years, and in this way increase still further the distance to which the method can be applied. In this respect, we are fortunate that the Sun's peculiar velocity is rather high. The problem is that the stars move as well. However, the stars move in all directions, so that the average velocity of all stars relative to a point at rest vanishes. The reference is the average velocity. By observing the apparent motion of a single star one can estimate its motion and distance. Hertzsprung's result was hampered by the inaccuracies of this technique, particularly given his small sample of only 13 Cepheids and imprecise knowledge of the motion of the Sun relative to the rest of the Galaxy. Yet, he was able to obtain the first distance estimate to the SMC, even if there was an error of about a factor of 6.

apprentice in an import–export business. However, he was attracted to navigation, and through it to astronomy and mathematics. Though self-taught, he very quickly excelled to the point that, at the age of 26 (in 1809), he was appointed director of Frederick William III of Prussia's new Königsberg Observatory and a professor of astronomy. But there was a problem. It was unheard-of for a German university to offer a professorship to a scientist who did not have a doctorate, or who had not graduated from a gymnasium. So Bessel approached the then high priest of mathematics, Gauss (1777-1855m), for help. Gauss met Bessel in Bremen in 1807 and in five hours of conversation discovered Bessel's qualities and recommended him to the authorities of the university. So solely upon Gauss' recommendation, a doctorate was conferred by the University of Göttingen, based on his achievements. This is just to let the reader know that there are exceptions (provided you can get a recommendation from Gauss).

Bessel had numerous scientific achievements to his credit, but for our story here the relevant discovery was his prediction in 1841 that Sirius, the brightest star in the sky, and Procyon, each have an unseen companion. This was an era in which celestial mechanics based on Newton's laws was accomplishing great feats. Bessel measured the positions of many stars very accurately, including Sirius and Procyon.[59] He observed that the path of these stars in the sky is not straight, but makes very small but noticeable oscillations.[60] Bessel guessed that Sirius and Procyon have invisible binary companions, and that each of the two stars revolves around the mutual center of gravity. The unavoidable conclusion from Newtonian mechanics was that, if a star performs a 'dance' of this kind in the sky, one can infer that it is due to an as yet undiscovered partner. In 1845 Le Verrier (1811–1877m) predicted in a similar way that there is another planet disturbing the orbit of Uranus, and this was of course discovered at the predicted place by Johann Gottfried Galle (1812–1910m) in 1846, constituting yet another great victory for Newtonian mechanics. It was the second planet to be discovered in modern times and the first to be discovered following a prediction.

In 1851, Peters[61] showed that the variable motion can be explained by assuming that Sirius moves in an elliptical orbit around some center of gravity with a period of 50.093 years. In 1862, the American telescope maker Alvan Clark (1804–1887m) discovered the companion star while testing a new 18 1/2 inch refractor for the Dearborn Observatory.[62] The new star was designated Sirius B (see Fig. 4.6). Clark was fortunate to observe the system when its components were at their largest separa-

[59] Bessel developed unique methods for measuring the position of stars very accurately. He made the first precise measurements of refraction of light in the Earth atmosphere, and in 1811 prepared tables of refraction which allowed astronomers to measure stellar positions with greater accuracy than ever.

[60] As a matter of fact, Bessel did not observe even a single full oscillation because the period of the system is 50.1 years, and he did not observe the stars for such a long time before he reached his conclusion.

[61] Peters, C.H.F., A.N., No. 748 (1851).

[62] The announcement of the discovery was made by G. Bond, director of the Harvard Observatory, in MNRAS **22**, 170 (1862).

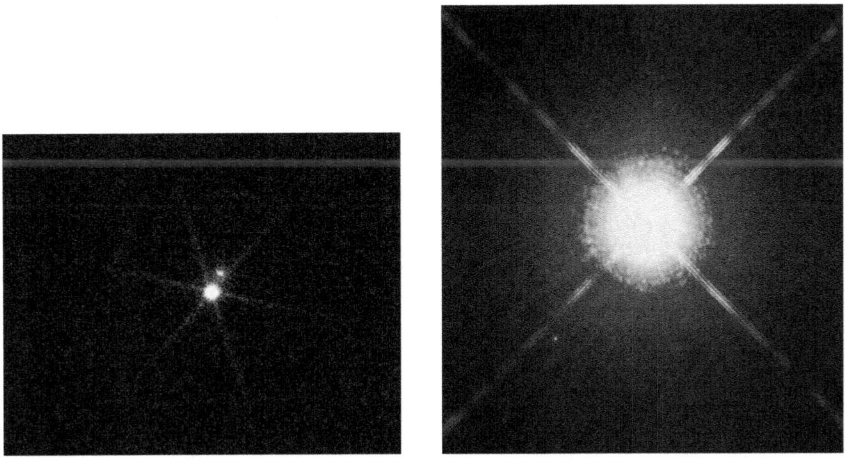

Fig. 4.6 *Left*: A unique image of Sirius-B taken by the Chandra X-ray satellite in October 1999, during the testing period. The bright source is Sirius B, which produces very low energy X rays. The dim source at the position of Sirius A may be due to ultraviolet radiation from Sirius A leaking through the filter on the detector. The exposure time was 14 hours. The Sirius system is 8.6 light-years away. Credit NASA/SAO/CXC. *Right*: This Hubble Space Telescope image shows the Sirius system. Sirius A is the bright star and Sirius B the dim one. The image of Sirius A was overexposed so that the dim Sirius B could be detected. The scale of the two images is very different. Credit: European Space Agency and NASA

tion. The maximum distance between Sirius A and B is about the distance from the Sun to Uranus, which is roughly 19.8 AU.[63]

For more than 50 years, Sirius B posed no problem to astrophysics. In 1915, the separation between Sirius A and Sirius B was again maximal, and Adams (1876–1956m)[64] reported how, after two years of failed attempts, he secured a spectrum of the companion of Sirius. The difficulty was that Sirius B is a faint star. Its luminosity is about one hundredth the luminosity of the Sun, and when placed near a very bright star like Sirius, it was impossible to see or take a picture of it. Adams used a special technique and the Mount Wilson 60-inch reflector (with a very long focus). His dramatic finding was that *the line spectrum of the companion is identical with that of Sirius in all respects so far as can be judged from a close comparison of the spectra*. Now imagine, the two stars Sirius A and B have the same spectrum and hence the same surface temperature. But Sirius A is about 10 000 more luminous! The only way this could happen would be if the surface of Sirius B were 10 000 times smaller than the surface of Sirius A, which would imply that the radius is 100 times smaller than the radius of Sirius A. That puts the radius of Sirius B at about 0.92 times the radius of the Earth. At the same time, analysis of the motion of Sirius

---

[63] AU stands for astronomical unit, which is half the size of the longer axis of the Earth's orbit around the Sun, whence 1 AU = 1.495978×10^{13} cm.

[64] Adams, W.S., PASP **17**, 239 (1915).

revealed that the mass of Sirius B should be only slightly less than one solar mass (0.98 to be precise). These two numbers provide the possibility of calculating the mean density of Sirius B, and the value obtained is 1.7 ton/cm^3! Or as Eddington put it,[65] *a density which is absurd*. To Eddington, Sirius B was:

> *[...] a strange object which persists in showing a type of spectrum entirely out of keeping with its luminosity, and hence may teach us more than a host which radiates according to the rule.*

At that time several similar stars were already known. The irony was that, in spite of Eddington's unsurpassed understanding of stars and their making, he did not read the message Sirius B was sending, and continued to assume that the stars along the cooling series are liquid, and that his recent mass–luminosity law was only valid on the ascending path, where the stars are gaseous.

These strange stars got the name white dwarf because their color was white, and they were very small, in fact, about the size of the Earth. (This time 'dwarf' was indeed a reference to size.) The important point for our story is that it became clear that not all stars fall into the two categories of Lockyer's scheme. Some of them are extremely strange. As such, the white dwarfs indicated a flaw in Lockyer's scheme. The white dwarfs troubled theoreticians, because all attempts to find the energy source for stars were applied also to white dwarfs, but failed, leading to various rather unconvincing conclusions.

4.23 Can We Already Guess What the Energy Source Might Be?

It was June 1919 when Russell was invited to the meeting of the Astronomical Society of the Pacific to review the (unknown) sources of stellar energy.[66] First, by comparing the radiation the Earth gets from the Sun, calculating the total energy output from the latter, and assuming also that it is at least the age of the Earth (a few billion years), Russell found that the gravitational supply of energy was short by a factor of a 1 000. He concluded, therefore, that the stars must have an enormous supply of energy which they manage to tap and radiate out into space.

It is interesting to see how physicists can draw conclusions about an unknown phenomenon, and it is instructive to examine later how nature agrees or disagrees with the predictions and demands of physicists – or how Nature can be kind or cruel to physicists' predictions. In this particular case, Russell claimed that the unknown energy source should respect the following conditions:

- It should generate a large amount of heat per gram of matter under the conditions prevailing in stars, and none under laboratory conditions or in the interior of the Earth, because none is observed in the latter two cases.
- It must be a stable process, so as to remain steady for many years.

---

[65] Eddington, A.S., *The Internal Constitution of the Stars*, Dover, NY (1959) p. 171.

[66] Russell, H.N., PASP **31**, 205 (1919).

- It must in some way be regulated so as to supply heat to each star at almost exactly the rate at which the star radiates heat into space. If the total energy produced is higher than the energy radiated away, the star will heat up, rather than cool gradually as implied by the HR diagram.
- The source is limited and eventually the star runs out of energy and dies.
- The supply is available even at late stages, because we see many cool stars which radiate energy into space.

All of the above demands were found to be correct when the solution was eventually found 20 years later, save the issue of stability.

Russell surmised that the *unknown process*, in his words, should reach equilibrium with the emitted radiation when the core of the star reached an *equilibrium temperature*. According to Russell, the unknown energy release takes place during the giant phase, when the star is gaseous. During this phase, most of the energy generated is stored in the star. In the dwarf stages, when the star goes down through the dwarf phase it *will still occupy more time than in the giant stages, since the rate of transformation of energy is so much smaller*. Russell claimed that his hypothesis explained the scarcity of red giants. The reason given was that the temperature of the giants is below the critical value for energy generation, whence they must contract very fast. But this is somewhat confusing when non-gravitational energy is released. Russell ended his address without giving any guess as to what the source of the energy could be. Even radioactivity was not mentioned.

4.24 Eddington's Response to Russell

A few months later, the October issue of The Observatory contained Eddington's response to Russell's address.[67] He reviewed the evidence against gravitational contraction and treated this hypothesis with the same disregard as he treated Ussher's hypothesis. Eddington mentioned that at one time radioactivity had been considered as a possibility, but that by the time of writing it had already been rejected. Still, he found it necessary to add evidence against the short time allowed by gravitational contraction. To this end he mentioned the recent geological dating by Lord Rayleigh and Shapley's[68] claim that:[69]

> *The similarity in the types and distribution of giant stars in globular clusters is at variance with the rapid evolution of this stage, since the light time from the most distant clusters is as much as 200 000 years.*

At this point Eddington reviewed the talk given by Russell, going over the five conditions and explaining them. The first line of reasoning provided by Eddington

[67] Eddington, A.S., Obs. **544**, 371 (1919).

[68] Shapley, H., PASP **30**, 283 (1918).

[69] Globular clusters are collections of stars which contain a million or more stars, all moving under the gravitational force of the cluster as a whole, and at the same time moving around the galaxy. The spherical shape of these clusters explains why they are described as globular.

led him to reject the idea that the stars can store the energy during the giant phase and release it during the dwarf phase. The reason he gave was that, if energy is pumped into the star, it expands and cools, not the opposite. Two new pieces of evidence were added by Eddington: the evolution prior to becoming a spectral class M dwarf and the stability (meaning constancy in time) of the pulsation period of the Cepheid variables. Later Eddington remarked that *Russell refrains from speculation as to the nature of this new source of energy*, and added that *his paper must inevitably set wild ideas traveling through our heads, as we try to contemplate the various possibilities*. He then tried to examine one of these speculations.

Eddington had the gut feeling that the answer should be in the $E = mc^2$, but he still could not find it. Take the entire Sun, for example. According to this relation, the total energy is 1.75×10^{54} erg. However, extracting all this energy *will not leave behind a dark star, but absolutely nothing*. This energy should suffice for the Sun for 15 billion years if it continues to radiate at the same rate:

> *But in any case, there is room here for a concealed source of energy which would serve for as long a period as anyone has asked for.*

Annihilation of electrons with the positive charges was a possibility, but Eddington pointed out that it would be sufficient *if only one thousandth of this energy is released*, implying that the demands from such a process were not difficult to contemplate. This was essentially Jeans' mass annihilation idea.

But why would such a process take place in stars? Eddington explained that, at high temperatures, the atoms are fully ionized and *leave the positive nucleus a free target*. For this reason, we do not see the process occurring on Earth. If a hydrogen atom is annihilated, Eddington calculated that the resulting frequency of the radiation liberated would be very high indeed, about $2.3 \times 10^{23}\ \mathrm{s}^{-1}$. At this incredibly high frequency *we know nothing that can absorb* the resulting photon, and it should escape from the star without interacting with the stellar material. But probably, by some unknown process, it would *suffer the usual scattering* and thus gradually be *brought to more tractable form*. In summary, at this time, Eddington believed in restricted annihilation, although he raised some basic problems which he could not solve.

4.25 The Components of the Nucleus

So far, the proton and the electron were known to be particles that compose the atom and the nucleus. However, it was not clear how the heavier nuclei were made up. The atomic weights of most nuclei and chemical elements are not products of one single number and hence do not appear to be made of the same fundamental unit. Every chemical element appeared to be different.

A major breakthrough came in 1919 when Aston[70] managed to significantly improve his mass spectrometer and published extensive research results on the masses of the chemical elements. Aston examined a series of chemical elements[71] and found that all these elements are composed of a mix of isotopes, all of which have integer atomic weights. On the other hand, Aston found hydrogen to have an atomic weight of 1.008 (relative to oxygen 16), in agreement with the findings by chemical methods. For He, Aston found an atomic weight of 4.007. Aston reached the conclusion that, except for the two light elements hydrogen and helium, all elements had, within the accuracy of his measurements, an integer atomic weight, a result he called the whole-number rule. The non-integer atomic weights found by the chemists were fortuitous and arose from a mixture of whole numbers, namely, if you mix two or more isotopes and then measure the atomic weight of the mixture, just as found in nature, you find a non-integer atomic weight. The implication for atomic structure was enormous. The insurmountable difficulty of explaining non-integer weights disappeared, because it became clear that the atom could be made from a single building block. In Aston's own words:

> *An elementary atom with mass m may be changed to one with mass $m+1$ by adding a positive particle and an electron. If both enter the nucleus, an isotope results.*

Aston's major discovery was therefore that all atoms were built out of the same standard building block.

But Aston faced a problem. He knew about $E = mc^2$ (although he wrote that this formula arose from the electromagnetic theory, and not relativity), which meant that the weights could not be additive. When two atoms are fused together, the mass of the new nucleus should be less than the sum of the masses of the individual atoms, *[...] but it only becomes so when the charges are relatively distant from each other.* This, according to Aston, appeared to be the case in light nuclei. In the case of the heavier atoms, Aston argued that the charges are very close, and *the law $E = mc^2$ is not effective*. But if the weights are an exact product of the same building unit, there is no room for the formula $E = mc^2$! Aston, though an experimentalist, did not give the error of his measurements. Fate would have it that the first generation of Aston's instruments could not see the effect of $E = mc^2$ on the nuclear weight, and hence he claimed incorrectly that it does not apply. Aston could not imagine that the nuclear binding energy affects the mass to a degree below the accuracy of his measurements! Only 8 years later, and not before the accuracy was improved, the effects of the binding energy were discovered with the mass spectrometer.

Aston's results won him the Nobel Prize just three years later. In giving the committee's arguments for awarding Aston the 1922 Nobel Prize for Chemistry, professor Söderbaum stated that:

> *All the masses so far measured [...] can be expressed by means of whole numbers in relation to oxygen 16 [...] and must be regarded as the expression of natural law [...] and has been named the 'whole-number rule'.*

---

[70] Aston, F.W., Nature **104**, 393 (1919); Phi. Mag. Ser. 6, **39**, 611 (1920); a precursor paper appeared in Nature **104**, 393 (1919).

[71] Oxygen, carbon, neon, chlorine, nitrogen, argon, krypton, xenon, and mercury.

In other words, Aston got the Nobel Prize for showing the unity of matter, by demonstrating that all elements were composed of the same units. In a sense, he confirmed Prout's hypothesis that the atoms of the elements are all made up of aggregations of larger or smaller numbers of atoms of the lightest known element, hydrogen. If Prout had been right, so claimed Söderbaum, then all elements should be exact multiples of hydrogen, and they are not. So Prout's hypothesis was confirmed with a twist: the fundamental unit à la Aston is the sum of a negative and a positive charge. There is no mention of $E = mc^2$ in Söderbaum's address! Careful reading shows that the argument given by Aston as to why all elements save hydrogen and helium have whole-number atomic weights was actually incorrect. If Aston's argument had been correct, then the deviation from the 'whole-number rule' should increase with the atomic number, and this was not observed by him, nor found in Nature. The impression one gets from reading Söderbaum's address is that he and the committee did not fully comprehend Aston's result – the results left no room for binding energy. A few year's later, Aston improved the mass spectrometer and discovered the behavior of the binding energy with mass, whereupon he definitely deserved the prize for this discovery, but doubtfully for the earlier discoveries, which on the face of it contradicted relativity.

4.26 Early Hints

Jean Baptiste Perrin (1870–1942, Nobel Prize for Physics in 1926) was interested in reactions between atoms and light, and published long papers on the subject[72] in 1919 and 1920. Perrin is mostly known for this research. However, in his attempts to demonstrate the wide applicability of his theory, he also discussed the energetics of radioactivity and solar energy, and drew Eddington's attention to it.

Perrin explained that all atoms are made of hydrogen and its electron. According to Perrin, who was apparently unaware of Aston's discoveries, this was the fundamental building block Prout[73] had in mind. If so, all atoms should have atomic weights which are integer products of the weight of hydrogen, in contradiction to

[72] Perrin, J., Ann. de Physique, II, 1919, p. 5, March–April issue, Revue de Mois **21**, 113 (February 1920).

[73] Prout, W., Annal. of Phil. **6**, 321 (1815). More accurately, Prout's conclusion from the data he presented was: *That all elementary numbers, hydrogen being considered as 1, are divisible by 4, except carbon, azote (the French name for nitrogen) and barytium (the old name for barium), and these are divisible by 2, appearing therefore, to indicate that they are modified by a higher number than unity or that of hydrogen. Is the number 16, or oxygen? And are all substances compounded of these two elements?* asked Prout in his paper. The statement that the atomic weight of carbon (12) is not divisible by 4 appears strange. Prout wrote: *Carbon. I assume the weight of an atom of carbon at 7.5. Hence the sp. gr. of a volume of it in a gaseous state will be found by calculation to be .4166, or exactly 12 times that of hydrogen.* Because he compared the specific densities of the elements in the gaseous form, he fell into the trap that gaseous carbon does not form a double molecule like hydrogen or oxygen (H_2 or O_2), whence the atomic weight thereby derived was off by a factor of 2.

the findings of the chemists. The explanation given by Perrin was that the elements found in nature are mixtures of isotopes of the same element, and the isotopes satisfy Prout's hypothesis.[74] However, there are some notable deviations from this rule, and Perrin brought, as an example, the difference between four hydrogen atoms and the helium atom. To explain this difference, he cited Langevin (1872–1946)[75] as the one who (correctly) interpreted Einstein's theory just after its publication. According to this theory and the interpretation of Langevin and Perrin, the meaning of $E = mc^2$ is that the energy which goes to combine the 4 hydrogens to helium is converted into mass, and for this reason the mass of the helium atom is smaller than the mass of 4 hydrogen atoms. This is the correct interpretation of the fact that the masses of heavy elements are not an integer product of a fundamental building block. However, Perrin did not have much experimental data, and for this reason erred significantly.

Perrin assumed that the particular behavior of hydrogen and helium continues all the way to the radioactive elements. So far so good, but what happens when you reach the radioactive elements? Perrin claimed that, contrary to common wisdom, the sign of the energy was wrong. If the radioactive transformation had been explosive, as people thought(!), it would have been accompanied with mass loss. But for the reasons he gave, the disintegration is endothermic, which means that it is accompanied by mass gain.

Perrin treated the radioactivity transformation as a chemical reaction in equilibrium, though he did not check to what extent this assumption was justified on Earth or in stars (it is not justified at all). As an example, Perrin discussed the formation of ozone from oxygen, writing $3O^2 \rightleftharpoons 2O^3$. He did not add the radiation to the reaction equation. Today this reaction is written as $3O^2 + \gamma \rightleftharpoons 2O^3 - Q$, where γ denotes the photon and $Q = W' - W$ the energy absorbed. W' and W are the binding energies of oxygen and ozone. In Perrin's words, W and W' are, respectively, the *quantities of light* ν and ν' which transform oxygen into ozone and ozone into oxygen. In other words, the mass, radiation and energy are all combined into a single equation.

If all reactions are in equilibrium, then the material and molecules we see on the Earth are the consequence of the conditions of radiation on the Earth. Under different conditions, other types of matter may exist or be more stable. Deep in stars, the conditions lead to other forms of matter. Matter transforms from one stable form to another depending on the conditions. Perrin now turned to discuss the history of the Earth and why Kelvin's theory was not correct. The explanation went as follows. Assume that, when the pressure is very small, we have a nebula composed of (observed) hydrogen, nebulium, and helium. At the temperatures of the nebula (about 12 000–15 000 K according to Fabry and Buisson), these are the stable elements and these temperatures are well below those at the center of stars. Hence one can reason, à la Perrin, that in the interior of stars the ultra energetic X-ray radiation gives rise to the formation of heavier and heavier atoms. This ultra energetic X-ray radiation does not escape from the star due to absorption by the outer layers.

---

74 Since Perrin did not cite Aston, it is not clear how he knew about isotopes.

75 Langevin, P., Journal de Phys. Theo. et App. **4**, 165 (1905).

Perrin, like the chemist Arrhenius before him, required gravitation to form the heavy elements by compression. In the next phase, the heavy elements disintegrate in stars and give back the gravitational energy they stored. The formation of heavy elements during the contraction converts electric energy into heavier atoms like radium. Hydrogen and helium convert into radium and similar atoms.

Among many other topics, Perrin, who attempted to explain almost everything on the basis of reactions in equilibrium, discussed the problem of radioactivity and solar energy. As for radioactivity, he suggested that *light provokes the radioactivity*. If so, the cosmic conditions can affect the duration of the radioactive element, in particular, in the interior of the Sun. Had the conditions on Earth been similar, we would have observed similar changes. Contrary to common opinion, Perrin claimed that radioactive transformations are very endothermic, which is to say that they absorb energy, and that the transformation takes place by extremely powerful rays which escape our perceptions:[76]

> *The atom of radium is not a packet of dynamite which can be exploded by means of a small excitation, but an extremely stable combination which can only decompose with the investment of an enormous energy.*

A simple observation concerning the behavior of the light elements is extrapolated to the heavy ones.

Perrin completely ignored the by now extensive work on stellar models by Eddington, Jeans, and others. He wrote several times in the article that it was easily explainable, but did not carry out any calculations. On the other hand, Eddington cited Perrin in his book, published some 6 years later, when he came to the idea of energy from the transmutation of hydrogen into helium. Eddington was quite generous in giving Perrin so much credit for the idea of stellar energy.

4.27 Eddington's Presidential Address

In August 1920, the British Association met at Cardiff and Eddington gave the presidential address.[77] The address can be considered as a milestone in the search for the energy source in stars, because here Eddington saw the light, as it were, after reading Aston's paper and reinterpreting Aston correctly. The address is also a good summary of the state of knowledge of stellar structure as Eddington saw it, exposing the stumbling blocks at that time in the move towards a complete theory of gaseous stars. The address was very long and extended over 18 pages of The Observatory.

In typical style, Eddington began by noting that, since astronomy had made such great progress in recent years, the most secret place in Nature was now *10 miles below our feet*, commenting on the fact that we know so little about the interior of

---

[76] My translation.

[77] Presidential address to Section A of the British association at Cardiff, 24 August 1920, Obs. **63**, 341 (1920).

the Earth. Yet he drew our attention to the stars, which are *cut off by more substantial barriers*. Even so, claimed Eddington:

> *Science has material and non-material applications to bore into the interior, and I have chosen to devote this address to what may be described as analytical boring devices – absit omen!*

This is the essence of theoretical astrophysics. (As a matter of fact, even today we know more about the interior of the Sun than we know about the interior of the Earth.)

Eddington recapitulated that *heat has two forms – the energy of motion of material atoms and the energy of ether waves*. The first is energy stored in the random motion of the atoms.[78] As for the energy stored in radiation, in an analogous way, Eddington assumed that the energy of radiation was stored in ether waves, and he considered his own discovery of the role of radiation in stars: *In the giant stars the two forms [of energy] are present in more or less equal proportions.*

In the giant phase, the star contracts and the generated energy is stored mostly in the form of *imprisoned radiant energy – ether waves traveling in all directions*. What we observe is just the radiant energy which leaks out. Somehow stars with predominantly *etheral energy* are not found, and Eddington guessed that these effects limit the masses of stars. Eddington pointed to the fact that Lane, Ritter, and Emden were concerned with how the energy is brought to the surface, while the true problem was how the star stores the energy generated by the contraction.

The new discovery by Eddington that the star is loaded with etheral energy implied that the energy flows by radiation and not by mass currents, as was supposed by Lane, Ritter, and Emden. For the historical record, something similar had been suggested by Sampson[79] years before Eddington, but it had not attracted any attention because of an error in the physics. The flow of radiant energy gave rise to radiation pressure which added to the gas pressure in supporting the star against gravitation. In this case, Eddington was left with two unknowns, namely the absorption properties of the matter and the molecular weight.

4.27.1 Predicting Saturation in the Absorption

Eddington compared his theoretical results with observations. In 1918, Eddington wrote down equation (4.1) for the first time, and it became his hallmark formula for stellar structure. All the ingredients were at his disposal from previous papers. So he

---

[78] Consider a gas in a box. The atoms inside the box move in all directions in a chaotic motion. The long time average velocity of this chaotic motion vanishes (average over the velocity of a single atom or average over the velocities of many atoms). However, the average of the velocity squared does not vanish and is a measure of the heat stored in the gas.

[79] Sampson, R.A., MNRAS **55**, 280 (1894). Sampson made several errors in his treatment. For example, he assumed that the emissivity of a mass element is proportional to the temperature and not to the fourth power of the temperature, as had been discovered by Stefan over a decade earlier. The reference is to an abstract of a longer paper.

applied the formula to derive the absorption coefficient from the observed mass and luminosity. He made a point of the fact that, at the predicted temperatures of a few million degrees, one expects the radiation to be in the form of *soft X rays*.[80] Soft X rays had been investigated by physicists in the laboratory, and hence the absorption properties of the matter were known. The catch was, as Eddington pointed out, that the absorption properties depend on the unknown composition and molecular weight:

> *In the extreme case, probably not reached in a star, when the whole of the electrons are detached from the atoms, the average weight comes down to about 2, whatever the material.*

Eddington was unable to calculate the detachment of the electrons from the atoms, and could not believe that in a star all the electrons might be detached.[81] So he found that, for a molecular weight of 2, the absorption coefficient is 10 g/cm^3, and for an infinite molecular weight, the absorption coefficient is 130 g/cm^3. Eddington claimed that these values agreed well with laboratory measurements of the absorption coefficient. So what was the snag? The comparison with the giants implied that, despite the broad range of temperatures and densities, the absorption coefficient remained practically the same, that is, the absorption coefficient did not change with temperature or density, a result he could not accept because it contradicted his physical intuition. Eddington thus hypothesized that, at high temperature, the absorption coefficient somehow approaches a limiting value, so that it remains practically constant over a wide range. He himself was surprised with this 'unavoidable' conclusion. How could there be such a limit? But he was right. The absorption coefficient does tend to a constant value.

The radiation inside the star is very intense, much more intense than radiation in the laboratory, where absorption properties are measured. So the atoms reach saturation and simply cannot absorb so much radiation.[82] At this point, Eddington mentioned another possibility suggested to him by Barkla,[83] namely that the detached electrons simply scatter the radiation, and it is this scattering which provides the limit Eddington was looking for. However, if Barkla was right, the limit should be 0.2 g/cm^3, and not somewhere between 10 and 130 g/cm^3. Even taking into account all possible factors, Eddington found that Barkla's suggestion was insufficient.[84]

---

[80] Soft X rays have wavelengths of 3–30 angstrom. Hard X rays are more powerful, and hence have shorter wavelengths.

[81] Eddington did not know the composition, and guessed that the star contains plenty of very heavy elements like uranium and iron. These elements hold the electrons so strongly that, under the most extreme conditions, they nevertheless retain a large fraction of their electrons. On the other hand, hydrogen and helium lose all their electrons at a rather moderate temperature (under 200 000 K). As Eddington did not yet know that the stars he observed were about 97–98% hydrogen and helium by mass, he considered the molecular weight of 2 as an extreme that would never be realized.

[82] We know today that this is wrong. The absorption coefficient depends on the temperature, the density, and the composition. The radiative flux must be determined together with the temperature, density, and ionization of the atoms, and not separately as imagined by Eddington.

[83] Charles Glover Barkla (1877–1944m) won the Nobel Prize in 1917 for his work on X rays emitted by various elements.

[84] Eddington was wrong. In 1902 Thomson [Phil. Mag. **4**, 253 (1902)] discovered the process known today as Thomson scattering, namely, the scattering of light by electrons, which is exactly

How do we know that the numbers were correct? Once the structure was known, the mean density could be calculated, and from it the pulsation period. So Eddington predicted for a gaseous star a period between 4 and 10 days, and was happy to note that the period of δ Cephei is 5 1/3 days. The discrepancy, argued Eddington, might be due to the fact that he had neglected the rotation. Since the stars rotate, the centrifugal force must be included in the calculations, and so far it had not been. If the truth be told, the situation was confusing, as the reader may wish to confirm.

4.27.2 The Prediction of Nuclear Energy and Transmutation of the Elements

In the last part of his address, Eddington turned to the energy source of stars:

> *What is the source of the heat which the Sun and stars are continually squandering? The answer given is almost unanimous, that it is obtained from the gravitational energy converted as the star steadily contracts. But almost as unanimously this answer is ignored in its practical consequences. Lord Kelvin showed that this hypothesis, due to Helmholtz, necessarily dates the birth of the Sun to about 20 000 000 years ago, and he made strenuous efforts to induce geologists and biologists to accommodate their demands to this time-scale.*[85] *I do not think they proved altogether tractable. But, it his among his own colleagues, physicists and astronomers, that the most outrageous violations of this limit have prevailed.*

And Eddington gave several examples of longer measured ages. Moreover, Eddington claimed that:

> *No one seems to have any hesitations, if it suits him, in carrying back the history of the Earth long before the supposed date of formation of the Solar System [...] Lord Kelvin's dates [...] are treated with no more respect than Archbishop Ussher's.*

Eddington used the constancy of Cepheid periods[86] to show that *the stellar universe proceeds at a slow majestic pace*, and of course stressed that *there is no room for a companion star*. As Eddington said:

> *Only the inertia of tradition keeps the contraction hypothesis alive – or rather, not alive, but an unburied corpse. A star is drawing on some vast reservoir of energy by means unknown to us. This reservoir can scarcely be other than the subatomic energy which, it is known, exists abundantly in all matter; we sometimes dream that man will one day learn how to release it and use it for his service.*

---

the process Barkla suggested to exist. In the meantime, the absorption coefficient due to radiation scattering by electrons had been calculated correctly, and Barkla was indeed right. The scattering process is called Thomson scattering and not Barkla scattering. A few years later, Stewart proposed the same idea again, and once again the idea was not ascribed to the proposer (see later).

[85] As we saw, Kelvin got an even shorter time scale, but Eddington cited Ritter's result.

[86] Chandler [AJ **24**, 65 (1904)] found a decrease in the period of 1/20 seconds per year, while Hertzsprung [AN **210**, 17 (1920)] found a decrease of 1/10 seconds per year. Thus, the observational evidence was that the evolution proceeds at 1/400 of the rate of what gravitational contraction required.

Unbelievable, but see later.

Finally, Eddington turned to Aston's recent discoveries, to which he devoted four pages. In contrast to Aston's explanation for the mass of helium, Eddington's interpretation was the correct one, as he identified the fact that the mass of a helium atom is less than the mass of four hydrogen atoms, the difference being just 1/120 of the mass, or 0.7%. Now, in contrast to what he had assumed before, Eddington contended that *mass cannot be annihilated and the deficit can only represent the mass of the electrical energy set free in the transmutation* of hydrogen into helium:[87]

> *If only 5% of the mass of the star consists initially of hydrogen, the total heat liberated will more than suffice for our demands. Is this possible?* pondered Eddington and argued: *If Rutherford could break down the atoms of oxygen in his lab, driving out an isotope of helium, then what is possible in the Cavendish laboratory may not be too difficult in the Sun.*

Eddington reached the landmark conclusion that stars generate energy by fusion of hydrogen. Energy generation and element fusion are the same process:

> *In the stars matter has its preliminary brewing to prepare the greater variety of elements which are needed for a world of life.*

Note the terminology 'needed for life'.

Eddington noticed that since radioactive elements release energy when they disintegrate, energy had to be invested in forming them. So the fusion of hydrogen into helium releases energy and the disintegration[88] of the heavy elements releases energy. Hence, there must be a nucleus which occupies the border between the two types of nuclei. In other words, Eddington predicted the shape of the mass formula or the binding energy of nuclei[89] which Aston found about a decade later, and for which von Weizsäcker and Bethe discovered a phenomenological expression more than 10 years later! This was a conclusion Perrin missed, and he consequently confused nuclear synthesis. Stars gain energy through fusion and invest energy in generating the heavy nuclei. Eddington explained that Aston's results were not sufficiently accurate to confirm his predictions, but that when the accuracy of the instruments had been improved, this would be the expected result. It took seven years for Aston to prove that Eddington was right:

> *If indeed the subatomic energy is set free in stars [...] it seems to bring a little nearer to fulfillment our dream of controlling this latent power for the well-being of the human race – or for its suicide.*

---

[87] The idea of the existence of nuclear forces was not yet born, and the Rutherford experiment showed that the nucleus is small and acts on the α particles via the Coulomb force. The energy of the α particles in the Rutherford experiment was too low to probe the nuclear force. For this reason, I suppose, Eddington used the term electrical energy.

[88] Eddington uses the wording that energy must be invested in generating the radioactive nuclei. We use the word 'disintegrate', because energy is released with the emission of a radioactive nucleus. The same is true when heavy nuclei undergo fission into two large nuclei.

[89] The mass formula is a phenomenological expression for the binding energy of nuclei as a function of the numbers of protons and neutrons in the nucleus.

How predictive, true, and profound Eddington was!

The last three pages of the address were devoted to the role of hypothesis and speculation as a driving force for scientific research. Here, we recall Eddington's grand tour of the theories of stellar energy. First, he assumed radioactive decay, then he adopted a variant of Jeans' mass annihilation as the source of energy, and finally, when he saw the atomic weights of hydrogen and helium, he grasped the right idea very quickly: subatomic energy. More than twenty years later, the explicit mechanism of what we call today nuclear energy was finally worked out. Eddington lived to see his hypothesis confirmed, since he passed away five years after Bethe's discovery of the CN cycle. An interesting and symbolic coincidence was that the obituary of Sir Norman Lockyer, whose theory of stellar evolution Eddington destroyed, appeared cover to cover in The Observatory[90] just after Eddington's address.

Eddington died in 1944, and Russell wrote the obituary in the Astrophysical Journal.[91] While Russell enumerated Eddington's contributions to theoretical astrophysics, he did not mention that Eddington predicted the correct source of stellar energy in his famous address in 1920, shortly after Russell's address at the Astronomical Society of the Pacific in 1919.

Success sometimes brings success. After using radiation and radiation pressure to obtain models of gaseous stars, it became rather popular to invoke radiation pressure in many phenomena, even if it was not relevant. Thus, Eddington[92] complained that:

> *Until recently the great possibilities of radiation pressure as an agent in cosmical phenomena were scarcely appreciated. Now, however, there is a tendency to go to the opposite extreme, and to invoke its aid almost too freely.*

A well known phenomenon still today. So Eddington demonstrated that radiation pressure is not important for the Sun, even on its surface.

4.28 Chemical Elements in Equilibrium

Excited by Eddington's 1920 address, Richard Tolman (1881–1948)[93] picked up on Perrin's idea of equilibrium in 1922, and hypothesized that the stars derive their energy from processes in equilibrium (in contrast to a process with a given rate and which goes only in one direction). Tolman noted that, in 1915, two chemists, Harkins and Wilson, had already had the idea of the conversion of hydrogen into helium[94] by a process in chemical equilibrium, and that this process releases energy even when in equilibrium. Following his predecessors, Tolman assumed that hydrogen converts into helium under conditions of equilibrium, writing $4H \rightleftharpoons He$. Interestingly, he remarked that he did not want to write the reaction as $4H^+ + 2e^- \rightleftharpoons He^{++}$

[90] Rolston, W.E., Obs. **43**, 358 (1920).

[91] Russell, H.N., Ap. J. **101**, 133 (1945).

[92] Eddington, A.S., MNRAS **80**, 723 (1920).

[93] Tolman, R.C., J. Am. Chem. Soc. **44**, 1902 (1922).

[94] Harkins, W.D., & Wilson, E.D., Am. J. Chem. **37**, 1367, 1383, 1396 (1915).

or as $4H^+ + 3e^- \rightleftharpoons He^+$, and so on, because *we consider spectroscopic evidence for the presence of an ionized hydrogen and unionized helium* in stars. Tolman suggested that this equilibrium reaction might *account for the magnitude of the radiation from the giant stars.* Indeed, if just the mass difference between the helium and the four hydrogen nuclei is released, then Tolman was right.

The results for the assumed temperatures that prevail in stars (10^6 K according to Tolman's estimate) were disappointing as they did not agree with the observations of hydrogen and helium on the surface of stars. Tolman found that the amount of helium relative to hydrogen was $1 : 10^{-30000}$, which is clearly nonsense. Consequently, the new problem was to explain how come hydrogen exists together with helium in stars. What actually went wrong was that the assumption of equilibrium between nuclei simply does not apply by many orders of magnitude, and if the assumption is nevertheless implemented, absurd results are found. Moreover, Tolman assumed that the stars are fully mixed, and hence that the surface composition reflects the composition of the interior. Tolman's paper, like the papers by Harkins and Wilson, was published in a journal for chemists and consequently went unnoticed by the astrophysical community, which never cited any of them.

Another series of papers which went unnoticed appeared six years later. These were by Suzuki,[95] in which he discussed the H/He equilibrium. He realized that in raising the temperature to about 10^9 K a fit with observations could be reached. However, none of the stellar models predicted such a high temperature to exist in stars. The inadequate equilibrium idea continued to float around for some time.

4.29 Eddington Discovers Kramers' Law

Bothered incessantly with the problem of the radiation absorption coefficient, Eddington decided to attack the question from first principles.[96] The calculation Eddington carried out was a mixed quantum and classical calculation. Eddington argued that the radiation energy depends on the temperature to the fourth power and the mean energy of a photon depends on the temperature to the first power. Hence, the number of photons absorbed is proportional to $kT^4/T = kT^3$, where k is the absorption coefficient.

On the other hand, the rate of absorption is proportional to the velocity, which in turn is proportional to the square root of the temperature.[97] We have now an equation for the absorption coefficient, namely, $kT^3 = aT^{1/2}$ or $k = aT^{-3.5}$. To obtain the constant a, Eddington assumed that:

> *When an electron encounters an ionized atom it will be captured if, and only if, it actually hits the nucleus of the atom.*

---

[95] Suzuki, S., Proc. Phys. Math. Soc. Japan **10**, 166 (1928); ibid. **11**, 119 (1929); **13**, 277 (1931).
[96] Eddington, A.S., MNRAS **83**, 32 (1922).
[97] Recall that the energy, which is $mv^2/2$, is equal to $3k_BT/2$.

From this simple, but incorrect hypothesis, he obtained the constant *a*. Eddington reversed the argument again and used the absorption coefficient *k* to determine the conditions of capture for an electron, and in this way to obtain a *fairly secure proof of the approximate truth of the hypothesis.*

Eddington found that the absorption coefficient is directly proportional to the density and inversely proportional to the temperature to the 7/2th power. The exact numerical coefficient in front of this formula was of secondary importance at that moment because, due to the lack of an appropriate theory, Eddington could not correctly calculate the number of free electrons or the probability for electron absorption by the atom. Nonetheless, Eddington got the correct dependence of the absorption coefficient on density and temperature. This absorption law is known in astrophysics as Kramers' law, after Kramers who discovered it a year later and gave the exact numerical value of the constant.

Eddington used the result to improve his stellar model. However, he quickly discovered that the problems did not disappear. He was unable to make accurate predictions without a better knowledge of the molecular weight of the matter, since the results were very sensitive to the exact value of the molecular weight. It was therefore necessary to improve the theory of the matter before any progress could be made. Eddington attempted[98] to do so but to no avail.

4.30 The Universe Has Uniform Composition

It was 1921 when Russell was invited to address the Pacific Division of the American Association for the Advancement of Science.[99] This time Russell discussed the discoveries made by means of the spectroscope, which allows one to unravel the compositions of the distant stars:

> *Long before the story reached the second chapter the main lesson of the stellar spectra was clear. The elements which we know on the Earth are to be found all thru the visible universe. Matter is of the same ultimate constitution everywhere.*

Elated by the results, Russell quoted a few lines from the poet Stedman:

White orbs like angel pass
Before the triple glass
That we may read the record of each flame,
Of spectral line and line
The legendary divine
Proclaiming them the same, and still the same,
The atoms that we knew before,
Of which ourselves are made: dust, and no more.

---

[98] Eddington, A.S., MNRAS **83**, 98 (1923).

[99] Russell, H.N., PASP **33**, 275 (1921).

Kirchoff, Bunsen, Huggins, and Lockyer had shown that the same elements exist on Earth and on stars. Russell went one step further, and demonstrated that the abundances on Earth and in the stars are the same.

He recapitulated the emerging atomic theory and the observed uniformity in stars. The simple stellar classification contained 99.5% of all stars, and there were just a few classes. Russell exposed the Lockyer theory and explained that:

> *Our Sun is denser than water and is evidently pretty well advanced in the process of cooling. It was probably once ten times brighter than at present and very likely more. Can we hope to understand the make-up of the hot interior of stars where the pressure is beyond imagination?*

Russell argued that we can learn from the laboratory and then use the 'analytic boring machine' of Eddington to 'see' into the interior. Next, the role of radiation pressure as discovered by Eddington was described. Russell explained the 'Eddington limiting luminosity', but without referring to it as such. He argued that the pressure of the radiation causes massive stars to break into smaller stars. For this reason, the most massive stars are less than 100 solar masses. What about the smallest mass? Here, Russell claimed that very small stellar masses, for all we know, may be abundant in space, but invisible. This was the first time that a prediction of dark matter was made, in this case stars that do not shine!

And for how long do the stars shine? According to Russell:

> *If the Sun had been fifty per cent hotter than it is now, all the oceans would have been heated to the boiling point, and all terrestrial life destroyed. If for a single century the Sun had been fifty per cent cooler, the whole surface of the Earth would have been ice-clad, and life must have perished.*

This argument was used by Russell to infer the long life of the Sun (several billion years) at the present rate of energy production. The total energy radiated during the lifetime of the Sun is colossal. So regarding the energy source:

> *The stars must posses an unimaginable source of energy which they can release slowly and so enable the Sun to keep shining for billions of years.*

At this point, Russell repeated the arguments he presented in his 1919 address. But now he knew about Aston's results. As a good experimentalist, he quoted the accuracy of Aston's results (an accuracy, as you may recall, that Aston himself did not provide) as one in a thousand. Russell explained how the fusion of hydrogen into helium could provide energy for the Sun for 100 billion years. However, the reference to Eddington was missing, along with the prediction that the synthesis of heavy elements requires energy, while the synthesis of the light elements releases energy.

Finally, a sobering point:

> *The constants which enter into Eddington's equation, from which these results have been deduced, are the most fundamental that we know; the constant of gravitation, the mass of the hydrogen atom, the velocity of light, and the Planck constant. We may therefore say that the masses of the stars, and hence their other properties that depend on the masses, are predetermined by the most general properties, not even of atoms, but of the structural units out of which atoms themselves are built.*

The reader should also refer to Russell and Webster,[100] where this idea was further expanded.

4.31 Hertzsprung Once More. The Observed Mass–Luminosity Law

In 1922, Hertzsprung made yet another seminal discovery. He considered the relation between the mass and brightness of stars found in a binary system.[101] As we know, the advantage with binary systems is the possibility of putting bounds on the masses of the components. Hertzsprung's list of stars contained 14 pairs and one case of a single star, our Sun. Hertzsprung plotted the data (see Fig. 4.7). Though small in number, the data was sufficient to discover that the luminosity of the star is a function of its mass. The units of the ordinate are again rather typically 'astro-

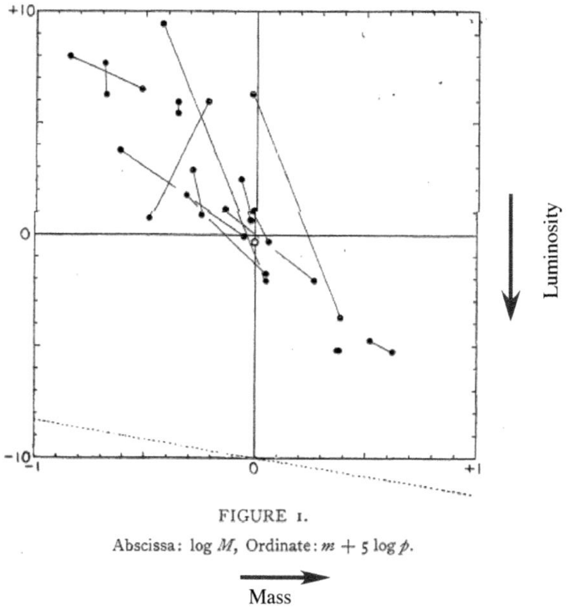

Fig. 4.7 The adapted original mass–luminosity diagram of Hertzsprung, as discovered in 1922. The *abscissa* is the logarithm of the mass in units of the mass of the Sun, and it runs *from left to right*. The *ordinate* is the absolute brightness, which is the magnitude of the star as it would be seen from the Earth if it were placed at a distance of 10 light-years. Note that the luminosity increases downward

[100] Russell, H.N., & Webster, D.L., MNRAS **82**, 181 (1922).

[101] Hertzsprung, A., BAN **43**, 15 (1923).

nomical' (the ordinate is the apparent magnitude plus 5 times the logarithm of the parallax, which is equivalent to the log of the intrinsic luminosity). The basic finding was that, the brighter the star, the greater its mass. The luminosity was almost proportional to the mass, and Hertzsprung noted that a more complicated relation described the data a bit better, but it was not clear to what extent the data really warranted such a correction. This was the first observational hint that there might be such a thing as a mass–luminosity relation, although it was not as simple a relation as Eddington had first obtained. Converted to normal units, Hertzsprung's result yields $L \propto M^{1.1}$.

A year later, Russell, Adams, and Joy[102] investigated the properties of double stars. In an effort to obtain better results, they collected data on 1 636 pairs of stars. However, much of the data was missing, and in particular many of the masses of the pairs were not known. Since not all the masses were known and since they wanted to include as much data as possible to render their result more accurate, they plotted the data as a function of the mean mass of the binary system. It was this assumption which fogged the data and hid the mass–luminosity law. So the authors reached the conclusion that:

> *It is obvious [...] that the mean mass of a binary pair is by no means a simple function of the spectral type. If the mass is plotted against type, the usual diagram shaped like a figure seven is obtained with the white dwarfs quite isolated. But if the masses are plotted against the absolute magnitude (log of the luminosity), all the stars – red giants and white dwarfs alike – fall into line, for the first time in the author's experience. It is evident that statistically considered, the mass of a binary system is a function of its absolute magnitude.*

There was no such figure[103] in the paper, and it is surprising that the white dwarfs were included in the mass–luminosity law, because they did not and were not expected to follow it. Even majestic observational discoveries could be confusing.

In the same year, Öpik (1893–1985)[104] carried out similar and more extensive research on wide binary stars. In contrast with Russell et al., Öpik concluded correctly that:

> *Components of close double stars cannot be regarded from a statistical point of view as representative of single stars; in counting them together we introduce consciously a non-homogeneity which can, e.g., considerably disfigure our conclusions on the luminosity curve of the stars.*

[102] Russell, H.N., Adams, W.S., & Joy, A.H., PASP **35**, 189 (1923).

[103] The figure in the paper is the spectroscopic parallax over the geometrical parallax as a function of the magnitude, not luminosity as a function of mass.

[104] Öpik, E.J., *On the luminosity curve of components of double stars*, Pub. de l'Observatoire Astronomique de l'Université de Tartu, No. 5, XXV (1923); Ibid. No. 6, XXV (1924). I discovered this forgotten paper in the library of the ARI, The Astronomisches Rechen-Institute, Heidelberg. The librarian kindly found the paper in the basement and to my dismay the pages were bound as printed in those days. Nobody had read the paper until my visit. The vigilant librarian looked for a special knife to cut the pages of the never read manuscript, then kindly xeroxed the paper for me, for which I would like to express my gratitude, and sent the original back to the basement.

4.32 Kramers Discovers His Law

In 1923, Kramers (1894–1952m), a Dutch physicist who collaborated with Niels Bohr for over ten years, attacked the problem of how X rays are absorbed. Kramers[105] is known for many contributions to atomic physics, but to astrophysicists he is mainly known for his seminal paper about the absorption coefficient of matter with regard to X rays, results which are crucial in the study of stellar structure.[106] While the formula $\kappa \sim \rho T^{-3.5}$ for the absorption coefficient was discovered by Eddington, it is known as Kramers' formula, after the person who put in the correct numerical coefficient. Kramers did not mention Eddington's result, probably because he did not read the astrophysical literature and hence did not know how important it was for the theory of stellar structure, and nor would he have known that Eddington discovered the formula but with the wrong constant.

Kramers calculated how matter absorbs X-ray radiation. Since Kramers was interested in down-to-Earth problems rather than cosmic ones, he assumed the material to be at room temperature, or more accurately, he did not calculate the influence of the temperature on the absorption coefficient. This was essentially done before by Eddington. The value of the absorption coefficient in the laboratory was crucial because it was the starting point for the calculations of the absorption coefficient of matter under stellar conditions.

In the same year, Stewart[107] picked up on an old proposal by J.J. Thomson[108] that electrons detached from atoms, referred to as free electrons, are the cause of the absorption coefficient in the solar atmosphere. The problem was the following: atoms absorb light only when bound electrons absorb the light and jump to a higher energy level. Accordingly, ionized gas cannot absorb radiation and should be transparent, a conclusion not borne out by observation. Stewart suggested that the free electrons scatter the light, whence an ionized gas would not be transparent.

So far it had been impossible to formulate Thomson's suggestion, but now that the Saha equation was available, Stewart decided to see the implications of Thomson's idea (see Sect. 4.35). It turned out that this effect led to the constant absorption coefficient that Eddington inferred must exist in giant stars, and that had been suggested by Barkla (who was not given the proper credit by anyone, save Eddington).

---

[105] This was during World War II, a period which A. Pais referred to as the 'dark days' in his book *The Genius of Science: A Portrait Gallery of Twentieth-Century Physicists*, Oxford University Press (2000). Pais, like the Frank family, went into hiding in a house in Amsterdam. Hendrik Anthony Kramers, who was his mentor, visited him there once a week, and during one of these visits (as recounted in the chapter on Kramers), the Gestapo raided the house and Pais hid in the attic behind a wall, with Kramers fronting. Pais was caught by the Gestapo in March 1945 and imprisoned, presumably slated for deportation and death. Kramers, known to the world for his work on electromagnetic dispersion relations, wrote to Werner Heisenberg, who was the head of the German atomic bomb program, on Pais's behalf, but Heisenberg replied that he could do nothing. However, a copy of the letter from Kramers was conveyed to a high Nazi official by Pais's friend, Tineke Buchter, and ultimately saved his life.

[106] Kramers, H.A., Phil. Mag. **46**, 836 (1923), communicated by N. Bohr.

[107] Stewart, J.Q., Nature **111**, 186 (1923).

[108] Thomson, J.J., Phil. Mag. **4**, 253 (1902).

In parallel, Lindemann[109] examined Eddington's premise that an ionized atom can capture an electron only if it hits the nucleus. Lindemann pointed out correctly that the atom lacks the electrons in the energy levels which are lower than the energy of agitation (kinetic energy in our terms) and not in the nucleus. Eddington's conclusion seemed therefore to be improbable, and his absorption coefficient wrong:

> *What he has shown,* criticized Lindemann, *is that one collision in 10 000 results in the recombination of an electron without the production of a new one by collision. Physically, it seems almost inconceivable that this should be true.*

4.33 Absorption in Strong Radiation Fields

The appreciation that the structure of stars depends critically on the complicated absorption coefficient gained ground in the early 1920s. To simplify the treatment of the absorption coefficient, Milne[110] suggested an important approximation which later became known as the two-level atom approximation. Consider an atom in the strong radiation of the star. After it has absorbed a quantum of radiation, the electron jumps to a higher level and the atom cannot absorb a new quantum of radiation unless the electron returns to the original state. Hence the absorption of radiation depends on the rate of decay of the electron or how fast the emission takes place. An immediate corollary of this correct argument concerned the question as to whether the absorption could saturate? An ionized atom cannot react to radiation unless it recaptures an electron. Thus it appeared that, in stars, absorption could reach saturation, whereupon absorption would effectively be limited in strong radiation fields.

4.34 Subatomic Energy in Difficulty

In 1924, Jeans[111] returned to the annihilation hypothesis. And so he wrote:[112]

> *Some years ago I put forward the suggestion that the energy of stellar radiation is produced by a secular decrease in the star's mass, the mechanism possibly being that positive and negative electron charges fall together and annihilate one another, their energy being transformed into radiation. The conjecture has gained enormously in probability since Eddington has shown that, as a matter both of theory and of observation, the radiation from a star is approximately a function only of the star's mass. It is now clear, both from theory and observation, that as a star's development proceeds, its mass must decrease.*

So Jeans set down the equations for stellar evolution with decreasing mass. The rate of mass loss is clearly given by L/c^2, where L is the luminosity of the star and c the speed of light. Jeans calculated that the Sun must lose 4.2 million tons per second,

---

[109] Lindemann, F.A., MNRAS **83**, 332 (1923).

[110] Milne, E.A., Hdb. d. Ap. Springer Pub., Berlin (1930).

[111] Jeans, J.H., MNRAS **85**, 2 (1924).

[112] In the original paper the idea was formulated as annihilation between a negative electron and a positive proton.

a very large number in absolute terms, but very small when compared to the mass of the Sun. The giant stars, where the energy is produced, are therefore decreasing in mass. This was not yet the idea that stars on the main sequence lose mass due to energy production by mass annihilation and evolve from the massive early type stars to the low mass late type stars. According to Jeans' views at that moment in time, the dwarf main sequence stars were cooling liquid stars and radiated the energy stored in the past.

4.35 Saha. Getting the Stellar State

The question as to what happens to atoms at high temperatures and pressures surfaced from the very first attempts to understand the stars. It was obvious that the state of the matter was crucial for the structure of stars.

Clo[113] investigated the stability of atoms against high temperatures. His experiment reached a temperature of 650 K, and he discovered very little breakdown of the atom due to the high temperature, which only aggravated the question. Could atoms preserve their stability as their kinetic energy increased? Clo's answer so far was affirmative.

It seemed implausible that atoms could survive the extreme stellar conditions as implied by Clo's experiment. It was an essential question to what extent the conditions in the stars, which cannot be mimicked here on Earth, do not lead to the destruction of atoms. Eddington showed that assuming that all elements lose their electrons, so that the star contains a soup of nuclei and electrons, yielded a better agreement between theory and observation. Hence, the problem became one of calculating the state of the atoms under conditions that could not be mimicked in the laboratory.

In 1919, Lindemann (1886–1957)[114] was engaged in a controversy with Chapman (1888–1970m) about the origin of magnetic storms. In the wake of this controversy, Lindemann derived the ionization formula for hydrogen for the first time, and discussed the possibility of the complete ionization of hydrogen in the solar chromosphere. However, he did not develop the formula any further and nor did he generalize it to other atoms.

Eggert (1891–1973)[115] was the first to assume that the ionization of atoms can be treated as a chemical reaction in equilibrium, that is, on the one hand the atom loses its electron and on the other the electron recombines with the atom, exactly as would happen in any chemical reaction $A + B \rightleftharpoons C + D$ in equilibrium. Eggert was a student of Nernst, who developed the statistical theory of chemical reactions.[116] Ionization of an atom can take place in many ways, so the calculation must include

[113] Clo, J.H., Ap. J. **33**, 115 (1911).

[114] Lindemann, F.A., Phil. Mag., December 1919, p. 540.

[115] Eggert, J., Phys. Zeit. **20**, 570 (1919).

[116] Nernst, W., *Die theoretische und experimentellen Grundlagen des neuen Wärmesatzes*, Verlag von Wilhelm Knapp, Halle (1918).

all possible processes. To circumvent such a lengthy calculation, which requires many unknown atomic data, Eggert assumed that the neutral state is in equilibrium with the ionized state, i.e., neutral $\rightleftharpoons$ ionized + e, so that he could directly apply Nernst's statistical mechanics to such a problem. The advantage of the method was that it automatically takes into account all possible processes. The Nernst theory of chemical reactions in equilibrium contained a single constant, namely, the energy involved in the decomposition. The problem with Eggert's derivation was the value of this constant, which he obtained in a rather artificial manner.

At this point the East Indian Mmegh Nad Saha (1893–1956m) entered the scene. At the time he was visiting Alfred Fowler at Imperial College in London, and later went on to Nernst's laboratory in Berlin. He suggested[117] that the uncertain constant in Eggert's formula could be eliminated by introducing what he called the ionization potential of the element. This is the minimum energy needed to remove an electron from the atom. By showing how to calculate this constant, Saha produced a formula that could be implemented confidently in situations where the experimental data were not available, or under extreme conditions. Saha quickly realized the huge effect the revised formula could have on the interpretation of stellar spectra, and stellar structure and composition, and quickly began to implement it.

For a while people called the formula the Lindemann–Saha equation[118] (forgetting about Eggert). Today Lindemann's name is omitted, and Eggert's name is not added either. As with the absorption coefficient, the one who fixed the last constant got the honor of having the formula named after him.

As soon as Saha published his paper, Milne[119] and Ralph Fowler[120] (1889–1944m)[121] realized its importance, and applied it extensively. Within a year, Saha[122] extended the application and managed to explain how the ionization changes along the spectral types of the main sequence, giving rise to the observed changes in temperature and composition. Furthermore, the new Saha equation was used by Cecilia Payne[123] (1900–1979) to derive for the first time the abundances of elements observed on the surface of stars. The possibility of using the observed spectral lines and the theory to derive actual abundances of elements on the surface of stars was finally open.

Shortly afterwards, in 1923, Noyes and Wilson[124] confirmed the Saha equation by comparing the theory with observations of very hot flames. The theoretical proof of Eddington's assumption that the matter in stars is in the form of a plasma was provided by the Saha equation.

---

[117] Saha, M.N., Phil. Mag. **40**, 472, 809 (1920).

[118] See for example, Von Engel, A., *Ionized Gases*, American Vacuum Society Classics, 1994, Springer Pub. p. 295.

[119] Milne, E.A., Obs. **44**, 261 (1921).

[120] Ralph Fowler and Alfred Fowler share a mountain on the Moon.

[121] Fowler, R.H., & Milne, E.A., MNRAS **83**, 403 (1923).

[122] Saha, M., Proc. Roy. Soc. London **99** A, 135 (1921).

[123] Payne, C.H., PNAS **11**, 192 (1925).

[124] Noyes, A.A., & Wilson, H.A., Ap. J. **57**, 20 (1923).

The new equation by Saha was reported by Milne[125] to the Royal Society as a major advance. Furthermore, Milne applied the Saha theory to explain many phenomena in stellar spectra, of which the most conspicuous example was the appearance of helium in stellar spectra. Helium has the highest known ionization potential (24.5 eV) and is seen spectroscopically only in stars with the highest surface temperatures. The old observation by Lockyer and Vogt, to the effect that helium is seen in early type stars but not in late type stars, was explained over 40 years later, when the Saha equation showed that, in stars like the Sun, the temperature is not sufficient to ionize the helium, whence it is not observed.[126] The opposite is true in the hotter stars. Milne was very enthusiastic about Saha's results, writing:

> *The value of a workable quantitative treatment of high temperature ionization in relation to its spectroscopic consequences can hardly be overemphasized, and Dr Saha is to be congratulated on the fruitfulness of the result.*

Yet Milne found an error in one of Saha's equations and corrected it.

4.36 The Theoretical Mass–Luminosity Law

Soon after Kramers had published his results, they were adopted by the astrophysical community and Eddington[127] published his second paper on the absorption coefficient, this time using Kramers' numerical values. However, the most important paper came later that year,[128] and in this paper Eddington carefully examined all the available data on stars and inferred from the observations of Hertzsprung, Russell et al., and others, the existence of a relation between the mass of the star and its luminosity. This was one of the most important results in stellar theory: the mass–luminosity law.

In principle, the mass–luminosity law depends on both the type of energy source and the absorption coefficient. The first generates the energy, and the second controls the way it is transferred to the surface. The mass–luminosity law is obtained from the equilibrium between these two features. But at the time Eddington wrote his paper, only the absorption coefficient was known (calculated with the wrong composition). So Eddington had to mix theoretical expressions with observations, namely the data on Capella, in order to calibrate the law (see Fig. 4.8). Capella was chosen because, as Eddington put it:[129]

---

[125] Milne, E.A., Obs. **44**, 261 (1921).

[126] The electron in a neutral helium can absorb radiation, jump to a higher level and emit radiation when it returns to the original state. But the emitted photon is not in the visible range and the classification is based only on the visible light, and in fact on only part of it.

[127] Eddington, A.S., MNRAS **84**, 104 (1924).

[128] Eddington, A.S., MNRAS **84**, 308 (1924).

[129] Capella vitae: As Capella played such an important role in the theory of stellar evolution, let us review some of its properties and history. Capella (the 'she-goat'), or α Aurigae, is the brightest star in the Auriga constellation. It is the sixth brightest star as seen from the Earth. In 1899, Campbell

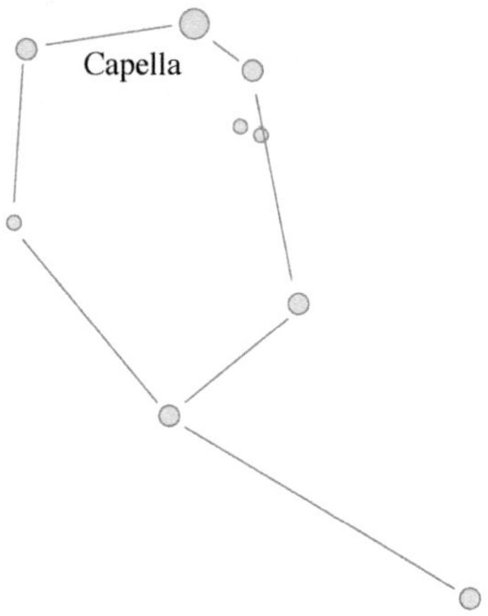

Fig. 4.8 The constellation of Auriga and its brightest star Capella, which served to calibrate Eddington's mass–luminosity law

It is the only (giant) star for which the required observational data reach a high standard of accuracy.

In view of what became known later about the complex structure of Capella, we can say in retrospect that Eddington could hardly have chosen a worse case.

[Ap. J. **10**, 177 (1899)] and Newall [MNRAS **60**, 2 (1899)] discovered independently that Capella is a spectroscopic binary, i.e., it exhibits two sets of spectral lines which move periodically with respect to one another. The two stars are so close to each other that it is impossible to observe them separately. The period of the binary is 104.022 days, which means that the two stars are quite close together, the distance between them being about a factor of ten greater than the radius.

Besides being the second brightest star in the Northern Hemisphere, the distance to Capella (42.2 lyrs) is known very accurately due to a combination of an interferometric method which provides the size of the orbit in degrees, and a spectrographic method which yields the orbit in kilometers. The star is apparently quite complex, as Otto Struve (1897–1963m) [PNAS **37**, 327 (1951)] complained that the spectra exposed by Capella were the most complicated he had ever seen. And yet the knowledge of the precise distance was not something that could be ignored.

The mass of the Aa component is $2.7M_\odot$ and the radius is $12.2R_\odot$, while the luminosity is $78.5L_\odot$. The primary is only slightly more massive than the secondary Ab, which has a mass of $2.6M_\odot$, a radius of $9.2R_\odot$, and a luminosity of $77.6L_\odot$. Both stars rotate rather fast, giving rise to complex atmospheric phenomena. The data in Eddington's time gave a mass of $4.18M_\odot$ for the brighter star.

The binary system of two giant stars has two faint companions labelled as Capella C and D. The former is a red dwarf star with mass $0.3 - 0.4M_\odot$, radius $0.56R_\odot$, and luminosity about $10^{-2}L_\odot$. The radius of the orbit is about 0.11 AU, or just 30 times the radius of Capella A. Capella D is also a red dwarf with mass 0.25–$0.3M_\odot$ and luminosity $5 \times 10^{-2}L_\odot$. The radius of the orbit of this star is 0.022 AU, or only 6.2 times the radius of Capella A. There is no doubt that this star disturbs the atmosphere of Capella A. While Capella served as a testing ground for many theories of stellar structure, one can hardly imagine a more complicated system for such a comparison between theory and observation [Heintz, W.D., Ap. J. **195**, 411 (1975)]. For comparison, the distance of the largest planet, Jupiter, from the Sun is 5.2 AU, which is 1.1×10^5 times the radius of the Sun, while the mass is about $9.7 \times 10^{-4}M_\odot$. Mercury, the nearest planet to the Sun, has an orbital radius of 0.47 AU, or $10^4R_\odot$, and a mass of $1.6 \times 10^{-7}M_\odot$.

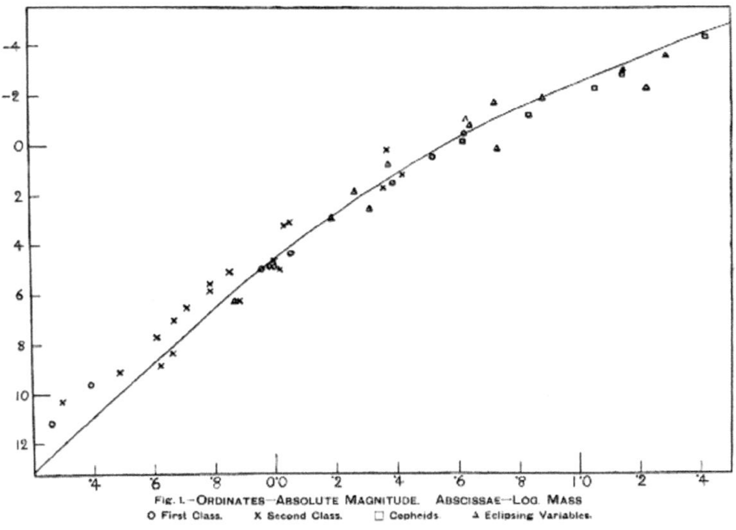

Fig. 4.9 The mass–luminosity law. The figure contains practically all the data available at the time. Altogether there are 36 stars. The *ordinate* is the logarithm of the luminosity, while the *abscissa* is the logarithm of the mass. From Eddington 1924

According to the observations, two values for the luminosity were found for every mass and effective temperature. As an example, for class G stars, see Fig. 4.10. There are dwarf stars with a brightness magnitude of 5, and giants with a brightness of magnitude 0 (which amounts to a difference by a factor of a hundred in the brightness). Hence, the brightness should be a double-valued function of the spectral type. In the case of the Sun, Eddington faced the following dilemma. A star of one solar mass and effective temperature of 5 860 K has two possible luminosities: (1) that of the present Sun, and (2) that of the Sun when it passed through the same temperature on the way from P to Q with a much larger radius than today, that is, when it was a giant. Since the Sun is now on the dwarf sequence and was supposed therefore to be liquid, and since the theory assumes gaseous stars, it is clear that Eddington tried to predict the luminosity of the Sun when it was a giant. But Eddington discovered that his calculation yielded the luminosity of the Sun today, as if it were not liquid! Eddington did not realize that the assumption of a liquid Sun was wrong, and claimed that:

> *If the theory gives the right luminosity of the wrong stars, it is presumably wrong.* Having said that, a few sentences later, he suggested that: *Even dense stars like the Sun are in the condition of a perfect gas and will raise the temperature if they contract.*

The first signs of the revolution to come were there, but not yet recognized.

After some algebraic manipulations and using the data from Capella, Eddington reached the following result:

$$L = \text{Const.} \times M^{7/2}(1-\beta)^{3/2}\mu^{4/5}T_e^{4/5} ,$$

where μ is the assumed molecular weight (which is the same for all stars) and T_e is the effective temperature. This was not yet a mass–luminosity formula because the effective temperature appears in the formula. Eddington had to eliminate the effective temperature to get the mass–luminosity relation. If we check the range of luminosity and temperature, we find that the luminosity changes by about a factor of 10 000, while the effective temperature changes by a factor of 25, so Eddington felt that he could assume, as a first approximation, that the effective temperature is constant, and in this way obtain the mass–luminosity law.

The comparison Eddington got between the semi-theoretical formula and Hertzsprung's observed data is shown in Fig. 4.9. The results were impressive. This agreement was reached despite the (unacceptable to many) tacit assumption that the energy source is spread uniformly throughout the star in such a way that energy generation is inversely proportional to the absorption coefficient.

What about the discrepancy in the absorption coefficient? Eddington did not discuss it, but stated that:

> *As regards composition, an unduly large proportion of hydrogen would make the star fainter; apart from that, not much effect is likely to be produced.*

As we shall see, this was another point Eddington missed.

4.37 Conceptual Difficulties

The success of the theory in predicting the observed mass luminosity law, even where it was not supposed to be valid, gave Eddington good reasons to reflect. What Eddington found was that the luminosity he got for one solar mass star agreed with point S on the dwarf sequence. So could a dense star like the Sun obey the perfect gas law?

Moreover, consider a star at point S (see Fig. 4.10). If the energy is derived from contraction, then as Eddington pointed out correctly, the evolution should be from point T to point T′. But if there is another source of energy, this is probably not the course of evolution. Moreover, if the stars evolve with constant mass, then Eddington suggested that the line QR is the locus of stars with different masses and not an evolutionary track of all stars. This was a revolution! As we know today, it is also the correct explanation!

How did Eddington justify his bold departure from the standard picture? The latter was based on the assumption that the luminosity did not depend on the mass of the star, so that all stars with all masses must evolve along the same track. This was of course, in contradiction with the observations of Hertzsprung and others. The statistics of stars showed that the stars along the dwarf sequence are on the average less massive than those along the giant branch. The explanation according to the standard picture was that the lighter stars cross the giant phase faster than the massive stars. But the contraction theory predicts the opposite.

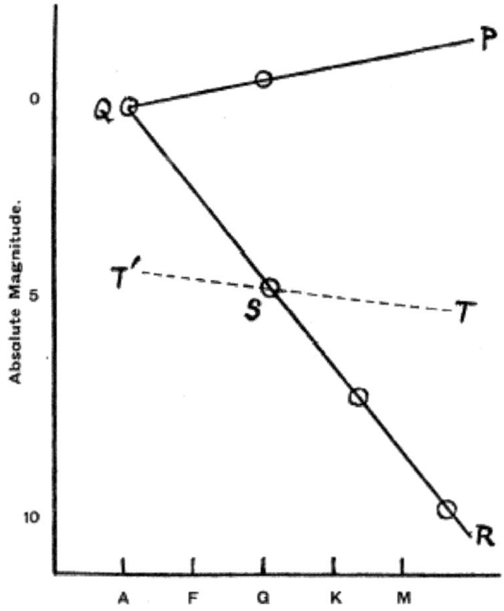

FIG. 2.

Fig. 4.10 The problems with the standard picture of stellar evolution as pointed out by Eddington in 1924

Eddington hypothesized that the atoms stripped of electrons, in the form of a plasma, behave like an ideal gas, but he was unable to calculate the behavior of the electric forces due to the non-neutral particles. Eddington realized the conceptual problems and wrote: *in view of these difficulties we do not feel able to attribute great weight to the individual result*, which he calculated for certain stars. However, the assumption of an ideal gas, against all odds, led to agreement with observation. Eddington came extremely close to our present day understanding when he stated that:

> *In applying the Lane–Ritter theory to stellar evolution we have been influenced by the false analogy between the mutilated stellar atoms and ordinary atoms: we can at least see that this analogy is unfounded and approach the problem again, free from this bias.*

This was just before Debye and Hückel[130] published their theory of electrolytes,[131] which has nothing to do with stars, and invented the very important concept of electron screening. Free electrons are attracted to the heavy ions and create a cloud around each heavy ion, called the Debye sphere. The cloud neutralizes the electrical force between the ions and allows the gas to behave almost like an ideal gas. This surprising behavior of the stellar matter was predicted by Eddington before the physics was really known, and was one of the controversial issues.

Eddington went ahead and examined the possibility that the evolution might be explained by mass loss, for example, due to annihilation of electrons and protons. This hypothesis seemed capable of explaining the standard evolution. So Eddington stated that there was:

> *[...] no need at present to contradict the current theory*, and that: *Our explanation cannot be developed in detail owing to ignorance of how the rate of generation of subatomic energy depends on temperature and density.*

4.38 Objections to the Mass–Luminosity Law

As one would expect, Eddington's derivation of the mass–luminosity law was rejected by many, and his basic assumptions did not go unchallenged. Few great discoveries in astrophysics have ever been accepted without attempted rebuttals. Jeans[132] criticized Eddington by claiming that: *When the problem is treated in a general way the surprising properties which Professor Eddington attributes to the stars disappear*. The physical idea was Jeans' assumption that *a star with given mass can always adjust the luminosity to radiate away what the energy source produces*. As a matter of fact, a tacit assumption by Jeans, which he did not specify explicitly, was that the rate of energy generation does not depend on the conditions inside the star or on the mass of the star. However, Jeans related the difference between his and Eddington's analysis to Eddington's simplifying assumption that β is constant throughout the star. This particular assumption by Eddington changes only the numerical constant in the mass–luminosity law, and not the basic behavior.

In contrast to Jeans' provocative introductory statements, his results did not differ that much from Eddington's. After eliminating the effective temperature, Jeans got $L \sim M^{4.77}$ for low mass stars, while Eddington got $L \sim M^{4.4}$, and $L \sim M^{1.3}$ for high mass stars, in contrast with Eddington's result of $L \sim M^{1.4}$. In view of the

[130] Debye, P., & Hückel, E., Phys. Zeit. **24**, 3 (1923).

[131] An electrolyte is a solution of an acid, or a salt, or a base, in water. These types of chemical compound have the property that the chemical bond is based on electrical attraction. Once placed in water, some of the molecules disintegrate into positive and negative ions. For example, NaCl, partially disintegrates into Na^+ and Cl^-. The properties of water molecules cause the weakening of the chemical bond. In solutions there are positive and negative charges, as in stars. There are, however, differences between stars and electrolytes which do not affect the application of the theory. In stars there are positive ions and negative electrons, and there are no water molecules in stars.

[132] Jeans, J.H., MNRAS **85**, 196, 394 (1925).

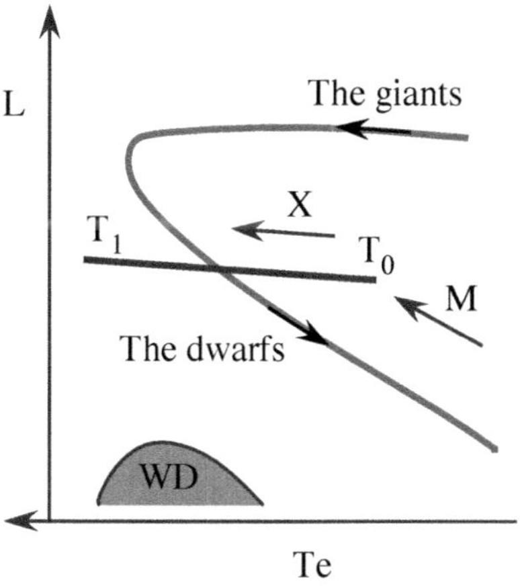

Fig. 4.11 The schematic evolution theory on the basis of Eddington's results

inaccurate data and vastly different assumptions about the energy production, the small difference in the mass–luminosity laws required an explanation, and it had not yet been provided.

Eddington[133] defended his theoretical derivation of the mass–luminosity law. In the introduction he stated that the luminosity depends mainly on the mass, with a small correction for the surface temperature. Moreover, he criticized Jeans' logic that the luminosity could accept any value irrespective of the mass of the star. Contrary to common sense, argued Eddington, *a star does not generate energy at a given rate but at a rate which it can modify by expansion or contraction. It is this flexibility which enables it to find a state of equilibrium*, so that if for some reason the energy generation increases or decreases, the star can restore the equilibrium.[134]

Soon after Eddington,[135] who loved to present his cases as paradoxes, came up with an extreme example which demonstrated his logic. Eddington assumed that all the energy of the star was generated in a point at the center, a point which had negligibly small volume. Hence, the star could not regulate energy production, because expansion of the star would not change the energy production. A point of vanishing volume remains a point. Eddington solved the problem and demonstrated that it led to unphysical results, viz., the density increased outward and similar unphysical consequences.

The exchange of 'compliments' is noteworthy:

> *In place of pointing out in his 'reply' any error in the analysis by which my equation (2) is obtained, Professor Eddington merely claims that he could have reached my result in a*

---

[133] Eddington, A.S., MNRAS **85**, 403 (1925).

[134] This argument immediately rules out radioactive decay as a source of energy in a gaseous star, because the rate of radioactivity does not depend on pressure and temperature.

[135] Eddington, A.S., MNRAS **85**, 408 (1925).

shorter way. His way is certainly shorter, but it disregards entirely the difficult question of the boundary condition at the star's surface.

Moreover, Jeans claimed that radiation pressure:

[...] is of comparatively little importance in the problem under discussion [...] the old-fashioned sphere of gas, in which radiation was left entirely out of account, still provides a remarkable good model of a star.

In other words, all Eddington had done and discovered in the study of stellar structure was of no importance and in fact wrong:

The various paradoxical theorems recently enunciated do not seem to me to correspond to anything in the facts of nature.

How blunt!

In his reply, Eddington was no less sharp and acerbic, showing that, for reasonable assumptions regarding the radiation absorption coefficient, the dependence of the luminosity on the surface temperature is slight. Eddington showed where Jeans was wrong, and claimed that:

Unfortunately mathematical analysis shows plainly that the common-sense view will not work. For if the given rate of generation of energy is greater than the rate of radiation L fixed by the formula, the energy of the star is increasing and it must expand.

Jeans claimed that it must contract. And Eddington then set out a fundamental fact about the physics of stars:

Since we find actual stars in equilibrium, we must take the view (which is perhaps not entirely opposed to common sense) that a star does not generate energy at a ***given*** *rate, but at a rate which it can modify by expanding or contracting. It is this flexibility which enables it to find a state of equilibrium.*

And Eddington was right. This sentence embodies Eddington's deep understanding of the making of stars. Few in the history of stellar structure understood it the way he did.

With this result, Eddington quenched the heated debate on the mass–luminosity relation, but just for one year.

4.39 The Writing Was on the Wall

In a note entitled *On the Relation Between the Mass, Temperature and Luminosity of a Gaseous Star*, Russell[136] confirmed that Eddington's results agreed with observations. However, argued Russell, if the mass–luminosity law did not contain the radius of the star, it meant that a star of a given mass and luminosity could have any radius, whereas this was not observed. Moreover, the stars are not found just anywhere on the HR diagram, but only in certain locations. Hence, there was a problem with this theory. Russell reiterated that:

[136] Russell, H.N., Nature, 8 August 1925.

All commentators agree that if the mass of a star remains nearly constant throughout its history, no comprehensive scheme of evolution appears to be possible.

He thus rejected Eddington's idea that the dwarf branch was a locus and not a track of evolution. The entire scientific community was against Eddington's theory, but he was right!

The most important conclusion, as can be seen from Russell's Fig. 4.12, where the central temperatures are calculated on the basis of Eddington's model, was that the central temperature is essentially constant along the main sequence (a name given by Eddington to the dwarf sequence). Hence, the main sequence could not be a cooling sequence. As Russell concluded correctly:

In the neighborhood of a temperature of about thirty million degrees, the rate of transmutation of matter into energy increases very rapidly.

This is the equilibrium temperature at which energy production equals the losses. Russell assumed that mass annihilation took place at this temperature, and consequently that stars lose mass and thereby move down the main sequence.

To account for the white dwarfs, Russell hypothesized that:

There exists a certain residue of refractory material, immune to transformation at a million degrees. As the main constituents become exhausted this will preponderate, and at last be almost exclusively present. If this residue were incapable of transformation, rapid gravitational contraction would ensue until even the ionized atoms were jammed close together. The considerable abundance of the white dwarfs per unit volume suggests, however, that further energy liberation changes occur and delay the last act.

What he meant by the 'last act' is not specified. As for the evolution, Russell rejected the subatomic energy and quoted Eddington as saying that *the diminishing brightness in the dwarf series is due to decreasing mass, and not to falling off in compressibility*. He concluded as follows:

On the other hand, the difference of mass between the giants and the dwarfs is now explained, and the white dwarfs – formerly most puzzling – now, thanks to Eddington, find an orderly place at the end of the sequence.

In short, in spite of the fact that the main sequence appears to be a locus, where stars of different mass have the same central temperature, the old idea due to Lockyer was still alive. By the way, note that Russell wrongly attributed the solution of the white dwarf puzzle to Eddington.

4.40 The Russell–Vogt Theorem

An important theorem in the context of the mass–luminosity theorem was proven by Vogt[137] in 1926 and a year later by Russell.[138] The theorem states that, for a

[137] Vogt, H., AN **226**, 301 (1926).

[138] Russell, H.N., Astronomy Vol. 2, p. 910, by Russell, N.H., Dugan, R.S., & Stewart, J.Q., Ginn & Co. Boston.

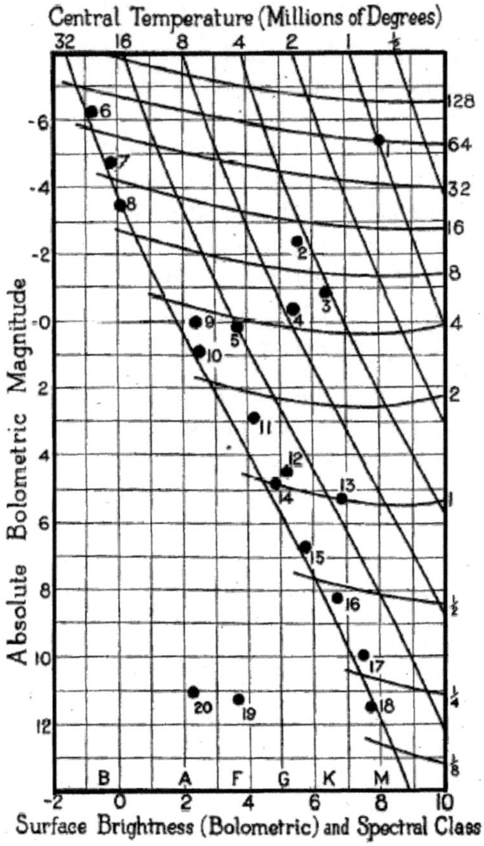

Fig. 4.12 The HR diagram with the central temperatures calculated according to the Eddington model. From Russell 1925. The points are stars for which the masses and absolute luminosities are known. The main sequence corresponds to a constant central temperature of 32 million degrees

given gaseous star, once the dependence of the composition and the energy source on density and temperature are given, the mass determines the luminosity. This is exactly what Eddington tried to prove from his models, of course, without precise knowledge of the energy source, but making assumptions about how it was spread through the star. The theorem is quite general,[139] and implied that there exists a mass–luminosity law for gaseous stars, in accord with Eddington's claims.[140]

[139] Chandrasekhar, S., *An Introduction to the Study of Stellar Structure*, Univ. Chicago Press (1939) p. 252.

[140] The theorem actually states that there is a solution for every mass and that it is unique. Both statements turned out later to be wrong. Here, we point only to the second part. In its present day

4.41 The Rosseland Mean: The Situation Worsens

The radiation emerging from stars extends over all wavelengths. Hence, the equation for the radiation emerging from the atmosphere of the star must include all wavelengths. And indeed, the equation Schwarzschild wrote down describes how radiation behaves at any wavelength. Furthermore, the absorption coefficient depends on the wavelength. The absorption of matter in the visible differs from the absorption of X rays. In addition to these two complications, the equation of radiative transfer is a complicated equation to solve, because what happens at a given point depends on what happens elsewhere. Photons are born at one point, but absorbed at another.

When Eddington applied the equation to the interior of the star, he realized the difficulty in solving the equation for the entire star and including all the frequencies. Ergo, Eddington proposed to replace the full radiative transfer equation with an average-over-all-frequencies equation. In other words, he chose to treat all wavelengths as a single wavelength. The average equation derived from the radiative transfer equation was called the diffusion equation for radiation, and this equation is still in use today to calculate the transfer of radiation inside stars. Even today's computers are not big and fast enough to allow more accurate calculations.

The averaging over frequencies process leading to the diffusion equation requires an average over the radiation absorption coefficient. Averaging of this type can be a very tricky matter if the resulting equation is to be a good representation of the real situation. We know today that, in some cases, it is even impossible. What Eddington did was to take the simplest average possible, which is known today as the Planck mean, i.e., a simple average with the Planck function as weight.

In 1924 there came a young Dane named Rosseland (1894–1985m),[141] who found that Eddington's averaging procedure was not the right one. The formula for averaging needed a major revision which is known today as the Rosseland mean.[142] So far so good, but after some calculations, Rosseland realized that:

> *The discrepancy between the observation and the theory found by Eddington will not be removed [...] on the contrary, the discrepancy [...] comes out essentially larger.*[143]

---

form, the theorem confirms that there are no close pairs of solutions. There can be two distant solutions, that is, two solutions with very different masses and luminosities.

[141] Rosseland, S., MNRAS **84**, 525 (1924). Rosseland's paper was communicated to the MNRAS by Eddington himself.

[142] In the Planck mean, one averages the absorption coefficient with the Planck function as weight. Consequently, a higher weight is given to the frequencies with high absorption of radiation. In the Rosseland mean, one averages one over the absorption (with the same weight as before). But since it is one over the absorption which is averaged, the frequencies where the radiation escapes more easily get a higher weight. Thus, if the absorption changes very much with frequency, as is the case, the differences can be extremely large.

[143] The reason for the dramatic difference is that Eddington's average emphasizes the maximal values of the absorption (where radiation is mostly absorbed), while the Rosseland average emphasizes the minimal values (where radiation can more easily escape).

Actually, as pointed out by Milne[144] the situation became even more complicated. When there are two absorbers like iron and calcium, for example, the Rosseland mean gives a result which is not the simple mean between the absorptions of pure iron and pure calcium and can be very different from both. The Rosseland mean of the absorption of two species is not the sum of the Rosseland means of each species alone. Consequently, Milne claimed that his new absorption coefficients were *enormously higher than those calculated by Russell and Stewart.*

Milne[145] thus recalculated the absorption coefficient and claimed to find differences from Eddington's result.[146] Eddington found a discrepancy in the form of a factor 1/8 when the star Capella was modelled. Milne found a discrepancy of 1/2. Milne's conclusion was interesting:

> *I conclude that there is no very grave reason to suppose the theoretical value of the absorption coefficient to differ from the astronomical value more widely than by a factor of 1/2. In fact, the discussion might be taken to imply that the stars are not principally made up of iron but rather some element like calcium or silver [...] whether this residual discrepancy of 1/2 is significant is difficult to say.*

Milne pointed to several uncertain assumptions in Kramers' calculation which might yield an inaccurate absorption coefficient. Moreover, Milne drew attention to the fact that the composition of the stars was poorly known. As a matter of fact, this is the first time that the solution to the discrepant absorption coefficient was proposed in terms of a composition other than iron. Milne also checked the possibility that Capella might be made of titanium or silver, but the agreement was poorer. It would take several years for Eddington to find out that Milne was right, and that the composition was indeed totally wrong.

Finally, Milne rejected the idea raised by Eddington that some of the absorption of the radiation might take place in the nucleus, and not only by the electron, as was considered hitherto. On this point Milne was right as well.[147]

4.42 Are the Stars Well Mixed?

Observation of sunspots indicated right away that the Sun rotates with a period of about 25.36 days.[148] The rotation of stars can be detected through the shape of the spectral lines. As early as 1877, Abney[149] and Vogel[150] had suggested such an effect

[144] Milne, E.A., MNRAS **85**, 750, 768, 979 (1925).

[145] Milne, E.A., MNRAS **85**, 750 (1925).

[146] The differences are due to technical reasons which we shall skip here.

[147] As an amusing note, Milne described in a footnote to his paper how an error he published with Fowler was frequently copied and quoted by many without too much reflection.

[148] Note, however, that the Sun does not rotate like a rigid body, and the rotation period changes with latitude.

[149] Abney, W. de W., MNRAS **37**, 278 (1877).

[150] Vogel, H.C., AN, No. 2141, **90**, 71 (1877).

on the spectral lines. However, the first real calculation of the shape of spectral lines in a rotating star is due to Fowler.[151] It soon became clear that most stars rotate, some of them fast and some slowly. The Sun is no exception, though it is a slow rotator.

In 1924, von Zeipel[152] published a stellar shaking theorem about rotating stars. The theorem stated that, in a rotating star in equilibrium and in which the energy is transferred by radiation, the energy generation must satisfy the following relation:

$$\varepsilon \propto 1 - \frac{\omega^2}{2\pi G\rho}, \tag{4.4}$$

where ε is the energy generation, ω the angular velocity, ρ the density, and G the constant of gravity. The formula appears to dictate a very bizarre situation in the star, namely, a connection between rotation, which is a macroscopic quantity, and energy generation, which is a microscopic quantity. No one expected such a relation. This sounded unthinkable to Eddington.[153]

The way out of the quandary suggested by von Zeipel and Eddington was to assume that the relation is never satisfied in stars, even though the stars continually try to satisfy it. In the attempt to satisfy the formula, the star initiates currents of heat and matter in order to reach the state required by the formula. These large scale currents, which extend over the entire star, are called meridional circulation, because they flow in the meridional plane. They are critical for our case here, because their existence raises the question to what extent the stars are fully mixed. Do we see on the surface of the star a special surface composition, or is what we see indicative of the composition throughout the star? Is the observed surface composition the primordial one, or has it changed since the formation of the star by mixing with the interior?

The result, although it shocked theoreticians, is not surprising after all. The star is in an equilibrium wherein the gas and radiation pressure and the centrifugal forces all help to balance the gravitational attraction. But the radiation pressure is closely associated with radiative transfer, which removes energy from the inside of the star and brings it to the surface to be radiated into space. So the radiation enters into the equation of hydrostatic balance in which rotation has now been introduced, and for this reason the existence of such a relation between the source of radiation and rotation should not be so surprising.

The stars try incessantly to reach the state which satisfies the equation, but they always fail. In the attempt to satisfy this equation, the star shifts mass and this gives rise to a flow of matter: the meridional circulation. The current flows along the meridian and brings matter from the inside out. The currents mix the star and bring to the surface the synthesized elements from the interior, according to this account of what happens inside stars.

---

[151] Fowler, A., MNRAS **60**, 579 (1900).

[152] von Zeipel, H., Seeliger Festschrift, 1924, p. 144.

[153] Eddington, A.S., Obs. **48**, 73 (1925).

It all boils down to how fast the currents can mix the star. Five years after von Zeipel shook the theorists, Eddington[154] succeeded in estimating the speed of the currents. Eddington's estimate was 2×10^{-4} cm/s. Vogt,[155] who had discussed the problem 4 years before Eddington, did not estimate the speed of the currents. Vogt just wrote down the equations, but his conclusion about the importance of the currents was identical to Eddington's. This velocity leads to a mixing period of just 10^7 yrs, and hence mixing became extremely important.

Could there be alternatives to currents? Randers (1914–1992)[156] examined this question. In particular, he examined what happens if the rotation is not uniform in the star, that is, if different parts rotate at different speeds.[157] Is there a particular rotation which minimizes the meridional circulation so that the star does not exhibit large currents, whereupon it might remain inhomogeneous? Gerasimovič (1889–1937m)[158] did indeed discover such a rotation law, but could not confirm that stars rotate according to his law. As an alternative, he raised the idea that the star might be extremely viscous, and that the viscosity might impede the currents. As a matter of fact, Rosseland[159] suggested precisely this idea claiming that *there are both observational and theoretical indications that the motion in the Sun is turbulent*, which implies a very high viscosity.[160] However, Biermann[161] argued that, on the contrary, the Sun is stable and no turbulent motion can take place. The survival of solar spots for many days indicated that Biermann was probably right.

Randers solved for the steady state solution of the meridional circulation and found that the period of the circulation was given by the square of the radius divided by the viscosity. If the viscosity was the normal one for the ideal gas in the Sun, the period of the meridional circulation in the Sun came out to be $P \approx 10^{13}$ yrs, provided that the Sun was not turbulent, and $P \approx 10$ yrs if it was turbulent, which implies a very well mixed Sun. But what would happen if there was no viscosity at all? Randers found that, in this case, there could not be a steady state solution, and the currents should therefore change continuously.

The end to this red herring came in 1950 when Sweet[162] showed that the actual velocities are of the order of 10^{-10} cm/s (which yields a mixing time of 10^{13} yrs, much longer than the lifetime of the stars), whereupon the meridional currents are unimportant. Stars rotate too slowly for the effect to be important. So how come Eddington made such a blunder, causing people to believe in full mixing for about twenty five years? A careful analysis of Eddington's calculation shows that he introduced a factor q into the calculation whose value he did not calculate, but merely

---

[154] Eddington, A.S., MNRAS **90**, 54 (1929).

[155] Vogt, H., AN, No. 5342, **223**, 229 (1925).

[156] Randers, G., Ap. J. **94**, 109 (1941).

[157] The technical term is differential rotation.

[158] Gerasimovič, B.P., Obs. **48**, 148 (1925).

[159] Rosseland, S., MNRAS **89**, 49 (1928).

[160] The flow becomes turbulent if the viscosity is sufficiently low. But if the flow becomes turbulent, the turbulence generates a high effective viscosity.

[161] Biermann, L., Zs. f. Ap. **5**, 117 (1932).

[162] Sweet, P.A., MNRAS **110**, 548 (1950).

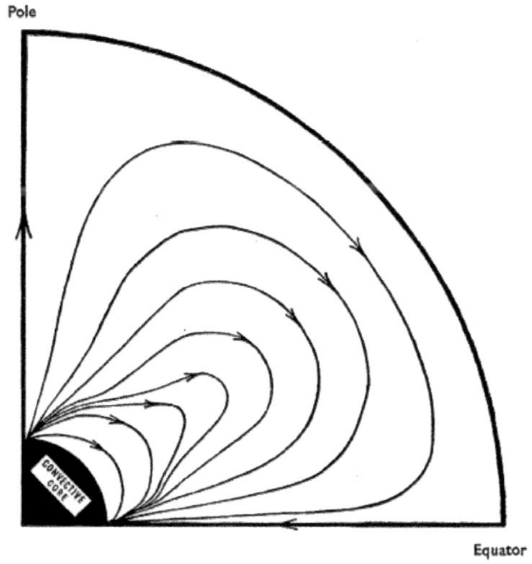

Fig. 4.13 The pattern of the circulation currents as solved by Sweet (1950), and shown to be much too slow to have any effect on mixing in stars

guessed. His guess was just a factor of 1 000 000 too high! The factor q is a dimensionless factor, depending on the deviations of the star from spherical symmetry due to the rotation. Sweet calculated this factor accurately.

Quite independently, Öpik,[163] not knowing about Sweet's work, reached a similar conclusion. He found that, for the solar rotation, Eddington's estimate was wrong by a factor of 1 000. The sequence of events is particularly interesting. Öpik submitted his paper to the MNRAS on 1 December 1949. On 28 August 1950, Sweet submitted his paper to the MNRAS. On 28 September 1950, Öpik's paper was accepted for publication (about 10 months after submission). On 13 October 1950, Sweet's paper was accepted for publication (about 6 weeks after submission). Sweet's paper was published in volume 110 of the MNRAS, while Öpik's paper was published in volume 111. Öpik was in the Armagh Observatory in Northern Ireland, while Sweet was in Glasgow, just on the other side of the Irish sea. Today the credit goes mainly to Sweet,[164] and Öpik is not mentioned at all in connection with meridional circulations,

Sweet put an end to attempts by many, for example, Wasiutynski,[165] to insist that stars are well mixed all the time. As we will see, the fact that stars are not mixed dramatically affects all the evolution after the exhaustion of hydrogen.

The von Zeipel theorem played into the hands of Jeans. Consider the above expression when the angular velocity tends to zero, that is, when the star rotates very

[163] Öpik, E.J., MNRAS **111**, 278 (1951).

[164] On August 2007, Sweet's paper had 140 citations while Öpik's only had 28.

[165] Wasiutynski, J., Astrophys. Norvegica **4**, 1946

slowly or not at all. In this case, the energy production must be constant. This is exactly what Jeans claimed all the time. This fact, however, did not prevent Jeans[166] from criticizing von Zeipel, but after an exchange of papers von Zeipel[167] convinced Jeans that his, as Jeans put it, *very surprising theorem* was in fact correct.

Today, the phenomenon of currents flowing along the meridian of the star is called the Eddington–Sweet circulation. The name refers to the fellow who discovered the flow and the fellow who showed that it was unimportant.

4.43 The Absorption Coefficient Dispute and the Mass–Luminosity Relation

The atomic physics needed to calculate the absorption coefficient was far from being developed in 1925, and yet Rosseland[168] recalculated the opacity from scratch, including the scattering of radiation by free electrons released by ionization of the stellar matter. This radiation scattering by electrons is called the Thomson limit. As Rosseland wrote, the weakness of the theory was that some of the numerical coefficients were off by 50%. Applying the new absorption coefficients, Rosseland checked whether this time the agreement with Capella could be improved, finding that:

> *The theory can conform with the requirements of Eddington's mass–luminosity relation only by assuming the core of a star to consist of elements of very large atomic numbers to a degree which surpasses what might be expected from the distribution of elements in the Earth crust or in meteors.*

Otherwise, the agreement with Eddington was reasonable. So the two theoretical calculations agreed among themselves, but not with the observations.

After a detailed calculation, Rosseland reached the conclusion that:

> *If the core of a star is in true hydrostatic (non-convective) equilibrium, and if it does not contain an enormous proportion of hydrogen, then hydrogen must be repelled from the core and strongly concentrate in the surrounding convective layer and at the surface.*

Rosseland assumed that the disintegration to positive ions and negative electrons would create a strong electric field which would have caused these effects. He was wrong. The electric field is not important at all in stars, because of the screening effect Debye and Hückel had discovered. Rosseland argued correctly that the electron force between the ions is much larger than the gravitational force, because the ratio between the Coulomb attraction/repulsion and the gravitational attraction is $e^2/m_{\mathrm{H}}G = 1.3 \times 10^{36}$, where m_{H} is the mass of the hydrogen atom, G the constant of gravity, and e the basic unit of electric charge. Hence, argued Rosseland, this time wrongly, when the hydrostatic equation which describes the balance between

[166] Jeans, J.H., MNRAS **85**, 333 (1925).

[167] von Zeipel, MNRAS **84**, 678 (1925).

[168] Rosseland, S., CMWCI **296**, 1 (1925); MNRAS **85**, 541 (1925); Ap. J. **41**, 424 (1925).

the forces acting on the star is written down, it must include the electric force. This was an error, and on the contrary, because the electric forces are so strong, one can consider them first and ignore the weak gravitational forces. In this case, one gets the theory of Debye and Hückel, which claims that the positive charges are surrounded by negative charges and each ion is neutralized within a very short distance (a distance equivalent to several times the distance between ions).

To sum up, the fine structure of the ions and electrons and the balance of the star can be safely separated. The result of Rosseland's incorrect assumption was that stars should have a very strong electric field, and that this field pushes the hydrogen from the core to the surface. In this case, whatever we observe of hydrogen on the surface is the maximum amount, and the interior should be devoid of hydrogen. But if all the hydrogen is on the surface and the interior contains only very heavy elements, one cannot reconcile the theory with observation. Thus, claimed Rosseland:

> *There are several facts indicating that hydrogen really is present to an abnormal amount in the stellar atmosphere,*[169] *but whether it is permissible to allow a sufficient amount of hydrogen to be imprisoned in the core of the star to bring about agreement with theory must provisionally remain an open question, which the calculation as regards the electrical state of the star would tend to answer in the negative.*

Thus, the incorrect theory and erroneous considerations prevented Rosseland from reaching the correct results on that occasion. The very same arguments were repeated by Eddington in his book.[170] However, when he realized that the consequences disagreed with the observations, he suggested that *the star may strive to reach the steady state dictated by the solution*, but the time needed may be extremely long, much longer than the age of the star.

Another issue that came up was diffusion of elements. If the star is gaseous, then the heavy elements tend to sink while the light elements tend to float. Diffusion thus increases inhomogeneity. Eddington estimated the effect, and found that the time scale is about 3×10^{13} years, whence it could be neglected.

4.44 The Mass–Luminosity Relation Again

Two years after the publication of the mass–luminosity law by Eddington, Jeans wrote to the Editor of The Observatory:[171]

> *I claimed in Nov. 1924 that my suggestion that the annihilation of matter provided the source of stellar energy had 'gained enormously in probability since Eddington showed that, as a matter both of theory and of observation, the radiation from a star is approximately a function only of the star's mass'. I wish to withdraw this claim now. Further study of the*

[169] Compton, K.T., & Russell, H.N., Nature **114**, 86 (1924).

[170] Eddington, A.S., *The Internal Constitution of the Stars*, p. 272. Eddington's main error was to include the gravitational force in the basic Debye–Hückel equation, and to miss the point that gravity can be completely neglected in this problem.

[171] Jeans, J.H., Obs. **49**, 60 (1926).

problem[172] *soon convinced me that Prof. Eddington's work did not prove this at all. It is determined, independently of the star's mass, by the rate at which energy is being generated inside the star. The argument I based on Prof. Eddington's work is therefore fallacious.*

Indeed, the simple-minded physical thinking would be that the luminosity depends on the energy generation and not on the mass or any other parameter of the star. Thus, Eddington's result was even more surprising! Jeans could not accept even the case in which Eddington's results were supposed to confirm his own calculation! The mass–luminosity law fell victim to the hypothesis of the energy source of the stars.

The most annoying feature in Eddington's theory remained the old assumption about the mode of stellar energy production. There never was any physical reason why that should be the case. Jeans[173] discussed the internal temperatures and densities of the stars and demonstrated what he considered to be the rather poor logic on the part of Eddington. Now that Kramers had come up with his results, and the particular way the absorption coefficient depends on temperature had become known, Eddington's assumption looked even more unphysical. Jeans felt that a more accurate calculation could be carried out with Eddington's assumption, claiming that:

So long as the true value of the absorption was unknown, such calculations might be regarded as legitimate speculation, but now that Kramers' formula for the absorption is available, they ought to be superseded by calculations based on this formula.

So Jeans calculated the structure of stars using Eddington's critical assumption, just in order to demonstrate how wrong the results of the theory were. The main conclusion Jeans drew, however, was that the central temperature *varies sufficiently to dispose of the suggestion advanced by Russell*[174] *that all stars on the main sequence have very approximately the same central temperature.* Note that Jeans carried out all these calculations assuming the stars to be gaseous, although he stressed in his book published two years later that the stars on the main sequence were liquid. The implication was that one had to abandon Russell's suggestion that the star's energy generation was a consequence of its central temperature attaining a certain definite critical temperature.

So what determines the main sequence? According to Jeans,[175] it had to do with the stability of the star when electrons are stripped from the atom. Thus the main sequence phase is reached when all electrons are stripped off the atoms. Until the main sequence phase, the atoms are gradually more and more compressed until all the electrons are removed, and then the star reaches the main sequence. If so, argued Jeans:

My suggestion required that atoms ceased to generate energy when they were stripped bare of electrons [and] the generation of energy in the central regions of the main sequence star would be very slight.

---

[172] Jeans, J.H., MNRAS **85**, 196, 394, 792 (1925).

[173] Jeans, J.H., MNRAS **87**, 36 (1926).

[174] Russell, H.N., Nature **116**, 209 (8 August 1925).

[175] Jeans, J.H., MNRAS **85**, 914 (1925).

The inner part is inert and energy generation should continue further out. The evidence, claimed Jeans, was that stars on the main sequence were[176] *unduly luminous for their masses*. Recalculating the models assuming the ideal gas law to prevail to very high temperatures and densities, Jeans found for the Sun a central density of 300 g/cm^3 and a temperature of 70×10^6 K, which he admitted were to a large extent arbitrary.

Eddington realized the annoying fact that his theory actually agreed better with Jeans' hypothesis that the stars convert mass into energy, and that their masses decrease with time, whence they move down the main sequence. He considered this case in his book as well, and found rather long lifetimes for stars. However, stellar statistics were not sufficiently good to distinguish between Jeans' and Eddington's assumptions. As a matter of fact, the last paragraph in Eddington's book is puzzling. Eddington wrote:[177]

> *Somewhere in the present tangle of evolution and sources of energy I have been misled; and my guidance of the reader must terminate with the admission that I have lost my way.*

Hard to believe, but true. Maybe it explains why Eddington did not return to the problem of the stellar energy source in the early 1930s, when new developments were about to show that he was right after all.

Vogt[178] brought up a very interesting argument. If the components of a binary system radiate away their mass, the ratio of the masses tends to unity as the stars get older (because the more massive star has a higher luminosity and hence loses mass faster, and when it reaches the mass of the companion, they keep their mass ratio fixed at unity). The observational data by Vogt confirmed this prediction, displeasing Eddington. Eddington summarized the arguments for and against Jeans' hypothesis with 6 arguments in favor of Jeans' hypothesis and only one argument against it, the fact that the energy release must depend on temperature, otherwise the star would become unstable. Jeans' hypothesis did not satisfy the sixth condition, and eventually failed completely.

4.45 How Pristine the Stars Remain

Stars live for billions of years, during which they move through the galaxy and encounter gas clouds, dust clouds, and so on. One would therefore expect stars to accrete some fresh material from the various clouds. Hence the following question popped up: how pristine does a star remain? *Accretion of mass by stars moving through the interstellar medium must in general be very much less than loss of mass by radiation*, estimated Eddington.[179] If stars do not mix and do not accrete mass,

[176] Jeans, J.H., MNRAS **85**, 199 (1925).

[177] Eddington, A.S., *The Internal Constitution of the Stars*, Dover Publ. (1930) p. 392.

[178] Vogt, H., Zeit. f. Phys. **26**, 139.

[179] Eddington, A.S., Obs. **49**, 193 (1926). Bakerian Lecture of the Royal Society.

CONTENTS

CONTENTS

Fig. 4.14 *Left*: The table of contents of Jeans' book entitled *Astronomy and Cosmogony* (1928). Separate chapters are devoted to liquid and gaseous stars. *Right*: The table of contents of Eddington's book entitled *The Internal Constitution of the Stars* (1926). There is no mention of liquid stars

then their surface composition is pristine, and provides evidence of the original composition. If on the other hand, the stars were fully mixed, one would expect to see the results of element synthesis on the surface. As Eddington's bad luck would have it, the stars are not mixed.

In 1926, Eddington published a summary in the form of a seminal book entitled *The Internal Constitution of the Stars*. The interesting points were:

- there was no mention of liquid stars (see Fig. 4.14), and
- he gave an extensive discussion of the major problems of molecular weight, absorption coefficient, and energy sources.

Of course, the role played by radiation in stars dominates the book.

4.46 A Devastating Argument

A severe blow to practically all energy generation hypotheses in gaseous stars was inflicted by Jeans in 1927,[180] when he proved a mathematical theorem that the stability of a gaseous star requires the energy mechanism to be insensitive to temperature. Actually, the first paper Jeans wrote about stability[181] contained an incorrect energy

[180] Jeans, J.H., MNRAS **87**, 400 (1927).

[181] Jeans, J.H., MNRAS **85**, 914 (1925).

equation. The error was pointed out by Vogt.[182] After correcting for the error,[183] Jeans' basic conclusion remained unchanged.

A star is supposed to be in an energy balance when the total energy produced is equal to the energy emitted into space. If for some reason the star generates more energy than it can emit, the extra energy will stay in the star and cause it first to heat up, and later to expand, in an attempt to lower the energy production. Hence, if the energy production does not decrease upon expansion of the star, an unstable situation arises whereby the star expands in an attempt to quench the energy production, but fails to overcome the accelerating energy production. On the other hand, if expansion of the star affects the rate of energy production and the production of energy decreases very quickly, then clearly the star has somehow managed to overcome the attempt of the energy production mechanism to escalate, and the star is stable. In summary, the nuclear reactions must be more sensitive to the temperature than anything else in the star. In particular, they must be more sensitive than the energy removal processes, like radiative transfer, for the star to be stable.

Jeans carried out a mathematical analysis and derived a mathematical condition for the stability of a gaseous star. At the end of the analysis, he invoked a simple physical argument for stability. This physical argument is almost identical to the one given in the last paragraph. However, the condition derived from the physical argument was the opposite of the condition derived with the extended mathematics. What was stable mathematically was unstable physically, and vice versa. In a footnote, Jeans explained that he had got the physical condition wrong in a previous paper[184] (the sign was wrong) and that Russell[185] and Eddington had also got the wrong sign. It is truly puzzling how Jeans himself did not notice the contradiction between the physical and mathematical conditions.

The paper was published, but somehow the mathematical error got lost in the many approximations and heavy applied mathematics, and nobody noticed it. Scientists used the 'accurate' but incorrect mathematical condition to rule out any energy mechanism! Maybe it just looked more rigorous than the simple and straightforward physical logic. After several years, during which the argument was regularly used to eliminate good ideas, it was simply ignored in the discussion about the energy mechanism of stars. About ten years later, the condition was corrected by Cowling (see later). Here we shall just note that, before this analysis was carried out, Jeans had shown that a gaseous star is stable only due to deviations from the ideal gas, which implied that Eddington's simple basic model was unstable! Looking back many years later, one is compelled to ask how come so many were misled and for so long? We can only guess that it was probably because of Jeans' scientific eminence that nobody dared to expose the blunder. However, that would be a poor excuse.

The application of the wrong theorem had a devastating effect for a good many years. The hypothesis of subatomic energy, as well as the possibility of mass annihilation, were ruled out because of the theorem. One consequence was that only liquid stars could possess subatomic energy, for example. Thus, Eddington's results for the

[182] Vogt, H., AN **232**, 5545 (1928).

[183] Jeans, J.H., MNRAS **88**, 393 (1928).

[184] Jeans, J.H., MNRAS **85**, 923 (1925).

[185] Russell, H.N., MNRAS **85**, 928 (1925).

Sun, the mass–luminosity relation, the theory of the main sequence, in short the entire theory, had to be dismissed. The amazing fact was that the theorem contradicted physical logic and was obviously wrong.

About two years later, Gerasimovič[186] showed that, if one includes the changes in ionization of the gas during the perturbed state, then all stars are stable, and Jeans' argument does not apply. However, ionization of elements is important only near the surface of the star, and hence in a very small region, so that the argument does not really eliminate the validity of Jeans' theorem.

As Jeans had concluded that practically all mechanisms for energy generation in gaseous stars were unstable, he revived the argument in favor of liquid stars, and even ventured the following new argument. About half the stars are in binary systems. A long list of mathematicians and astrophysicists, notably Poincaré and Darwin, had investigated the formation of a binary system via fission. Back in 1917, Jeans[187] had shown that fission, the splitting of a given star into two stars, could take place only in stars whose interior was incompressible, namely, liquid stars. As about one-third of the stars in the sky are binaries, *which have almost certainly been formed by fission*, Jeans concluded that this *direct evidence of observational astronomy pronounces in favor of the stars being liquid rather than gaseous structures.* While the mathematics was correct, we would argue the other way round today. Since fission requires the stars to be liquids, and they are not, this means that we have to look for another mechanism for the formation of binaries. Jeans pointed out in a footnote that Milne[188] had claimed to prove that gaseous rotating stars of high mass would behave like a liquid. But, claimed Jeans, Milne's treatment was invalid because of incorrect assumptions. A similar criticism of Milne's result was expressed by Vogt.[189]

The problem of the stability of gaseous stars was resolved by Cowling (1906–1990).[190] He overlooked the fact that Jeans had conflicting conditions and dismissed Jeans' criteria as *not very accurate*.[191] The accurate condition is that the nuclear reactions must be very temperature sensitive to secure the stability of the star. What Cowling effectively did was to show that the high temperature sensitivity of the nuclear reactions in the cores of stars leads to convective cores, and then Jeans' stability condition (Jeans assumed radiative transfer) became irrelevant. We note that, although in 1935, when Cowling wrote his important stability paper, the details of the subatomic energy were not known, the general temperature dependence of nuclear reactions in stars was in fact known (proportional to a high power of the temperature and to the density).

[186] Gerasimovič, B.P., Astronomy **15**, 347 (1929).

[187] Jeans, J.H., Phil. Trans. **218**, 209 (1917). Bakerian Lecture.

[188] Milne, E.A., MNRAS **83**, 141 (1923).

[189] Vogt, H., AN **229**, No. 5480, 125 (1927).

[190] Cowling, T.G., MNRAS **94**, 768; MNRAS **96**, 42 (1935).

[191] An examination of Jeans' lengthy and almost impenetrable mathematics yields that Jeans' basic error was in the simplification of the form of the perturbation used to check the stability. Cowling was right in stating the 'inaccuracy' of the derivation, but the perplexing issue is the blindness of the author, and subsequently the many who applied his mathematical stability criteria, to the erroneous mathematical stability condition, which contradicted his own physical argument.

4.47 Reaching the Limits of Knowledge

By 1927, the problem of the disagreement between the theoretical and observed phase relation of the light and velocity curves in Cepheids was still unsolved. Eddington[192] admitted to having no idea how the problem could be cured. In Eddington's language, the results were hostile to his hypothesis, and he admitted that there must be something missing from the theory. At this stage, Eddington confessed that he dropped the problem, and it remained unresolved until years later, when massive computer simulations including the best available approximations for the absorption coefficient became available.

What else could go wrong? Eddington[193] reexamined the state of gaseous stars in view of recent progress in the theory of electrolytes, in particular Debye and Hückel's recent theory which had already been applied to stellar matter by Rosseland[194] and by Fowler and Guggenheim.[195] This time Eddington treated the Debye–Hückel theory correctly, and did not include the gravitational force in the discussion of the electric force. Consequently, using this theory, he vindicated his assumptions. It should be mentioned that there was a discussion in the physical literature regarding the extent to which this theory was correct for solutions, and several alternative theories were proposed. Years later, it became clear that the Debye–Hückel theory is indeed the correct one for electrolytes and stars.

Another fundamental question regarded the extent to which ions tend to cluster and form crystals or solids. If the answer was in the affirmative, that they did tend to do so, then clearly the stars were liquids with all the implications for Eddington's theory. However, investigation returned the negative answer, according to which these effects could not produce liquid cores in stars:

> *Will the stellar gas crystallize?* asked Eddington, and replied: *It is not contemplated that the whole star would form a crystal, but single elements or possibly alloys might form crystals which would no doubt have a certain optimum size.*

The physical question is essentially whether a crystal has more or less energy than the ions and electrons. Eddington noted that his analysis was good only for stars with high Z elements, and was not valid for stars composed mainly of hydrogen. But the effects were totally negligible in stars with large amounts of hydrogen. Years later, it would become clear that Eddington's analysis was relevant to white dwarfs,[196] and he essentially proved that the ideal gas assumption for gaseous stars (and the Sun) was excellent.[197]

---

[192] Eddington, A.S., MNRAS **87**, 539 (1927).

[193] Eddington, A.S., MNRAS **88**, 352 (1928).

[194] Rosseland, S., MNRAS **84**, 720 (1924).

[195] Fowler, R.H., & Guggenheim, E.A., MNRAS **85**, 939, 961 (1925).

[196] Kovetz, A., & Shaviv, G., A. & A. **8**, 398 (1970).

[197] In the case of white dwarfs, the ions behave as an ideal gas in hot white dwarfs. The electrons follow the equations for a Fermi–Dirac gas. As the white dwarf cools, the ions arrange themselves in special crystals, while the electrons remain uniformly distributed, because of the high pressure. See next chapter.

4.48 Russell Again. Discovery of the Most Abundant Element in the Solar Atmosphere

Recall that Russell's statement about the solar composition, made almost a decade earlier, had been based on superficial evidence, so this time in 1929, he returned to analyze, in a very long paper (72 pages),[198] the composition of the solar atmosphere. The paper was groundbreaking, and showed how Russell combined his expertise as a spectroscopist and as an astronomer when he laid down some of the basic principles of the spectroscopic analysis of stellar spectra. By analyzing the binding energy of electrons in atoms, he was able to explain why the spectral lines of certain elements were observed, while the others were not. He found that only elements in which the binding energy of the electron was less than 5 eV were observed, apart from hydrogen, which holds its electron more strongly. Fifty six elements were identified in the solar atmosphere and six compounds. The most astounding result was that six elements, Na, Mg, Si, K, Ca, and Fe, contribute 95% of the whole mass. The mean atomic weight was found to be 32. *The abundance of the non-metals, and especially hydrogen, is difficult to estimate from the few lines which are available* claimed/concluded Russell. The surface composition of the Sun, as found by Russell, is given in Table 4.1. The most dramatic result was the preponderance of hydrogen (by volume, rather than by mass). In Russell's words: *The calculated abundance of hydrogen in the Sun's atmosphere is almost incredibly great.*

This was the first observational indication that something was wrong with the composition that a 'liquid Sun' would have implied. Actually, the molecular weight seemed to be the lowest possible. In particular, Russell discussed the *significance of the 'absences'*. In other words, he considered what could be concluded if a certain element was not observed. Did that mean that it did not exist? This part would turn out to be the most important, because most of the matter in the Sun, which is in the form of hydrogen and helium, is not observed![199]

Russell stressed that the temperature of the solar atmosphere was not high enough to ionize certain elements, and that was why they were not seen. Russell realized that, at the surface temperature of the Sun, the kinetic energy of the atoms is about 1/2 eV. Consequently, atoms which hold their electrons with energy of the order of 1/2 eV or less would lose their electrons in collisions between the atoms. On the other hand, atoms which hold their electrons with energy much greater than 1/2 eV would not be affected and would not show any spectral lines, whence they would not be observed! Along the main sequence, as the surface temperature of the stars rises from about 2 000 K to about 60 000 K, one expects to see different elements in stars with different surface temperatures, and this nicely explains the different spectral lines and compositions observed in stars with different spectral classes/surface temperatures. The chemical elements along the periodic table hold their electrons with different energies, and this is the reason for their different chemical properties,

[198] Russell, H.N., Ap. J. **70**, 11 (1929).

[199] Helium is only observed during eclipses, when the corona of the Sun becomes visible. Hydrogen shows weak diffuse lines which are not sufficient to determine its abundance.

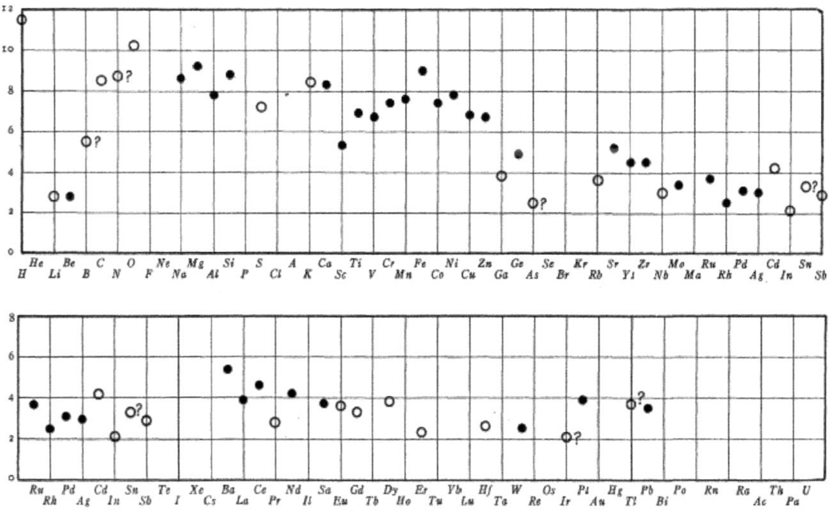

FIG. 3.—Values of log Q, where Q represents the total mass of the atoms or molecules of an element per unit area of the sun's surface.

Fig. 4.15 The relative abundances of the elements, showing that elements with even atomic weight are more abundant than elements with odd atomic weight. From Russell 1929

and the reason why stars with different surface temperatures show spectral lines of different elements. In particular, if two stars of different surface temperatures show different spectral lines of elements, it does not mean that the stars have different compositions. The surface temperature is determined by the mass of the star which has no direct connection to the composition.

Russell discussed limits on the abundances of elements not observed in the Sun, the most important of which was helium. Figure 4.15 shows the abundances of the elements as a function of atomic number. The astonishing thing here was that, as a rule, the even atomic number elements are more abundant than the odd atomic number elements. Russell hypothesized that: *The abundance of an element is probably a function of yet unknown properties of the structure of the atomic nucleus.* Years later this hypothesis would be proven correct. An independent confirmation of this fact was made by Elsasser a couple of years later.[200]

Russell confirmed the results obtained before by Miss Payne,[201] who used a method devised by Milne. Maybe because hydrogen is rare on the Earth, Payne assumed that the intensity of the hydrogen lines on the Sun was caused by some abnormal behavior, and was not a result of high abundance.

[200] Elsasser, W., Nature **131**, 764 (1933).

[201] Payne, C., *Stellar Atmospheres* (Harvard Observatory Monographs, No. 1) Cambridge, Mass. (1925) p. 184.

Table 4.1 Abundances of the elements in the Sun

Element	By volume	By mass
Hydrogen	60 parts	60
Helium	2 ?	8
Oxygen	2	32
Metals	1	32
Free electrons	0.8	0
Total	65.8	132

The first solid results of high hydrogen abundance in the Sun were obtained by Unsöld[202] three years later, using his own method for analyzing stellar spectra. Unsöld's results were substantiated by McCrea[203] who used measurements of hydrogen lines appearing in the upper part of the solar atmosphere (the chromosphere),[204] where the temperature is higher than in the photosphere. Despite earlier results about the hydrogen abundance in the solar atmosphere, Russell was very careful and discussed in detail the extent to which the huge amount of hydrogen was real or a consequence of uncertainties in the calculations, because he derived the startling hydrogen abundance from what he did not see.

Next Russell compared the abundances in the Sun with those of meteorites and the crust of the Earth, and concluded that the abundances of the metals in the Sun most closely resemble the abundances in meteorites, rather than those in the crust of the Earth. Russell added the conclusion (now known to be wrong) that:

> *It is probable that the Earth and the meteorites were formed by condensation from matter ejected from the Sun, as first suggested by Chamberlain and Moulton.*

The ejected matter would have been hot, so that all volatile elements would have been lost from the final condensed Earth (and meteorites).

The art of deriving the abundances from the spectra and the theory of stellar atmospheres was in its early days and far from perfect. As a consequence, slightly different numbers were found for the various abundances by applying different methods. However, the basic conclusion that the mostly unseen elements in the solar

[202] Unsöld, A., Zeit. f. Phys. **46**, 778 (1928).

[203] McCrea, W.H., MNRAS **89**, 483 (1929).

[204] The chromosphere is a thin layer of the solar atmosphere just above the photosphere. The chromosphere is more visually transparent than the photosphere. The name comes from the fact that it has a reddish hue, as the visual spectrum of the chromosphere is dominated by the deep red H_α spectral line of hydrogen. The chromosphere is hotter than the photosphere. It is heated by small shock waves propagating from the convection zone of the Sun. The photosphere, from where most of the solar radiation seen by us emerges, has a temperature between 4 000 and 6 400 K, but the chromosphere has a temperature as high as 20 000 K. The Sun does not have different layers, and the temperature runs smoothly. At first the temperature decreases outward and then, when the surface becomes transparent, starts to heat up, to reach temperatures of about 2 million degrees. Since different emission emerges from different parts depending on the temperature, it is customary to attribute different names to different regions.

spectra, namely, hydrogen and helium, were in fact the most abundant in stellar atmospheres was universally accepted.

4.49 Two Birds in a Single Shot

The discrepancy between the derived absorption coefficient for all stars (about a factor of 10, assuming the Eddington model to be correct) defied all attempts to resolve it. Various researchers tried to solve the problem by accounting for different effects in atomic physics[205] but to no avail. The real solution came with Gaunt[206] who noticed an inconsistency in Kramers' calculation of the absorption coefficient for stars.[207] As a result, a correction factor had to be introduced, the Guillotine factor, which cuts the absorption coefficient (who else save Eddington could have invented such a name?). With the assumed composition of stars at this time, Eddington found that: *It may be stated at once that the observational evidence does not support these factors.*

Starting from about 1930, Bent Strömgren (1908–1987) entered the game by investigating various stellar models. In his first publication,[208] he already argued that all the explanations given so far were wrong: Sugiura's[209] calculations overestimated the absorption coefficient by a factor of 8 or 9, and a long explanation was given as to why Biermann[210] was wrong. Jeans[211] had adopted a new equation of state that led to incompressibility (which meant that the stars were liquids), but this was *pure empiry without connection with present theoretical physics*, claimed Strömgren. He then argued that recent results on the stellar atmosphere by McCrea,[212] Russell,[213] and Unsöld[214] showed that the amount of hydrogen in the atmospheres of stars is 30% by mass and above.

After extensive calculations with stellar models, Strömgren reached the breakthrough conclusion that:

---

205 Sugiura [Sugiura, Y., J. de. Phys. **8**, 113 (1927)], Nishina and Rabi [Nishina, Y., & Rabi., J., Verh. d. D., Phys. Ges. **9**, 8 (1928)], Reiche [Reiche, F., Zeit. f. Apstrophys. **53**, 168 (1929)], Stobbe [Stobbe, M., Ann. d. Phys. **7**, 661 (1930)], Sauter [Sauter, F., Ann. d. Phys. **9**, 217 (1931)], Hall and Oppenheimer [Hall, H., & Oppenheimer, J.R., Phys. Rev. **38**, 57 (1931)], Roess and Kennard [Roess, L.C., & Kennard, E.H., Phys. Rev. **38** (1931)] to name but a few of the attempts.

206 Gaunt, J.A., Proc. Roy. Soc. A **126**, 654 (1930).

207 It had to do with taking into account the fact that the atoms lose their electrons in the ionization process, and hence are unavailable for absorption of radiation.

208 Strömgren, B., Zeit. f. Astrophys. **4**, 118 (1932).

209 Sugiura, Y., Sci. Pap. Inst. Phys. Chem. Res. No. 339 (1931).

210 Biermann, L., Zeit. f. Astrophys. **3**, 116 (1931).

211 Jeans, J.H., *Astronomy and Cosmogony*, Cambridge University Press (1928).

212 McCrea, W.H., MNRAS **89**, 483 (1929).

213 Russell, H.N., Ap. J. **70**, 11 (1929).

214 Unsöld, A., Zeit. f. Physik **46**, 765 (1928).

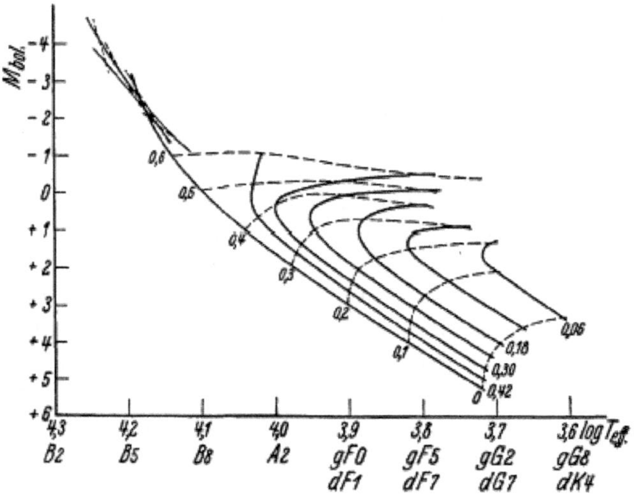

Fig. 4.16 The Hertzsprung–Russell diagram according to Strömgren, where the hydrogen abundance affects the exact position in the diagram. The *continuous lines* are lines of constant hydrogen, while the *broken lines* are lines of constant stellar mass. We see that the main sequence corresponds to a line of high constant hydrogen abundance of about 42% by mass. The x axis is the logarithm of the effective temperature and the spectral type, and the y axis is the logarithm of the luminosity of the star

> *It is seen that the hydrogen abundance [throughout the star] is roughly constant for the three stars considered.*

The value was about 33% by mass. Thus the amount of hydrogen in stars was much higher than had been previously thought. Moreover, the amount of hydrogen correlated with the position in the Russell diagram,[215] as can be seen from Fig. 4.16. This is a very interesting figure, which appeared about twenty years too early. If you consider a constant mass, then you see that, as the hydrogen abundance decreases, the star moves to top right, i.e., towards larger radii. In the early 1950s, such a graph was used to interpret the evolution of stars off the main sequence.

The large amount of hydrogen in the star was not accepted right away. What prevented the acceptance of such high hydrogen abundances in stars was the 'common knowledge' that large amounts of hydrogen lead to high radiation pressure, and this was assumed to cause instability in the star, even for modest amounts of hydrogen.

A short and very interesting abstract was read by Menzel[216] at the Pasadena meeting of the American Astronomical Society in June 1931. In this abstract, Menzel announced that the factor of 10 discrepancy in the absorption coefficient in stel-

[215] Strömgren did not call it the Russell–Hertzsprung diagram. This was, however, corrected in the forthcoming papers.

[216] Menzel, D.H., PASP **43**, 358 (1931).

lar interiors could be removed if one assumed the star to consist almost entirely of hydrogen. The agreement was achieved mainly by lowering the molecular weight (now that the star had become pure hydrogen), which in turn reduced the central temperature. As Menzel remarked:

> *In view of the probable predominance of hydrogen in stellar atmospheres, there seems to be some justification for the assumption that a star consists almost entirely of the element, though it has usually been tacitly assumed that the observed abundance was almost entirely a surface phenomenon, arising from the low atomic weight and the natural tendency of hydrogen to float to the top of the atmosphere.*

Menzel had discovered that the most abundant element in the universe is hydrogen, although he hesitated to state this explicitly, and pointed to the misconception that what we see on the surface of a star may not represent the entire star.

Menzel's important discovery went unnoticed by Eddington, and in a paper entitled *The Hydrogen Content of the Stars*, Eddington[217] returned to the open problem, with the admission:

> *When I first found the discrepancy I showed that the calculated and observed luminosities would agree with the standard star Capella, if the material contained 20% of hydrogen.*[218] *But I did not at the time think that the abundance of hydrogen was the actual explanation of the discrepancy.*

At this time, the discrepancy in the absorption coefficient stood at a factor of 10 for massive stars and a larger factor for the Sun and low mass stars. Eddington admitted that, when he had found the discrepancy 8 years earlier and triggered the debate on the validity of his models of gaseous stars, he did not believe that hydrogen could be the solution. What caused the change in Eddington's views? Just that all other explanations failed. Improvement in stellar modeling removed questions like how well the stellar material was mixed. Hence, Eddington suggested using the position in the HR diagram to determine the hydrogen content in stars. Eddington's idea was not new, as Russell and others had applied it before.

It was about twelve years after Eddington had had the idea of hydrogen synthesis in stars, when the possibility that dwarf stars should contain a lot of hydrogen probably crossed his mind with the appreciation that it solves the absorption coefficient problem. But it took several years before Eddington, Strömgren, and Russell realized that stars contain so much hydrogen, and in this way discovered that hydrogen is the most abundant element in the universe.

Moreover, when asked to explain his results at the meeting of the Royal Astronomical Society,[219] Eddington made the following point. The luminosity of the star is extremely sensitive to the hydrogen content. A change in the hydrogen content from 0 to about 80% changes the predicted luminosity of the star by a factor of about 600. But we know that the mass–luminosity relation depends on the hydrogen content and is sensitive to it, and hence it turns out that:

[217] Eddington, A.S., MNRAS **92**, 471 (1932).
[218] Eddington, A.S., MNRAS **84**, 114 (1924).
[219] Eddington, A.S., The Obs. **55**, 125 (1932).

- The hydrogen content in stars hardly varies from one star to another. For unknown reasons, all stars contain about the same amount of hydrogen.
- Hydrogen, so discovered Eddington, is the most abundant element in the Cosmos.

This result had a profound impact, and implications, not all of which had yet been spelled out. For example, one of these was that the star should get energy by building up from the abundant hydrogen, and not by breaking down rare heavy atoms. The original matter out of which stars form must have been uniform all over the Cosmos. Eddington told the audience how Strömgren[220] had communicated his results to him, and that the two independent results agreed with one another. Thus, Strömgren and Eddington solved the absorption coefficient controversy and discovered the preponderance of hydrogen in stars, and consequently in the Universe as a whole.

It is interesting to note that, in contrast to what was observed in all other stars, a very low amount of hydrogen was found in white dwarfs, and Chandrasekhar noted that, if a molecular weight of 2 was assumed, then no hydrogen at all could have existed in white dwarfs.[221] No interpretation of this fact was given.

4.50 The Hydrogen in the Sun

Of particular interest was the Sun. Calculations showed that the amount of hydrogen in stars had two solutions (see Fig. 4.17), that is, two possible abundances of hydrogen were consistent with the observed luminosity. One solution was that the Sun was made mostly of hydrogen, about 99.5%, with traces of other elements:

> *The other solution, which rightly or wrongly I have assumed to be the more probable, gives approximately 33% hydrogen in the Sun, Capella, Algol and Krueger 60.*

What we see in Fig. 4.17 is a violation of the simply formulated Russell–Vogt theorem, which would state that there is only one composition for which the Sun would have just the solar luminosity.

A few months before the publication of the paper, the neutron was discovered. Eddington did not know about the properties of neutrons, and considered it to be a new element. Consequently, he showed that, even if it did exist in stars as an element, it would not change his conclusions about the hydrogen content.

This time it appeared that the discoveries of Strömgren and Eddington were accepted by the scientific community. Russell[222] hailed this breakthrough:

> *We have already seen that the interior of the star probably contains about 30% of hydrogen by weight. In the outer parts of the star a large fraction of the remainder is atoms of carbon, nitrogen, and oxygen.*

220 Strömgren, B., Zeit. f. Astr. **4**, 118 (1932).

221 Chandrasekhar, S., Zeit. f. Astrophysik. **3**, 302 (1931).

222 Russell, H.N., JRASC **27**, 411 (1933). First Maiben Lecture before the American Association for the Advancement of Science, given at Atlantic City on 3 December 1933.

Similarly, the mass–luminosity law for gaseous stars was accepted. No word about liquid stars appeared in the literature after this discovery. After Jeans' 1928 stability discussion, the idea of liquid stars had practically disappeared from the literature.[223]

The recognition that the light elements are the most abundant, while the heavy elements are rare diminished support for the idea of radioactive elements being the energy source. Russell discussed the possibility that more massive radioactive elements might exist in stars and supply the energy, claiming it to be very speculative. (He did not know about trans-uranium elements and why the periodic table ends, but this lack of information did not change his conclusion.) He was ready to accept matter annihilation, but claimed (correctly) that it requires much higher temperatures than those we think exist in stars (calculated from models which reproduce the observed luminosity). By now he knew that there were neutrons, and that the nuclei of atoms are composed of protons and neutrons. Since more progress had been made in the discussion of how heavier atoms can be built up out of hydrogen, his favorite theory for the energy source became the construction of heavier and heavier nuclei by absorption of protons.

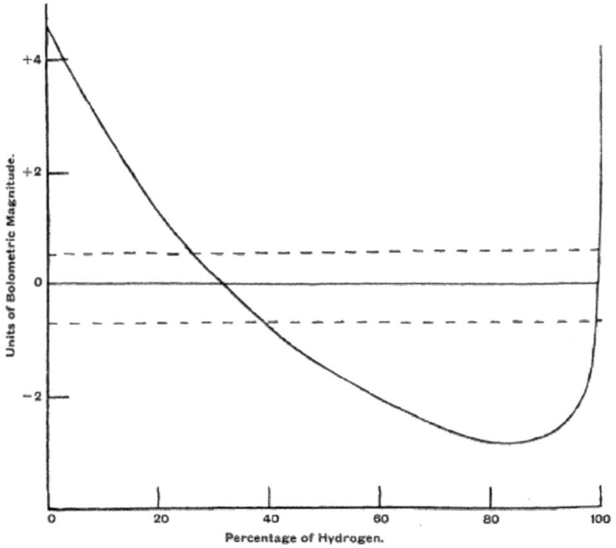

Fig. 4.17 The predicted luminosity of the Sun as a function of the assumed hydrogen content (Eddington 1932). Of the two solutions, Eddington preferred the one with lower hydrogen abundance

[223] A comment is due. One finds the claim in various places that Cecilia Payne discovered that hydrogen is the most abundant element in the Universe. In fact, Payne discovered that hydrogen is the most abundant element on the surface of the Sun (and stars). What Strömgren and Eddington showed was that hydrogen is the most abundant element throughout the Sun and stars. The difference is fundamental.

Russell calculated what should be the temperature of the core of the Sun to produce the observed luminosity. (He assumed that the core of the Sun is just the innermost 10% of the mass.) The result was 15 million degrees, amazingly close to the present day estimate of 15.5 million degrees. As more massive stars have a higher luminosity, he estimated that it is sufficient to increase the temperature to 40 million degrees to generate the energy produced in the most massive known stars. In doing so, he confirmed Eddington's hypothesis that the main sequence is the locus of all stars converting light elements into heavier ones.

The bells were ringing out for the revolutionary conclusion that hydrogen is the most abundant element in stars (and the Universe), but nobody heard them. Maybe because it was not a clean sweep.

4.51 First Signs of Finer Details

In spite of the Russell–Vogt theorem, the sensitive eye of Struve discovered that the spectra shown by stars were not fully covered by the standard Harvard system. In 1933, Struve[224] discussed the extent to which the spectrum of a normal star was fully described by the temperature and the pressure, or in other words, the extent to which two parameters might be sufficient to describe the spectrum uniquely. *If the pressure and temperature of two stars are identical, is it clear that the spectra are the same?* By posing such questions, it became clear that there were additional factors in the making of a star, which were not included in the theory. However, these were considered as being of secondary importance and largely ignored.

By 1933, the following facts became accepted by the community:

- The stars are gaseous.
- The Jeans stability condition could be safely ignored.
- The mass–luminosity relation is valid both theoretically and observationally.
- The most abundant element in the Universe is hydrogen, and next in abundance is helium. All stars have about the same composition.
- Mass annihilation as a source of energy for stars was losing support. On the other hand, subatomic energy, starting from hydrogen and building up, was gaining support.
- The main sequence is the locus of stars using the same energy source and at roughly the same phase of their evolution.
- Stars are thoroughly mixed. If the subatomic energy is generated deep within the cores of stars, we expect to see the signs at the surface.

[224] Struve, O., Ap. J. **78**, 735 (1933).

4.52 The Cowling Model

While the Eddington model was investigated and implemented extensively, Cowling decided to deviate from the by then classical assumptions of Eddington's model. In particular, Cowling, as a student of Milne, was unhappy with Eddington's assumption about the distribution of the energy production in stars.[225]

Cowling[226] decided to check the other extreme, that is, he imposed the assumption that all the energy is generated at the center of the star. In this sense the model is the opposite to Eddington's. The extremely large energy generation in a very small region causes a breakdown of the radiative transfer that Eddington worked so hard to introduce. The extreme energy released cannot be removed by radiation alone, and the energy transfer becomes unavoidably unstable against convection. But convection means that mass currents can carry the energy. When you heat water in a kettle using a very low burner, you find that the water remains calm until almost boiling. The hot bottom heats the water and the heat is transferred in the water by conduction. However, if you turn the burner to maximum heat, the energy flux is too high for conduction to carry it away. Then energy transfer by conduction breaks down and turbulence develops rather early, so that the energy is carried by turbulent motion: convection. Cowling reasoned that a point source, in which all the energy is generated in a very small volume, necessarily leads to a breakdown of radiative transfer and establishes a region where the energy is transported outward by convective currents. Mathematically, it means that the star possesses two different zones: an inner zone where the energy is transported by mass currents, and an external zone where the energy is transported to the surface by radiation, from where the energy is radiated into space.

Furthermore, Cowling assumed a uniform composition throughout the star. After completing the model, Cowling realized that it was impossible to deduce any mass–luminosity law from such a model.[227] As the mass–luminosity relation was very important, the model was not widely favoured in its early years. However, when Bethe discovered the exact form of the energy release via the CN cycle, the Cowling model became very popular, at the expense of Eddington's model.

---

[225] Eddington's assumption was as follows. Define a function η which is given by $L_r/M_r = \eta L/M$, where L_r is the luminosity produced within radius r, and M_r is the mass out to that radius. L and M are the total luminosity and mass of the star, respectively. Obviously, η varies from unity on the surface of the star to some unknown *but not very large value at the center. The form of* η *depends on the unknown law of liberation of subatomic energy.* If k is the radiative absorption coefficient, Eddington assumed that ηk was constant throughout the star. (See Eddington, A.S., *The Internal Constitution of the Stars*, Chap. 6.)

[226] Cowling, T.G., MNRAS **91**, 92, 472 (1930).

[227] The reason is that the convective currents are so powerful in carrying the energy that, no matter what the energy flux is, the temperature slope is practically the same. This is in contrast to a radiative energy flux, which depends on the temperature slope.

Chapter 5
From Chemistry to Dying Stars

5.1 The Problem with the Existence of Different Atoms

With all the new atomic physics, it was not clear why there are chemical elements and why they differ from one another in their chemical properties. And neither was it clear why there is a periodic table. The principle of minimal energy claims that all systems tend to the lowest energy state. Hence in a system with many electrons, all electrons tend to the lowest energy state and, if allowed, all electrons will end up in the same ground state energy. Hence, all elements would behave in a very similar way. But this is not the case in Nature. There is a reason why there is a limit to the capacity of energy states, so that in a system of many electrons, not all the electrons find their way to the lowest energy state.

5.2 The Road to the Pauli Principle. The Multi-Electron Atom

A year after Bohr published his successful model, which accurately predicted all the observed spectral line series of hydrogen, Rydberg[1] discovered that the total number of possible electronic states in the $n's$ principal quantum state is $2n^2$, and that all states are filled in the noble gases. The factor of 2 in front was not understood at all. Somehow, the electrons came in pairs, or there was some symmetry which provided a place for just two electrons on each site. Rydberg however, did not provide any explanation as to why this should be so.

---

[1] Rydberg, J., Phil. Mag. **28**, 144 (1914).

G. Shaviv, *The Life of Stars*, DOI 10.1007/978-3-642-02088-9_5,

5.3 The Chemists Have It Differently

In an attempt to explain how atoms combine with one another to form compounds, Parson (1889–1970)[2] came up with the idea that the electrons in the atoms were arranged with cubic symmetry and that the electrons were at rest at the corners of the cube. The chemists did not seem to accept Bohr's model, which required perpetual motion of the electron. Lewis (1875–1946m),[3] for example, suggested what he called the cubical atom, which was an extension of Parson's model (see Fig. 5.1):

> *The atom is composed of the kernel and an outer atom or shell, which, in the case of the neutral atom, contains negative electrons equal in number to the excess of positive charges in the kernel, but the number of electrons in the shell may vary during chemical change between 0 and 8.*

The fact that there are at most 8 electrons in the last electronic level triggered the idea of electrons in the corners of a cube. Moreover, Lewis hypothesized that:

> *The atom tends to hold an even number of electrons in the shell, and especially to hold eight electrons which are normally arranged symmetrically at the eight corners of a cube.*

But how can the electrons 'stay put' in the corners of a cube? To overcome this problem, Lewis assumed that:

> *Electric forces between particles which are very close together do not obey the simple law of inverse square which holds at greater distances.*

Actually, Lewis hypothesized that the electrical attraction turns into repulsion at short distances. If Lewis was right, it meant that Bohr's assumption of the Coulomb force governing the atomic structure was wrong. As a matter of fact, Lewis argued with Bohr, claiming that his theory was simpler, because Bohr's ad hoc assumption of quantum theory, namely that the energy must be quantized, was not required by him (Lewis) to explain the spectral lines. However, Lewis did not calculate what his theory predicted the spectral lines should be.

Lewis was not alone in considering the idea of cubical atoms. Similar models were developed by Kossel (1888–1956)[4] and Langmuir (1881–1957m).[5] Kossel was a physicist but his theory of the molecular bond made him a favorite among chemists.

Landé (1888–1975)[6] did not accept that the electrons at the corners of the cube were at rest, and developed a mathematical theory for the motion of eight electrons distributed with cubic symmetry in atoms. Still the puzzle was the supposedly simpler helium atom, which has just two electrons and is more stable than all other

---

[2] Parson, A.L., Smithsonian Inst. Publ. Miscel. Collections **65**, no. 11 (1915).

[3] Lewis, G.N., **35**, 72 (1916).

[4] Kossel, W., Ann. Physik **49**, 229 (1916).

[5] Langmuir, I., **41**, 868 (1919); see also Proc. Nat. Acad. Sci. V, 252 (1919).

[6] Landé, A, Zeit. f. Physik **2**, 83, 380 (1920).

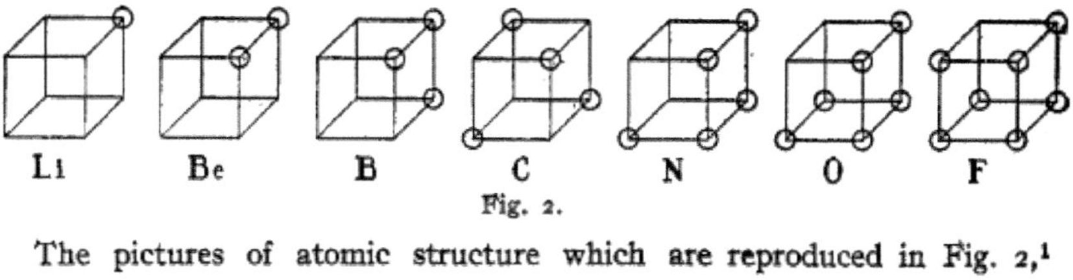

Fig. 5.1 Lewis' 1916 picture of the outer electronic shell of the atom. The figure is from Lewis' paper. The fact that the last electronic shell contains 8 electrons is explained by assuming they are fixed at the corners of a cube

atoms. Landé postulated that the two electrons move in coupled orbits without giving any reason or mechanism to explain why these states should exist, work, and be stable.

Langmuir[7] developed the 'octet theory of valence' by assuming that the radii of the consecutive electronic shells varies as $1 : 2 : 3 : 4 : \ldots$, so that the area varies as $1^2 : 2^2 : 3^2 : 4^2 : \ldots$. He then postulated that each shell is divided into equally sized cellular spaces, so that the first shell contains 2×1^2 locations, the second cell contains 2×2^2, and the nth shell contains $2 \times n^2$ cells. Hence the number of possible cells varied as the area of the shell. Each cell contained at most two electrons. Why only two electrons could reside in the same shell was not explained, although Langmuir suspected that there must be some hidden reason.[8] In a way, Langmuir had already noticed that:

> *Every pair of electrons is in a cell. There cannot be more electrons than cells.* He went on to assume that: *Electrons contained in the same cell are nearly without effect on each other, but the electrons in the outside layer tend to line themselves up.*

Most importantly, Langmuir identified the noble gases as the atoms with 'closed' shells, i.e., shells with no room for any further electron, and hence the most stable atoms. Because of the fact that the outermost shell never contains more than 8 electrons, Langmuir called the theory the octet model. The rest of the very long paper (67 pages long) was devoted to explaining how the bond between atoms took place and how molecules formed.

As Langmuir wrote:

> *The remarkable stability of the pair and the octet is not explainable on the basis of Bohr's theory. [...] The electrons in atoms are coupled together in a rather complex manner, which seems quite inconsistent with the ordinary properties of the electron.*

But it was not explainable by his theory either. The electrons, claimed Langmuir[9] were in their most stable positions and moved only within certain limited regions or

---

[7] Langmuir, I., The octet theory, J. Amer. Chemical Soc. **41**, 868 (1919).

[8] Langmuir referred to the reason as a hidden two-fold symmetry.

[9] Langmuir, I., The octet theory, J. Amer. Chemical Soc. **41**, 868 (1919).

'cells' within the atom. The motion of each electron was restricted à la Langmuir to its cell. Langmuir won the 1932 Nobel Prize for Chemistry for his research on surface phenomena. The octet model, though published in many papers, was not mentioned in his Nobel lecture (1932).

5.4 The Helium Atom

Even the simplest atom after hydrogen, the helium atom, posed problems. Indeed, Bohr himself tried to extend his hydrogen model to helium but failed.[10] Bohr's model for helium predicted an ionization energy of 28.8 eV, while the experimental value is 25.5 ± 0.25 eV.[11] This difference should be contrasted with the highly accurate prediction of the hydrogen ionization energy. Sommerfeld himself[12] stated that his model did not agree with experiment.

In 1921, Langmuir[13] discussed the helium atom. He explained that Bohr's atomic model was unsatisfactory because it predicted too large an ionization potential and did not agree with the optical and magnetic properties of helium. So he considered two models (see Fig. 5.2). In the first model, the two electrons were assumed to move in *two separate parallel circular orbits*. This model was found by Langmuir to

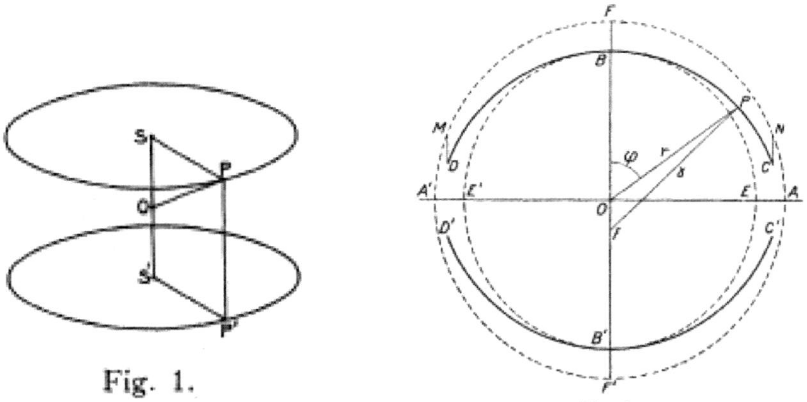

Fig. 1.
Double-circle model for the helium atom.

Fig. 2.
Oscillating or semi-circular model for the helium atom.

Fig. 5.2 Langmuir's two models for helium in 1921, as a demonstration of how scientists imagined atoms with small numbers of electrons a few years after the birth of Bohr's quantum theory

[10] Bohr, N., Phil. Mag. **26**, 488 (1913).

[11] Frank, J., & Knipping, P., Phys. Zeit. **20**, 481 (1919).

[12] Sommerfeld, A., *Atombau*, 1st edn., Braunschweig (1919) p. 70.

[13] Langmuir, A., Phys. Rev. **17**, 339 (1921).

be unstable: one has to add energy to the atom to prevent the electron from escaping. In the second model, *each electron is assumed to oscillate back and forth along an approximately semi-circular path in accord with classical mechanics.* The resulting ionizing energy agreed well with the experimental value.

Even the chemists did not accept Langmuir's model, and criticism of the Langmuir theory was expressed by Bury (1890–1968),[14] who pointed out several inconsistencies in the octet theory. In particular, Bury rejected Langmuir's postulate that more than two electrons cannot share a cell. As a matter of fact, he repudiated the idea of electron cells, and instead offered an electronic configuration which is actually the one accepted today, in order to explain the periodic table.

5.5 The Physicists' Approach. Was It Any Better?

Bohr's semi-classical theory of the atom was a theory of a single electron. Arnold Sommerfeld (1868–1951m) immediately generalized the Bohr theory to include the effects of the special theory of relativity. In particular, the circular orbits assumed by Bohr were generalized (because of relativistic effects) to elliptical orbits (like the planetary orbits). In 1918,[15] Sommerfeld published a model of the multi-electron atom which combined Nicholson's idea of a ring of electrons with the elliptical orbits. The model, which was called 'ellipsenverein' (see Fig. 5.3), claimed that the motion of the electrons in their respective elliptical orbits is correlated in such a way that the symmetry of the ring is preserved. Moreover, the symmetry is preserved irrespective of the number of electrons in the ring.

While Sommerfeld's theory gained support from experimentalists, it had problems in predicting the X-ray spectra Moseley and others had obtained. The basic problem was how to treat the innermost electrons, and to what extent they shield the outer electrons (completely or partially) from the positive charge in the nucleus. The inability to calculate the shielding correctly meant that parameters had to be introduced, whose values could be tuned to obtain agreement with experiment. Various prescriptions were proposed,[16] but the patching up method could not continue forever,[17] as more and more results had to be explained by the resulting theory. Various attempts were made, such as assuming the atom to be three-dimensional, in which case the ring idea with all its advantages from the standpoint of radiation losses broke down.[18] Actually, the collapse of the planar atom came when Born and Landé[19] demonstrated that crystals made of flat atoms would not have the compressibility properties observed experimentally.

---

[14] Bury, C.R., J. Am. Chem. Soc. **43**, 1602 (1921).

[15] Sommerfeld,A., Phys. Z. **19**, 297 (1918).

[16] Debye, P., Phys. Z. **18**, 276 (1917); Vegard, L., Phil. Mag. **35**, 293 (1918).

[17] Reiche, F., & Smekal, A., Ann. Phys. **57**, 124 (1918).

[18] Smekal, A., Phys. Z. **22**, 400 (1921).

[19] Born, M., & Landé, A., Verh. dt. Phys. Ges. **20**, 202 (1918).

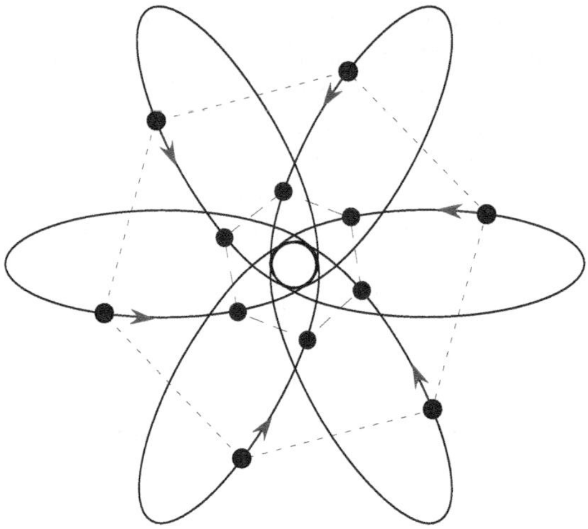

Fig. 5.3 Sommerfeld's 1920 'ellipsenverein' model

Bohr's theory, which explained the different series of spectral lines, led researchers in X-ray spectroscopy to concentrate on identifying the energy levels where the electrons reside. Bohr's basic idea was that, whenever an electron jumps from a high level to the lower levels, it gives rise to a series of spectral lines. Consequently, an extensive search was launched for the energy levels and resulting spectral lines. It was not long before Sommerfeld,[20] and independently Kossel,[21] produced a picture in which the main energy levels were designated by K, L, M, N, ..., with the electrons of the heavy elements occupying these levels. The position of each level was found by spectroscopy, and not from theory. The patchy theory described above was unable to predict the wavelengths of the spectral lines. The amazing thing was that no one had yet raised the question as to why the electrons should fill the levels, and why the electrons do not all drop down to the lowest level. Instead, they were asking only how they fill the energy levels.

It is interesting to point out that Sommerfeld, considered to be the high priest of German physics, had a unique position in science and was extremely influential. Sommerfeld had 28 students and practically all rose to become dominant figures in physics.[22] Consequently, his model, though not very helpful, was treated with great respect.

[20] Sommerfeld, A., Zeit. f. Phys. **1**, 135 (1920).

[21] Kossel, W., Zeit. f. Phys. **1**, 119 (1920).

[22] His students were in order of graduation: Debye, Hopf, March, Lenz, Ewald, Lande, Epstein, Herzfeld, Lang, Burmeister, Fues, Pauli, Wentzel, Heisenberg, Guillemin, Heitler, Unsöld, Bethe, Thüring, Fröhlich, Urban, Franz, Welker, Seebach, Waldmann, and Apfelbacher.

5.6 Bohr and the Periodic Table

It took Bohr 9 years to formulate his idea of how the periodic table is arranged,[23] and to establish the 'aufbau' principle,[24] specifying how the electrons are arranged in atoms. In the Nature paper, Bohr discussed the difficulties with the previous pictures, and explained why basic modifications of the theory should be introduced. As a matter of fact, Bohr was looking for:

> *[...] configurations and motions of the electrons which would seem to offer an interpretation of the variations of the chemical properties of the elements with the atomic number as they are so clearly exhibited in the well-known periodic table.*

Bohr criticized all theories which assume that the electrons are in groups placed at equal angular intervals (like Sommerfeld's theory):

> *All such theories involve, however, the fundamental difficulty that no interpretation is given why these configurations actually appear during the formation of the atom through a process binding the electrons to the nucleus, and why the constitution of the atom is essentially stable in the sense that the original configuration is reorganized if it should be temporarily disturbed by external agencies.*

Bohr pointed to the difficulty that an electron in an external group may have an orbit which brings it closer to the nucleus than an electron from an inner 'group', whence the electrons from different groups may seem to penetrate each other. This is so because, once Sommerfeld's relativistic corrections to Bohr's theory had been introduced, the possible orbits of the outer electrons included very elliptical ones, so elliptical that the minimum distance to the nucleus was shorter than the radius of the circular orbit. Bohr's main contribution in this context, was the aufbau principle, which states that, in the transition from one element to another, the new electron enters an unoccupied state with the lowest energy. Because each shell, which is characterized by a principal quantum number n, can have several substates and orbital quantum numbers, there are cases, such as the $n = 4$ case, where the next n includes a few states lower in energy than the previous n levels. As a consequence the electrons do not fill the $n = 3$ level first and then proceed to the $n = 4$ level, but fill first the $n = 4$ states that are lower in energy, and only when these levels are full do they continue to enter the $n = 3$ states (see Fig. 5.4).

However, the aufbau principle did not explain why the new electron enters into a new level and does not descend to the lowest state. In other words, it did not explain why the presence of an electron in a lower state prevents the next electron from entering the same level. Did quantum levels have finite capacity? After all, a classical system can contain any number of particles with the same energy.

---

[23] Bohr, N., Zeit. f. Phys. **9**, 1 (1922); Nature (March 1921).

[24] The German term 'aufbauprinzip' means the 'building-up principle'.

Tabelle 1. Ursprüngliches Bohrsches Schema der Edelgaskonfiguration.

Element	Atom-Nr.	Anzahl der n_k-Elektronen														
		1_1	2_1	2_2	3_1	3_2	3_3	4_1	4_2	4_3	4_4	5_1	5_2	5_3	6_1	6_2
Helium	2	2	—	—	—	—	—	—	—	—	—	—	—	—	—	—
Neon	10	2	4	4	—	—	—	—	—	—	—	—	—	—	—	—
Argon	18	2	4	4	4	4	—	—	—	—	—	—	—	—	—	—
Krypton	36	2	4	4	6	6	6	4	4	—	—	—	—	—	—	—
Xenon	54	2	4	4	6	6	6	6	6	6	—	4	4	—	—	—
En.anation	86	2	4	4	6	6	6	8	8	8	8	6	6	6	4	4

Fig. 5.4 The arrangement of the electrons in shells, as found by Stoner in 1924 and confirmed by Pauli

5.7 Stoner. Getting Closer

The correct electron arrangement in atoms was found by Edmund C. Stoner (1899–1968)[25] in 1924. On the basis of optical spectra, Stoner attempted to find the arrangement of the electrons in the various levels. Stoner reviewed previous attempts to find the distribution of the electrons, and showed that none of the proposed schemes actually worked:[26]

> *It is remarkable*, stated Stoner, *that the number of electrons in each complete level is equal to double the sum of the inner quantum numbers as assigned.*

The electrons appeared to come in pairs which occupy the same quantum states. Stoner's distribution of electrons was the one we know today, and as Stoner had already shown, it explained the chemical and physical properties in the periodic table. In this distribution, the electrons come in pairs, and no more than two ever occupy the same quantum state. However, Stoner went one step further, and characterized the states of the electrons by two numbers,[27] the first being identical to Bohr's principal quantum number n, and the second taking values from 0 to $n - 1$. Stoner noticed that each electron has another l value.

5.8 The Pauli Principle

One of the strangest physical principles, and yet crucially important for the structure of stars and atoms, was discovered by Wolfgang Pauli (1900–1958m) in a series of papers which culminated in the discovery and formulation of what is known

[25] Stoner, E.C., Phil. Mag. **48**, 719 (1924). The paper was communicated to the Phil. Mag. by R.H. Fowler.

[26] Bohr, N., Zeit. f. Phys. **9**, 1 (1920); Landé, A., Zeit. f. Phys. **16**, 391 (1922): ibid. **24**, 88 (1924); ibid. **25**, 96 (1924); Bohr, N., & Coster, D., Zeit. f. Phys. **12**, 342 (1923); de Broglie, L., & Danvillier, A., Jour. de Phys. **6**, 1 (1924).

[27] Stoner and Pauli used the notation k_1 and k_2. Years later these numbers became n and l.

today as the Pauli exclusion principle (hereafter PEP). While in Munich,[28] Pauli had long discussions with Sommerfeld about the meaning of the series of 2, 8, 18, 32, ..., for the numbers of electrons in the atomic shells, as discovered by Rydberg. Pauli's interest in the problem arose in 1922 when he met Bohr for the first time. Bohr lectured in Göttingen on his new theory to explain the periodic system of the elements.

As soon as Bohr had come up with his model of the multi-electron atom, the burning question was: Why do the electrons in the atom not all fall down to the lowest energy level? Or again, why are they not attracted by the nucleus with the result that they simply fall into it? As a matter of fact, Bohr had already discussed this problem, but could not find a satisfactory solution. A hint as to what was going on came when a strong magnetic field was applied to the atom. So far, it was known that all electrons in a given shell possess the same energy. However, when a magnetic field was applied to the atom, the various substates within each shell got different energies. Very soon Pauli realized that electrons immersed in a strong magnetic field have different quantum numbers, and still do not descend to a lower state. Yet he had no idea why this should be so.

In 1923, Pauli returned to the University of Hamburg. The lecture he gave to obtain the title of 'privatedozent' was on the periodic system of the elements, with the disappointing conclusion that the problem of closed electronic shells had no explanation. The only thing that was clear was the connection with the multiplet structure of the energy levels. According to a popular notion at the time, the belief was that non-vanishing angular momentum had something to do with the doublet splitting. But it was just a guess.

In 1924, Pauli published some arguments against this point of view, in particular the idea that a quantum number can have just two values. At that time, the following key remark by Stoner was published:

> *For a given value of the principal quantum number the number of energy levels of a single electron in the alkali metal spectra in an external magnetic field is the same as the number of electrons in the closed shell of the rare gases which corresponds to this principal quantum number.*

According to Pauli, it was this comment by Stoner which led him to the idea that:

> *The complicated numbers of electrons in closed subgroups are reduced to the simple number one, if the division of the group by giving the values of the four quantum numbers of an electron is carried so far that every degeneracy is removed.*[29] *An entirely non-degenerate energy level is already closed, if it is occupied by a single electron. States in contradiction with this postulate have to be excluded.*

---

[28] Pauli, W., Nobel lecture, 1945.

[29] Meaning that no two electrons have the same 4 quantum numbers.

The general principle was finally formulated in Hamburg in the spring of 1925.[30] In simple terms, in a given system of many electrons, no two electrons can have the same quantum numbers. However, this statement did not reflect the exact truth.

Only three quantum numbers[31] were known for the electron in the atom at the time Pauli announced the principle, and these numbers were not sufficient for the principle because they led to the following rather strange formulation of the principle: *no more than two electrons can occupy the same state*. Pauli therefore followed Stoner, and introduced a fourth quantum number (m_1 in his notation), which could assume just two values, and in this way each electron was identified by four quantum numbers. Once the idea of four quantum numbers was in the air, the principle became that two electrons cannot occupy the same state, i.e., no two electrons can have four identical quantum numbers.

The idea of the exclusion principle was hard to swallow for many reasons. For example, how could one electron know about the state of the other? There was no classical analogue to the idea. The Pauli exclusion principle is a genuine quantum phenomenon. And what about the meaning of the fourth, apparently arbitrary, quantum number, hypothesized by Pauli in order to obtain a nice version of the principle?

Shortly after the publication of Pauli's paper containing the PEP, Uhlenbeck (1900–1988) and Goudsmit (1902–1978)[32] realized that the anomalous Zeeman effect becomes simple to understand if one assumes that the electron has an additional quantum number with dimensions of angular momentum and with a value of 1/2 in atomic units. The basic idea of Uhlenbeck and Goudsmit was that the newly invented quantum number is associated with the intrinsic spin of the electron. How a point particle could have intrinsic spin was another strange problem. At first, even the extraordinarily innovative Pauli had strong doubts about the intrinsic angular momentum picture of Uhlenbeck and Goudsmit. He only became an adamant supporter of the spin idea after Thomas' calculation,[33] in which he adopted the Uhlenbeck and Goudsmit idea of intrinsic angular momentum, yielded the correct magnitude for doublet splitting in a magnetic field. The belief in the connection between the idea that the electron has an intrinsic angular momentum, as if it were rotating about its own axis, and the observation that the level was split into two levels, was strengthened when Bohr demonstrated that the spin of the electron cannot be measured by classically describable experiments, and must therefore be considered as an essentially quantum mechanical property of the electron.

---

[30] Pauli, W., Zeit. f. Phys. **31**, 765 (1925). The title of the paper, viz., *On the Connexion between the Completion of the Electron Group in an Atom and the Complex Structure of Spectra*, does not give even the slightest hint of the extremely important and far-reaching new principle described there.

[31] Three quantum numbers were associated with the dynamic properties of the electron, viz., the energy, and two of the three components of the angular momentum. The newly invented quantum number did not have any *obvious dynamic property to quantize*. It came as an invention to *simplify the formulation of the principle*. However, it had dramatic consequences.

[32] Uhlenbeck, G.E., & Goudsmit, S., Naturwissenchaften **47**, 953 (1925); Nature **117**, 264 (1926).

[33] Thomas, W., Zeit. f. Phys. **34**, 586 (1925).

The history of the fourth quantum number of the electron was quite acrimonious. In 1921, Landé[34] introduced a factor g, the so-called Landé factor, as a correction for the energy of the electron in a magnetic field. This hand-waving correction was introduced to get around the problem that it was impossible to describe the energy of the electron with only three quantum numbers. The half-valued quantum number was also introduced by Heisenberg.[35] It seems that the idea of a new two-valued quantum number had crossed Heisenberg's mind earlier, but that it had been rejected by Sommerfeld.[36]

At the beginning of 1925, Ralph Kronig (1904–1955) suggested that the fourth quantum number might have something to do with the self-rotation of the electron. Pauli, in his frequently rejecting mood, dismissed the idea on the grounds that the rotation of the electron would have to be so fast that it would violate the special theory of relativity. Consequently, Kronig refrained from publishing his idea.[37] Finally, Uhlenbeck and Goudsmit, following the advice of their mentor Paul Ehrenfest (1880–1933), published the idea in a journal of secondary importance. This time the idea was accepted, mainly because Thomas had succeeded in explaining a complicated discrepancy between Uhlenbeck and Goudsmit and the unpublished Kronig calculation on the one hand, and experimental results on the other. A year later, Uhlenbeck and Goudsmit published the discovery in Nature.

It was in 1927 that Pauli accepted the idea of the self-spin of the electron and formalized the theory. As late as 1940, Pauli finally established the connection between the spin of elementary particles and their statistical properties (see later). The spin quantum number was difficult to accept, because it was the first quantum number which arose from observation but not from theory, and had no classical property associated with it. Much later, further quantum numbers were introduced without any classical analogue to describe elementary particles.

The two principles, the aufbau and the Pauli principles, were sufficient to explain the variation in the chemical properties and most physical properties across the periodic table, but not some details of levels containing two electrons. The discussion of these fine details involves two additional rules due to Hund.[38] The Hund rules discuss the order of the energy levels, that is, they specify when a level with a higher quantum number has an energy lower than the energy of a level with a smaller quantum number. Unlike the previous two principles, which are obeyed without exception, there are exceptions to the Hund rules, in particular in heavy elements.

The PEP has incredibly far-reaching consequences in most disciplines. But let us note the rather long time it took the Nobel committee to reach the conclusion that this amazing and fundamental principle deserved the recognition of the Nobel Prize. Admittedly, it was no easy matter to accept such an imaginative principle.

---

[34] Landé, A., Zeit. f. Phys. **5**, 231 (1921).

[35] Heisenberg, W., Zeit. f. Phys. **8**, 231 (1921).

[36] Pais, A., *Inward Bound. Matter and Forces in the Physical World*, Clarendon Press, Oxford (1986).

[37] Tomonaga, S., *The Story of Spin*, University of Chicago Press (1997).

[38] See Herzberg, G., *Atomic Spectra and Atomic Structure*, Dover Publ., p. 135; Hund, F., Zeit. f. Phys. **33**, 345 (1925); ibid. **34**, 296 (1925).

When discussing Pauli's contributions to physics, it is impossible to refrain from commenting on his personality. Pauli had a perfectly unparalleled way of expressing himself that knew no mercy. He was well known for using the expression *it is not even a nonsense*, and he honestly admitted that: *I have indeed mistakenly considered something right to be wrong, but never considered something wrong to be right.*

5.9 The Unique Behavior of Particles

Prior to the announcement by Pauli of his principle, S.N. Bose (1894–1974) attempted to derive the Planck distribution from a new set of statistical assumptions. Attempts to derive the Planck distribution of photon energies by statistical methods were carried out by Debye in 1910.[39] On the other hand, Einstein[40] derived the Planck distribution of photon energies emitted by a black body by assuming equilibrium between matter and radiation. Against this background, Bose[41] came up with a new idea. The basic new concept was that, in contrast with electrons, many massless photons could occupy the same energy state. The paper was published before Pauli announced his principle for electrons. After Bose succeeded in reproducing the Planck distribution with the above assumption, he sent the paper to the English Philosophical Magazine, which rejected it. He then sent the paper to Einstein in Berlin. Einstein realized the implications of the paper, translated it, and sent it to the prestigious German Zeitschrift für Physik journal[42] for publication. And indeed, with Einstein's recommendation, the paper was published.

At the end of the three page paper, it was noted that the translator was Einstein himself. Next came a short note by the translator saying (in German) that:

> *The derivation of the Planck formula by Bose is an important advance. The method here can be used to apply the quantum theory to ideal gases, as I intend to show elsewhere.*

The second paper by Bose was received by Einstein just 5 days later and was nine pages long.[43] Einstein also translated this paper. However, Einstein added a page at the end of the paper, in which he asserted that the basic assumptions applied by Bose in the derivation of Planck's law were not applicable to the case discussed in the paper, and he provided two arguments. (The first argument had to do with the classical limit and the second with behavior at low temperatures.)

About a year later, Einstein[44] had apparently changed his mind. He extended Bose's analysis to particles with mass and predicted the condensation of such a gas, i.e., at very low temperatures, all particles descend to the lowest energy level, a

[39] Debye, P., Ann. de Phys. **33**, 1427 (1910).

[40] Einstein, A., Phys. ZS **18**, 121 (1917).

[41] Bose, S.N., Zeit. f. Phys. **26**, 178 (1924). Submitted 2 July 1924. The paper was reprinted in English by the J. Astrophys. Astr. **15**, 3 (1994).

[42] Bose, Zeit. f. Phys. **27**, 178 (1924).

[43] Bose, S.N., Zeit. f. Phys. **27**, 384 (7 July 1924).

[44] Einstein, A., Sitzungsberichte der Preussischen Akademie der Wissenschaften, 8 January 1925.

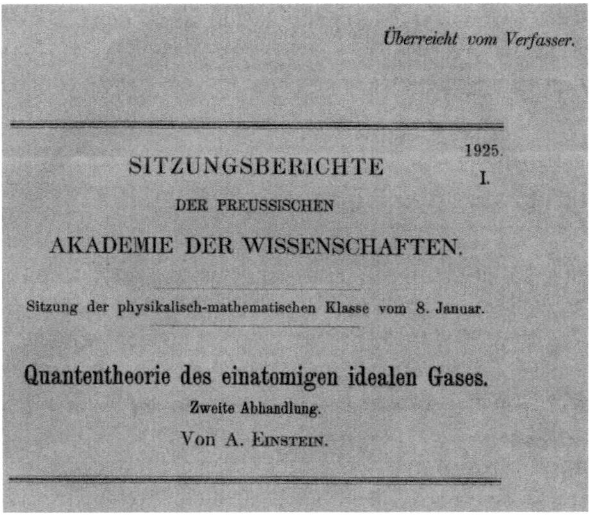

Überreicht vom Verfasser.

SITZUNGSBERICHTE 1925. I.

DER PREUSSISCHEN

AKADEMIE DER WISSENSCHAFTEN.

Sitzung der physikalisch-mathematischen Klasse vom 8. Januar.

Quantentheorie des einatomigen idealen Gases.

Zweite Abhandlung.

Von A. Einstein.

Fig. 5.5 The front page of Einstein's report to the Prussian academy on the extension of Bose statistics to particles of finite mass, and the prediction of condensation

phenomenon only demonstrated experimentally in 1995.[45] Einstein used the term unsaturated gas (ungesättigten gas). Today the phenomenon is called Bose–Einstein condensation, although Bose did not predict it. Some people claim that this was Einstein's last great discovery. In the same paper, Einstein tried to apply his new theory to electrons in metals, only to discover that it does not provide an explanation.

Interestingly, the two papers by the indian Bose published in a prestigious German scientific paper gave rise to some dispute in the German physical society.[46] As is well known, Philipp Lenard (1862–1947) who wrote a book on *Deutsche Physic* (German Physics) and Johannes Stark (1874–1957) were supporters of *Aryan physics* long before the Nazi party came to power. As early as 1915, Stark[47] had labeled Arnold Sommerfeld as the *energetic executive secretary of the Jewish and philo-Semitic circle of mathematicians and theoretical physicists*. In 1924, Max von Laue wrote to Sommerfeld and Wien, then the chairman of the German Physical Society, and Scheel, the secretary of the society and the editor of the journal, that the paper by the indian Bose had given rise to an argument over the extent to which this German journal should be open to non-Germans. The editor Scheel admitted

[45] Anderson, M.H., Ensher, J.R., Matthews, M.R., Weiman, C.E., & Cornell, E.A., Sci. **269**, 198 (1995).

[46] Singh, R., Current Science **81**, 1489 (2001).

[47] Forman, P., & Hermann, A., in *Dictionary of Scientific Biography*, Scribner and Sons Pub. New York (1975) p. 525.

that publishing Bose's paper was a mistake. My guess is that this might have been the reason why Einstein agreed to translate and send for publication the second and longer paper, about which he had some reservations concerning its correctness. To pacify his professional conscience he added his two critical comments.[48]

5.10 The Behavior of a Collection of Particles

Enrico Fermi (1901–1954m) was bothered by the fact that the ideal gas equations, in particular the expression for the heat capacity at constant volume, did not satisfy the law due to Nernst (1864–1941m),[49] according to which absolute zero temperature cannot be reached in a finite number of steps. When Fermi saw the papers by Stoner and Pauli, he set out to determine to what extent the application of the new principle to the molecules of an ideal gas would yield an expression that did satisfy Nernst's general principle. Interestingly, in Fermi's paper, there was no mention of electrons, the particles to which the Pauli principle applied. Eventually, Fermi[50] discovered what is known today as the Fermi–Dirac[51] statistics. Fermi and Dirac immediately grasped the far-reaching implications of the PEP for gases of particles which obey it, like electrons. With his extraordinary physical intuition, Fermi derived his results directly, while Dirac with his superb mathematical skills derived the general theory of the behavior of quantum particles, and derived both Fermi's result (which he apparently did not know about) and the Bose–Einstein result as special cases of his general theory.

Fermi ended his paper by showing that, with his new theory in which the quantum correction was included, the equations for the ideal gas did in fact satisfy the laws of thermodynamics, and in particular the phenomenological Stern–Terode formula for the entropy. Classical physics was not sufficient to explain the consistency of ideal gases with thermodynamics. So, while searching for geese, Fermi found swans!

The discoveries of Fermi and Dirac caused two revolutions. Elementary particles, like protons or electrons, are considered in classical physics to be different from one another. Each molecule or atom or electron has its own identity, and it is possible to distinguish between two atoms of the same element or two electrons. In quantum

[48] The opening statement of Einstein's addendum was: *Ich halte Boses Hypothese über die Wahrscheinlichkeit der Strahlungselementarvorgänge aus forgenden Gründen für nicht zutreffend.*

[49] From purely thermodynamic considerations, Nernst found that one cannot reach the absolute zero of temperature in a finite number of steps, and that the minimum of the entropy occurred at absolute zero. Fermi had discussed the question in previous papers, but the problem remained. The discovery of this law, which is called the third law of thermodynamics, won Nernst the 1920 Nobel Prize for Physics.

[50] Fermi, E., Rend. Acc. Lincei, Ser. 6, **3**, 145 (1926); Zeit. f. Phys. **36**, 902 (1926). Paper submitted to the journal on 24 March 1926.

[51] Dirac, P.A.M., Proc. Roy. Soc. A **112**, 661 (1926). The paper was submitted to the journal on 26 August 1926. Dirac (1902–1984) made no reference to Fermi's paper (in Italian), which had preceded his by just 5 months, but there was a reference to Bose's and Einstein's papers (in German).

theory, all particles of a given type are alike and one cannot distinguish between them. This is a crucial difference between classical and quantum physics, and it affects the way the particles are distributed. In an atom, an electron in a high energy state does not descend to a lower state if an identical electron already occupies this level. It would descend if a different particle occupied the level. Or, in Dirac's own example:

> *Denote by (mn) the state of the atom in which one electron is in the state labeled m and the other in the state n. The question arises whether the two states (mn) and (nm) which are physically indistinguishable [...] are two different states or one?*

And Dirac concluded that the two alternatives (mn) and (nm) *count as only one state*. In classical physics the two states are counted as two different states. As simple and trivial as it sounds, this fact had a profound impact on the final result. Using this fundamental inability to distinguish between particles, Dirac showed that it leads either to particles which satisfy the Pauli principle or to particles which can occupy the same state.[52] Quantum theory does not allow other alternatives.

Why should the counting of states be so crucial? The fundamental assumption of statistical mechanics is that all possible states of a gas are equally probable. So if (nm) and (mn) are the same state, the probability of both would be p. But if these states are different, the probability would be $2p$.

The theory of metals and the theory of a gas of electrons, for which the new Fermi–Dirac statistics had overwhelming ramifications, was not mentioned by either Fermi or Dirac. It was Sommerfeld who applied the new statistics to the theory of metals, and introduced the idea that the free electrons in a metal constitute a Fermi gas.

The Pauli principle was first verified for the electron in atoms. Fermi and Dirac separately generalized the Pauli principle to any system. Furthermore, it follows from Dirac's formalism that particles must obey either Bose–Einstein statistics or Fermi–Dirac statistics.

5.11 Why the Difference in Behavior? Why Does PEP Apply?

Is there a more basic reason for the peculiar behavior of particles? Elementary particles like electrons, protons, neutrons, or even atoms composed of all the above, can have spin. However, the angular momentum (the mass times the radius times the velocity) can take only integer values or half-integer values of a certain basic quantity $\hbar/2\pi$. So electrons, protons, and neutrons have a spin of 1/2, while photons have spin 1. The helium atom has spin 0. In 1940, while in the USA, Pauli published[53] one of his most brilliant papers, in which he proved that the demands of the special theory of relativity and quantum mechanics dictate the behavior of particles. Pauli

---

[52] At about the same time, this very same result was also obtained by Heisenberg [Heisenberg, W., Zeit. f. Phys. **38**, 411 (1926)].

[53] Pauli, W., Phys. Rev. **58**, 716 (1940).

found that just the requirement that the energy of free particles (namely the kinetic energy) should be positive compels particles with half-integer spin to obey Fermi–Dirac statistics and PEP, while a similar more delicate condition compels particles with integer spin to obey Bose–Einstein statistics. In Pauli's opinion:[54]

> *The connection between spin and statistics is one of the most important applications of the special theory of relativity theory.*

5.12 Why Is There Chemistry? Why Does Matter Have Its Bulky Form?

The classical and quantum picture of the electron and proton is of a charged point mass particle. As early as 1911, Jeans[55] had found a problem with the assumptions of a point particle and the Coulomb force. He reasoned that if the Coulomb force behaves as one over the distance squared, then point mass particles would experience an infinite force as they approached one another. Finding this result implausible, he assumed that the Larmor model of the atom,[56] which requires a modification of the Coulomb law at short distances, solved this problem. But we know that this model had many other problems and consequently was rejected. Moreover, Rutherford's experiment proved that the Coulomb law holds at least down to a distance of 10^{-11} cm.

So why does matter have its finite size and not simply collapse to a point? What prevents this? The question of the source for the stability of matter was not asked again for about 15 years, until it was raised again by Ehrenfest in 1931.[57] Ehrenfest also asked why atoms are so big? The question was asked at about the time the same problem was discussed and solved in astrophysics (see later).

After Ehrenfest and Jeans had raised the above question, it was abandoned for about 35 years, until it was revisited by Fisher and Ruelle.[58] However, the final answer was given by Dyson and Lenard.[59] If the Pauli principle had not existed, all electrons in all atoms would fall to the lowest level and the energies of these electrons would have been much greater than in the present case, because the lowest level is closer to the nucleus. There would be no difference between the elements except for increasing ionization energy, which would vary as the square of the electric

[54] Pauli, W., ibid. conclusions.

[55] Jeans, J.H., *The Mathematical Theory of Electricity and Magnetism*, Cambridge Press, 2nd edn. (1911) p. 168.

[56] Larmor, J.J., *Aether and Matter*, Cambridge Press 1900. Larmor suggested that a molecule consists of rings of fast-moving electrons. However, the model did not solve the problems it was designed for, and actually created additional ones.

[57] Dyson, F., J. Math. Phys. **8**, 1538 (1967) cites Ehrenfest, P., *Collected Scientific Papers*, Klein, Ed. North-Holland Pub. (1959) p. 617.

[58] Fisher, M.E., & Ruelle, D., J. Math. Phys. **7**, 260 (1966).

[59] Dyson, F.J., & Lenard, A., J. Math. Phys. **8**, 423 (1967).

charge of the nucleus. There would be no periodicity in the sequence of the chemical elements, and the chemical reactions would require more and more energy, the heavier the element is.

Chemical reactions with iron, for example, would need energies in the keV to form molecules like iron oxides or hemoglobin. It is due to the Pauli principle that, as Z increases, the electrons in the last shell are further and further away from the more and more highly charged nucleus. This situation proceeds in such a way as to keep the ionization energy almost constant (relative to the enormous change in the binding of the lowest energy state, which varies by a factor of 10 000 between hydrogen and uranium). So to a large extent, the higher charge of the nucleus is compensated by driving the valence electrons, those that participate in the chemical reactions, further away from the nucleus. The result is that the binding energies of the valence electrons are generally less then a few eV. One can say that Bose–Einstein condensation is exactly the phenomenon that would have occurred if the electrons did not obey PEP. All electrons would descend to the lowest level.

If the energy of a system cannot decrease below a certain limit, we say that the system is stable against collapse. The first proof that a collection of electrons and protons is stable was given in 1967 by Dyson and Lenard. Lieb[60] extended the theorems by Dyson and Lenard to include more cases. However, the essence remained the same, namely, that it is the PEP which secures the existence of matter as we know it today.

In general, the binding energy E_b responsible for holding the system together must be smaller than the rest mass energy, and in most cases even much smaller ($E_b \ll E$). Hence, from special relativity, we know that the absolute minimum of the binding energy is the rest mass energy. If so, what happens in systems not bound by PEP?

This is exactly what happens in gravitation. According to the law of gravity, any mass which does not experience a counter- pressure can collapse to a point. The Schwarzschild solution to Einstein's equations of general relativity, already discovered in 1916, shows that, before the mass shrinks to a point, it forms a black hole of radius $R = 2GM/c^2$, and all information stops coming out of the mass. As there is no PEP or an equivalent principle for gravity, there is nothing to prevent a star from contracting forever. The existence of PEP for electrons implies that, up to a certain stellar mass, one can expect the electrons to be able to provide enough pressure to resist gravity, but beyond that mass, nothing can help and gravity will win in the end. The massive star will continue to shrink unless something happens to prevent it from collapsing.

The consequences of the relentless attraction of the gravitational force can be observed in globular clusters (see Fig. 5.6). These are giant objects which contain around a million stars. There are about 150 globular clusters in our galaxy, and they move around the plane of the galaxy. At the same time, each star moves under the mutual gravitational attraction of all the other stars in the cluster. From time to time, a star 'evaporates', i.e., it escapes from the cluster and, as a consequence, the core

---

[60] Lieb, E.H., Phys. Rev. **48**, 553 (1976).

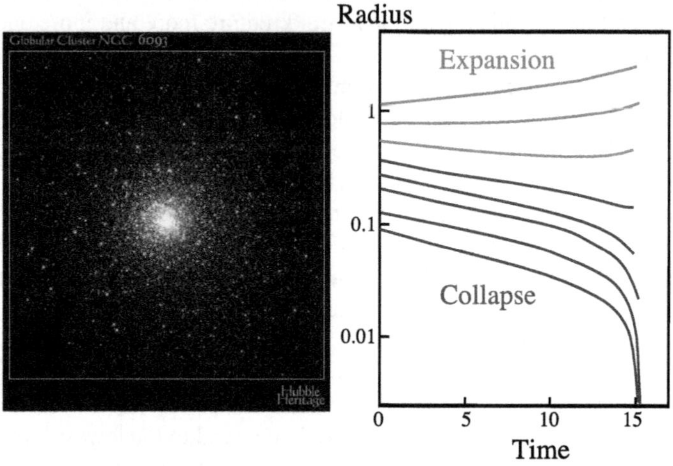

Fig. 5.6 *Left*: Globular cluster number 6093 in the New General Catalogue. (Credit, NASA, Hubble Heritage). *Right*: The result of a simulation of the evolution of a globular cluster with time. As time goes by, the central parts collapse and the radius contracts to vanishing values. At the same time the envelope expands. The results are adapted from Joshi, K.J., Rasio, F.A., Ap. J. **540**, 969 (2000)

of the cluster contracts and the envelope expands. The process continues until the density of stars in the center reaches such high values that the stars destroy each other, or collapse to form the object whose existence Schwarzschild had predicted, namely, a black hole. Indeed, black holes have been discovered in the center of several globular clusters. (For a review, see for example Kormendy et al. 1995.[61])

5.13 Observational Limits on the Pauli Principle

No new phenomenon which may lead us to doubt the absolute validity of the PEP has yet been discovered. On the other hand, various upper limits to possible violations, or perhaps it would be better to say deviations, from the universality of the principle have been set in various experiments. Because of the dramatic consequences that a violation of PEP would have, it should be easy to discover even minute deviations. One of the most recent results[62] set an upper limit for the probability of violation of the principle at 4.5×10^{-28} for electrons in an atom (of copper).

[61] Kormendy, J., & Richstone, D., ARA&A **33**, 581 (1995).

[62] Bartalucci et al. The VIP Collaboration, `quant-ph/0605047 v1` (4 May 2006).

5.14 Eddington's White Dwarf Paradox

Eddington[63] pointed out to a paradoxical situation in his famous book. A star is an object in a balance between the gravitational pull inward and the gas pressure acting outward. As the star contracts, the gravitational pull increases, and as a consequence the temperature and the density of the gas must increase so as to counterbalance the increase in the gravitational pressure. At the same time, the star continues to lose energy from the surface. How can this be? Part of the gravitational energy goes into heating the gas (to create the counterbalance) and the rest is radiated away. So stars are unique objects in that they lose energy all their life and as a consequence heat up! And conversely, stars cannot cool! As Eddington pointed out, to die by cooling, the star must lower its temperature and hence reduce its gas pressure, and in order to stay balanced must decrease the gravitational pull, which it can do only by expansion or by having some extra source of energy which nobody had yet thought of. In Eddington's words:

> *We can scarcely credit the star with sufficient foresight to retain more than 90% in reserve for the difficulty awaiting it. [...] Imagine a body continually losing heat but with insufficient energy to grow cold!*

The paradox shook the scientific community, because it implied that stars cannot die by cooling and must heat up forever!

5.15 Cracking the Paradox: Ralph Fowler 1926

Ralph Howard Fowler (1889–1944m)[64] was a leading physicist who made contributions to statistical mechanics and astrophysics. In 1925, with Guggenheim,[65] he worked out the properties of stellar material assuming that the gas in stars behaves like an ideal gas. Fowler also made contributions to the theory of stellar spectra.[66]

Dirac's paper containing the derivation of what had since been called Fermi–Dirac statistics was communicated by Fowler to the Royal Society on 26 August 1926. On 3 November, Fowler communicated a paper of his own,[67] in which the application of the laws of the 'new quantum theory' to the statistical mechanics of assemblies consisting of similar particles was systematically developed and incorporated into the general scheme of the Darwin–Fowler method.[68] And by 10

---

[63] Eddington, A.S., *The Internal Constitution of the Stars*, Dover Publ. (1926) p. 172.

[64] Fowler served as a lieutenant in the Royal Marine Artillery and was seriously wounded during the battle of Galipoli, a battle in which Moseley lost his life. Fowler married Eileen, Rutherford's only daughter.

[65] Fowler, R.H., & Guggenheim, E.A., MNRAS **85**, 939, 961 (1925).

[66] Fowler, R.H., MNRAS **85**, 970 (1925); JRASC **18**, 373 (1924); Obs. **47**, 216 (1924).

[67] Fowler, R.H., Proc. Roy. Soc. A **113**, 432 (1926).

[68] C.G. Darwin and R.H. Fowler [Phil. Mag. **44**, 450 (1922)] introduced a sophisticated and relatively simple method to calculate the thermodynamic properties of an assembly of particles.

December his paper entitled *Dense Matter* was read before the Royal Astronomical Society.

Fowler solved Eddington's paradox by applying Fermi–Dirac statistics to matter at high densities. What Fowler found[69] was that the ultimate state was one in which the star can be considered as a gigantic atom with the only one possible configuration. The star, if it is devoid of energy sources, can reach zero temperature, while the pressure generated by the compressed electrons would be large enough to balance the weight of the stellar layers attempting to collapse inward due to the gravitational pull.

Fowler assumed that all atoms were stripped of their electrons and that the electrons roam freely over the entire star, as was shown previously by Eddington. The whole star *is strictly analogous to one gigantic molecule in its lowest quantum state.* By *lowest quantum state*, Fowler meant that all states are occupied as in an atom on the Earth. Is this plausible? Can the electrons move freely between the compressed nuclei? And what about the stripped nuclei? As a matter of fact, the same problem exists in metals. The atoms in metals lose their last one or two electrons. These electrons are free to move inside the metal, irrespective of how big the piece of metal may be. In a way, this is the essence of the electron theory of metals developed first by Drude (1863–1906m),[70] then later expanded by H.A. Lorentz (1853–1928m),[71] and finally extended by Sommerfeld[72] in 1928 to include quantum statistics. Note that Fowler's idea of viewing the entire star as a single system, or like a piece of metal (but with all atoms stripped of their electrons) in which Fermi–Dirac statistics plays the dominant role, preceded the application of Fermi–Dirac statistics to metals.

A word of an explanation is in order. When we discussed the electrons in an atom, it was clear that there are certain energy states and no two electrons can occupy the same state. But what about electrons bound only to remain within some region V, where V is the huge volume of a star. Are there also 'states' in a bounded space? Indeed, there are states in any system bounded by whatever volume, from a piece of metal right up to a star, even though the physical dimensions are so different. Consider the sea of electrons moving around in the volume of the star. Since the space is bounded, quantum theory asserts that the electrons are restricted to occupying certain discrete states. Once particles are bound in some way, they cannot have just any energy, but are restricted to definite energy states.[73] A gas in which all possible states are occupied is called a degenerate Fermi–Dirac gas. On the other hand, a gas of bosons, as the particles which obey Bose–Einstein statistics are called, in which all particles reside in the lowest energy state, is called a condensed Bose–Einstein gas.

---

[69] Fowler, R.H., MNRAS **87**, 114 (1926).

[70] Drude, P., Ann. der Physik **1**, 566 (1900); ibid. **3**, 369 (1900); ibid. **7**, 687 (1902).

[71] Lorentz, H.A., Proc. Acad. Amst. **7**, 438, 585, 684 (1905).

[72] Sommerfeld, A., Zeit. f. Phys. **47**, 1 (1928).

[73] The number of cells in a volume V with energy E less than $E = p^2/2m$ is equal $V(4\pi/3)p^3/h^3$, where p is the momentum of the particle and is E the energy.

The situation is quite amazing. The temperature of a gas (or the entropy) reflects the number of states the system can be in. The higher the temperature or the entropy, the more states the system can be in. Here we find, according to Fowler's new idea, that white dwarfs are in the single lowest possible state, namely, all particles fill all the energy levels (cells), exactly like the electrons in an atom. The gravitational force, which pushes the white dwarfs into this state, appears to act in the opposite direction to thermodynamics. As time goes by and the star cools to the state of a white dwarf, it reaches the most ordered state, which also has the state of lowest entropy.

Chandrasekhar, in his obituary to Fowler,[74] described this discovery as *among the more important of the astronomical discoveries of our time.* Indeed, Fowler's application of the Pauli exclusion principle in the form of Fermi–Dirac statistics changed the theory of stellar evolution forever. In Eddington's language, Fowler allowed stars to die by cooling.

5.16 Pokrowski. A Limit on the Mass of a Collapsed Star

A rather surprising paper appeared in 1928, written by a Russian author by the name of Pokrowski.[75] Pokrowski assumed that the maximum density of the matter in the star would be obtained when all atoms had lost their electrons and the nuclei touched each other. Provided that the nuclei could not be compressed, as was found later to be the case, this should be the maximum density that matter could be in. This state is known today as nuclear matter. Pokrowski estimated this density to be $4 \times 10^{13\pm1}$ (see also Darwin[76]). Assume now a star with mass M and uniform density equal to the maximum density. It is simple to calculate the energy required by a particle of mass m on the surface of the star to escape to infinity.[77] Since the maximal density is fixed, there exists a stellar mass for which the energy needed to escape exceeds the rest mass energy $E = mc^2$, and hence no energy/particle can leave this star, and it cannot be observed. Pokrowski claimed that, for masses above the limiting mass (and having the limiting density), energy cannot leave the star. According to Pokrowski's calculations, this mass is $30.29M_\odot$.

Pokrowski's calculation was based on Newtonian mechanics which is not valid for such a large gravitational force. In a way, Pokrowski essentially repeated the centuries old calculation by Laplace, who discussed the idea that the limiting state of a star is reached when it is so dense that light cannot escape from it. Such an object is known today as a black hole. Fourteen years after the discovery of general relativity by Einstein, and twelve years after Schwarzschild discovered his solution to the general theory of relativity, there was no justification for carrying out a calculation

[74] Chandrasekhar, S., Ap. J. **101**, 1 (1945).

[75] Pokrowski, G.I., Zeit. f. Phys. **49**, 587 (1928).

[76] Darwin, C.G., Proc. Phys. Soc. London **39**, 359 (1927).

[77] The energy E_{esc} is given by $E_{\mathrm{esc}} = M^{3/2}G(4\pi\rho_{\mathrm{m}}/3)^{1/3}$, where ρ_{m} is the maximal density.

which completely ignored general relativity. Moreover, Pokrowski formulated his result by stating that: *The strong gravitational field curves the space around the star in an extraordinary way*, which is the language of the general theory of relativity. So it is plausible to assume that Pokrowski knew about general relativity and yet still published an incorrect calculation. Furthermore, there was no reference to Fowler's seminal work. On the contrary, Pokrowski adopted Eddington's assumption for stars on the main sequence, namely, that the stars behave like ideal gases.

5.17 Anderson Expands on Pokrowski's Idea, but Changes the Reasons

Hardly a year after the publication of Pokrowski's 3 page paper, Wilhelm Anderson[78] from Tartu university in Estonia,[79] took Pokrowski's idea a bit further. Repeating a calculation without the new general theory of relativity, Anderson argued as follows. The luminosity that the star radiates is equivalent to mass, so when the star radiates into space, it decreases its mass. He thus calculated how much mass a star loses as a function of the original mass before it reaches the limiting density.[80] For example, if the initial mass is $334M_\odot$, about $0.55M_\odot$ of the stellar mass is radiated before the star reaches the limiting density, and when the initial mass is $4.82\times10^7M_\odot$, the final mass is $370M_\odot$, so that the amount radiated away is $1-10^{-6}=0.999999$ of the initial mass. Hence, concluded Anderson, the final mass of a star must be smaller than $370M_\odot$.[81]

Anderson then criticized Eddington's claim that the gravitational contraction energy is insufficient to support the Sun for billions of years. The contraction, claimed Anderson, can be so high that it can easily supply all the energy the Sun needs in its lifetime. Anderson was right from the point of view of the energy balance. In nuclear transmutation of hydrogen into helium, about 0.007 of the rest mass is converted into energy. So if gravitational contraction can supply the entire rest mass, it should be able to supply a small part of it. However, Anderson did not carry out a calculation of the lifetime of the Sun, and did not refer to the necessary changes in the radius of the Sun, had it really derived its energy from contraction. As a matter of fact, except for references to Pokrowski, Eddington, and Heyl (for the value of the constant of gravity), Anderson chose to ignore all previously published results. After sending the paper for publication, Anderson became aware of Stoner's paper[82]

78 Anderson, W., Zeit. f. Phys. **55**, 386 (1929).

79 The address on the paper is Dorpat, the historical name of Tartu, the second largest city in Estonia.

80 The resulting formula is $1/M_\mathrm{i}-1/M_\mathrm{f}=1.2263\times10^{-24}/M_\mathrm{i}^{1/3}$, where M_i is the initial mass of the star and M_f is the final mass.

81 During the calculation, Anderson noted that Pokrowski had made a numerical error of $(5/6)^{3/2}=0.76$ in the mass.

82 Anderson's paper reached the journal on 23 February 1929, while Stoner's reached Tartu at the end of April, 1929.

(see below) and remarked correctly in *a note added in proof* that Stoner ignored the change in the mass of the electron due to special relativity, and hence that his results were correct only for small stellar masses. Anderson was right about this point.

5.18 Stoner Again

At this point Stoner[83] entered the picture once more, publishing a sequence of papers in which the idea of a limiting mass evolved gradually. By now he was aware of Pauli's principle, and of course of Fowler's work, which he cited and applied. In the first paper, Stoner developed the idea that there might be a limiting density, not due to nuclei losing all their electrons, but due to the 'jamming up' of the electrons, which had to obey Fermi statistics. This effect does not depend on the size of the atom, or whatever is left out of it under the immense pressure in the star. Thus the idea was basically that there exists a limiting density which was smaller than the one assumed by Pokrowski and Anderson. Stoner mentioned Jeans' stellar stability theory (which had not yet been shown to be wrong) that a star cannot be stable if it satisfies the ideal gas laws. Hence, the matter in a stable star had to be in a liquid state. Stoner quoted from the newly published book by Jeans[84]:

> *In the white dwarfs, atoms are mainly ionized down to their nuclei [...] it is their jamming, rather than that of the nuclei, which results in the departure from the gas laws which ensure the stability of the star.*

So Stoner set out to calculate the revised limiting density, now caused by the Pauli exclusion principle. He adopted the new theory of Fowler and assumed that the mean molecular weight of white dwarfs is 2.5. Next, he calculated the gravitational energy and the energy in the gas. To simplify the calculation he assumed that the density in the star was uniform and did not change from the high density in the center to vanishing density on the surface. If the star behaves like a liquid, this assumption is logical, as a liquid can hardly be compressed. But it was this assumption of constant density which caused him to lose priority in the discovery.

As the star contracts, the density rises, and as a consequence the energy of the electron rises because of the exclusion principle. We can see this in the following way. From the uncertainty principle in one dimension $\Delta x \Delta p \sim \hbar$, so as the 'living space' Δx for an electron decreases, its momentum must increase. But this increase in momentum requires energy. So to compress the gas of electrons requires energy, and as the density increases, more and more energy is needed to raise the electrons to higher and higher energy levels in order to satisfy the uncertainty and exclusion principles. This raises the question as to whether the gravitational force can always win over? In fact, Stoner found the critical density beyond which the gravitational pull no longer has the power to provide the required energy to the electrons, whence

[83] Stoner, E.C., Phil. Mag. **7**, 63 (1929).

[84] Jeans, J.H., *Astronomy and Cosmogony*, Cambridge Press, 1928.

no further contraction is possible. The resulting density (for a molecular weight of 2.5) was found to be:

$$\rho = 3.85 \times 10^6 \left(\frac{M}{M_\odot}\right)^2 \text{ g/cm}^3 . \tag{5.1}$$

Stars that reach this density cannot contract anymore, claimed Stoner, so they cannot extract energy from the gravitational field and consequently they do not shine. They are dark and have zero temperature. They are dead stars. Indeed, all stars are doomed to die when they reach this limiting density, and this is the end of stellar evolution.

The comparison with observations was excellent. The mean density of Sirius B is 5×10^4 g/cm^3, while for Eridani B it is 9.8×10^4 g/cm^3, van Maanen's star has a mean density in excess of 10^5 g/cm^3, and Procyon B has a mean density of several thousand g/cm^3. If the mass of Sirius B is $0.85M_\odot$, then according to Stoner the maximum density should be 2.77×10^6 g/cm^3, while Eridani B with a mass of $0.44M_\odot$ should have a maximum density of 7.48×10^5 g/cm^3. Since the temperatures of the stars are not yet zero, it appeared that the observed densities agreed nicely with the predictions of the theory. Moreover, if the density in the star is equal to the maximum density everywhere, one gets the minimum radius of the star, and this can be compared with observations. Indeed, the minimal radius of Sirius B was calculated to be $0.0075R_\odot$, while observation yielded $0.03R_\odot$. For Eridani B, the minimal theoretical radius was $0.011R_\odot$, while the observed value was $0.018R_\odot$. Every star has a minimal radius, and it cannot contract beyond this radius.

Stoner was happy with the results because the electron gas in which all the energy levels are occupied is practically incompressible. Even the strongest gravitational force cannot compress it. In other words, it behaved like a liquid and hence satisfied Jeans' condition for the stability of stars. On the other hand, Stoner mentioned that his results had no effect on the difficulties Jeans' condition implied for the stability of ordinary main sequence stars. This statement was published three years after Eddington's seminal book. No reference to Pokrowski, whose paper was published well before, or to Anderson who published his paper roughly at the same time. Both papers were published in the prestigious German Zeitschrift für Physik.

5.19 Anderson

Soon after the semi-critical paper on Pokrowski's limiting density, Anderson[85] published an analysis of the state of the electron gas in white dwarfs, in which he criticized Stoner's treatment of the problem. Anderson's most important contribution was to note that, as the density increases and the electrons are driven to higher and higher energies, they quickly reach the point where the velocity of the electrons is close to the speed of light, whereupon special relativity can no longer be igno-

[85] Anderson, W., Zeit. f. Phys. **36**, 851 (1929).

red. Indeed, at a density of 10^6 g/cm^3, the kinetic energy of the electron is already 0.28 of its rest mass energy. The main effect of special relativity is that the velocity of the electrons cannot reach the speed of light. Thus, as the speed of the electrons increases towards c, the energy needed to compress the gas quickly increases to infinity, and it is clear that the gravitational force will be unable to continue to compress it. The effective mass of the electron is given by $m = m_0\sqrt{1-v^2/c^2}$, where m_0 is the mass of the electron at rest.[86] The inclusion of special relativity turned out to be crucial.[87]

Anderson explained that, when the electrons move at speeds close to the speed of light, the rest mass of the electrons is negligible compared with the effective mass and can therefore be neglected. This is correct, and the result is an equation that is identical to the equation for a gas of photons, namely, the pressure of the electrons behaves like radiation pressure. Here, Anderson referred to the theory of Louis de Broglie,[88] which assumed that the photon is a particle with an extremely small mass, about 10^{-50} g. This is wrong. The photon has no mass, and for this reason moves at the speed of light.

5.20 Stoner Responds

Shortly after Anderson's paper was published, Stoner[89] criticized his mathematical treatment, but accepted the basic idea that the role of the special theory of relativity is crucial. The main criticism Stoner made was on the accuracy of the approximations Anderson applied, and not on the idea that relativity is important. Stoner found a way to carry out the calculation accurately (by using the energy of the particles, rather than the mass). Again, Stoner assumed a mean molecular weight of 2.5. The new results are displayed in Fig. 5.7, where Anderson's approximate results are also shown. The idea of a limiting mass appeared for the first time in these two papers, and the approximate values for it are very close to the accurate one. The effect of the

---

86 The kinetic energy of the electron is given by $E_k = (m-m_0)c^2 = m_0c^2\left(1/\sqrt{1-v^2/c^2}-1\right)$.

87 The Newtonian gravitational pressure can be shown to behave as $P = a(M)\rho^{4/3}$, where $a(M)$ is a function which depends only on the mass of the star. On the other hand, at zero temperature, the pressure of an electron gas without any relativistic effects behaves as $P = b\rho^{5/3}$, where ρ is the density and b a numerical constant. So as the density increases, the pressure of the electron gas increases more and more quickly, and will always be able to balance the gravitational pressure. However, when relativistic effects are taken into account, the pressure of the electron gas goes as $P = c\rho^{4/3}$, hence exactly like the gravitational pressure. The constants b and c depend only on the properties of the gas and do not refer to any feature of the star. The coefficients do not depend on the temperature, because we assume zero temperature. So the question as to who wins, the electron gas pressure, in which case we have a stable star, or gravity, in which case the star collapses to a point (becomes a black hole), depends on the coefficients $a(M)$ and c. Substituting in $a(M) = c$ yields the expression for the limiting mass. For $M \le M_{\text{limit}}$, the gas wins, and for $M > M_{\text{limit}}$, gravity wins.

88 de Broglie, L., Phil. Mag. **47**, 447 (1924); Ann. de Phys. **3**, 79 (1925).

89 Stoner, E.C., Phil. Mag. **9**, 944 (1930).

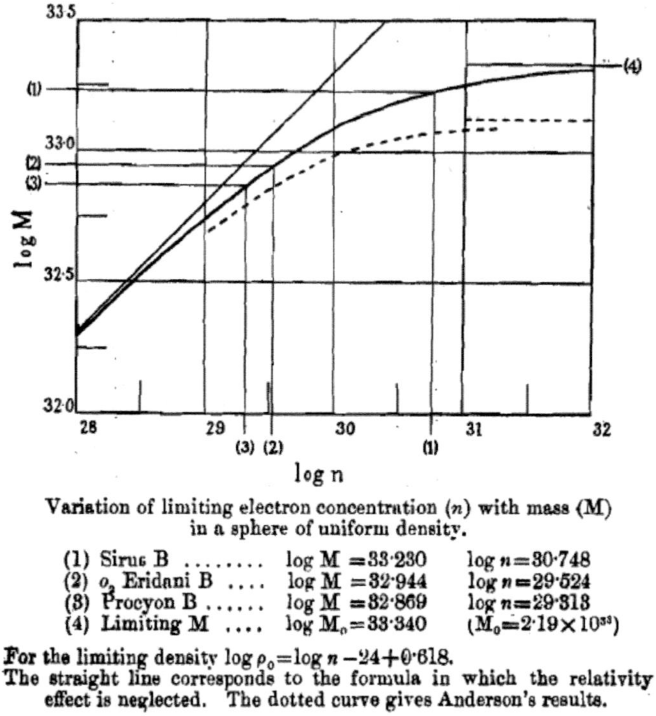

Fig. 5.7 The original curve from Stoner (1930), in which the limiting mass appears. The *horizontal axis* is the logarithm of the number of electrons per centimeter cubed, and the *vertical axis* is the mass of the star in grams. The *broken line* is Anderson's result. The limiting mass is $1.10M_{\odot}$ for a mean molecular weight of 2.5. The *straight line* is the result obtained when special relativity is ignored

special theory of relativity is clearly seen. Without incorporating relativity, the curve obtained was the straight line, which shows no signs of 'saturation' or of tending to a finite mass. It is due to relativity that the curve bends and tends to a finite mass.

It is worth mentioning why a molecular weight of 2.5 was adopted. Stoner assumed that white dwarfs, which are at the end of their stellar evolution, are composed of lead, and the mean molecular weight of fully ionized lead is $207.2/(82+1) = 2.50$. Jeans derived a molecular weight of 2.6, because he assumed that the matter in dwarf stars was composed of uranium. Stoner also remarked that these stars do not satisfy Eddington's mass–luminosity relation, and that the *radiation emitted by these stars does not decrease steadily with the mass, as had been suggested* by Jeans.[90] These stars do not behave like normal stars.

Stoner also discussed the energy source of the white dwarfs. He brought up Jeans' hypothesis that:

---

[90] Jeans, J.H., *The Universe Around Us*, Cambridge Pub. p. 310.

The energy generation in stars is to be traced to electron–proton annihilation occurring in hyper-uranium atoms as a result of one extra nuclear electron of the atom falling into the nucleus.

Stoner argued that, since the atoms are almost all stripped of their extra nuclear electrons, the rate at which such electrons fall into the nucleus, annihilate protons, and produce energy is very small, in agreement with the low luminosity of these stars. Only in the very outer layers of the stars where conditions were less extreme would the atoms still keep their electrons, so energy would only be generated in the outer layers, contradicting the assumptions needed to derive the mass–luminosity law. Still, Stoner pointed out that:

It does nonetheless remain peculiar that Eridani B, which approaches the condensed state, should have a much higher surface temperature than Sirius B, and that it should generate more energy per gram.

This clearly signalled the failure of the theory of the white dwarf energy source, and Stoner even noticed it, but he did not draw the inevitable conclusion.

Stoner did not discuss what happens to stars which are more massive than the limiting mass. Do they contract forever? Later, Stoner attempted[91] to improve the estimates of the limiting mass by taking into account the density distribution. To that end, he applied the model polytropes from Emden's 1907 *Gasgugeln* monograph. Actually, the polytropic hypothesis amounts to assuming that the pressure varies as ρ^γ, where γ is called the polytropic index. The pressure of the condensed electron gas varies as $\rho^{5/3}$ at low densities and as $\rho^{4/3}$ at high densities. The effect of special relativity is to reduce the power of the pressure dependence on density by just 1/3. It was this change in the exponent that would be the subject of the fierce and emotionally charged controversy that arose between Chandrasekhar and Eddington.

As a matter of fact, Stoner and Tyler managed to solve the case of low density, but just missed the idea of assuming an ideal star in which the polytropic index is everywhere equal to 4/3, as dictated by the special theory of relativity. Interestingly, in two papers published in 1932, Stoner[92] discussed the pressure dependence on the density, and in particular what happens between the low and high density limits. Both papers were communicated to the journal by Eddington. In other words, Eddington communicated papers which included a result he objected to. Moreover, Stoner ended the paper with an acknowledgment to Eddington for proposing the problem of the 'upper limit'. One may suspect that Stoner's cardinal contribution to the theory of white dwarfs is not widely recognized by astrophysicists because of its publication in the Philosophical Magazine, a journal not so frequently read by them.

---

91 Stoner, E.C., & Tyler, F., Phil. Mag. **11**, 986 (1930).

92 Stoner, E.C., MNRAS **92**, 651, 662 (1932).

5.21 Chandrasekhar. The Final and Accurate Answer

Chandrasekhar (1910–1995) met Sommerfeld in 1928 during Sommerfeld's[93] trip to India, and heard his seminar on the new theory of metals and Fermi–Dirac statistics. He even got the galley proofs of the new article from Sommerfeld.[94] Here, we witness the opposite sequence of events to what happened to Fowler, namely, the theory of metals was applied by Chandrasekhar to stars. At this time Chandrasekhar decided to go to England and not Germany, although the intentions of Sommerfeld's visit to India were to strengthen relations between German and Indian science. The decision might have been affected by the language barrier. This preference for England over Germany had a major impact on Chandrasekhar's life in the coming years.

The story has it[95] that, at the age of 19, Chandrasekhar worked out the limiting mass of white dwarfs while on the boat from India to England, work that earned him the Nobel Prize in 1983.[96] The basic difference between Stoner's limiting mass expression (which Chandrasekhar apparently was not aware of while on the boat) and Chandrasekhar's was that the latter[97] included a better model for the density distribution in the star, and consequently led to a more accurate value for the limiting mass. Indeed, the first result for the limiting mass obtained by Chandrasekhar was $0.91M_{\odot}$. Later Chandrasekhar compared his result with Stoner's and concluded that:

> *The agreement between the accurate working out, based on the theory of the polytropes, and the cruder form of the theory is rather surprising.*

There was not a word about what would happen to stars more massive than $0.91M_{\odot}$ in Chandrasekhar's two page paper.[98] It is amusing to note that Chandrasekhar was Fowler's PhD student in Cambridge and got the degree in 1933, so his prize-winning limiting mass paper was published while he was still a graduate student.

Chandrasekhar's short paper about the limiting mass was published in the American Astrophysical Journal, although the most important astrophysical literature on the subject of stars was published at that time in the Monthly Notices of the Royal Astronomical Society. It should be noted that Chandrasekhar used to write long and comprehensive papers, so the paper on the limiting mass was exceptionally short by

[93] Arnold Sommerfeld, the high priest of German science, was an adamant supporter of Indian physics, and the trip in September–October of 1928 was documented by him in Zeitwende **5**, 289 (1929).

[94] Parker, E.N., National Academy of Sciences, Biographical Memoirs.

[95] Parker, E., Chicago University, Chandrasekhar's Obituary.

[96] His uncle, C.V. Raman, won the Nobel Prize for Physics in 1930.

[97] Chandrasekhar, S., Ap. J. **74**, 81 (1931). See also Phil. Mag. **11**, 592 (1931).

[98] Chandrasekhar assumed like Stoner that the mean molecular weight was 2.5, and at the time did not express the explicit dependence of the limiting mass on the mean molecular weight.

his standards.[99] One can only wonder why Chandrasekhar chose this venue for his seminal contribution.

In 1934, Chandrasekhar[100] summarized the physical state of the matter in the interior of stars by distinguishing between matter which obeys the ideal equation of state, dense matter which obeys the equation $P \sim \rho^{5/3}$, and ultradense matter which obeys the equation $P \sim \rho^{4/3}$. A limiting mass is obtained only for the ultradense case. So Chandrasekhar classified the stars according to the mass. The very massive stars satisfy Eddington's equation, and the matter in them remains in the ideal gas state. The matter in these stars depends only marginally on the Pauli principle. On the other hand, the small masses were divided again into two classes. For stars with mass less than $(1.74/\mu^2)M_\odot$, where μ is the mean molecular weight, the relativistic effects never become dominant, and the density in a star of mass $M < (1.74/\mu^2)M_\odot$ never exceeds $6.301 \times 10^5 \mu^5 (M/M_\odot)^2$ g/cm^3.

Then came the white dwarfs. For white dwarfs with masses that are smaller than $(3.822\mu^2)M_\odot$, relativistic effects never play a role. White dwarfs in the mass range $1.743\mu^2 M_\odot$ to $6.623\mu^2 M_\odot$ reach a density in which relativistic effects play a dominant role. Finally, matter in stars with masses $M > (6.623/\mu^2)M_\odot$ always obeys the ideal gas law. As for their fate, Chandrasekhar entered the land of speculation. He relied on Steensholt[101] and Sterne,[102] who had shown that the star could be stable only if the transmutation of the elements took place in chemical equilibrium.[103] But here was a problem. For fusion to be in equilibrium, the temperature must be very high, much higher than any temperature imagined to exist in stars. So Chandrasekhar speculated that, as the density approaches the critical density, the behavior of matter changes in an unknown way.

In summary, what Stoner and Chandrasekhar proved was that cold stars are stable for masses smaller than the limiting mass, while more massive stars are apparently unstable.

In 1935, Eddington[104] published his first straight attack on the idea that special relativistic effects are important to the theory of white dwarfs. One may wonder what triggered Eddington's reaction, and why he was so upset, to put it mildly, with Chandrasekhar's result. Maybe the answer can be found in the introduction to his paper:

---

[99] In the same issue of the Ap. J., there was an obituary to Michelson (contributor to the Michelson–Morley experiment which led to special relativity), and the preceding paper to Chandrasekhar's is by Edwin Hubble and Milton Humason, entitled *The velocity–distance relation among extra-galactic nebulae*, the first experimental evidence for the Big Bang. This is a rare conjunction of ground-breaking papers.

[100] Chandrasekhar, S., Obs. **57**, 93 (1934).

[101] Steensholt, G., Zeit. f. Ast. **5**, 140 (1932).

[102] Sterne, T.E., MNRAS **93**, 736 (1933).

[103] To say that a reaction is in chemical equilibrium means that, while $A+B$ goes to $C+D$, the process also goes backwards at the same time, i.e., $C+D$ goes to $A+B$. The notation is therefore $A+B \rightleftharpoons C+D$. When the reaction is in equilibrium, the temperature and to a lesser extent the density determine which way the reaction will go. At sufficiently high temperatures, the product $C+D$ decomposes back into the original components $A+B$.

[104] Eddington, A.S., MNRAS **95**, 194 (1935).

> *Using the relativistic formula, he [Chandrasekhar] finds that a star of large mass will never become degenerate, but will remain practically a perfect gas up to the highest densities contemplated. When its supply of subatomic energy is exhausted, the star must continue radiating energy and therefore contracting – presumably until, at a diameter of a few kilometers, its gravitation becomes strong enough to prevent the escape of radiation. This result seems to me almost a reductio ad absurdum of the relativistic formula. It must at least rouse suspicion as to the soundness of its foundation.*

In other words, Eddington did not believe in the physical reality of the Schwarzschild solution, exactly like Einstein, who refused to accept it as physical possibility (see later). So, because he did not believe in what we call today black holes, he turned the argument round to conclude that, if Chandrasekhar's theory led to the formation of black holes, then it must be wrong. One may guess that Chandrasekhar knew about Eddington's basic reasons for the objection to his results, and for this reason refrained from predicting the fate of a massive star in his communication to the Royal Astronomical Society (February 1934), instead speculating that *the nature of the interaction between the nuclei changes at high density.*

Eddington set out to look for flaws in the derivation of the result $P \sim \rho^{4/3}$ for electrons moving with velocities close to the speed of light. He raised a series of technical questions, and one fundamental one. Let us discuss the latter. Fowler's basic assumption was that the electrons released from the atom in the star move freely throughout the entire volume of the star.[105] The derivation assumed the (paradoxically correct) assumption that, as the density rises, the electrons move more like particles in a box. The surrounding nuclei, stripped of all their electrons and at high density, do not affect the motion of the electrons, and consequently the latter can move very long distances without colliding with either nuclei or other electrons. This is exactly what happens in metals, and results in the excellent heat conductivity of metals.

Why is this so? Recall that electrons occupy all the possible states. Now, when an electron collides with a nucleus and as a result wants to change its course and move to another state, it finds that this state is already occupied, since all states are occupied in a degenerate gas. The result is that the electron cannot move to the 'new' state, so remains in the initial state. The effect of the Pauli principle on the motion of the electrons is to cause them to move like free particles. This is what Eddington could not accept, so he attacked the idea, claiming that this 'idealization' or approximation was basically wrong. In his words, *the condition that the electron must go into an unoccupied state can only operate if the electron is being added to a distribution already present.*[106] Note that Fowler did discuss this point, and came to the conclusion that the assumption, no matter how incredible it might sound, was basically correct.

---

[105] The motion of an electron in the star resembles the motion of a particle in a box. So long as the particle does not collide with the wall, it moves as if it is free. When it hits the wall, the wall exerts a force on the particle which keeps the particle inside the box. When we write 'free electron', we mean like a particle in a box. The electron moves throughout the entire star, but it is prevented from leaving it.

[106] To be precise, Eddington used the term 'half-cell' rather than 'state'. It is surprising, because this was about 10 years after the discovery of spin.

Moller and Chandrasekhar[107] immediately responded to Eddington's attack. Actually, it was no wonder Moller and Chandrasekhar could respond so quickly, since they were *indebted to Sir Arthur Eddington for allowing us to see a manuscript copy of his paper.* As a consequence, the two papers appeared in the same issue of the Monthly Notices of the Royal Society.

Just one volume later, the MNRAS carried Eddington's reply.[108] Again, mostly technical, but this time including a statement that the exclusion principle had been *abundantly verified* for electrons in the atom:

> *Undoubtedly there exists a generalization of it applicable to large assemblies of particles* [he meant stars] *but the generalization cannot be of the form assumed by Moller and Chandrasekhar, which conflicts with the uncertainty principle.*

Eddington accepted Pauli's principle for atoms, but rejected its extension by Fermi and Fowler to larger systems. Nobody else doubted the validity of the Pauli principle in stars. Moreover, this very statement contradicted Eddington's earlier statements in 1916 about the validity in stars of the laws of physics discovered on Earth.

In 1936, Chandrasekhar recruited Rudolf Peierls (1907–1995), a leading nuclear physicist, to write a note on the derivation of the equation for a relativistic gas.[109] This time the paper was communicated to the MNRAS by Chandrasekhar. Peierls discussed Eddington's contentions that the behavior of the gas in the star might depend on the shape of the volume containing it. Peierls admitted that the solution was obvious, but *in view of the controversy, it is perhaps worthwhile to give a proof.* An acknowledgment to Chandrasekhar appeared at the end, although it was Chandrasekhar who had to thank Peierls.

5.22 What Determines Stellar Masses?

Several theories of nature combine to create the limiting mass. These are the theory of gravitation, quantum mechanics (via the Pauli and uncertainty principles), and the special theory of relativity. Gravitation is controlled by the gravitational constant G, quantum theory by the Planck constant h, and special relativity by the speed of light c. Gravitation needs a mass to act upon, so let us assume that the relevant mass is the mass of a proton m_{p}. Using these constants, the simplest expression with physical dimensions of mass is the cosmic mass defined as:

$$M_{\mathrm{cosmic}} = \left(\frac{ch}{G}\right)^{3/2} \frac{1}{m_{\mathrm{p}}^2} = 29.246 M_{\odot} \,. \tag{5.2}$$

It follows from Chandrasekhar's analysis that the limiting mass of a white dwarf is given by

---

[107] Moller, Chr., & Chandrasekhar, S., MNRAS **95**, 673 (1935).

[108] Eddington, A.S., MNRAS **96**, 20 (1931).

[109] Peierls, R., MNRAS **96**, 780 (1936).

$$M_{\text{limit}} = 0.196702 M_{\text{cosmic}} \,. \tag{5.3}$$

In the previous chapter, we mentioned Eddington's quartic equation, the equation that relates β = gas pressure/radiation pressure to the mass of the star. Since β is dimensionless, there must be a constant with the dimensions of mass in the equation. We can call this mass the Eddington mass, and the value is:

$$M_{\text{Eddington}} = 4\pi \left[\left(\frac{k_{\text{B}}}{\mu m_{\text{p}}} \right)^4 \frac{3}{a} \frac{1-\beta}{\beta^4} \right]^{1/2} \frac{1}{(\pi G)^{3/2}} \frac{1}{(\mu m_{\text{p}})^2} C_{\text{n}} M_{\odot} \,, \tag{5.4}$$

where k_{B} is the Boltzmann constant, a is the Stefan–Boltzmann constant, and C_{n} is a numerical constant that depends on the density distribution in the star. Eddington derived his equation without any reference to quantum theory, and hence Planck's constant does not appear in the expression for the mass. However, it can be shown[110] that $a = 8\pi^5 k^4/15c^3h^3$, and if we assume the same density distribution as Chandrasekhar assumed,[111] then Eddington's mass becomes

$$M_{\text{Eddington}} = 0.617511 \left(\frac{ch}{G} \right)^{3/2} \frac{1}{(\mu m_{\text{p}})^2} \,, \tag{5.5}$$

so that

$$M_{\text{limit}} = \frac{0.196702}{\mu^2} M_{\text{cosmic}} \,, \qquad M_{\text{Eddington}} = \frac{0.617511}{\mu^2} M_{\text{cosmic}} \frac{\beta - 1}{\beta} \,. \tag{5.6}$$

What we have shown here is essentially how the fundamental constants of physics, i.e., the speed of light, Planck's constant, and the constant of gravity, determine the mass of a star whose basic building unit is the proton.

It is incredible that, in spite of the fact that the electrons are the particles which supply the pressure against the gravitational pull in the case of the white dwarf and the absorption of radiation in the case of main sequence stars, the mass which appears in the formula is the mass of the proton and not the mass of the electron. The protons contribute the mass of the star, while the contribution of the electrons to the mass of the star can be neglected. It is the Pauli principle, a purely quantum effect that does not depend on the mass of the identical particles, and with it, special relativity, that are so essential to the structure of collapsed stars, effectively fixing the masses of stars. When the energies of the particles are very high relative to the

---

[110] This was shown by Einstein when he discussed the equilibrium between mass and radiation.

[111] Chandrasekhar assumed a polytrope of index 3. Eddington's basic model assumed also a polytrope of index 3. A polytrope of index n means that the pressure–density relation is $P = K\rho^{(n+1)/n}$, where K is a constant. Note that Eddington's model for stars on the main sequence assumed exactly the same density distribution as Chandrasekhar assumed for the cool white dwarfs, where relativistic effects play a dominant role. The reason is simple. In Eddington's model, the pressure is mainly radiation pressure, and when the electrons become relativistic, i.e., when they move with speeds close to the speed of light, their rest mass plays no role and can be neglected, so that they resemble the photons of the radiation field.

rest mass energy m_0c^2, the rest mass is no longer important. The same is true in neutron stars, where the density reaches 10^{15} g/cm^3 and the neutrons, which also obey the Pauli principle, move at relativistic speeds.

In the case of the Eddington mass, the processes of radiation absorption controlled by the electrons do not affect the fundamental mass. While Eddington's mass was calculated without any reference to the quantum theory, the Planck constant nevertheless cropped up. This is due to the fact that the Planck constant is required to describe the radiative processes participating in the absorption of radiation in stars. Hence, the two fundamental masses which govern stellar evolution can be expressed in terms of the same constants of nature.

5.23 1930. Milne's Attack on Eddington's Stellar Structure

The rivalry between Milne and Eddington reached a new peak when in 1930 Milne[112] staged a long vilification of Eddington's and also Jeans' stellar theories, presenting an alternative theory for the structure and evolution of stars. Milne disagreed with the *current theory*, and added in a footnote that by this term he meant the theory of Sir Arthur, to avoid writing everywhere in the article 'according to Eddington'. Milne explained that, at the time of writing, there were two theories of stellar structure. Regarding the first, due to Jeans, he asserted that:[113]

> *It accounts for the existence of giants, dwarfs, and white dwarfs, but only at the cost of ad hoc hypotheses quite outside physics. It assumes stars to contain atoms of atomic weight higher than that observed on earth, and it assumes them to be relentlessly disappearing in the form of radiation. [...] I think that it is true to say that the majority of astronomers do not accept this theory.*

On the other hand, claimed Milne:

> *The theory of Eddington does not claim to account for the observed division of stars into dense stars and stars of ordinary density, nor does it establish the division of ordinary stars into giants and dwarfs.*

Milne's verdict about Eddington's theory was that:

> *Closer consideration of the actual formula used by the theory shows that it scarcely bears out the claims made for it by its originator.*

In particular, and this was the crucial point made by Milne:

> *The claim to establish the mass–luminosity law from mere equilibrium considerations cannot, however, be sustained for a moment.*

---

112 Milne, E.A., MNRAS **90**, 17, 678 (1929); ibid. **91**, 4 (1930). A summary was published in Nature (January 1931) p. 16.

113 This was written after Jeans had abandoned his mass annihilation theory because of his condition for stellar stability and suggested very massive radioactive elements that do no exist otherwise in Nature. See later for the suggestion by Nernst.

By 'equilibrium' Milne meant both the mechanical equilibrium (the pressure of gravity is balanced by the pressure of the gas) and the thermal equilibrium (all the energy generated in the star is radiated away).

Milne elucidated the reasons for these harsh statements. Consider a star with an energy source. A star in thermal equilibrium is a star which adjusts its energy generation to the energy radiated away. We can look for a star having any mass M and any luminosity L because, if the energy production is not equal to L, *the star can adjust itself to suit any arbitrary* L. This means that M and L do not depend on each other and one should be able to find stars with any combination of M and L. But, as we know today and as Milne knew at the time, this is not what is observed in Nature. The observed existence of a mass–luminosity relation implies, à la Milne, that the properties of the energy source must be taken into account to sort out from all possible pairs M, L only those that satisfy the mass–luminosity relation. Eddington's idea of determining the mass–luminosity relation from equilibrium considerations alone appeared to Milne to be *a philosophical blunder*. Furthermore, Milne argued that:

> *It is unphilosophical to assume that the interior of a star is perfect gas.* Why? Because *the knowledge of the interior is forever unattainable, or we should be able to infer it from the observation of the outer layers.*

Milne alleged to have shown that it is impossible to have a gaseous star in a steady state and that the core must be exceedingly dense and hot. The stars must either have an extremely dense core or be 'collapsed', and this division corresponds to the separation between 'ordinary stars' and 'white dwarfs'. This, according to Milne, was not a consequence of any special new hypothesis, but followed naturally from the method of analysis. All previous analyses of the structure of stars, according to Milne, were fallacious! The division of ordinary stars into giants and dwarfs would appear to be less fundamental and not to indicate any special difference in structure. Even Stoner was wrong, according to Milne, with his maximum density, because Stoner assumed that the only source of energy in white dwarfs was gravitational contraction and:

> *[...] as soon as the possibility of an internal supply of subatomic energy is admitted, his 'maximum density' condition becomes invalid.*

As for the source of energy, Milne believed in Jeans' hypothesis of matter annihilation in the very dense core of the star, although Jeans himself had abandoned this idea by then. In white dwarfs, he stressed, the temperature may reach 10^{10} K or higher.[114] Such high temperatures had never appeared in any previous theory. The basic idea was that the mutual destruction of a proton by an electron is a reversible process, matter $\rightleftharpoons$ radiation, i.e., in the dense core, the protons and the electrons annihilate to produce radiation, while at much lower densities, the balance shifts towards the protons and electrons. In equilibrium, the number of annihilations per

114 In a footnote to his paper, Milne noted that *the temperatures are sufficiently high for the synthesis of radioactive elements.* But it was not clear what he meant. Was he assuming that the radioactive elements were synthesized in white dwarfs?

second is equal to the number of syntheses of protons and electrons by absorption of radiation, whence at each density there is a definite concentration of matter, like water and water vapor:

> *If the temperature increases, the concentration of matter would decrease to preserve equilibrium, for the annihilation process is an 'exothermic' one in the sense of chemistry.*

A word of explanation is due. When an equilibrium process takes place in a closed volume like the one above, one can control it by raising the temperature or the pressure, and in this way push the process in one direction or the other, while the energy is conserved, since nothing leaks out. However, if for example the radiation leaks out, as in a star where radiation is slowly but continuously lost into space,[115] the delicate balance between the radiation and the matter is lost, and more matter annihilates to compensate for the energy lost in the form of radiation.

Milne published his series of papers a short time after Atkinson and Houtermans[116] had proved that sufficient amounts of energy to support the solar luminosity are released by the transmutation of hydrogen into helium at temperatures as low as 4×10^7 degrees. So, according to Eddington, there was no reason to maintain the idea of mass annihilation as an energy source for stars.

In summary, Milne's theory aspired to:

- explain the existence of all types of stars including the white dwarfs,
- explain the energy generation,
- eliminate the problem of the stellar absorption coefficient.

So what did Milne assume in order to get such strange and provocative results? While Eddington assumed that the absorption coefficient and mass determine the luminosity, Milne supposed that the mass, the luminosity, and the absorption coefficient were completely independent. The three could have arbitrary values, claimed Milne. If this were so, then Milne claimed that:

> *The difficulty with the discrepancy between a supposed astronomical value of the stellar absorption coefficient and the value predicted by pure physics is not encountered.*

This was his attempt to solve the nagging problem of the absorption coefficient in stars. In his analysis, the radius of the star is determined by the mass, luminosity, and absorption coefficient. This was before Eddington and Strömgren had solved the absorption coefficient problem by discovering that hydrogen is the most dominant element in stars. The observational comparison of stellar masses, luminosities, and radii should determine the absorption coefficient, although it could not be observed directly. Milne criticized what he referred to as the current theory:

> *It endeavours to calculate the luminosity from steady state considerations only, i.e., from general physics without any knowledge of the nature of the energy generating process. It does so by making a hypothesis about the state of aggregation of the whole mass (the perfect gas hypothesis) and then claiming to set up a relation between mass, luminosity, and absorption coefficient for any distribution of energy sources.*

[115] Relatively slowly means that the time it takes the radiation to empty the energy store is long.

[116] Atkinson, R.d'E., & Houtermans, F.G., Zeits. f. Physik **54**, 656 (1929).

On the other hand, Milne claimed that:

> *Mass, luminosity, and absorption coefficient must be in nature three independent variables capable of having arbitrary prescribed values, as far as steady state considerations go.*

And just in case the reader failed to understand which of Eddington's results Milne was referring to, Milne added a footnote in which he explicitly stated that he was referring to the mass–luminosity absorption coefficient which appeared in Eddington's book. The formula at the focus of this controversy was (4.1), which was the basis for Eddington's estimates of the absorption coefficient. The amazing result that the radius disappeared from the formula outraged Milne.

5.24 Condensed Models

Who was right? As a matter of fact, both Milne and Eddington were right, but for reasons that became clear only decades later. A close examination shows that Milne allowed for several possibilities that were summarily eliminated in Eddington's theory. Eddington implicitly assumed a homogeneous star, but this need not be the case. We have no proof that the stars are uniform. Eddington assumed that L and M were independent and the star homogeneous. As a consequence, he found what the absorption coefficient should be. On the other hand, Milne assumed that L, M, and κ were independent, and hence obtained many more possibilities. Indeed, Milne found his theory to yield non-homogeneous models, or what are called today condensed models. The star may hide a very dense core surrounded by an extensive envelope (see Fig. 5.9).[117]

Milne's basic results are summarized in Fig. 5.8. Each star of a given mass and absorption coefficient has two critical luminosities L_0 and L_1 such that there cannot be a steady-state stellar configuration with L greater than L_1. Eddington's homogeneous model with an ideal gas is possible only for $L = L_0$, and when L is between L_0 and L_1, the steady-state configuration is composed of a very dense and high-temperature core surrounded by a tenuous envelope, a model known as the 'centrally condensed configuration'. The core of these configurations has a very small radius, but because of the tenuous envelope, the radius is huge. For stars with luminosity less than L_0, the only possible configuration is that of a white dwarf. Milne succeeded for the first time in inventing a theory that unified the models for the three so vastly different stars: the dwarfs (main sequence), the giants, and the white dwarfs. Figure 5.8 shows what happens. A homogeneous star cannot exceed the luminosity L_0, but a non-homogeneous star, a condensed one, can reach the maximum luminosity L_1. For stars to reach the maximum luminosity, they must be non-homogeneous.

[117] You can think about this as follows. Choose the mass and luminosity and follow Eddington to find the absorption coefficient, assuming the star to be homogeneous. If you now assume the same mass and luminosity as the resulting absorption coefficient, you find according to Milne a homogeneous solution. Now change the absorption coefficient arbitrarily. Clearly, there is no homogeneous solution to the new problem. The only possible solution is an inhomogeneous model. You trade the assumption of homogeneity for the freedom to chose the absorption coefficient.

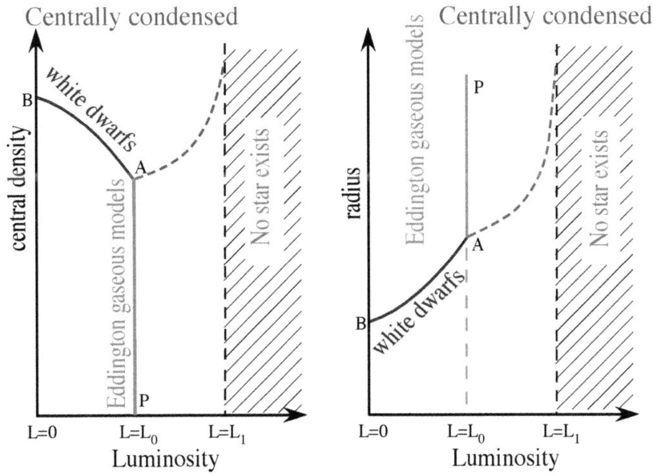

Fig. 5.8 The possible solutions to the stellar structure equations for a fixed mass and absorption coefficient and any luminosity L, according to Milne. *Left*: Central density as a function of the luminosity. *Right*: Radius of the star as a function of the luminosity. Adapted from Milne's paper in Nature

The luminosity L_0 is obtained from a solution of Eddington's quartic equation (see footnote 29 in the last chapter).

The luminosity L_1, which is given by Milne as $L = 4\pi GM/\kappa$, is known as the Eddington limiting luminosity, in the sense that no star with mass M and absorption coefficient κ can be in a state of mechanical equilibrium and radiate more than this formula predicts. The irony is that the first time that the Eddington limit appeared in the literature as a limit was in Milne's paper, where he wrote that $L < L_1 = 4\pi cGM/\kappa$. Eddington used this formula to estimate the absorption coefficient. While it is a special case of Eddington's general formula,[118] the first to apply it as a limit was Milne in an attempt to destroy Eddington's theory. In this way Milne actually helped in naming the result after Eddington. In the classical application of the Eddington limit, one substitutes the constant Thomson scattering limit for the absorption. This is supposed to be the lowest possible absorption coefficient. However, Eddington did not mention Thomson scattering at all in his book, and the chapter on opacity[119] just ignored this extremely important phenomenon. On the other hand Eddington repeated his incorrect and justifiably criticized theory of nuclear capture of radiation.

Milne concluded by stating that:

[118] In general, β appears in the formula, but in the expression for the maximum possible luminosity, L_1 the parameter and β is set to zero as the most extreme case.

[119] Eddington, A.S., *The Internal Constitution of the Stars*, Cambridge Univ. Press (1926). See chap. IX *The Coefficient of Opacity*.

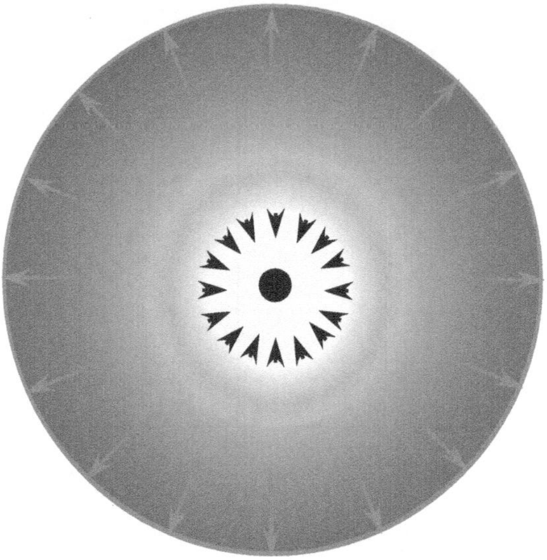

Fig. 5.9 The consequence of hydrogen burning in the cores of stars is the contraction of the core and the expansion of the envelope. The star tends towards Milne's centrally condensed configuration

> *It is not possible to infer from the observed masses, luminosities, and temperatures that the interiors of stars are necessarily composed of a perfect gas; and it is not possible to deduce the value of the absorption coefficient for the stellar interior.*

The observed correlation of luminosity with mass had to depend upon the intrinsic physics of the energy generation and could not be deduced à la Eddington. Milne was already aware of Kramers' absorption coefficient law, but did not investigate it. In a way, Milne was right, because unless you assume the model to be homogeneous, Eddington's procedure does not apply. Here is a concrete example, not provided by Milne. The consequence of hydrogen burning is the conversion of four hydrogen nuclei into one helium nucleus. The idea gas pressure is proportional to the number of particles, so as the latter decreases, the core contracts under the gravitational weight of the other layer, and increases in density and temperature as shown in Fig. 5.9. The density distribution is shown in Fig. 5.10. There is a very dense core and a very extended, low density envelope. The contraction of the core to supply the required pressure releases gravitational energy (the star sinks deeper into the gravitational potential well). This energy is mostly absorbed by the envelope, which thus expands.

The Milne–Eddington controversy dragged in several additional players. Jeans had retired by then,[120] but Larmor (1857–1942m)[121] and Woltjer (1891–1946m) joined the fray. In Larmor's first paper,[122] he reviewed Eddington's and Milne's results and concluded that:

Perhaps not much stress should be laid on the deduction. The formula is regarded probably by its author as essentially an empirical result.

The author he was referring to was Milne. In other words, he was not convinced by the derivation. For this reason, he applied the term 'empirical' to what Milne considered to be theoretical. On 22 March, Milne replied and argued again that, without an assumption about the energy source, the problem could not be solved. Apparently, Larmor remained unconvinced, and in the next paper he drew attention to what was assumed to take place on the surface of the star (what mathematicians call the boundary conditions). Larmor insisted, and explained[123] why Milne was nevertheless wrong. After sending the paper for publication (on 13 February), he had a discussion with Milne, who convinced him that he, Milne, was right. So Larmor added a post scriptum (dated 27 February) in which he expressed gratitude to Milne for convincing him in a face to face discussion.

In the second paper[124] Larmor discussed first Sects. 91–93 of Eddington's book, which elaborated on just what one has to assume to take place on the surface of the star when one attempts to solve the equation of stellar structure. Larmor, in a typical understatement, claimed that it was *a treatise in places easier to read than to digest*, and went on to demonstrate the problems. When Larmor later came to discuss Milne's theory, he admitted that *it is very confusing to a reader*, and so set about clarifying it. However, the reader was left confused as to who was right. Eddington responded to Larmor,[125] attempting to explain his viewpoint. Larmor[126] acknowledged that he had overlooked a few facts, as Eddington pointed out in his reply, but remained unconvinced by Eddington. The paper was sent for publication on 4 August 1930, and before publication Larmor felt a need to add a post scriptum (28 August) stating that:

One notes that on both sides of the discussion the range of the analysis is conditioned by the same assumption, namely, the ideal gas law for deep seated matter [...] but it is only on Sir Arthur Eddington's side that reason is found for adhering completely to these rather daring postulates.

---

[120] Jeans took an early retirement in 1929 and dedicated himself to the writing of several very successful popular books. To name but a few titles: *The Universe around Us* (1929), *The Mysterious Universe* (1930), *The Stars and Their Courses* (1931), *The New background of Science* (1933), *Through Space and Time* (1934). Jeans is famous as a mathematician, theoretical physicist, astrophysicist, and science popularizer.

[121] Larmor, J., The Obs. **53**, 249.

[122] Larmor, J., Nature (22 February 1930) p. 273.

[123] Larmor, J., The Obs. **53**, 113 (1930).

[124] Larmor, J., ibid., p. 167.

[125] Eddington, A.S., The Obs. **53**, 208 (1930).

[126] Larmor, J., The Obs. **32**, 676 (1930).

In short, Larmor adhered to Milne's view, notwithstanding the fact that it lacked proper justifications. The discussion between Woltjer and Milne had to do with the modeling of the corona of the Sun. However, many of the arguments raised above entered into this discussion as well.

It is interesting that, in the same issue of the Monthly Notices of the Royal Society in which Milne's 51 page paper was published, Cowling (1906–1990),[127] who was a student of Milne, published a piece of work about Eddington's model, saying that:

> *The model proposed by Eddington [...] is the only model which has been investigated with any attempt at completeness: and in the discussion of this model there have been important gaps which are only now being filled.*

In contrast, Cowling presented his own model in which the energy was generated at the center as a point source.[128] Cowling demonstrated that many of Eddington's assumptions, like uniform generation of energy in the star, lead to unacceptable physical results, such as an infinite mass for the star. In doing so, Cowling arrived at the conclusion that:

> *No evidence has been found for a belief that to a given luminosity corresponds a unique mass.*

In other words, according to Cowling, there was no theoretical mass–luminosity law. But he was wrong.

Some ten years later, Cowling[129] summarized the past decade of stellar structure theory, and explained that *it was all about an unresolved discrepancy between the astronomical and physical values of the absorption coefficient of stellar material.* Milne claimed that there was no problem with the absorption coefficient (because he was free to choose whatever value he wanted), and it was only the model that was defective, but as Cowling explained, Milne did not supply an alternative model in which the difficulty was resolved. As early as 1942, when the review was published, Cowling was already a leading astrophysicist and had published a landmark book.[130]

The relations between Eddington and Milne can be inferred from the style of the writing. For example, Milne wrote:

> *The slightest speck of meteoritic dust falling on a star built in Eddington's model would cause it either to collapse or to develop a central condensation.*

Or again:

> *To change the surface absorption coefficient and then omit to investigate the layer of changed absorption coefficient is scarcely to deal with the problem.*

---

[127] Cowling, T.G., MNRAS **91**, 92 (1930).

[128] This was not a new idea. Eddington himself had examined such a model as a limiting case. See Eddington, A.S., MNRAS **85**, 408 (1925). No credit was given to Eddington.

[129] Cowling, T.G., The Obs. **64**, 224 (1942).

[130] Chapman, S., & Cowling, T.G., *The Mathematical Theory of Non-Uniform Gases*, Cambridge Univ. Press (1939).

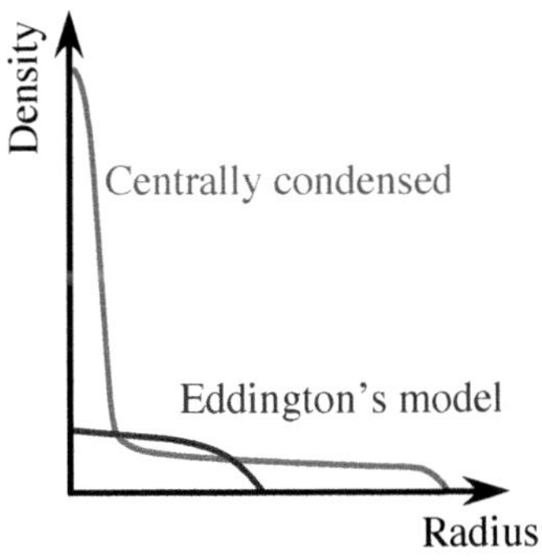

Fig. 5.10 The structure of a simple Eddington model and Milne's centrally condensed model

Milne could simply have said *Eddington's model is unstable*, but he clearly could not help the note of sarcasm.

Eddington's response to Milne's criticism was no less poisonous. He published a summary in the German periodical Zeitschrift für Physik.[131] The wording in this German publication was, however, milder. First, Eddington pointed out that his theory developed in 1916 had been unchallenged until Milne published his work. To our mind, this is a poor argument. Maybe before Milne came on the scene, no one had been clever enough to expose the error! Next, Eddington[132] wrote that:

> *Milne's recent papers contain a great number of attacks on the current theory. It would be absurd to take up every challenge in detail; but I have replied to one which he emphasizes most in his original onslaught on L_0 that it is sensitive to the conditions in the photospheric layers*[133] *of the star. Some general remarks may be permitted. A theory is said to be wrong (a) if there is a fault in the mathematical or logical deduction from the premises assumed, or (b) if the actual conditions differ more widely than was anticipated from those assumed, so that the theoretical model fails to represent the actual object to the degree of approximation intended. Generally, an author is presumed to guarantee the correctness of his theory as regards (a). As regards (b), he advocates it with varying degrees of confidence and readjusts his views from time to time as new evidence is obtained; sometimes indeed he takes no responsibility for (b). [...] In stellar constitution accusations of mistakes under (a) have continually been made during the past 14 years and have seldom been admitted or withdrawn. [...] It is impossible to avoid the conclusion that [...] the progress of this subject is being unnecessarily handicapped by the publication of a most unusual amount of careless deductions.*

Eddington did not completely dismiss Milne's theory and rephrased it as follows. Milne assumed that the stars were made of a white dwarf core surrounded by a tenuous envelope. As such, this was a model whose correctness should be examined by comparison with observation. As to Milne's philosophy, Eddington dismissed it

---

[131] Eddington, A.S., Zeit. f. Astrophysik **3**, 129 (1931).

[132] Eddington, A.S., MNRAS **90**, 279; Ibid. 808.

[133] The photosphere of the star is the layer from which photons can escape to the outside world. Since the star is gaseous, it is not a rigid surface, but a layer.

completely. In years to come, it would turn out that both Eddington and Milne were right. Eddington's model applies only to homogeneous stars, and as soon as the stars have burnt their hydrogen, they become non-homogeneous (with a helium core and a hydrogen envelope), adopting the configuration of a condensed core (hot white dwarf) with a tenuous envelope.

These facts only became clear at the beginning of the 1950s, but by then Eddington and Milne were no longer alive. Recall that, at the time of the controversy between Milne and Eddington, it was thought that the stars were made of heavy elements with very little hydrogen. In this respect, Milne's models came too early. The puzzle at that time was the structure of main sequence stars, or even young main sequence stars, and these were quite well represented by Eddington's models. Years later, when post main sequence phases of evolution were being investigated, the truth of Milne's model became evident.

It turns out that giants are stars that have completed the fusion of hydrogen into helium in their core. In this fusion, four protons are converted into one helium nucleus. The pressure of an ideal gas depends on the number of particles. But nuclear fusion decreases the number of particles, whence the ability of the star to resist gravity decreases. The result is a contraction of the core which releases gravitational energy. The latter is then pumped into the envelope, causing it to expand. In this way, giant stars form, and they can be nicely represented as condensed configurations. The latter have enormous densities at the center and extremely low densities near the surface, and they have very large radii. These large dimensions conceal the extreme conditions at the center.

The irony is that Eddington thought at the beginning that his theory explained the gaseous giant stars and not the dwarf (main sequence) stars. As it turned out, Eddington's theory explains the dwarfs and Milne's theory explains the giant stars.

5.25 Independent Derivation of the Limiting Mass

In 1932, the young Lev Davidovitch Landau (1908–1968m)[134] denounced Milne's proof that a star consisting throughout of classical ideal gas cannot exist, and thus defended Eddington's view.[135] Milne's idea, as described by Landau, was that, if L and M are chosen completely arbitrarily, there is no guarantee that such a star can actually exist. That is, of all possible luminosities and masses, only certain combinations correspond to actual stars. Landau claimed that Milne reached this conclusion because he assumed the absorption coefficient to be constant throughout the star. Furthermore, Landau claimed that this assumption was made for mathematical convenience and had nothing to do with reality. Under this limiting assumption, the radius of the star disappears from the relation between L, M, and R. Moreover, any

[134] Landau won the Nobel prize in 1962 for his development of a mathematical theory of superfluidity. However, no lecture was delivered because of a serious accident, from which Landau never fully recovered.

[135] Landau, L., Physik. Zeits. Sowjetunion **1**, 285 (1932) [Sov. Phys. **1**, 285 (January 1932)].

real absorption coefficient leads to a relation between L, M, and R, and in this way is *exempt from the criticism levelled against Eddington's mass–luminosity relation.*

So Landau proposed to overcome this problem by *methods of theoretical physics*, a statement reminiscent of Eddington's motto years earlier. What Landau did was to derive the equation for the structure of the star (the equation of hydrostatics) from thermodynamics, rather than from dynamical considerations, which is the usual way.[136] Assuming the cold (vanishing temperature) gas to obey Fermi–Dirac statistics, and without mentioning what gas particles obeyed these statistics, Landau deduced theoretically that the gas should abide by exactly the equation used by Eddington for his gaseous stars (supported by radiation) and Chandrasekhar for his white dwarf (supported by a gas of electrons),[137] demonstrating that a true result in physics can be derived in more than one way.

Given the physics that enters the theory, Landau's critical mass naturally turns out to be

$$M_{\text{crit}} = \frac{1}{20} M_{\text{cosmic}} = 1.5 M_{\odot} \,, \tag{5.7}$$

where the mass of the particle which provides the gravity was assumed to be twice the mass of the proton.[138] In Chandrasekhar's case, the particles which provide the pressure are the electrons, while the particles which provide the mass are the protons. Hence, Chandrasekhar's result was more general, and included the molecular weight, which tells one how many particles provide the pressure for every particle that provides the gravity. Landau did not specify what particles were supposed to obey the Fermi–Dirac statistics, nor did he specify which particles provided the mass. However, he did assume that they were the same particles, and for this reason the mean molecular weight did not appear in his expression.

If we assume as Landau did that the proton mass is 2, we get $M_{\text{crit}} = 1.5 M_{\odot}$. At the time Landau published his paper, he did not know about the existence of neutrons. So if we take the oxygen nucleus, for example, the mass of the nucleus is 16 a.m.u., while the atomic number is 8, and we find that the mass of the proton must be 2.

Landau reached the conclusion that a mass larger than $1.5 M_{\odot}$ should collapse to a point. But he noted that:

> *As in reality such masses exist quietly as stars and do not show any such ridiculous tendencies, we must conclude that all stars heavier than* $1.5 M_{\odot}$ *certainly possess regions in which the laws of quantum mechanics are violated.*

What led Landau to draw such a far-reaching conclusion? He went on to say:

---

[136] The proof can also be found in the famous series of lectures on physics: Landau, L., & Lifshitz, M., *Statistical Mechanics*, Pergamon Press (1958) p. 340. The relevant chapter in the book discusses only neutron stars, so Oppenheimer & Volkov (1939) and Oppenheimer & Snider (1939) are cited, but not Chandrasekhar or Landau himself.

[137] In all these cases, it is a polytrope of index $n = 3$.

[138] There is a typographical error in Landau's paper, as he wrote that *we get an equilibrium state only for masses greater than a critical mass.* It should be 'less than', as can be understood from the continuation.

> *As we have no reason to believe that stars can be divided into two physically different classes according to whether the mass is greater or smaller than* M_{crit}, *we may suppose that all stars possess such pathological regions.*

In retrospect, it is difficult to understand why the assumption of the violation of quantum theory by stars, which contradicts Landau's very first premise about the use of theoretical physics, was easier to accept than the assumption that stars may have different courses of evolution depending on their mass. Or was it a tacit resentment of the collapse to a black hole? Should we apply Landau's famous assertion that *cosmologists are often wrong but never in doubt* to this case?

Landau then changed the subject and discussed the energy source of the stars. He rejected the suggestion of annihilation of electrons with protons because (a) such annihilation was never observed and (b) the electron and the proton exist in the nucleus at the same time (this was before the discovery of the neutron!) and they do not annihilate each other:

> *It would be very strange*, argued Landau, *if the high temperature did help, only because it does something in chemistry.*

Following Bohr's idea of the violation of energy conservation as a solution to the β decay problem, Landau claimed that:

> *We are able to believe that the stellar radiation is due simply to a violation of the law of energy, a law, as Bohr first pointed out, that is no longer valid in the relativistic quantum theory, when the laws of ordinary quantum mechanics break down, as is experimentally proved by continuous ray spectra and made probable by theoretical considerations.*[139]

Landau expected this phenomenon to occur when the density became very high. The reference to the paper by Landau and Peierls is interesting. In this paper, the authors elaborated on Bohr's idea about the conservation of energy in the nucleus and argued that the uncertainty principle was not valid in the relativistic domain, and consequently that ordinary quantum theory did not apply in the nucleus where special relativistic effects are important.

Despite what Landau wrote in the introduction, he concluded by supporting Milne's basic theory, namely, that the central region of the star must consist of a core of highly condensed matter surrounded by matter in an ordinary state. He then argued that:

> *If the transition between these two states were a continuous one, a mass smaller than the critical one would never form a star because the normal equilibrium state (without pathological regions) would be quite stable. As far as we know, it is not the case, and we must conclude that the condensed and non-condensed states are separated by some unstable state in the same manner as a liquid and its vapour.*

Before leaving this section, we note that the paper, which preceded the discovery of the neutron, did not mention neutrons, protons, or electrons, and as a matter of fact, did not mention any particles at all.

---

[139] Landau, L., & Peierls, R., Zeit. f. Phys. **69**, 56 (1931).

The paper was sent for publication in February 1931 (from Zurich) and without any mention of Chandrasekhar's[140] or Stoner's[141] discoveries of a critical mass, made only shortly before. The only paper Landau mentioned was his own article with Peierls.

In 1931, Chandrasekhar[142] extended his research in two directions. In a paper communicated by Milne, he expanded on Milne's theory of collapsed objects and attempted to explain the structure of white dwarfs. At the end of this paper, Chandrasekhar gave a table in which he distinguished the fate of the low mass stars and the high mass stars, just the point Landau had rejected. This is one of the first occasions on which the fate of stars was considered as a function of their mass. In parallel, he worked on his theory of white dwarfs. It so happened that the paper on Milne's composite models (which did not work so well in any case) came out just before Chandrasekhar submitted his paper about the critical mass of white dwarfs.

In December 1932, Russell[143] gave the First Maiben Lecture before the American Association for the Advancement of Science. The topic was *The Constitution of the Stars*. It is interesting that, in the discussion on white dwarfs, Russell attributed the understanding of white dwarfs to Milne:

> *The white dwarfs have, within the last few years, changed their role from most perplexing to the best-understood class of stars. The present theory of their nature (which we owe to Milne) is the second notable triumph of the application of general physics to stellar constitution.*

All the new results were attributed to *Milne and his colleagues at Oxford.* The Cantabrigian Chandrasekhar was not even mentioned. At the end of his talk, Russell discussed the energy source: *stellar energy synthesis or annihilation of atoms*. Based on indirect evidence, he concluded that:

> *There is then no room for doubt that the synthesis of heavier elements out of lighter ones and hydrogen may actually occur within the star.*

This, however, was not the energy source Milne assumed to operate in stars.

Just two years later, Chandrasekhar[144] reached the dramatic conclusion that:

> *It is necessary to emphasize one major result of the whole investigation, namely, that it must be taken as well established that the life history of a star of small mass must be essentially different from the life history of a star of large mass. For a star of small mass the natural white dwarf stage is an initial step towards complete extinction. A star of large mass cannot pass into the white dwarf stage, and one is left speculating on other possibilities.*

The theory of how stars without energy sources die was discovered before the energy source of the living stars was found. Stars die either as a white dwarf or in another way.

---

[140] Chandrasekhar sent his paper to the American Astrophysical Journal on 12 November 1930, and it was published in the July 1931 issue.

[141] Stoner's paper in the English Phil. Mag. **9**, 944 (1930), was sent for publication in December 1929. Stoner's first paper was sent for publication a year earlier.

[142] Chandrasekhar, S., MNRAS **91**, 456 (1931). Published 13 March 1931.

[143] Russell, N.H., JRSAC **27**, 375 (1933).

[144] Chandrasekhar, S., Obs. **57**, 373 (1934).

Chandrasekhar's[145] last paper on the limiting mass with the new and rigorous derivation of the limiting mass for white dwarfs came in 1935. First, Chandrasekhar removed any reference to the radiation (symbolically, because introducing the radiation was Eddington's main achievement). Next, came the question: what happens to masses above the limiting mass? What Chandrasekhar had hesitated to state in the previous paper,[146] he dared to write this time: *Configurations of greater mass must be composite* (this referred to Milne's models), and *these composite configurations have a natural limit [...] zero radius*. In a footnote, Chandrasekhar added that:

> *In the previous paper this tendency of the radius to zero was formally avoided by introducing a state of 'maximum density' for matter, but now we shall not introduce any such states, namely for the reason that it appears from general considerations that, when the central density is high enough for marked deviations from the known gas laws to occur, the configuration then would have such small radii that they would cease to have any practical importance in astrophysics.*

In other words, Chandrasekhar did not believe at that time in the reality of what we call today black holes. However, Chandrasekhar changed his mind years later.[147] In his concluding remarks, he stated that the white dwarfs are *the limiting sequence of configurations to which all stars must tend eventually*. How the more massive stars would do this was not explained. Last but not least, in an appendix, Chandrasekhar gave a reference to Landau's 1932 paper, pointing out that it gave the same law for the pressure of the gas, but not mentioning that Landau had independently obtained a critical mass.

5.26 Erupting Stars

An eruptive star is a star which suddenly and without any known prior signal increases its luminosity by a large amount. The nature of these stars was an enigma for many years. Today we know that there is a large variety of them. Among the different types we find the supernova (plural supernovas or supernovae, hereafter SNs) and the novas. In the case of a nova, the brightening can be by a factor of 10^5, and in the case of SNs, the brightening reaches a factor of 10^{10}. A SN can be as bright as an entire galaxy which contains 10^{10}–10^{12} stars. The physics of SNs will be discussed in the chapter on the death of massive stars. Here, we describe how it was gradually recognised that SNs are the end product of the evolution of certain stars. At the time our story begins, around the turn of the nineteenth century, the difference between SNs and novae was not known, and all eruptive stars were assumed to be identical. The recognition that the eruptive stars compose a highly non-uniform group came after much time and effort and solved many astrophysical conundrums.

---

145 Chandrasekhar, S., MNRAS **95**, 205 (1935).

146 Chandrasekhar, S., MNRAS **91**, 456 (1931).

147 Chandrasekhar, S., *The Mathematical Theory of Black Holes*, Oxford University Press (1998).

Table 5.1 Discovered supernova and SN remnants in the Milky Way

Year	Duration [days]	Remnant name	Distance [light-years]
185	140	MSH 14-63	3 040
386	90	G 11.2-0.3	16 000
393	210	–	–
1006	240	PKS 1459-41	4 500
1054	540	Crab	6 400
1181	185	3C58	8 300
1604	330	Kepler	14 000
1752	480	Tycho	7 400
–	–	Cas A	9 000
–	–	G 292.0+1.8	11 500
–	–	RCW 103	10 500

Table 5.1, based on Strom,[148] summarizes the known SNs that have been observed in our galaxy. The overall number of SNs per galaxy is about one SN per century per galaxy of the size of the Milky Way. Comparing the known SNs in our galaxy with the expected number, we may deduce that human civilization has missed a good number of SNs that are likely to have occurred in the Milky Way. Some SNs may have happened on the other side of the galaxy, behind very opaque clouds of dust, and hence gone unobserved, and some SNs probably did not leave a stellar remnant to be detected centuries later. Note that radio observation would have detected the remnants, if they existed. In three cases, we see the remnants of suspected SNs and do not have an estimate for the date when the star erupted. The name in the table contains the name of the catalog and the number in that catalog, or the astronomer who detected the SN.

5.27 The Observations of How Stars Perish

The most famous of all supernova remnants is the Crab nebula, discovered by John Bevis in 1731. At the same time, Charles Messier (1730–1817m) was interested in comets. Recall that this was about a century after the advent of Newton's mechanics, which beautifully explained Kepler's laws, and there was widespread interest in cometary orbits. The basic question was: are the comets periodic, i.e., are they bound to the Solar System, or do they move in space and get captured by the Sun? Consequently, Messier was looking for the predicted return of the Halley comet. During this search, he discovered this nebula, apparently independently, on 28 August 1758. It was a faint object and he misidentified it as a distant comet. Shortly

[148] Strom, R.G., A&AL **288**, L1 (1994).

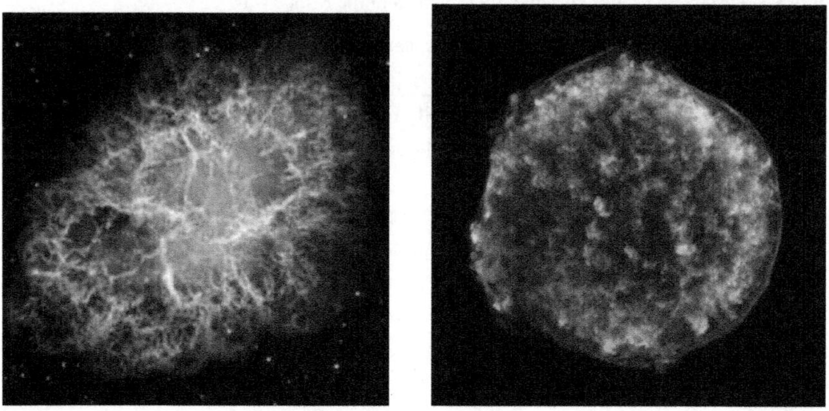

Fig. 5.11 *Left*: Probably the most famous supernova remnant in our galaxy, known as the Crab Nebula, M1, or NGC 1952. This is the remnant of the supernova observed by the Chinese in the year 1054. No record was found in Western archives. Credit: NASA, Hubble Space Telescope. *Right*: The Tycho 1572 supernova remnant. The supernova was discovered by Tycho. The present picture is from the Hubble Space Telescope. *Colors* refer to different chemical elements. Credit: NASA, Hubble Space Telescope

afterwards, he realized his misidentification (the faint object did not move in the sky relative to the fixed stars), and decided to catalog the faint objects that were not pale, distant comets, but which occasionally led him astray. In this way, the famous Messier catalog was born. The first object in this catalog, denoted M1 (M for Messier) was the nebula shown in Fig. 5.11. The name Crab nebula was attributed by William Rosse (1800–1867m) in 1844.

Edouard Biot (1803–1850) was a sinologue and the son of the physicist Jean-Baptiste Biot. Together they had a particular interest in Chinese astronomy and science,[149] and published separately catalogues based on the Chinese chronicles. So it was natural for them to search the Chinese astronomical records for the strange eruptive variable stars. Biot summarized his search with a list of such objects which included the suddenly brightening stars in the years 1054, 1572, and 1604. These records turned into a gold mine in the 1930s.[150]

It was already self-evident from the first spectra, that the Crab nebula was very special. Vesto Slipher (1875–1969m) reported in 1915[151] that the Crab nebula had a very strange spectrum[152] that did not resemble the spectrum of any other nebula.

[149] Biot, E.C., *Note sur la connaissance que les chinois ont eue de la valeur de position des chiffres*, Journal Asiatique **8**, 497 (1839).

[150] There is no record in Western archives of the nova in 1054. As for the two others, the 1752 event was observed by Tycho Brahe, while the one in 1604 was observed by Kepler. It seems that, during the Middle Ages, people in Europe were not interested in what was going on in the sky.

[151] Slipher, V., Nature **95**, 185 (1915).

[152] The nebula exhibits the N_1, N_2, and N_3 lines of nebulium, as well as two lines of hydrogen. This in itself was not such a mystery. But the puzzle was that, while each line was double, with a

In 1919, Sanford (1883–1958m)[153] reported his observations of the Crab nebula. These indicated that the spectrum was continuous and crossed by bright lines. The light from the nebula was made up of a relatively more continuous spectrum for the brightest part of the nebula than for the other regions adjacent to it. Sanford succeeded in measuring the velocities in the nebula, and discovered that different parts were moving with quite different velocities. Some parts were receding with velocities up to 1 000 km/s, while others were moving towards the observer with velocities up to 1 600 km/s. Sanford dismissed Slipher's explanation that the lines appeared doubled because of the Stark effect (spectral lines can be split by an electric field). No explanation for the continuum radiation was given.

In 1921, Lampland (1873–1951m)[154] noticed time variations in the brightness and changes in the shape and structure of the Crab nebula, based on observations carried out since 1913. So the nebula had changed in less than ten years (see Fig. 5.11). The observation was important because many other observed nebula did not show such variations. Lampland noticed the changes against the background of Slipher's announcement in 1915 that the spectrum was *most extraordinary* and did not resemble that of any other nebula. Simultaneously, Lundmark (1889–1858m) examined the list provided by Biot, and using comparisons with other evidence, listed the strange objects and started to correlate their locations in the sky, as reported by the Chinese, with the locations of known nebulas.

The hottest topic in astronomy in the early 1920s was the question of whether the amorphous objects we call nebulas are inside our own galaxy or further away, like isolated islands in the Cosmos. The telescopes of the day were not powerful enough to resolve stars in distant galaxies, even in the Andromeda galaxy, which is our closest neighbour galaxy in the Northern Hemisphere. (The largest telescope at this time was the Hooker 100-inch telescope on Mount Wilson, California, which became operational in 1917.)

Two personalities stood at the center of the Great Debate: Shapley and Curtis. The details of this discussion, though extremely interesting, would carry us too far from the present story. Here we simply note that one of the issues was whether the 1885 nova, also known as S Andromeda,[155] in the Andromeda nebula, was a regular nova or not. This nova appeared very bright and, if considered as a regular nova, it would have implied a very small distance to Andromeda. Curtis, who supported the idea that Andromeda was a distant galaxy, had observed the nova in Andromeda, and clearly S Andromeda was a very disturbing exception for his thesis. So he made a

separation of about 40 Å, the separation between the lines varied across the nebula. For this reason, no measurements of the radial velocity were carried out by Campbell, W.W., & Moore, J.H., Pub. Lick Obs. **13**, 134 (1918).

[153] Sanford, R.F., PASP **31**, 108 (1919).

[154] Lampland, C.O., PASP **33**, 79 (1921).

[155] The S stands for 'second nova' in the Andromeda nebula. The nova was discovered on 20 August 1885, by E. Hartwig from the Dorpat Observatory in Estonia. This is the only known supernova in the Andromeda galaxy. The remnants were discovered by Fesen, R.A., Hamilton, A.S.J., & Saken, J.M., ApJL **341**, 55 (1989).

big mistake and decided to ignore it, retaining only the regular and fainter novas.[156] Shapley, his opponent, decided the opposite. The removal of this extraordinary nova from the sample did not help Curtis, and he lost the debate, although he was right!

In 1921, Lundmark[157] was interested in *what has become of ancient novae which must exist in thousands in the heavens*. He suspected that *former novae were planetary nebulae and the Wolf–Rayet stars.*[158] So he compiled a list of all known 'new stars' and, using Biot's list, identified the 1054 eruption as having occurred *southeast of* η *Tauri but nearby*.[159] He also provided the celestial coordinates. In a footnote, he wrote that *the object is near NGC1952*, which is the Crab nebula or Messier 1. He almost, but not quite, identified the nebula with the results of the nova eruption.

Hubble (1898–1953m)[160] was interested in settling the controversy about the locations of the nebulas, and in 1922 continued his impressive research on nebulas, on the way to resolving the Great Debate. Hubble added NGC1952 (the Crab nebula) to his list of diffuse nebulas, but noted the fact that this nebula was special in that it had a continuous spectrum, whence he did not consider it to be a distant, unresolved galaxy. The nebula was a mystery to him which he was unable decipher, and he was quite confused and disturbed.

A year later, Lundmark[161] noted that:

> *The existing data indicates that the novae return to their original conditions after a very short time*, but then: *For Tycho Brahe's nova the amplitude is at least 18.7 [way above the average brightening of a nova] as no star in or near the rather accurately known position is brighter than magnitude 13.7.*

In non-astronomical units, that meant a brightening by at least a factor of

$$2.512^{(18.7-13.7)} = 100\,.$$

Lundmark also drew attention to the fact that the nova S Andromeda of 1885 stood out by its behavior and brightness, thus defending Curtis' decision not to include it in his analysis. Moreover, Lundmark noted the following problem. If the nova S Andromeda had been a regular nova, then the distance to the Andromeda nebula would have been 4 300 000 light-years, while on the other hand:

> *If, as several astronomers think, the distance of the nebula is of the order of 20 000 light-years, then the conclusion must be that the novas in that system are of quite another class from those in our galaxy [...]. For the present it may not be possible to decide which of the two possibilities is the right one.*

---

[156] Curtis, H.D., J. Washington Acad. of Sc. **9**, (1919); Lick. Obs. Bull. **300**; Bull. NRC **2**, 194 (1921).

[157] Lundmark, K., PASP **33**, 225 (1921).

[158] Wolf–Rayet stars are massive stars which are losing mass by means of a very strong stellar wind. These stars are very hot, with surface temperatures in the range 25 000–50 000 K.

[159] The star η Tauri (known also as Allcyone) is the brightest star in the Pleiades cluster, and is the third brightest star in the Taurus constellation. The distance to the star is about 440 lyrs.

[160] Hubble, E.P., Ap. J. **56**, 162 (1922).

[161] Lundmark, K., PASP **35**, 95 (1923).

Lundmark was close, but too respectful of previous distance estimates.

However, he gradually began to suspect that there might be two classes of nova that differed immensely in their maximum brightness. Hints started to appear that erupting stars did not constitute a homogeneous class. Lundmark observed that some parts of Nova Aquilae[162] changed their velocities by 600 km/s in just three days, and raised the question: *Should we not characterize such a huge change in the dimensions of the star during so short a time by the word explosion?* Indeed, Lundmark had guessed correctly. It should be pointed out that distance measurements were problematic in those days, and consequently it was extremely difficult to compare the brightness of different novas. This problem caused Lundmark to misidentify several supernovas as novas.

In May 1925, Hubble[163] announced his identification, using the powerful 100-inch telescope, of Cepheids in the Andromeda nebula, and was able to measure the distance to the nebula accurately. In this way he resolved the Great Debate. The Great Andromeda nebula became the Great Andromeda galaxy. Curtis was right. Interestingly, Hubble applied Shapley's results for the period–luminosity relation for Cepheids to find that Shapley was wrong in the Great Debate.[164]

In June 1925, Lundmark investigated[165] the connection between *the motions and the distances of spiral nebulae*. The attempt to measure the distance to the spiral nebulas by means of the direct geometric parallax failed. The nebulas are too far away to succumb to such a simple method. So Lundmark had to rely on proper motion, i.e., a change in the coordinates of the nebula as time went by. Lundmark expected a correlation between the distances and radial velocities of the nebulas:

> *Few reasons were given for the opinion that the measured Doppler shifts of the lines are due either to motions in the non-relativistic sense or to motions and certain effects consequent to the general theory of relativity.*

Lundmark published Table 5.2, which hints at some as yet unclear relation.

Considering the observed novas in Andromeda, Lundmark independently reached the conclusion that Hubble had published just a few months earlier. Lundmark even argued that, once the Cepheids could be measured in Andromeda, the distance would be accurately known. This was exactly what Hubble did, using the largest telescope available at the time. Lundmark did not know about Hubble's discoveries at the time of writing. In his earlier paper, Lundmark found that the distance to Andromeda was 1 400 000 light-years, while Hubble got 930 000 light years. However, this very large distance puzzled Lundmark, who looked for particular reasons why he had obtained such an unusually great distance. As he could not find any error in

[162] The nova in the Aquila constellation erupted on 8 June 1918 and was the brightest nova since Kepler's in 1604.

[163] Hubble, E.P., Popular Astronomy **33**, April 1925; The Obs. **48**, 139 (1925). An abstract was presented at the 33rd meeting of the AAS.

[164] Note that even here there was a problem with the data. Although at that time Hubble did not know that there were two types of Cepheid variables, and hence got an incorrect distance, even the wrong distance was sufficiently great to end the Great Debate.

[165] Lundmark, K., MNRAS **85**, 865 (1925).

Table 5.2 Lundmark's findings regarding the radial velocity of nebulas

Class of objects	Mean radial velocity [km/s]	Number of objects
Globular nebula	727	11
Early spirals	647	18
Late spirals	396	6
Magellanic clouds	217	2

his analysis, he hypothesized in his conclusion that: *It is quite possible that we have to deal with two distinct classes of novas: the 'lower class' and the 'upper class'*, as he chose to call them.

Another interesting conclusion by Lundmark was: *There is every reason for the view that novae are not new stars.* Indeed, we know today that novas are dying white dwarfs, on which some hydrogen-rich material is poured by a neighboring star. Moreover, Lundmark wrote:

> *We find that the Nova S Andromeda at the maximum reached the huge magnitude of* −16. *One may hesitate to accept such a luminosity. I think that we have an analogous case in the famous Nova B Cassiopaeiae of 1572.*

That was the Tycho Brahe nova.

The first attempt to correlate the Crab nebula with a nova was made by Hubble[166] as early as 1928. Hubble realized that the unusual velocities measured in the nebula impled that it was expanding, and if this was so, it should be possible to calculate when the expansion had started. He found that the expansion had begun roughly 900 years earlier. This meant that, if the Crab was a consequence of an explosion, that explosion must have taken place around the year 1028. Indeed, Hubble cited the Chinese annals, which recounted the appearance of a 'new star' in the year 1054, at approximately the same location in the sky. Hubble was able to correctly identify the Crab with an actual nova eruption, but did not discover that this nova was in fact a supernova.

Support for the evidence derived from the Chinese Chronicles came in 1934, when the Japanese astronomer Iba published a series of articles about the history of Japanese astronomy,[167] discussing ancient Japanese records of 'visiting stars'. Apparitions of 'strange stars' occurred in the years 877, 891, 1006, 1054, 1166, and 1181 A.D. (this is a partial list). None of these were observed in the West (during the Middle Ages) or even by the arabs. *That of 1054 appeared in the region of* ζ *Tauri and was as bright as Jupiter*, recounted Iba.

We must digress for a short while here to discuss a momentous event in nuclear physics and its impact on our theme. The neutron, the missing neutral particle which together with the proton composes the atomic nucleus, was discovered by Chadwick

[166] Hubble, E.P., ASPL **1**, 55 (1928).

[167] Iba, Y., PA **42**, 243 (1934).

in 1932.[168] In April of that year, Eddington[169] was asked to describe his discovery that hydrogen is so abundant in stars before the Royal Astronomical Society. At the end of his short presentation, he explained that, while he was carrying out his calculation:

> *[...] a new element, the neutron, which may be of great cosmical importance, was announced. But I do not think it will alter things very much, even if it should turn out to be of high abundance.*

Well Eddington could not have imagined all the ramifications of the discovery of the neutron for the whole of nuclear physics, nor the way it would revolutionize the theory of stellar evolution. Reality can surprise even the most imaginative minds.

Hardly two years after the discovery of the neutron, in 1934, Baade (1893–1960m) and Zwicky (1898–1974m) published two seminal papers. In the first,[170] they coined the terms 'common novae' and 'supernovae' (rather than Lundmark's lower and upper classes).[171] The common novas reach a maximum brightness of about $20\,000L_\odot$. The supernovas, on the other hand, emit much more energy. As a matter of fact, Baade and Zwicky calculated that, in about 25 days, a typical supernova releases more energy than the Sun releases in 10^7 years! This means that the brightness of a supernova is about 6×10^6 times the brightness of the Sun.

Relying on Lundmark,[172] they claimed that the abnormal light of the 1572 nova (Tycho's nova) implied that the explosion was a supernova. The authors noted that the remnant gases had apparently been discovered, but that no star had been identified inside those gases. In addition, nothing was known about the initial state of the supernova. Using the Einstein relation $E = mc^2$, they estimated that the energy released in a supernova would be equivalent to the conversion of about $6M_\odot$ to energy! They thus concluded correctly that *the phenomenon of a supernova represents the transition of an ordinary star into a body of considerably smaller mass.* But if the original star had a mass less than $100M_\odot$ say, and if it converted $6M_\odot$ into energy, *this energy is very close to the annihilation energy!* Consequently, a significant part of the star must have been converted into radiation, inferred Baade and Zwicky.

In a previous paper, submitted less than two months before,[173] this time about the idea that cosmic rays are produced by SNs, Baade and Zwicky added at the end that:

> *We have tentatively suggested that the SN process represents the transition of an ordinary star into a neutron star. If neutrons are produced on the surface of an ordinary star they will 'rain' down towards the center, if we assume that the light pressure on neutrons is nearly zero. This view explains the speed of the star's transformation into a neutron star.*

---

[168] Chadwick, J., Nature (27 February 1932) p. 312. The full paper is: Proc. Roy. Soc. A. **136**, 692 (1932).

[169] Eddington, A.S., MNRAS **92**, 471 (1932).

[170] Baade, W., & Zwicky, F., PNAS **20**, 254 (1934). Sent for publication 19 March 1934.

[171] Zwicky, F., Rev. Mod. Phys. **12**, 66 (1940). Here, Zwicky recounts that: *Baade and I first introduced the term 'supernova' in seminars, and in a lecture on astrophysics at the California Institute of Technology in 1931.*

[172] Lundmark, K., Kungl. Svenska Velensk. Handlingar **60**, No.8 (1919).

[173] Baade, W., & Zwicky, F., Phys. Rev. Lett. 77 (1934). Submitted to the journal on 28 May 1934.

In an accompanying paper,[174] they proposed that cosmic rays might be emitted by (rather than accelerated by) SNs, and that:

> *Mass may be annihilated in bulk. By this we mean that an assembly of atoms whose total mass is M may be lost in the form of electromagnetic radiation and kinetic energy as an amount of energy* E_T *which probably cannot be accounted for by the liberation of the known nuclear packing fraction.*

It seems that Zwicky was already aware of his reputation for expressing unorthodox views, so the authors added:

> *We are fully aware that our suggestion carries with it grave implications regarding the ordinary views about the constitution of stars and therefore will require further careful studies.*

The argument regarding the radius of the collapsed object was right, provided that all the energy we see in a SN is due to gravitation. However, there was no explanation as to why it should be a neutron star and not, say, a proton star. At that time, there was no definition of what a neutron star should be. The only calculations published about the state of matter under extreme conditions were those of Sterne a year earlier, and he found protons and electrons to be the last state. Furthermore, this was two years before Hund suggested the idea of the neutronization of matter under high pressure. On the basis of an argument for why it should be a neutron star that they never provided, Baade and Zwicky are credited today with identifying the supernova as a transition to a neutron star. It should be pointed out that Milne[175] suggested that novas might be due to the collapse of a star to form a white dwarf. Thus the idea of a collapse was not completely new. However, once the energy balance involved in a SN was calculated, Baade and Zwicky realized that the collapsed star must be much smaller for the gravitational energy to provide the energy for the explosion. How this takes place is another issue that is not understood even today.

In 1938, Lundmark raised the following provocative question in the title of a conference paper:[176] *Was the Crab Nebula Formed by a Supernova in 1054 A.D.?*. No detailed follow-up was published.

An explanation of why the SN is a collapse to a neutron star was only provided in 1938, when Zwicky[177] returned to discuss some consequences of the hypothesis that certain stellar cores are composed mainly of neutrons. He explained:

> *It must be emphasized that we here use the term neutron star simply to designate a highly collapsed star, the average density of which is of the order of the density of matter existing inside of ordinary atomic nuclei. When we therefore speak of the neutron composition of such a star this does not necessarily mean neutrons in the ordinary sense. It must be taken rather as a short designation for an extended state of matter of nuclear density in which every region whose linear dimensions d are larger than about* $\delta = e^2/mc^2$ *is necessarily electrically neutral, where e and m are the charge and the mass of the electron and c is the velocity of light.*

---

[174] Baade, W., & Zwicky, F., Proc. Nat. Acad. Sci. **20**, 259 (1934).

[175] Milne, E.A., The Obs. **54**, 140 (1931).

[176] Lundmark, K., pobv. conf. (1938) p. 89.

[177] Zwicky, F., Ap. J. **88**, 522 (1938); Phys. Rev. **55**, 726 (1939).

It is not clear why Zwicky introduced the term $\delta = e^2/mc^2$ which is the classical radius of the electron. It is derived by equating the electrical energy of the electron (how much energy must be invested to put the charge on the electron), with the rest mass energy $E = m_0c^2$, and ignoring quantum theory altogether (this is the reason for the name 'classical' radius). However, at these distances and energies, quantum mechanics governs matter, and the description of the electrons as classical particles is not valid.

Further, the critical mass is due to relativistic effects and these were ignored by Zwicky. What Zwicky apparently meant was that the entire star was a giant nucleus. Recall how Fowler described the entire star as a superatom. Now we reach the densities at which the entire star can be considered as one giant nucleus. However, from Zwicky's observational point of view, what mattered was the radius of the star, which dictated the amount of gravitational energy released. Indeed, the observed energy released by the SN demanded such a small radius and high densities for the final configuration if the energy for the explosion was due to gravitation. It appears that Zwicky did not mean a neutron star in the sense we take it today, and the fortunate/accidental use of the term 'neutron star' led people to believe that he should be credited with the idea of the neutron star.

Zwicky briefly described some of the properties of neutron stars, as well as new observations of supernova which tended to support the neutron star hypothesis. And so contended Zwicky:

> *According to present knowledge, cold neutron stars represent the lowest energy that matter may assume without being completely transformed into radiation.*

Indeed, at that time the idea of collapse to a black hole was not yet known, as this was a year before Oppenheimer's and Volkoff's famous papers, which indicated that a collapse to a black hole could actually happen. He also asserted that:

> *According to the general theory of relativity, a limiting mass of stars exists for every given density.*[178] *At this limit,* claimed Zwicky, *the energy liberated because of gravitational packing is* $0.58Mc^2$*, where m is the mass of the star.*

Zwicky claimed that this result was derived in a discussion with Tolman and would be communicated in a joint paper. No such paper could be found in the published literature. A star which reached the Schwarzschild limiting configuration, explained Zwicky, *must be regarded as an object between which and the rest of the world practically no physical communication is possible.*

Soon after Schwarzschild published his sensational paper about the solution of Einstein's equation (to be discussed later), he published a second paper in which he assumed the star to be incompressible so that the nagging question of what happens to the matter when it shrinks to vanishing size is eliminated. It was as if Schwarzschild himself did not believe that his solution led to infinite densities and hence

[178] Here Zwicky cited the second paper by Schwarzschild, in which he discussed the solution he found to Einstein's equation assuming the star to be incompressible. According to special relativity there cannot be incompressible matter, but this assumption simplifies the calculation. The speed at which information is transmitted in incompressible matter is infinite. But according to the special theory of relativity, information cannot propagate faster than light.

represented an unphysical solution. So if one assumes the matter to behave in this way, one finds that there is a maximum density. The radius of the star must be larger than $R_{\text{Schw}} = 2GM/c^2$, which is known as the Schwarzschild radius. For the Sun, for example, the Schwarzschild radius is 2.86 km which means that, if the Sun contracts to a radius smaller than this, light and information will be unable to come out of the star and tell the rest of the universe what is going on inside. If the star has a mass M, the mean density is then the mass divided by the volume, so that the maximum density[179] becomes $\rho_{\text{limit}} = (c^2/G)/\left(8\pi R^2_{\text{Schw}}/3\right)$, and for a mass like that of the Sun, this density is 7.232×10^{11} g/cm^3.

Consider the following question: when is the gravitational energy of a star equal to its rest mass energy? The rest mass is given by $E = mc^2$. The gravitational energy of a star is given by Gm^2/R times a numerical factor of the order of unity which depends on the exact density distribution in the star, a refinement we leave aside here. By equating the rest mass with the gravitational mass, we get $R = Gm/c^2$. Thus, the radius $R_{\text{Schw}} = 2Gm/c^2$ has the significance that, if a star contracts to this radius, half the rest mass energy is released.

However, Zwicky did not follow Schwarzschild, and gave an expression for the limiting mass as follows. Let m_{p}, m_{n}, and m_{e} be the masses of the proton, the neutron, and the electron, respectively. According to Zwicky the limiting mass is given by

$$M_{\text{L}} = \alpha R^{3/2} m_{\text{n}} = \alpha 91.04 M_{\odot}\,, \quad \text{where } R = \frac{e^2}{G m_{\text{p}} m_{\text{e}}} = 2.267 \times 10^{39} \tag{5.8}$$

is the ratio of the electrical to the gravitational attraction between an electron and a proton, and α is a constant of the order of unity whose value was not specified by Zwicky. This large number is an expression for the strength of the electrical force between the elementary particles relative to the gravitational attraction between them. According to Zwicky, the limiting mass did not depend on the special theory or relativity, because c does not appear, and nor did it depend on quantum theory, because the Planck constant does not appear.

Several years earlier, Landau and Chandrasekhar derived the limiting mass in which special relativity and quantum theory not only play a dominant role, but are the very reason for the existence of a limiting mass. Overlooking these fundamental results was unthinkable. Electromagnetic and gravitational theories alone do not lead to a limiting mass. Note the following strange feature in Zwicky's expression. The term R is dimensionless, since it is the ratio between two forces. The dimension of mass comes from the mass of the neutron. If R is dimensionless, one can raise it to whatever power one wants, and not necessarily 3/2. So where did the 3/2 come from?

Zwicky gave two numerical examples, and the second example corresponds to a neutron star of mass $31.7M_{\odot}$. (Recall that, according to Landau, the limiting mass was $1.5M_{\odot}$.) Zwicky ended his paper with a comment on how strange the nuclear

[179] Zwicky found incorrectly that $R_{\text{Schw}} = 74$ km for the Sun, with a critical density equal to 1.2×10^{12} g/cm^3.

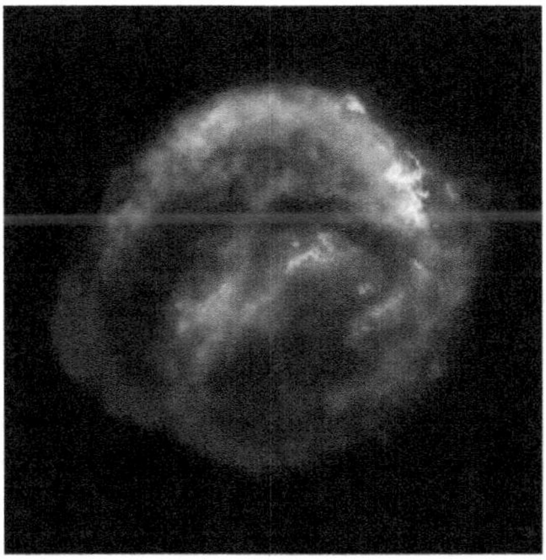

Fig. 5.12 The Kepler 1604 supernova remnant. The supernova was discovered by Kepler. The present picture is from the Hubble Space Telescope. Credit: NASA, Hubble Space Telescope

reactions inside a neutron star might be. However, neutron stars are the end of all nuclear reactions. No nuclear reactions can take place under such conditions. Zwicky commented that: *The derivation of these results, which was obtained in a discussion with Tolman, will be communicated in a joint paper.* But there is no such paper to be found in the SAO/NASA Astrophysical Data System.

Obviously, Zwicky got the idea of neutron stars for the wrong reasons. According to Rosenfeld,[180] when the news about the discovery of the neutron reached Copenhagen, Bohr and Landau were spending the evening together and started a discussion about the possible consequences of the existence of a neutron. It was then that Landau suggested the idea of a cold neutron star. However, there is no publication to that effect. Moreover, the spin of the neutron was not known at that time. Even in the book *Statistical Mechanics*, by Landau and Lifshitz, where the 'neutron sphere' is discussed, there is no mention of Landau's or Chandrasekhar's papers, let alone who discovered the limiting mass. The only references are to Oppenheimer and Volkoff and Oppenheimer and Server from 1939 (to be discussed next). Incidentally, the critical (maximal) mass Landau and Lifshitz derived in their book is $0.76M_{\odot}$. We note that the idea of a critical stellar mass depends on the fact that the particle which provides the pressure must obey Fermi–Dirac statistics.

When Chadwick announced his discovery (the paper was sent for publication on 10 May 1932), the spin of the neutron was not known, and hence it was not possible to determine whether the newly discovered particle obeyed Fermi–Dirac statistics, and consequently whether it implied a limiting stellar mass. About four months later, in August 1932, Bacher and Condon[181] analyzed the available data on the structure

[180] Rosenfeld, L., Astrophys. & Gravitation, 1974, in Proc. 16 Solvay Conf. on Phys., Univ. de Bruxelles, p. 174.

[181] Bacher, R.F., & Condon, E.U., PRL **41**, 683 (1932).

of light nuclei theoretically, and considered the following possibilities for the spin of the particles in the nucleus, in contrast to their spin as free particles (it was not clear at that time whether particles preserved their spin). The three possibilities were proton spin 0 or 1/2, neutron spin 0, 1/2, or 1, electron spin 0 or 1/2, and α particle spin 0. Had it been found that the neutron had an integer or zero spin even when free, this would have implied that no neutron star could exist. Bacher and Condon concluded that the most probable hypothesis for the spins of the particles in the nucleus was proton 1/2, neutron 1/2, and electron 0. Free particles could have different spins. Shortly afterwards, it became clear that the nuclear-bound and free particles had the same spin, i.e., spin is an intrinsic property that does not vary in time and does not depend where the particle is (the particle carries its spin with it wherever it moves). Thus, Bacher and Condon showed that neutrons obey Fermi–Dirac statistics, and lead to a stellar limiting mass.

Chadwick believed in the idea that the neutron was composed of a proton plus an electron, and hence was satisfied with the result that the neutron is more massive than the proton, but not as massive as the sum of the proton plus the electron masses. The difference is the binding energy of the neutron. According to Chadwick's calculation the binding energy of the neutron was about 1–2 MeV, or about one thousandth of the mass-energy of the proton. This is a large number because the rest mass of the electron is only $m_e c^2 = 0.51$ MeV. How the proton and the electron are bound to form the neutron was another question, at which Chadwick made a guess. The system of the neutron, composed of a proton and an electron, is not stable and disintegrates into its components when free. The neutron is stable in the nucleus only if all other possible lower states are occupied, or there is no lower state. If there is a vacant lower energy state into which the neutron can decay, then it undergoes a β^- decay, where the neutron in the nucleus disintegrates into a proton and an electron which leaves the nucleus.

To get a better grasp of the general thinking in those days, let us quote Rutherford 1933:[182]

> *It is believed that the two primary units comprising the nucleus are electrons and protons, but there is strong evidence that secondary more complex units may be formed by the combination of protons and electrons. The most important secondary unit is the α particle, consisting of a combination of 4 protons and 2 electrons, and recent evidence for the existence of a neutron – a close combination of a proton and an electron – has been obtained. Both of these secondary units may form an essential part of the nuclear structure.*

5.28 The Transformation of Matter into Neutrons

Although the scenario according to which a supernova is a collapse of the star towards a neutron star was already in the air, it was not clear at all how the nuclei, protons, and electrons could transform into neutrons. The only certainty was the total energy released in the transition. No mechanism was known or even suggested. It

[182] Rutherford, E., JRAS **27**, 155 (1933).

was known that the neutron is more massive than a proton plus an electron, yet how the transformation of ordinary matter to neutrons could take place was an entirely different question. It was clear from radioactive β decays that such a transformation could take place inside the nucleus, but would a sea of protons or nuclei, plus a sea of electrons transform under any conditions into neutrons? The answer was given in a long paper by Hund (1896–1997),[183] who was the first to predict the state of matter when the temperature and density become extreme.

Hund discussed the following equilibrium process:

$$\text{proton} + \text{electron} \rightleftharpoons \text{neutron} + Q\,,$$

and he cited Sterne[184] to justify the assumption that this reaction is in equilibrium in stars. This is not at all a trivial assumption, and for it to be valid, the reactions must be sufficiently faster than the rate of stellar evolution. Hund extended the possibilities and included equilibria like

$$\text{nucleus} \rightleftharpoons Z\,\text{neutrons} + Z\,\text{protons}\,, \quad \text{nucleus} + Z\,\text{electrons} \rightleftharpoons 2Z\,\text{neutrons}\,.$$

This was not an error. Since Hund did not know the difference in mass between the neutron and the proton very accurately (at this time, the mass of the neutron was given as 1.0085 ± 0.0005, while the mass of a proton plus an electron was 1.0080) his calculation was quite inaccurate.[185]

Hund found that, as the density increases and the electrons become more energetic and degenerate, the equilibrium moves towards the more massive particles, namely, the neutrons. At a high density, the free neutrons cannot decay into protons and electrons, because the electrons fill all the available energy states. When the electrons occupy all the energy states up to an energy $E_{\text{k}} = (m_{\text{n}} - m_{\text{p}} - m_{\text{e}})c^2$, the electron emitted by the decaying neutron, which has at most a kinetic energy E_{k}, has no vacant cell to go to because of the Pauli principle, and the neutron cannot decay. On the contrary, a proton which meets such an energetic electron prefers energetically to convert into a neutron. Thus, here on Earth, where the density of electrons is negligible, the free neutron is unstable, but when the environment is full of electrons, the stable state is a neutron. The most stable state of matter thus depends on the environment.

The idea of a possible equilibrium between neutrons and protons was first discussed by Flügge (1912–1997)[186] in his doctoral thesis. He realized that, when the temperature is low,[187] the system will be on the left-hand side, i.e., it will comprise a proton and an electron. But when the temperature is high, the balance shifts to

---

[183] Hund, F., Erg. d. exacten Naturwis. **15**, 189 (1936).

[184] Sterne, T.E., MNRAS **93**, 736, 767, & 770 (1933).

[185] Today's values are $m_{\text{n}} = 1838.69 m_{\text{e}}$ and $m_{\text{p}} = 1836.16 m_{\text{e}}$, so that $m_{\text{p}} + m_{\text{e}} < m_{\text{n}}$, and the kinetic energy available in the decay is $E_{\text{k}} = (m_{\text{n}} - m_{\text{p}} - m_{\text{e}})c^2 = 1.53697 m_{\text{e}} c^2 = 0.78$ MeV.

[186] Flügge, S., Veröffentlichungen der Universitäats-Sternwarte Göttingen, No. 31, VeGobe, **3**, 2 (10 March 1933).

[187] Low here means low relative to $T = Q/k_{\text{B}}$.

the right and the system takes the form of a neutron. Flügge even worked out the equation for a star in which the above reaction takes place. To formulate this equation, he assumed that the radiation pressure did not act on the neutrons because they were neutral. Further, he assumed that the neutrons behaved as an ideal gas, like the protons and the electrons (at low pressures). Finally, he ignored the general theory of relativity and assumed that Newtonian mechanics was valid even for dense stars. However, his main interest was in the effect of the neutrons on the nuclear reactions, and so he missed what Hund discovered a couple of years later.

5.29 The Neutron Star

In 1938, a year before nuclear fusion in stars was discovered, Landau suggested[188] as a model for the energy source in stars that each star might have a neutron star in the core. The neutron star was the collapsed core of the star à la Milne. Landau forgot his criticism of Milne's model and conceived of an identical model to Milne's, with a neutron star core instead of Milne's white dwarf. The gravitational field of the neutron star was extremely strong, so the envelope would be pulled inward by the neutron star, and the contraction of the envelope, à la Kelvin–Helmholtz–Ritter, would then supply the energy of the star.

Could a neutron core form inside a star, or as an isolated star? And if so, what would be the minimal mass for which a neutron state would be more favorable than an electronic state (which led to Chandrasekhar's white dwarf case). Landau calculated that the energy needed to transform one gram of matter into neutrons was about 7×10^{18} erg, and the key question then was whether or not gravitational contraction could supply this energy. Landau found that, when the stellar mass exceeded $0.05M_{\odot}$, the neutron sphere became energetically favored. If one takes into account the fact that neutrons obey Fermi–Dirac statistics, then the minimal mass for a neutron star becomes $0.001M_{\odot}$, whence there is no energy problem in forming such a star. But that in itself does not mean that such a star would actually form.

The gradual raining down of matter on the core releases plenty of energy to account for the stellar energy. The core grows gradually, and its size varies from one star to another. However, there remained two basic problems inherent to this picture:

- In 1932, Landau had shown that such a neutron core could exist in stars more massive than $1.5M_{\odot}$. If this were so, what then was the energy source of less massive stars, like our own Sun?
- If the mass of the core had to be larger than a certain number, it was clear that it could not form gradually. Hence, the unavoidable conclusion (which did not appear in Landau's paper) was that it could not form in a slow and gradual process, but could only come about dynamically, say in a supernova explosion.

[188] Landau, L., Nature **141**, 333 (1938).

5.30 Oppenheimer: The Collapse to a Neutron Star

During the years 1938–1939 Oppenheimer, (1904–1967) and his associates published three seminal papers explaining how neutron stars could form, what their maximal mass would be, and what would happen in stars with masses above the limit. In 1938, Oppenheimer and Serber[189] carried out an analysis of the stability of stellar neutron cores, the idea put forward by Landau. With the growing knowledge of nuclear reactions, argued Oppenheimer and Server:

> *It has grown clear that such reactions must take place in stellar interiors, and that, on the basis of a standard Eddington model, reactions must occur which can account in order of magnitude for the radiation of the lighter stars [...] Nevertheless, it has become clear that these reactions could in no way account for the enormously greater radiation of such stars as Capella.*

Hence, they concluded that Eddington's model did not apply.

If Eddington's model fails for Capella, what about Landau's model of a condensed neutron core? The issue addressed by Oppenheimer and Server in this case was the extent to which such a star would be stable. According to the authors, Landau's estimate that the minimal mass should be $0.001M_\odot$ was wrong. A revised calculation[190] yielded $M_\odot/6$. The conclusion was that the existence of such a core *would involve a complete breakdown of Eddington's model.* However, they were unable to assess the stability of such a model, since the detailed interaction between two neutrons was not yet known. Finally, given the uncertainty involved in the nature of the nuclear forces, they concluded that the knowledge available to them at that time precluded neutron cores in stars with masses comparable to that of the Sun.

Two complementary and jointly published milestone papers appeared in 1939, one by Tolman[191] and the other by Oppenheimer and Volkoff.[192] Both dealt with the question of the existence of neutron stars. Oppenheimer and Volkoff found that the maximal mass for a neutron star was $0.75M_\odot$, in agreement with Tolman. Neutron stars with masses less than about $0.1M_\odot$ were found to be unstable, as the neutrons disintegrate into protons and electrons (the density was not sufficiently high to prevent the decay of the neutrons). But they found that normal stars with masses under $1.5M_\odot$ would not collapse to a neutron star. So Oppenheimer and Volkoff concluded that: *It is unlikely that static neutron cores can play any great part in stellar evolution.* As for very massive stars, they continue to contract forever. Of course, Oppenheimer and Volkoff had to make certain assumptions about the properties of matter at these extreme densities, but as time has shown, various corrections to the assumed behavior of the matter have changed the estimates for the mass of the neutron star only within the range from $0.75M_\odot$ to about $3M_\odot$, but never beyond it, so that the problem has remained. In particular, as the authors warned, the

[189] Oppenheimer, J.R., & Serber, R., PRL **54**, 540 (1938).

[190] Oppenheimer and Server assumed constant density, exactly like Stoner. However, they argued that a more accurate calculation would change the result by no more than 10%.

[191] Tolman, R.C., Phys.Rev. **55**, 364 (1939.

[192] Oppenheimer, J.R., & Volkoff, G.M., Phys. Rev. **55**, 374 (1939).

densities in the limiting models exceeded the densities of the nucleus and thus represented uncharted territory, where experimental knowledge was meager and one had to rely on unverified assumptions.

It is interesting to note to whom Oppenheimer and Volkoff attributed, or did not attribute, the priority for the idea of a neutron star. The possibility of such a thing was attributed to Gamow:[193]

> *[...] who hypothesized that in sufficiently massive stars, after all the thermonuclear sources of energy, at least for the central material of the star, have been exhausted, a condensed neutron core would be formed.*

As for Landau's derivation of the limiting mass, Oppenheimer and Volkoff criticized the result as being obtained on the basis of Newtonian gravitational theory, while in this case general relativistic effects cannot be ignored. Furthermore, Zwicky was not mentioned at all. This fact infuriated Zwicky, who used to claim his priority on every possible occasion.[194]

Back to back with the paper by Oppenheimer and Volkoff was Tolman's paper, in which he gave his impressions of the conversation with Zwicky, whence he wrote:

> *My own present interest in solutions of Einstein's field equations for static spheres of fluid is specially due to conversations with Professor Zwicky of this institute, and with Professor Oppenheimer and Mr. Volkoff of the university of California, who have been more directly concerned with the possibility of applying such solutions to problems of stellar structure. Professor Zwicky in a recent note [Ap. J.* ***88****, 522 (1938); see also Phys. Rev.* ***54****, 242 (1938)] has suggested the use of Schwarzschild's interior solution for a sphere of fluid of constant density as providing a model for a 'collapsed neutron star'. He is making further calculations on the properties of such a model, and it is hoped that the considerations given in this article may be of assistance in throwing light on the questions that concern him.*

Apart from here, Tolman did not cite any paper by Zwicky. This presumably explains why a joint Zwicky–Tolman paper never saw the light of day.

In the last paper on neutron and collapsing stars, Oppenheimer and Snyder[195] discussed the fate of a collapsing star:

> *When all thermonuclear sources of energy are exhausted, a sufficiently heavy star will collapse. Unless fission due to rotation, the radiation of mass, or the blowing off of mass by radiation, reduce the star's mass to the order of that of the Sun, this contraction will continue indefinitely.*

This was the first calculation that indicated that a massive star might collapse at the end of its evolution to what we call today a black hole. Oppenheimer and Snyder did not say that the star would collapse to a point, as worried those who objected to black holes, but that *the contraction will continue indefinitely*. The force of gravity always wins. Nuclear energy just served to halt the collapse, in the star's race towards an unspecified end. The collapse to a 'mathematical point' poses logical problems, so Oppenheimer and Snyder explained that it was impossible to reach an infinite

[193] Gamow, G., *Atomic Nuclei and Nuclear Transformations*, Oxford, 2nd edn. (1936) p. 234.

[194] I was personally present on two such occasions during seminars in Caltech.

[195] Oppenheimer, J.R., & Snyder, H., Phys. Rev. **56**, 455 (1939).

density in a finite time. In other words, the collapse would take forever, so that the conceptual problem would never arise (actually they sought a solution that satisfied this condition to begin with).

Today the Tolman–Oppenheimer–Volkoff (TOV) limit is an upper bound to the mass of stars composed of neutron-degenerate matter, or in short neutron stars. The limiting mass is equivalent to the Chandrasekhar limit for electron-supported white dwarf stars. The TOV limit is estimated to be approximately 3–5$M_\odot$. Clearly, with progress in nuclear and elementary particle physics, we may learn about other types of particles that can form at very high densities and create objects that might escape the perpetual collapse to a 'mathematical point'. As the properties of the more exotic, hypothetical forms of degenerate matter are even more poorly known than those of neutron-rich nuclear matter, most astrophysicists assume, in the absence of evidence to the contrary, that the TOV limit for neutrons is in the range 1.5–3$M_\odot$ for any kind of particle which obeys Fermi–Dirac statistics.

Theoretical predictions at this time therefore suggested that the stars end up either as white dwarfs, or as neutron stars, or in an infinite collapse. Nobody paid attention to the fact (or the argument) that, with the predicted scenario, no synthesized elements could escape from the star. This was therefore a one-way street for matter. Interstellar gas condensed to stars, which ended their life in one of three states, all of which would bury the products of nucleosynthesis forever.

5.31 Back to Observations

In 1938, Baade established the differences between novas and SNs as two distinct phenomena, and brought to bear further arguments to show that the 1054 A.D. event was indeed a supernova. However, he was not fully convincing. It was only in 1939 that Mayall (1906–1993),[196] in a paper entitled *The Crab Nebula: A Probable Supernova*, accumulated enough data to identify the Crab nebula with the Chinese records of 1054. The beginning of the article is amusing:

> *The year of our Lord 1054, when Omar Khayam was a small boy, and the Battle of Hastings still twelve years in the future, an unknown Chinese astronomer, perhaps weary and sleepy after working all night, was astonished to see a strange and brilliant new star appear in the graying eastern sky just before sunrise.*

Mayall drew support from Iba's report, which confirmed the Chinese accounts. On the other hand, so claimed Mayall, the Chinese Chronicles had remained unknown in the West until 1921. Apparently, Biot's studies, published in French, were overlooked by Mayall, as he did not mention them.

The best evidence was supplied by the measured expansion velocities, which could be extrapolated backward to show the moment at which the expansion started. Mayall ended his paper with the conclusion:

---

[196] Mayall, N.U., ASPL **3**, 145 (1939).

It may be said that the identification of the Crab nebula as a former supernova possesses a degree of probability sufficiently high to warrant its acceptance as a reasonable working hypothesis. Bearing this in mind, together with the fact that supernovas are very rare and puzzling objects, about which little is known, perhaps we can appreciate why the Crab nebula is one of the most interesting objects in the sky.

Indeed, Mayall was quite right. The paper by Lundmark[197] was cited. However, according to Mayall, Lundmark merely pointed out that the position of the Crab nebula agreed with the Chinese records, *but he did not, in the absence of other data, suggest any closer relation between the two objects.*

5.32 Novas and Supernovas Are Not the Same Thing

The evidence that SNs and novas are not a homogeneous group of phenomena was beginning to accumulate, and by 1940 Zwicky published a long paper[198] in which he classified novas and SNs. Interestingly, in Table 1 of Zwicky's paper, the 1054 eruption was defined as a nova, although Mayall and Baade had already classified it as a SN. Zwicky explained that the expansion velocities (about 1 300 km/s, while most SNs have an expansion velocity of about 5000 km/s) were too low for a SN. Incorrectly identifying the source of the continuum radiation as thermal radiation from a hot star, Zwicky attempted to derive the temperature of the remnant star, and got a temperature in excess of 133 000 K, much higher than is known to exist on the surface of a star. SN 1054 was indeed unique, and its peculiar spectrum confused the astronomers. Zwicky also made the important observation that there were SNs which left a remnant, and those for which a remnant star was not discovered.

5.33 Clinching the Identification

J.J.L. Duyvendak (1889–1954) was a well known Dutch sinologist and a good friend of the astronomer Oort. So Oort (1899–1992) asked his friend to examine the Chinese and Japanese chronicles to see whether more data was available about the 'guest star' of 1054, beyond what had already been unearthed. And indeed, Duyvendak found an additional reference, and wrote to Oort about it. He even published the discovery.[199] Duyvendak discovered that Biot's translation to the French 'à la fin de l'année' was erroneous, as were the Chinese Chronicles themselves, because

[197] Lundmark,K., *Was the Crab Nebula Formed by a Supernova in 1054 A.D.?*, pobv. conf. (1938) p. 89.

[198] Zwicky, F., Rev. Mod. Phys. **12**, 66 (1940).

[199] Duyvendak, J.J.L., PASP **54**, p. 91. Duyvendak wrote to Oort about it, and Oort communicated the result to Mayal and Baade. Soon afterwards, Oort resigned from the Leiden University in protest for its nazification. The resignation gave him trouble because his university superior, Herztsprung, did not approve, and Oort had to spend the war hiding in a small village.

the number of the lunar months cited in dating the event was wrong. The irony was that a sinologist had to correct a physicist and Chinese astronomers. An extremely important new piece of data was that the guest star was visible to the naked eye for more than a year. Actually, Duyvendak discovered records which indicated that the guest star was visible from 4 July 1054 to 17 April 1056.

The closing chapter in the story of the identification of the Crab nebula with a remnant of the 1054 supernova takes place in 1942. On the basis of Duyvendak's findings, namely that the 1054 nova was bright for close to two years, Mayall and Oort[200] claimed that the 1054 nova was indeed a SN, and one of the brightest on record. They knew about the expansion velocities measured by Duncan. So by comparing photographs taken several years apart, they were able to find the distance to the nebula from the differences in the images, and found a distance of 4 100 lyrs. Since the historical claim was that the new star was as bright as Venus, they were able to derive the actual brightness of the SN. Mayall and Oort also discovered some inconsistencies in the Chinese chronicles and tried to rectify them. However, the greatest problem they faced came from Duncan's recent results (of expansion velocities), which implied that the outburst happened in 1172 A.D. Even when the authors corrected them by treating the measurements slightly differently, the discrepancy could not be reduced to less then 84 years.[201] The discrepancy was explained as being due to the assumption of uniform expansion, while the expansion was probably faster at the beginning. Examining the paper carefully, one finds that Mayall and Oort's argument was based on elimination. The huge brightening observed by the Chinese could have been a comet, an ordinary nova, or a SN. By elimination, they were left with the SN hypothesis as the best explanation.

5.34 There Is More Than One Type of Supernova

Classification of the different types of supernova was first performed by Minkowski (1895–1976m).[202] He noticed that the SN spectra can be divided into two well defined and homogeneous classes, which he called Type I and Type II. Minkowski's samples contained 14 SNs, out of which 9 were Type I and the rest Type II. In this paper, hardly two pages long, he succeeded in establishing that there is probably more than one way to a SN explosion. The spectra of Type II SNs resemble to a large degree the spectrum of a nova, while the spectra of Type I SNs do not resemble the spectrum of any other heavenly body. Minkowski remarked that the spectra of Type I SNs was a complete mystery. Only two bands of oxygen could be identified in them. Furthermore, the intensity distribution remained unexplained, and did not agree with that of a black body.

[200] Mayall, N.U., & Oort, J.H., PASP **54**, 95 1942.

[201] There was also a discrepancy with the results of Deautsch, A.N., & Lavdovsky, V.V., cited by Mayall & Oort, who found an age of 785 ± 140 years.

[202] Minkowski, R., PASP **53**, 224 (1941).

In 1942, Baade[203] returned to the Crab nebula. He agreed with Mayall and Oort[204] about the identification of the Crab with the 1054 SN, and added that this was a Type I SN. In a note, he specified that, as he used it, the term SN referred to Type I SNs. Type II SNs have luminosities between those of ordinary novas and Type I SNs, and appeared, according to Baade, to be closely related to ordinary novas. In any case, during an outburst, they exhibit essentially the same phenomena as ordinary novas. The lack of any reliable method to estimate the distances to novas and SNs misled Baade here, and caused this confusion between novas and SNs. It should be mentioned that Baade[205] devised a special method to measure the distances to supernovas and novas. However, the method only works if the radiation emerges from a well defined surface.

5.35 Is There Any Remnant?

After predicting with Zwicky that a SN is a collapse to a neutron star, Baade was interested in his 1942 paper in *the final state of a supernova*.[206] He thus began to look for the central star in the Crab nebula, finding it to be a faint double star, and assuming that it was the radiation from this object that excited the nebula. However, the star was very strange. In Baade's own words: *Curiously enough, all attempts to identify the star in this way have been indecisive*. The stars appeared to be abnormal. Baade's search for possible fainter stars as candidates for the SN remnant were in vain.

Using some clever tricks, Baade managed to classify the north star of the binary as an F or G type star, but all attempts to classify the south star failed: *What made the case puzzling was the fact that no lines could be seen in the spectrum*. Next Baade checked the velocities of the two stars and concluded that the north star was just a background star, leaving the abnormal star as the only viable candidate *to excite the nebula*, although it was not at the exact center of the nebula. Despite all his precautions, Baade was not happy and wrote that *there remain serious doubts whether this agreement is significant*.

In the same journal issue that published Baade's paper and just after Baade's article, there appeared the article by Minkowski,[207] in which he referred to the south star as the stellar remnant of the supernova. While he agreed in principle with Baade, Minkowski stressed that the velocity measurements were far from conclusive. Minkowski also realized that there was no way to find the temperature of the star. The

---

[203] Baade, W., Ap. J. **96**, 188 (1942).

[204] Mayall, N.U., & Oort, J.H., PASP **54**, 95 (1942).

[205] Baade, W., AN **228**, 359 (1926).

[206] In a footnote he pointed out that: *In the following discussion, the term 'supernova' always refers to a supernova of Type I. Supernovae of Type II, with luminosities intermediate between those of ordinary novae and supernovae of Type I, appear to be closely related to the ordinary novae. In any case, during an outburst, they present essentially the same phenomenon as common novae.*

[207] Minkowski, R., Ap. J. **96**, 199 (1942).

radiation from the star was not the usual stellar radiation, close to that of a black body, which explained Zwicky's strange result.

5.36 The Discovery of Neutron Stars

For years the predictions and the claims of the existence of neutron stars made during the 1930s were just part of the saga of astrophysics. But a great victory came in 1967, when radio pulses with highly precise periodicity were discovered by Hewish, Bell, Pilkington, Scott, and Collins.[208] This earned Hewish the 1974 Nobel Prize for Physics.[209] The first four objects, later named pulsars, had a periodicity of around 1 second, which indicated to the authors, as they suggested in the abstract to their paper, that:

> *Unusual signals from pulsating radio sources have been recorded at the Mullard Radio Astronomy Observatory. The radiation seems to come from local objects within the galaxy, and may be associated with oscillations of white dwarfs or neutron stars [...]*

The idea that the pulsars are rotating neutron stars was suggested independently by Gold[210] and by Pacini,[211] although the two authors were both at Cornell at that time. Note that the two theoretical explanations were published before the observations. The observers kept the results for a long time before publishing, giving rise to a flood of rumors about the discovery. The rumors spread over the entire scientific community, giving rise to various speculations and attempted explanations/models/predictions.

Neutron stars and white dwarfs are two possible final states of stars. These two categories of stars are perfect examples of the Kelvin–Helmholtz–Ritter gravitational contraction cooling stars. Of course, the above authors never dreamt of such stars as these, although their hypothesis was in fact realized by them. The end point for lone white dwarfs and neutron stars is a cold dead star.[212] Since the cooling time of white dwarfs is about the age of our galaxy, we do not expect there to be too many unobserved cold white dwarfs. Neutron stars cool much faster, so our galaxy may contain many cold dead neutron stars.

---

[208] Hewish, A., Bell, S.J., Pilkington, J.D.H., Scott, P.F., & Collins, R.A., Nature **224**, 472 (1969). At first, because of the small size of the emitter and the regularity in the signals, the hypothesis that this could be a signal from an extraterrestrial civilization was seriously considered, and the objects were given the nicknames LGM 1 for Little Green Man 1.

[209] Hewish got the Nobel Prize in 1974 with the radio astronomer Martin Ryle.

[210] Gold, T., Nature **218**, 731 (1968).

[211] Pacini, F., Nature **219**, 145 (1968).

[212] The end point for a close binary system containing a neutron star or white dwarf is quite different. Due to energy loss via gravitational waves, where the binary acts as a broadcasting antenna for these gravitational waves, the two stars approach one other and eventually merge to form an object with mass above the Chandrasekhar or the TOV limit, respectively. This object may collapse as a SN or become a black hole.

The first serious suggestion concerning the nature of the Type I SN spectra was made by Shklovskii, who, in 1953, identified the radiation from the Crab nebula as synchrotron radiation.[213] Synchrotron radiation is emitted by electrons moving at speeds close to the speed of light through a magnetic field. The name stems from the fact that such radiation was first detected from synchrotrons.[214] Synchrotron radiation has a unique dependence on wavelength, and is thus easy to identify. The idea of synchrotron radiation was first applied to a cosmic object by Alfven and Herlofson,[215] who recognized the importance of the discovery by Elder et al. for our understanding of the Sun. It took three years for Oort and Walraven[216] to confirm Shklovskii's hypothesis by discovering that the light from the Crab nebula is polarized, which is the signature of synchrotron radiation. However, it was still not known where the high energy electrons and the magnetic field needed to generate this radiation actually came from.

Soon after the discovery of the first four pulsars, it was natural to ask whether there might be one in the Crab nebula. The first search was made for an object with a period of about one second, but nothing was found. Staelin and Reifenstein[217] discovered radio signals from two sources near the Crab nebula. In the first report they could not identify a periodicity, the hallmark of pulsars. However, shortly afterwards, Cocke, Disney, and Taylor[218] discovered a periodic signal in the visible light, with a periodicity of 33 milliseconds![219] The remnant star of the SN was discovered to emit strong non-thermal radiation, which explained why astronomers were misled for a long time. Baade passed away in 1960, and could not enjoy, as Minkowskii did, the identification of his bizarre star with a neutron star.

5.37 Eddington's Objection to the White Dwarf Theory

If the reader has been wondering whether Eddington had mellowed with time, or even better, was eventually convinced by Chandrasekhar or Landau, she/he need only note that Eddington went to the trouble of sending his message again. In 1939,

---

[213] Shklovskii, I.S., Dokl. Akad. Nauk. SSSR **90**, 983 (1953).

[214] Synchrotron radiation was detected at General Electric in 1946 by Elder, F.R., Gurewitsch, A.M., Langmuir, R.V., Pollock, H.C., Phys. Rev. **71**, 829 (1947).

[215] Alfven, H., & Herlofson, N., PRL **78**, 616 (1950).

[216] Oort, J.H., & Walraven, T., BAN **12**, 285 (1956).

[217] Staelin, D.H., & Reifenstein, E.C., Science **162**, 1481 (1968).

[218] Cocke, W.J., Disney, M.J., & Taylor, D.J., Nature **221**, 525 (1969).

[219] The present theory claims that the energy radiated away by the remnant is supplied by the rotating neutron star. Consequently, the neutron star should slow down and the period lengthen. This is indeed what is observed. The change in the period has even been measured. The very short period of 33 milliseconds implies a very young, fast-rotating neutron star, embedded in the expanding remnants. Neutron stars rotating with a period of one second are considered to be old and significantly slowed down stars, whose nebula had already dispersed to the point of being unobservable. At the time these observations were made, one needed imagination to search for such a short period as one in the millisecond range.

Eddington[220] returned to the white dwarf problem, discussing the hydrogen content in white dwarfs and of course rejecting the recent advances:

> *In 1929 Stoner and Anderson independently put forward a modified equation [...] The modification is, however, fallacious [...]*

Eddington was not convinced, and did not change his view from the one expressed in earlier papers.[221] Eddington's return to the problem he had attacked previously was prompted by *one of the more recent attempts to defend it* by Chandrasekhar,[222] who published a summary of his theory in the form of a book.

Eddington just stressed once again that the electrons in the star cannot be treated as free particles, and resorted to observation. The question was whether white dwarfs contain hydrogen. Assuming the wrong equation, Eddington derived a central temperature for white dwarfs which was four to five times higher than that of main sequence stars. If that had been so, the release of subatomic energy, as recently discovered by Bethe,[223] should be very large (some 10^{16} times higher!), while we observe the white dwarfs to be low luminosity stars. Eddington thus faced a dilemma, and looked around for possible ways out.

Chandrasekhar assumed in his book that the hydrogen content of Sirius B was about 50% by mass. Eddington agreed that the amount of hydrogen in WDs was not negligible. This time they were both wrong! But the problem that really bothered Eddington was the measured gravitational redshift of Sirius B. According to Einstein's theory of general relativity, a photon emerging from the gravitational potential well of a star experiences a shift towards the red. The photon loses energy trying to climb the potential well on its way out. In 1925, Adams[224] attempted to measure the gravitational redshift of Sirius B and got about 19 km/s, much to Eddington's satisfaction.[225] Such a small redshift implied a large radius and low mean density. In 1928, Moore[226] confirmed Adams' measurement. The mass of Sirius B is $0.98M_\odot$, as measured quite accurately using Kepler's law. The radius was known from the surface temperature and luminosity. Hence the predicted redshift should have been 50 km/s and not the observed 19 km/s. This discrepancy plagued Eddington as he looked for various alternative explanations, but failed to find one. So he wrote:

> *Though the large hydrogen content may not be established with the ideal security which we should desire in a fact on which far-reaching conclusions are to be based, it must, I think, be admitted that the evidence for it is difficult to shake.*

---

[220] Eddington, A.S., MNRAS **99**, 595 (1939). Published in October 1939.

[221] Eddington, A.S., MNRAS **95**, 194 (1935); ibid. **96**, 20 (1935). See also the book by Eddington, entitled *Relativistic Theory of Protons and Electrons* (1936) Sect. 13.5.

[222] Chandrasekhar, S., *Study of Stellar Structure*, Chicago (1939) p. 366. This is the famous book by Chandrasekhar.

[223] Bethe published several papers, first on the pp chain and then, in March 1939, on the CN cycle. However, these papers were submitted in September 1938 and were apparently known to Eddington. Although he did not cite Bethe, he mentions his results.

[224] Adams, W.S., Proc. Nat. Acad. Sci. **2**, 382 (1925).

[225] Eddington, A.S., *Stars and Atoms*, Clarendon Press (1927) p. 53.

[226] Moore, J.H., PASP **40**, 256 (1928).

Eddington fell into his own trap. Eddington used to tout that you should not believe an observation that the theory does not explain/predict. In this case, both Adams' and Moore's results were heavily contaminated by light from the luminous Sirius A, and this contamination confused the observations. It was not until 1971 that Greenstein, Oke, and Shipman[227] succeeded in measuring the redshift from Sirius B and found the value 89 ± 16 km/s. The implication of the larger gravitational redshift is that the radius of Sirius B is much smaller and cannot contain any hydrogen. I am not sure that Eddington would have been happy with this observational result. Eddington knew about Bethe's 1939 discovery of the CN cycle as an energy source for stars, in which the carbon and nitrogen act as catalysts, but did not know about the direct proton–proton reaction which was discovered in the mid-1950s, and this may be the reason for his peculiar conclusions.

Back in the early 1940s, Eddington faced the problem of how to explain observations of white dwarfs. The first alternative, viz., *liberation of subatomic energy is in some way inhibited*, can happen if there is no hydrogen, for example, or no cȧtalyst (C or N, as needed for the CN cycle). However, since hydrogen is observed on the surface of white dwarfs,[228] Eddington concluded that this might imply that there were no catalysts to induce the nuclear reaction.[229] Since this is a rather a strange consequence, Eddington concluded that there had to be a process which eliminated carbon, or nitrogen, or both. This was obviously a contrived solution.

The second alternative was that the initial gas cloud did not contain any catalysts and that gravitational contraction continued until the star reached white dwarf densities. Only then would transmutation of the elements begin, but this time explosively. As a consequence, the star would expand to become a main sequence star. In other words, the white dwarf state preceded the main sequence state, a solution which creates many problems with the giants and the main sequence! In short, it was a solution that solved one problem by creating several bigger ones.

Eddington did not carry out any calculations (the entire paper was simply a hypothesis, without any calculation whatsoever to back it up), and was surprised by the fact that, of the white dwarfs known at the time, only one showed spectral lines of hydrogen, implying that his hypothesis did not even have minimal support from observation. In particular, he could not even reconcile Bethe's new theory with a single hydrogen-containing white dwarf.[230]

In 1940, Eddington published his last paper on the physics of white dwarfs,[231] repeating his arguments as to why the Stoner–Anderson formula was wrong:

---

[227] Greenstein, J.L., Oke, J.B., & Shipman, H.L., Ap. J. **169**, 563 (1971).

[228] Hydrogen is observed on the surface of white dwarfs. But the 'surface' happens to be an extremely thin layer, and this says nothing about the hydrogen content deep in the interior of the star. The atmosphere of a white dwarf is just few meters thick.

[229] In Bethe's theory of the nuclear energy source of stars, carbon and nitrogen act as catalysts which accelerate the reaction while remaining unchanged.

[230] The two major classes of white dwarf are class DA, where the surface contains hydrogen, and class DB, where no hydrogen is observed, but helium is.

[231] Eddington, A.S., MNRAS **100**, 582 (1949).

> *A formula established empirically in certain conditions,* claimed Eddington, *is extended to conditions in which it has not been verified by a procedure known as 'the principle of induction', or less euphemistically as 'blind extrapolation'. Such extrapolation, though often leading to progress, is fairly sure to break down sooner or later [...]*

Once again, Eddington betrayed his own principles about the universal validity of the physical laws.

But this criticism from the eminent Eddington did not deter investigators from implementing the Anderson–Stoner formula, as a well recognized quantum formula, in the still more extreme conditions of white dwarf matter. Eddington explained his different results for the pressure as being due to the fact that:

> *[...] there is a preferential time direction defined independently by the cube of matter under consideration, namely, the time direction with respect to which the matter surrounding the cube is at rest.*

It was not clear what it meant in the context of white dwarfs, and it has not become any clearer with the passage of time. Eddington did not give Chandrasekhar the honor of one of his attacks, and did not cite him. Stoner, who was attacked by Eddington, did not bother to reply. His last paper on the subject was published in 1939.[232] He apparently felt the issue was closed.

In 1939, Chandrasekhar published a long review[233] on stellar structure under the title *The Internal Constitution of the Stars*, which was exactly the title of Eddington's book. The review contained a chapter on white dwarfs, and surprisingly only Fowler and Pauli were mentioned in the section on white dwarfs. Strikingly absent were Anderson and Stoner. The accurate expression for the maximal mass of a white dwarf appeared for the first time, and it was $M = 5.75\mu^{-2}M_{\odot}$.

Eddington died in 1944, convinced he was right. In 1983 Chandrasekhar won the Nobel Prize for solving the problem of white dwarfs and for his contributions to understanding stellar structure. In his Nobel lecture he described the role of the limiting white dwarf mass in stellar evolution. No references to Stoner or Landau were given. Chandrasekhar was well known for his rigorous treatments of physics, so not bothering to supply the relevant citations is something puzzling to say the least. Not only that but, in the same year, Chandrasekhar published a book entitled *S. Eddington: The Most Distinguished Astrophysicist of His Time* (Cambridge University Press, 1983).

5.38 No Escape for Massive Stars

We return to Eddington's paradox. As we have seen, stars with masses less than the Chandrasekhar limiting mass and the TOV limiting mass can cool and end their life as cold dead bodies. What about the more massive stars? They are subject to

---

[232] McDougall, J., & Stoner, E.C., Phil. Trans. Roy. Soc. London Ser. A **237**, 67 (1939).

[233] Chandrasekhar, S., Proc. Amer. Phil. Soc. **81**, No. 2, 153 (1939).

Eddington's paradox, and were doomed according to Oppenheimer and Snyder, to contract forever.

5.39 Newtonian Black Holes versus Einsteinian Black Holes

Long before the events we have just been discussing, Pierre-Simon Laplace (1749–1827m),[234] who is known to astronomers for his theory of the nebular origin of the Solar System, considered the condition that material particles would not be able to leave a sufficiently compact dense star. Laplace found that, if the radius of a star was $R = 2Gm/c^2$, the escape velocity from the star would be equal to the speed of light, and he thus concluded that this object would appear dark, since light would not be able to emerge from it. But Laplace was not very accurate, even by the standards of Newtonian physics. When the velocity of the particle is below the escape velocity, the particle moves radially, decelerates until it reaches a maximum height, and then falls back again, exactly like a stone thrown up in the air by someone standing on the Earth's surface. When the speed is exactly equal to the velocity of escape, the particle moves up, slowing down all the time, but eventually reaching infinity after an infinite time. So an object with the radius Laplace mentioned could in fact be observed everywhere in the universe, provided the observer was within a finite distance from it. It was just that the time it would take the particle to reach the observer might be extremely long.

5.40 Schwarzschild. Solving the Einstein Equations

Very shortly after Einstein's publication[235] of his series of papers on the general theory of relativity, Schwarzschild[236] managed to get a full and accurate solution of the equations.[237] In this paper, Schwarzschild was mainly interested in the solution outside the object, and hence was not bothered with the fact that his solution became

---

[234] Laplace is immortalized on the Moon by having the mountain chain Promontorium Laplace named after him. Laplace was considered by many as the French Newton. Laplace, P.S., 1795 Astronomy Guide.

[235] Einstein, A., Sitzungsberichte der Königlich Preussischen Akademie der Wissenschaften zu Berlin, Phys.-Math. (1915) p. 778; ibid. (1915) p. 799; ibid. (1915) p. 831. The title of the last paper, which was published on 18 November is: *Explanation of the perihelion motion of Mercury from the general theory of relativity.*

[236] Schwarzschild, K., Sitzungsberichte der Königlich Preussischen Akademie der Wissenschaften zu Berlin, Phys.-Math. (13 January 1916) p. 189. It took Schwarzschild less than two months to find the sensational solution.

[237] According to general relativity and Schwarzschild's solution to the problem of motion around a given mass, every particle, or light signal, moves so that the quantity s given by

$$ds^2 = -(1-2Gm/rc^2)^{-1}dr^2 - r^2 d\theta^2 + (1-2Gm/rc^2)dt^2$$

Fig. 5.13 Karl Schwarzschild

infinite for the radius $R = 2GM/c^2$. As a matter of fact, he was interested in solving the problem Einstein had posed in his last paper, that is, he wanted to derive the motion of Mercury's orbit in space.

As is well known, the planets go around the Sun in ellipses which are close to circles, but not quite. If one considers only the motion of one planet around the Sun, neglecting the existence of all other planets, the orbit is fixed in space.[238] However, when the effect of all the planets is taken into account, it is found that they disturb each other's motion, and as a consequence the orbits do not stay fixed in space, but rotate. This phenomenon is called the advance of the perihelion (the perihelion is the point in the orbit where the planet is closest to the Sun). Observations indicated that Mercury's orbit rotates by about 575 seconds of arc per century. The effect of all the other planets amounts to 532 arcsec per century, leaving 43 arcsec unexplained. The discrepancy between Newton's theory of gravity and observational results had been known since the 1840s, when LeVerrier made accurate calculations of the perturbations in the Solar System. LeVerrier, whose thinking led to the discovery of Neptune on the basis of its perturbations to Uranus, naturally advocated the existence of a missing planet near the Sun. The mysterious planet was called Vulcan, but never discovered. Various other conjectures were put forward to explain this discrepancy. Among these was the idea that the Sun has a rotating core, and so is not spherical, or the suggestion that the elusive ether applied its drag to the planet, as well as many other theories.

---

and measured along the path between two points, has the maximum possible value. The reader will see from the expression that this simple result can cause a problem when the coefficient of dt^2 vanishes.

[238] The classical formulation of Kepler's first law says that the paths of the planets about the Sun are ellipses with the center of the Sun located at one focus. These elliptical orbits are fixed in space.

Einstein was only able to find an approximate solution, while Schwarzschild managed to obtain an exact solution. Furthermore, Schwarzschild was able to demonstrate that he had discovered the only solution. The prediction of the new theory was 42.9 arcsec, which was a great victory for Einstein's theory. No specially contrived circumstances such as an unknown star had to be assumed.

The particular radius appearing in Schwarzschild's solution, which eventually became known as the Schwarzschild radius, attracted the interest of physicists and astronomers. If the radius of the star is much larger than its Schwarzschild radius, the effects of general relativity are negligible, and vice versa. In the case of the Sun, $R_{\odot} = 690000$ km, while $R_{\text{Schw}} = 2.94$ km, and hence there is no problem. As a matter of fact, the ratio $2.94/690000 = 4.2 \times 10^{-6}$ is nothing but the ratio between the strengths of effects introduced by Einstein's general relativity and those of the Newtonian theory of gravity for the Sun.

A month after publishing his first paper, Schwarzschild submitted his second paper,[239] and three months later he died at the age of 42. In this paper Schwarzschild got the solution inside the Schwarzschild radius, and since he assumed an incompressible fluid, there is no wonder he got a maximum density and a maximum pressure. According to the results of the second paper, a heavenly body compressed to a radius smaller than the Schwarzschild radius will not let anything escape from it, not even light. This radius is also called the event horizon, because for a distant observer this radius specifies the horizon beyond which (or into which) the observer cannot see. It is easy to estimate the average density of such a body. If we compress the entire Sun to a radius of 2.94 km, we get a density of 6.2×10^{15} g/cm^3, which is slightly higher than the density of the nucleus. On the other hand, if we consider an entire galaxy with mass of about $10^{12} M_{\odot}$ inside its Schwarzschild radius, we get a mean density of 6.2×10^{-9} g/cm^3, which is much less than the density of air (about 0.001 g/cm^3). Thus, the fact that an object lies entirely within its Schwarzschild radius does not imply that it has some phenomenal density.

For historical justice, it should be mentioned that Droste[240] found the solution to Einstein's equations at roughly the same time as Schwarzschild discovered it. When Lorentz communicated his results to the academy, Droste learnt that Schwarzschild had preceded him by a few months, and that the two results agreed.

Schwarzschild's solution contains a factor $1/(1-2Gm/rc^2)$ which seems to imply that, for $r = 2Gm/c^2$, there is a singularity and that the solution therefore ceases to be physical. Schwarzschild was not so bothered with this problem, because it played no role in the case of the Sun and Mercury. However, there remained the problem that it might pose a problem for collapsed objects. This question was dealt with by various authors,[241] who found that it resulted purely from the particular choice of coordinates, and so was not real. On the other hand, the result has a very profound consequence. If we equate the rest mass energy ($m_0 c^2$) with the gravita-

---

[239] Schwarzschild, K., Sitzungsberichte der Königlich Preussischen Akademie der Wissenschaften zu Berlin, Phys.-Math. (24 February 1916) p. 424.

[240] Droste, J., Ned. Acad. Wet. S. **19**, 197 (1917), communicated to the academy by Lorentz, who was Droste's PhD supervisor, on 27 May 1916.

[241] Kruskal, M.D., Phys. Rev. **119**, 1743 (1960).

tional energy of mass m at a distance r, we find that, at the Schwarzschild radius, the rest mass energy is equal to twice the gravitational potential energy. This means that an object falling into a black hole should release about half its rest mass energy when it crosses the Schwarzschild radius.

It is astonishing that in the obituaries for Schwarzschild published shortly after his untimely death, the now famous solution to Einstein's equations which provides the basis for the modern theory of black holes was not mentioned at all.[242] Apparently, it took the scientific community some time to appreciate this gigantic discovery, or perhaps the authors of the obituary just wanted to avoid the problem of the singularity.

In 1927, Lemaitre[243] expanded on Schwarzschild's solution as part of the preparation for his PhD thesis, and essentially used the solution (with a constant density) to describe the entire universe. So in cosmology the reader may hear talk of the Lemaitre–Schwarzschild solution. Here again the interest was in large radii, so that the problem of the singularity did not play any role.

For a long time the fate of collapsed matter was a big question. We have quoted Eddington, who did not like the idea. However, his opponents in the white dwarf controversy did not think much differently.[244] Landau was also worried about the fate of massive stars and suggested that quantum theory might break down at some point in these stars.

How far can the collapse go? Does the density reach infinity? Very likely it does not. At sufficiently high densities, quantum effects are predicted to enter the problem, and the general theory of relativity must then undergo modifications. Such a theory is not yet available. For our present purposes, it now appears that anything which falls into a black hole is first destroyed, and second, never comes out.

5.41 Einstein and Black Holes

Einstein himself did not like the idea of black holes. In 1935, Einstein and Rosen[245] discussed the extent to which the properties of the matter might be able to prevent the collapse of matter to a point, and in this way remove the difficulty with such 'unphysical' solutions.

Einstein and Rosen found that, by introducing a small change into the hypothesized properties of matter, they could obtain a theory in which such problems did not arise. But this was not sufficient for Einstein, and in 1939,[246] he loudly expressed his resentment. First, he claimed that there was a problem with Schwarzschild's solution. Schwarzschild had calculated the gravitational field produced by

[242] The Obs. **39**, 336 (1916); Hertzsprung, E., Ap. J. **45**, 285 (1916).

[243] Lemaitre, G., Ann. Societe Scientifique de Bruxelle **47**, 49 (1927); MNRAS **91**, 490 (1931).

[244] Chandrasekhar, S., Obs. **57**, 373 (1934).

[245] Einstein, A., & Rosen, N., Phys. Rev. **48**, 73 (1935).

[246] Einstein, A., Annals Math. **40**, 922 (1939). Submitted on 10 May 1939.

an incompressible fluid. However, according to special relativity, there is no such thing as an incompressible fluid, because it would mean an infinite speed of sound, and such speeds are prohibited by special relativity.[247] So Einstein asserted that: *Schwarzschild's argument is not convincing.*

The alternative, to recalculate the problem with the assumption of a compressible fluid, as is appropriate, turns out to be very complicated. So Einstein's idea was to calculate the gravitational field of a collection of small gravitating particles which move freely under the influence of the gravitational force produced by all the other particles. Einstein solved this problem, as well as a few more, and showed that, as the system contracts, no situation emerges in which light cannot escape from the system. Even in the very extreme case, the system always (in his examples) reached a finite density, and did not proceed beyond it. The various examples Einstein treated led him to the conclusion that *Schwarzschild singularities do not exist in physical reality*. The reason Einstein gave was that:

> *Matter cannot be concentrated arbitrarily. And this is due to the fact that otherwise the constituting particles would reach the velocity of light.*

A particle cannot reach the speed of light without investing an infinite amount of energy, which is of course implausible. Did Einstein anticipate the eventual merging of quantum theory and the general theory of relativity? It appears unlikely, because he also had reservations about quantum theory. See the paper by Einstein, Podolsky, and Rosen.[248]

Oppenheimer and Snyder's paper was submitted on 10 July 1939, two months after Einstein's paper was published, and it did not contain any reference to Einstein, only to Tolman. Apparently, Oppenheimer did not know about Einstein's paper denouncing the black hole.

How far can the collapse go? If the general theory of relativity is valid right down the line, then Penrose[249] and Hawking[250] have shown that the collapse will go on forever, reaching ever greater densities. However, at a certain point, quantum effects become important, and a theory which combines general relativity and quantum theory is needed. Such a theory is not yet available.

5.42 Observations of Black Holes

In recent years, astronomers have discovered small black holes with masses similar to the mass of the Sun, and supermassive black holes with masses 10^7–$10^8 M_\odot$. The solar mass black holes have masses in excess of the TOV (plus a safety margin

[247] More accurately, special relativity does not allow the transfer of information faster than the speed of light.

[248] Einstein, A., Podolsky, B., & Rosen, N., *Can quantum-mechanical description of physical reality be considered complete?* Phys. Rev. **47**, 777 (1935).

[249] Penrose, R., PRL **14**, 55 (1965).

[250] Hawking, S.W., PRL **17**, 444 (1966).

to cover our inaccurate knowledge of the limit). Supermassive black holes have masses of a million solar masses or more, so there is no question of misjudgment or incorrect identification. We do not know at the moment whether there are black holes of intermediate masses, and we do not know why they should not exist. Similarly, we have no evidence that much smaller black holes do not exist. It is just a question of a not so simple observation.

The first strong evidence for a solar mass black hole came in 1972,[251] and two years later a candidate for a supermassive black hole was identified.[252] In 1970, the UHURU satellite was launched with X-ray detectors on board. The most exciting discovery of the UHURU satellite was a rapid variable, consisting of irregular bursts that came from the Cygnus X-1 source. These rapid irregular variations on a time scale of a few tenths of a second implied right away that the source had to be an extremely small stellar object. When the radio astronomers searched for this object, they discovered a weak variable radio source with no outstanding features. However, the radio astronomers were able to identify the position of the source accurately enough to allow the identification of the optical counterpart of the X-ray source.

Cygnus X-1 (the strongest X-ray source in the Cygnus constellation) was identified as a spectroscopic binary system with a period of 5.6 days. The visible companion of this system is a blue supergiant with a mass of at least $12M_{\odot}$, and the companion has to have a mass of at least $3M_{\odot}$. The latter is thus too massive to be a white dwarf or a neutron star, and is in all likelihood a black hole.

A black hole is discovered through its effect on its environment, e.g., in a binary system when one of the 'stars' is a black hole. The classical Kepler laws continue to be valid for the two masses (provided they are sufficiently far apart). So by application of Kepler's third law, an estimate for the masses can be derived, and if one of them turns out to be above the TOV mass, it is highly probable that the corresponding object is a black hole. As a rule, supermassive black holes are discovered at the center of galaxies. Here again, existence is inferred from the gravitational effect on the surrounding stars.

The existence of black holes closes our space of possibilities for the fate of stars. However, the observational discovery of black holes has not convinced all astronomers. As late as 1977 (after the observational discovery of black holes), Öpik[253] wrote:

> *The theories made in connection with the concept of black holes have undoubtedly been interesting and stimulating, leading us into unexplored possibilities and vistas of the physical world. At present, however, it would be more important to concentrate on a search for real objects, instead of expanding a theory which may or may not prove wrong.*

Why did he express himself in this way? Because he claimed that:

> *Black holes are apparently not possible, by reason of an extension of the same relativistic theory which has been the basis for formulating their existence. However, as a new and strange outlook on the possibilities hidden in the depth of the Cosmos, their discussion*

[251] Bolton, C.T., Nature **235**, 271 (1972). Webster, B.L., & Murdin, P., Nature **235**, 37 (1972).

[252] Balick, B., & Brown, R.L., Ap. J. **194**, 265 (1974).

[253] Öpik, E., IrAJ **13**, 125 (1977).

> *(perhaps somewhat overdone) has enriched our understanding of the laws of Nature, even if the final outcome turns out to be negative as it seems.*

Öpik thought that black holes were a mere academic exercise, and he was not the only one to think so.

Chapter 6
The Solution to the Stellar Energy Problem

Historically, astrophysics discovered how the stars die before it deciphered the way they manage to live. In the previous chapter, we almost reached the year 1940, ignoring the developments in nuclear physics. The reason is that the final state of the stars does not depend that much on the nuclear processes. We now return to the mid-1920s to discuss the evolution of nuclear physics, which had been confirmed to hold the key to the stellar energy supply.

By 1920, Eddington had already struck upon the idea that the fusion of hydrogen to helium is the energy source of the stars, but he still had no idea how that could work. Recall that the neutron had not yet been discovered and the helium nucleus was assumed to contain 4 protons and 2 electrons, to allow for its net charge of two. Furthermore, no nucleus with atomic mass 2 or 3 was known to exist in Nature. Hence, the only way this hypothetical fusion could occur would have been for 4 protons and 2 electrons to come together and for the 2 electrons to be absorbed, a very complicated process and highly improbable. And what was more, nobody had any idea how to calculate such a process. The relevant theory had not yet been discovered. The 6 particles have to meet at the same point and the same time for the process to work. If only two particles meet, then since no system composed of two protons exists, the two particles are liable to separate before the third particle has a chance to arrive, so the whole encounter would lead nowhere. Eddington could hypothesise about the end result, but he was unable to suggest how fusion should actually take place. Most of the required physics was not yet known, and another twenty years or so would be needed, for the development of quantum theory and nuclear physics.

In view of the fundamental difficulties with the fusion hypothesis, it was no wonder that, in 1925, Jeans would republish his annihilation theory (this time revised).[1] Once more, the idea of mass annihilation came prematurely, because much of the required physics was not yet known. In particular, it was several years before Dirac came up with the first relativistic quantum theory, and the idea of the existence of antiparticles. So Jeans calculated the photon resulting from a proton–electron colli-

[1] Jeans, J.H., Nature, 12 December 1925.

G. Shaviv, *The Life of Stars*, DOI 10.1007/978-3-642-02088-9_6,

sion using conservation of energy, but disregarded the fact that one must also satisfy conservation of momentum.

Even without knowing the details of the physical process, Jeans' idea could be shot down very easily, and indeed, a few weeks after Jeans' republication in Nature, Hughes and Jauncey[2] published a paper in which they proved that Jeans' process did not comply with 5 required physical conservation laws. As correct as Hughes and Jauncey's paper was, and despite the fact that it remained unrefuted, it was largely ignored by theoretical astrophysicists.

As for Jeans himself, a few years later he analyzed the stability of stars and concluded (wrongly) that none of the above explanations for the energy source of stars could operate in a stable manner. Consequently, he suggested that very heavy radioactive nuclei might be the required energy source.

6.1 The Quantum Revolution

The years 1924–26 saw dramatic progress in physics with the work of Heisenberg (1901–1976, Nobel laureate in 1932.) and Schrödinger(1887–1961m, Nobel laureate in 1933 with Dirac),[3] establishing the foundations of non-relativistic quantum theory and explaining the atomic structure of the elements.

The quantum theory developed by Heisenberg and Schrödinger did not satisfy Einstein's special theory of relativity. It was clear that the theory needed corrections, and these were discovered in 1928 by Paul Dirac (1902–1984[4]), who succeeded in formulating the quantum theory in a way that satisfies the requirements of special relativity. In doing so, Dirac discovered that his equations have two symmetric solutions, like the quadratic equation $x^2 = 4$, which has two possible solutions $x = 2$ and $x = -2$. One solution of Dirac's equation corresponds to positive energy, while the other corresponds to negative energy. The existence of two solutions raised the question as to whether all solutions could be realized physically. If the negative energy states were physical, then positive energy electrons would be able to jump to one of the negative energy states with the emission of radiation. One possibility Dirac faced was to assume that the negative energy states were not physical and hence should be disregarded.

However, Dirac chose another dramatic and imaginative option. He postulated in 1928 that the second solution corresponds to a symmetric particle which has the

---

[2] Hughes, A.L., & Jauncey, G.E.M., Nature, 6 February 1926.

[3] The year the nazis came to power, Schrödinger left his chair of theoretical physics in Berlin, where he had succeeded Max Planck, and moved to Dublin in Ireland. Heisenberg, who had high regard for Einstein and his theory of relativity, was not a member of the national socialist party. He even faced problems due to his liberal views about physics and against the notion of Jewish physics advocated by the Nazi party. Later, during WWII, he apparently compromised his views and became the chief theoretician of the Nazi atomic program. This difference may explain why one physicist was honored with a crater on the moon while the other was not. Craters are named at least three years after death. The Nobel Prize is only awarded to living scientists.

[4] It is an annoying fact that no lunar crater has yet been named after Dirac.

same mass as the electron, but the opposite charge. Moreover, to prevent the electron from falling into these low energy states, all of them had to be filled with particles. This was a far-reaching conjecture as it implied that all particles have a symmetrical particle and that the vacuum is filled with particles with negative energies. In particular, the electron had to have a symmetric particle which possessed the same mass but positive charge. Furthermore, when the electron meets the symmetrical particle, they are expected to annihilate each other and emit two photons.[5]

The symmetrical particle to the electron was called the positron. Later it became known as the antiparticle of the electron. When a particle meets its antiparticle, they annihilate each other. In this way the idea of antimatter was born, made up of particles which annihilate upon collision with their symmetrical counterparts. From a physical point of view, the fact that Dirac got two solutions was interpreted as a symmetry of matter. Along with any particle of matter, there exists an antiparticle of matter. So why is our world composed of matter and not of antimatter? Well, at some point in the early history of the Universe, the symmetry between matter and antimatter must have been broken, so that more matter was formed than antimatter. As early as 1933, Dirac speculated in his Nobel lecture about a world composed of antimatter. Note that it was the merging of Einstein's relativity theory with Schrödinger's quantum theory that led directly to this symmetry.

The reader is asked indulgence for this short description of how and when the quantum theory was developed, and finally understood, without mention of the other possibilities considered at the time. The subject is simply too extensive to be covered in the context of the present book.

It was only 4 years before a particle with the mass of an electron but with positive charge was observed by Anderson in 1932 (1905–1991,[6] Nobel Prize in 1936), while examining the reactions of cosmic rays showers. The cosmic rays passed through a gas chamber and a lead plate. This was surrounded by a magnet to distinguish between charged particles whose tracks are bent by the magnetic field. Anderson called his particle the positron. He also suggested renaming electrons as negatrons, but the suggestion did not appeal to physicists. The positron was the first evidence of antimatter. Beside being a victory for Dirac's theory, it was a boon to the idea of mass annihilation as the stellar energy source.

---

[5] One of the consequences of energy and momentum conservation is that two photons (and not just one) are emitted in the annihilation of an electron and a positron.

[6] Anderson, C.D., The positive electron, Phys. Rev. **43**, 481 (1933). The Anderson crater is named after John August Anderson (1876–1959), an American astronomer, who succeeded in measuring the radius of the binary orbit of Capella directly [Ap. J. **51**, 263 (1920)], and not after C.D. Anderson.

6.2 We Have Already Seen this Movie

As late as 1928, there was still no acceptable solution for the stellar energy source. So Andrews[7] proposed a new element as energy source. The reasoning was somewhat strange. Andrews had noticed that the atomic numbers of the six noble gases are helium 2, neon 10, argon 18, krypton 36, xenon 54, and radon 86, and that these fit the following series: $2(1^2+2^2+2^2+3^2+3^2+4^2+\cdots)$. So what about the next term with atomic weight $Z=118$, he wondered. This element should be heavier than uranium, so it had to be radioactive and release energy. However, no such element exists on Earth. Andrews named the hypothetical element hypon, as if giving it a name would make it more real!

In a way, this was a variation on Jeans' idea of radioactive decay as the source of energy (although no citation of Jeans was given). In this respect, extrapolation from the noble gases was not needed, and only confused the basic idea that he assumed some element beyond uranium to supply the stellar energy. Moreover, in contrast to what had been shown and proven before, Andrews claimed that *the super radioactivity of hypon is regulated by pressure*. So why was such an element needed? Because this particular radioactive element *can exist only under the stabilizing influence of great pressure, and if this pressure falls below a critical minimum, the hypon instantly explodes with intense violence*. Andrew had to assume this in order to explain why the hypothetical element did not exist on Earth. Cernuschi's idea,[8] which came some ten years later, was a follow-up of Andrews'.

6.3 Light at the End of the Tunnel

A major breakthrough in our understanding of radioactive decay came in 1929 when Gamow (1904–1968m),[9] then in Göttingen, and Gurney and Condon,[10] then in Princeton, discovered the phenomenon of quantum tunneling, whereby particles in a potential well can escape even when classical physics predicts that they are bound.

The sequence of publications is interesting. Gurney and Condon first published a short note in the issue of Nature that came out on 22 September 1928,[11] in which they discussed qualitatively how a quantum particle might escape from a potential well. Gamow saw these short publications and decided to write to Nature about his results. In this note in Nature, he referred to the qualitative results discussed by Gurney and Condon, and went on to report his own results, which by then (i.e., by

[7] Andrews, W.S., Scientific Monthly **27**, 535 (1928).

[8] Cernuschi, F., Phys. Rev. **56**, 450, (1939).

[9] Gamow, G., Zeit. f. Phys. **51**, 204 (1928), accepted for publication 2 August 1928, but submitted 26 July 1928.

[10] Gurney, R.W., & Condon, E.U., Phys. Rev. **33**, 127 (1929).

[11] Gurney, R.W., & Condon, E.U., Nature, 22 September 1928. Reports of the work were also presented at the National Academy of Sciences meeting of 20 November 1928, and the American Physical Society meeting on 1 December 1928.

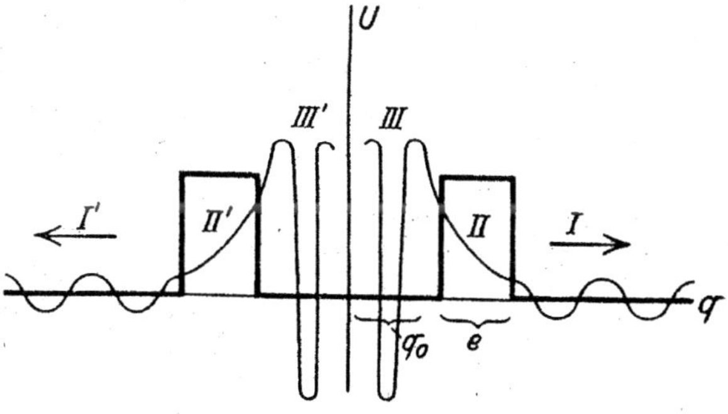

Fig. 6.1 The simplified nuclear potential well, as assumed by Gamow to represent the complicated nuclear plus Coulomb potentials. The wave function of the particle is not attenuated inside the nucleus or outside it. However, we see the exponential decay inside the potential

the time the Nature letter was published) had already been published in the Zeitschrift für Physik. Gurney and Condon submitted their full paper including all the mathematical analysis to the Physical Review on 20 November 1928, and the paper was published in February 1929. However, two weeks before submitting the paper to the Physical Review, they received the issue of Zeitschrift für Physik containing Gamow's paper. The authors had independently had the same idea and discovered the phenomenon called barrier penetration or tunneling (see Fig. 6.1).

The motivation of all the physicists involved was to explain natural radioactivity, and in particular, why certain nuclei are not stable and decay, and why in some cases it takes billions of years for a heavy nucleus to disintegrate, while in others it takes only a fraction of a second. Why should one atom from a collection of atoms suddenly disintegrate, and the others follow, but each at a different time? This problem appeared to have nothing in common with stars, let alone their energy source. However, it is difficult to overestimate the importance of the tunneling phenomenon in so many disciplines, and above all in astrophysics. This discovery essentially instigated nuclear astrophysics. It is amusing to note that Condon, who as a matter of fact discovered how nuclear reactions can take place in stars, believed several years earlier that mass annihilation was the source of stellar energy.[12]

In October of 1928, Gamow and the young Houtermans,[13] then still in Göttingen, published an extension of Gamow's theory to radioactive decay. They mention Gurney and Condon's paper in Nature, but apparently did not know about the Physical Review paper. Shortly after the discovery of barrier penetration, Houtermans met Atkinson and the idea of applying the tunneling discovery to stars was born. Atkinson, who was an English guest resident in Germany, married a German woman who knew Fritz Houtermans from an earlier meeting. So when the Atkinsons and Hou-

[12] Condon, E., Proc. Nat. Acad. Sci. USA **11**, 125 (1924).

[13] Gamow, G., & Houtermans, F.G., Zeit. f. Phys. Vol. XX, 496 (1929).

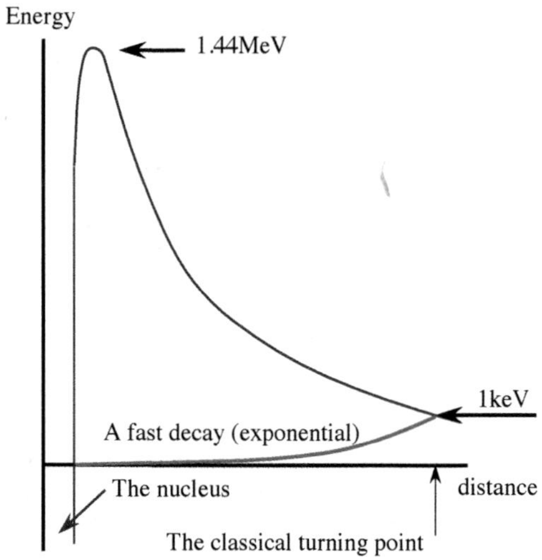

Fig. 6.2 The Coulomb barrier between two approaching protons, and the energies involved. The probability of a proton with energy smaller than the peak energy of the Coulomb barrier actually entering the classically forbidden region decreases as the wave–particle approaches the nucleus. The *red curve* gives the probability of entering the potential

termans moved from Göttingen to Berlin-Charlottenburg to accept new jobs, the two families met and the collaboration started. Atkinson (1893–1981) and Houtermans (1903–1966)[14] were quick to apply the tunneling effect to nuclear reactions in stars, and the paper was submitted on 19 March 1929.

The temperature in the core of a star is determined by stellar models. More precisely, if a mass of the size of the Sun is compressed into a sphere with the radius of the Sun, and if it behaves like an ideal gas, then the temperature of the gas is determined right away by the condition of hydrostatic balance, i.e., the pressure of the gas counteracts the gravitational pull. This is the temperature at which the gas pressure balances gravity. Hence we may say that the temperature of the gas is dictated by the gravitational field. On the other hand, the temperature of the gas is another expression for the energy of the particles composing the gas. More accurately, the mean energy of the particles in the gas is given by the Boltzmann constant times the temperature.[15] At a temperature of 1.5×10^7 K, a temperature at which hydrostatic equilibrium is reached, the mean kinetic energy of protons is about 1 keV. On the other hand, the peak of the Coulomb repulsion between two protons is about 1.44 MeV, which is more than 1 000 times greater than the mean kinetic energy.

[14] Atkinson, R.d'E., & Houtermans, F.G., Zeits. f. Physik **54**, 656 (1929).

[15] The mean kinetic energy of the gas particles, irrespective of their mass or charge (i.e., for protons and electrons alike) is given by $E_{\text{kin}} = 3k_{\text{B}}T/2$, where k_{B} is the Boltzmann constant and T the temperature.

Hence, according to classical physics, the protons cannot overcome the mutual repulsion, and thus repel each other, whence there is no chance of reaction. The closest distance at which two protons with an energy of 1 keV can approach each other, which is called the classical return distance, is much greater than the size of the nucleus. We see in Fig. 6.2 that the nucleus resides deep inside a very small but very deep potential well, while the Coulomb repulsion acts over a much greater distance than the size of the nucleus. The consequence is that a classical particle with insufficient energy to overcome the potential barrier is generally repelled far away from the nucleus, and is prevented from any nuclear interaction.

6.3.1 Tunneling: The Secret of Stellar and Biological Time Scales

Tunneling is so important that we must digress to explain it in more detail. Consider the simplest potential, as shown in Fig. 6.3, and consider the motion of a particle with kinetic energy E less than the height V of the potential. Let the particle come from infinity in region A. According to classical physics, the particle moves with constant velocity from right to left. At point a, the total energy E is less than the potential energy V, and hence the particle cannot go beyond point a in its motion to the left, and must bounce back. For this reason the professional jargon used to describe such a potential is 'potential barrier'. A classical particle can never go in

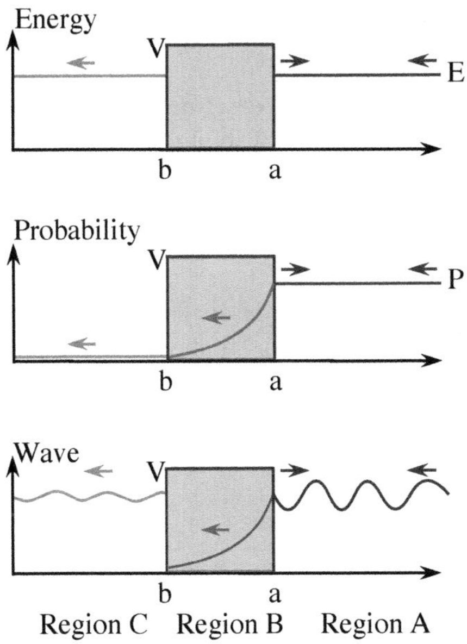

Fig. 6.3 Tunneling through a simple square potential

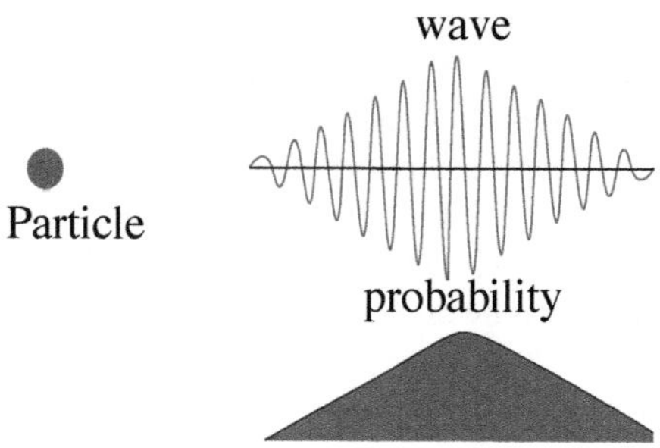

Fig. 6.4 The elementary particles have dual properties: particles and waves

the region between points a and b where the potential exists, because the potential V is greater than the total energy, and its kinetic energy would therefore be negative. But there are no limitations for a particle with energy $E > V$.

In the microworld, there is duality. An elementary particle can behave as a particle (an object with a well-defined location) or wave (an object occupying an extended region), as shown in Fig. 6.4. The picture seems confusing. Sometimes the 'elementary particle' behaves as a particle located in a certain region, and sometimes it is not clear where it is, since it is a wave. The connection between the two pictures is made via Heisenberg's uncertainty principle, which states how accurately the location and the momentum of a particle can be determined. If Δx is the uncertainty in the location of the particle and Δp is the uncertainty in the momentum, then $\Delta x \Delta p \sim \hbar$. The elementary particle has properties which are defined only when it interacts with another system. In our case, the wave property is exposed when it interacts with the potential barrier.

In quantum theory, each particle is described by a wave function. The square of the wave function represents the probability of finding the particle in a certain region. Returning to Fig. 6.3, in region ab, where the classical particle cannot go at all, even the wave has some difficulty. As a result, it decays (exponentially) with distance. The greater the value of $(V - E)$, the faster the decay of the wave in the forbidden region ab.

Once the wave reaches the other side of the potential, it exits and the simple 'free' particle state is restored. The energy of the particle is conserved, so it emerges with the same kinetic energy it had before it entered the potential. However, since the wave decayed during the penetration 'under the potential', the number of particles which manage to tunnel through the potential, without being reflected by it, is small.[16]

[16] The strength of the wave is described by the amplitude, that is, the height of the wave peak. The number of particles is given by the amplitude squared. Mathematically, we can understand the phenomenon as follows. The particles behave like a wave. The particle–wave cannot propagate in the region where V is greater than E, and hence is reflected. If the amplitude of the wave in region ab is set equal to zero it implies a discontinuity at point a (the wave exists on one side and does not exist on the other side). But waves cannot be discontinuous, with sharp boundaries. There are

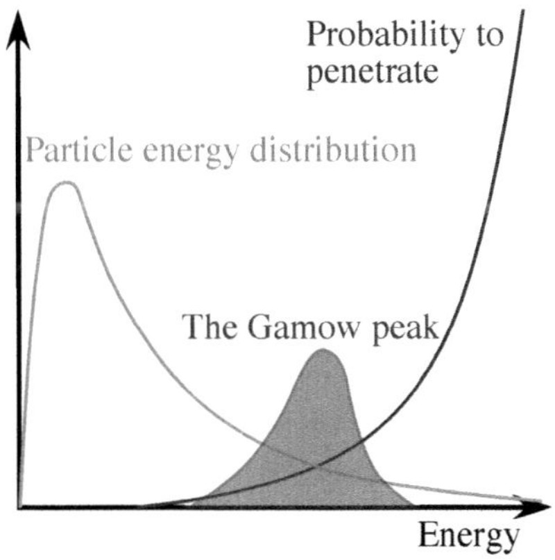

Fig. 6.5 The product of the probability for tunneling, which increases exponentially with the energy, and the number of particles, which decreases exponentially, gives rise to the Gamow peak

6.3.2 Tunneling Under Stellar Conditions

The particles in the gas have a mean energy of $3kT/2$, but some of the particles have more energy and some less. In short, there is a distribution of energies. So Atkinson and Houtermans realized that one has to average the higher probability of penetrating as a function of energy with the lower probability of finding particles with high energy. In the Sun, the mean energy of the particles is only 1 keV, but the relevant particles for penetration are those with an energy of about 5 keV, even though they are fewer in number than those with energy 1 keV. (The ratio between the number of particles with 5 keV to those with 1 keV is 0.0067!)

Since the probability of penetration increases with energy, while the number of energetic particles decreases with energy, it is clear that there exists an optimal energy, called the Gamow peak energy.[17] What this means is that the majority of penetrations occur at this energy.

It is clear from the acknowledgment to this important paper that the authors consulted Gamow, who was in the picture all the time. Gamow, Gurney, Condon, Atkinson, and Houtermans discovered jointly (part of) the reason for the long lifetime of stars and how they release their energy. It is interesting to note that the extended Physical Review paper by Gurney and Condon was published in February, while Atkinson and Houtermans submitted their paper in March of that year and were unaware of it (no reference to it). However, mail and information in those days were much slower than today, and hence it seems reasonable to assume that the full extent of Gurney and Condon's work was not known to the European researchers.

---

no discontinuities in physics. The square well potential is a mathematical simplification of a very steep potential or a strong force. The solution is therefore a decaying wave to the left of point a.

[17] In honor of the discoverers' mentor.

Shortly after Gamow's paper appeared, von Laue (1879–1960) (Nobel Prize for Physics 1914)[18] and Kudar[19] independently suggested that the light elements might be formed through the inverse process of α-decay. Unlike Houtermans, however, they found values for the probability per unit time that were much too low for the process to occur, even under stellar conditions.

Checking the possible rates of all available nuclear reactions, Atkinson and Houtermans discovered that significant probabilities of penetration exist only in proton collisions with the first 7 elements. That is, the lifetime of these elements against absorption of a proton turned out to be less than a few billion years. The penetration probability decreases quickly with increasing nuclear charge Z, because the potential barrier is too high for the conditions prevailing in stars. Thus they concluded correctly that, under the conditions in the Sun (a temperature of 40 million degrees was estimated at that time), proton reactions with heavy nuclei are totally negligible. On the other hand, they did not discuss the proton–proton reaction, probably because deuterium was not yet known to exist, so that the supposed resulting nucleus was not known to exist in Nature.

Was the issue of stellar energy settled? Atkinson and Houtermans admitted at the end that they did not yet have a solution to Eddington's red giant paradox. Once again, the attempt to find a theory which explained all types of stars continued to impede progress.

6.4 The Paradox of the Giants

And who else but Eddington could be behind this paradox? In his book, published a year before the Houterman and Atkinson paper, Eddington had set out the following red giant paradox[20] by comparing the giant Capella with the dwarf Sun:

- Capella releases 58 ergs/g/s compared with 1.9 ergs/g/s released by the Sun.
- The density of the Sun is 620 times the density of Capella.
- The temperature of the Sun at corresponding points is 4.3 times higher than the temperature of Capella, assuming Capella to be homogeneous like the Sun.

In short, the Sun is hotter and denser than Capella but releases less energy. This does not agree with the general prediction that the rate of energy release should increase with temperature and density. There had to be another incorrect assumption somewhere.

[18] von Laue, M., Zeit. f. Physik **52**, 726 (1929).

[19] Kudar, J., Zeits. f. Physik **53**, 61, 95, 134 (1929).

[20] Eddington, A.S., *The Internal Constitution of the Stars*, p. 297.

6.5 A Snag?

Soon after Houtermans and Atkinson published their paper, Fowler and Wilson[21] came up with a different approach to the problem. Recall that Gamow had been interested in the probability that an α particle in a radioactive nucleus would escape through the potential barrier. Atkinson and Houtermans used exactly the same formalism to get the probability for penetration through the potential. They did not ask what happened to the particle once it was inside the nucleus. They very likely realized that, since penetration and escape have the same small probability, once the particle is inside the nucleus, it has a much higher chance of reacting with the components of the nucleus than of escaping.

Fowler and Wilson took a different approach. They considered the potential well of the nucleus in a simplified form (a square well potential so that the calculation was much easier – but this fact was not essential) and asked for the probability that the incoming particle would settle into a stable energy state inside the nucleus. This is the correct way to do the complete calculation, and this is what the theory of nuclear physics does. They applied the conservation of (angular) momentum and energy to calculate the probability that the incoming particle would be captured and end up in a stable energy state. A priori, this probability is smaller than what Atkinson and Houtermans calculated. But as Fowler and Wilson noted, there is one exception, when the energy of the incoming particles coincides with the energy of a bound state. In physics, this is called a resonance. In this case, the rate of the reaction can be many orders of magnitude bigger.

As for the other case (when there is no resonance), Wilson concluded that the numbers found by Atkinson and Houtermans:

> *[...] were based on some speculation. If the calculations are carried out properly, it is found that the value given by these authors for the rate of transformation is incorrect.*

What were the consequences? Wilson required very high temperatures for the nuclear reactions to take place. Only Milne's model led to temperatures of the order of 10^{11} K, as required by Wilson. Hence Wilson concluded that:

> *[...] some modification of the theory of stellar interiors, such as proposed by Milne, is essential.*

In short, Eddington's model was out, and Milne's was in. Recall now the controversy that was raging at the time over the white dwarfs, between Eddington on the one hand, and Chandrasekhar, Milne, and Fowler on the other. Eddington must have had his hands full, what with all these controversies and 'proofs' that nuclear fusion was impossible!

So who was right? As late as 1933, Steensholt[22] checked the calculation by Wilson and Atkinson. Steensholt made a convincing job of this, solving the hydrostatic

---

[21] Fowler, R.H., & Wilson, A.H., Proc. Roy. Soc. **124**, 493 (1929), and shortly afterwards, Wilson, A.H., MNRAS **91**, 283 (1931). The paper was communicated by Fowler and Wilson, and Wilson merely summarized them in the paper.

[22] Steensholt, G., ZA **5**, 140 (1932).

balance equations of a star with Atkinson's model for nuclear reactions under stellar conditions to find a good agreement. Moreover, he found that Wilson's claim that one needs temperatures beyond the range of those expected in stars according to Eddington's model, in order for the proton to penetrate, was not supported by his calculations. Indeed, he found that temperatures in the range 10^7–10^8 K were sufficient. Thus Wilson's implications that nuclear reactions could not play a significant role in the stars was disproved. The controversy teaches us how delicate the situation was. If just one assumption is goofed, the results change so as to render the entire phenomenon negligible.

But this was not the end of the story. A year later, in 1934, Steensholt[23] returned to the problem and investigated the stability of the star with the energy source suggested by Atkinson, finding it to be unstable. There was no mention of Jeans' general conclusion, but a different criterion devised by Rosseland was used. However, the same destructive result was obtained. It thus looked as though no form of nuclear energy could be the source of energy for stable stars! There was, however, one 'small' caveat in Steensholt's calculations: *our calculation cannot claim high numerical accuracy*. On the other hand, this did not prevent Steensholt from claiming that he had established his result with a sufficiently large margin of safety for it to be beyond reasonable doubt.

What really happened was that the stellar models used had been oversimplified, and indicated an instability even when the models were actually stable. A year later,[24] Steensholt revisited the problem and concluded that:

> *Our analysis with idealized models does not necessarily rule out the possibility of the existence in Nature of stars with an energy generation that tends to set up a peculiar density distribution characteristic [needed for stability].*

So in the end, the calculations were not in fact sufficiently accurate to destroy Atkinson's nuclear energy models. It is amazing how finely tuned the nuclear reactions and the time scale must be in the stellar model!

6.6 Creation and Annihilation of Matter in Stars. A Brief Comeback

As Eddington's ideas about subatomic energy were still making no progress at the end of the 1920s, a new idea surfaced: creation of matter in stars. Years later, Bondi and Gold[25] and Hoyle,[26] in their theory of steady-state cosmology, would reinvent the idea of matter creation in the Universe. In 1929, Gerasimovič and Menzel[27] from

---

[23] Steensholt, G., ZA **7**, 373 (1933).

[24] Steensholt, G., ZA, 56 (1934).

[25] Bondi, H., & Gold, T., MNRAS **110**, 607 (1950).

[26] Hoyle, F., MNRAS **108**, 372 (1948).

[27] Gerasimovic, B.P., & Menzel, D.H., PASP **41**, 145 (1929).

Harvard published an award-winning essay entitled *Subatomic Energy and Stellar Radiation*,[28] in which they put forward the idea of proton creation inside stars.

The starting point for Gerasimovič and Menzel was the assumption that subatomic energy is released in the star in an equilibrium process. Here is their reasoning and their example. Consider a covered vessel containing water and vapor. Two processes take place inside the vessel. Water molecules evaporate from the water and join the vapor. At the same time, water vapor condenses and returns to the water. In the steady state, the same number of molecules leaves the water as returns to the water, and the amount of water and/or vapor does not change. This dynamic equilibrium means that the process and its reverse take place at the same time, while the total amount of any constituent does not change. Many processes behave this way, and the general name for them is indeed 'dynamic equilibrium'. The amount of water vapor changes only if the temperature and pressure change. The sum of water and water vapor stays the same, even when the temperature and density change. If the cover has a hole, then some of the vapor escapes. To compensate for this, the rate of evaporation increases. The total rate of evaporation is now equal to the rate of escape (depending on the size of the hole in the cover) plus the previous rate of evaporation in equilibrium. What an outside observer would see is only the escaping vapors and not the internal equilibrium flux of evaporating molecules.

What Gerasimovič and Menzel hypothesized was that subatomic energy release behaves in the same way. Thus they claimed that:

> *It is incorrect therefore to say that Capella generates 58 ergs per gram per second, when this is just the apparent excess of disintegrations over the reverse process.*

They introduced the concept of apparent generation, as opposed to true generation. At the same time as matter is converted into radiation in the annihilation process, the reverse process takes place, i.e., the creation of matter out of radiation. They claimed that the idea was *a necessary consequence of the fundamental conception of the energy supply*. But why this should be so was not explained, and the present author is at a loss to provide a better explanation.

By some simple calculations, they claimed to produce a theory which formed a bridge between Eddington and Russell. In particular, so they stressed, arranging the stars according to the mean density is equivalent to *some sort of evolution, for the density should increase with age*. Instead of thinking that *as a star grows older, the sources of energy approach exhaustion*, Gerasimovič and Menzel favored:

> *[...] an alternative view, that the relative number of reverse processes increases with density and the excess energy available for radiation accordingly decreases, thus producing an apparent 'exhaustion'.*

Unlike Jeans who put forward the idea based solely on energy considerations, the authors also had a detailed mechanism. The suggested energy–matter generation mechanism was as follows:

---

[28] Awarded an A. Cressy Morrison Prize in 1928, by the New York Academy of Sciences.

> *The interaction of free electrons and radiation results in the production of an excess of highly energetic quanta and high speed electrons. These excesses (over Planck's and Maxwell's discributions) do not accumulate permanently, but are spent in the formation of energy and matter. The protons are annihilated by the high speed electrons they themselves have created. An atom is a sort of a loaded gun that can be fired only by a very fast electron.*

In their summary, they wrote that:

> *We have dealt with some new aspects of the problem of stellar energy, trying to avoid, as far as possible, needless speculation. [...] The new hypothesis is unavoidably tentative and still very far from the ideal theory.*

By 1931, the only theories left in town were nuclear transmutation and mass annihilation. Menzel[29] returned to the annihilation/creation of matter hypothesis as a source of stellar energy. After an attempt to combine the theory with cosmology and the role of matter annihilation in the Universe, he fell back on Dirac's theory. With the acceptance of Dirac's theory, mass annihilation became a 'legal assumption', since for the first time there was a theory that predicted the existence of such a phenomenon, and above all, provided an unambiguous way to calculate it. This meant that there was actually a way of comparing theoretical results with observation.

Following Dirac, one postulates that the vacuum is a state in which all negative energy states are full and all positive energy states are empty. A single electron must then be in a positive energy state. If all negative energy states are filled save one, it is called a hole, and it represents a positively charged particle. Annihilation occurs when the single electron in the positive energy state falls into the hole, releasing the rest mass energy of the positron and the electron in the form of two photons. The energy release is then the mass of the electron plus the mass of the positron, or twice the mass of the electron.

Analogously, creation of matter takes place when an electron in a state of negative energy jumps to one of positive energy, creating a single electron with positive energy and a hole. Oppenheimer[30] used Dirac's theory to calculate the mean lifetime of matter when such a process is possible, and found 10^{-10} second, i.e., Dirac's theory predicted that this transition would take place almost instantaneously. But such a disappearance of matter had never been observed, as Menzel quite correctly claimed, and he argued that Dirac's theory should therefore be abandoned.

At this point Menzel returned to the stellar problem, and proposed matter annihilation (although he claimed that Dirac's theory was not the right one to calculate the phenomenon). Menzel made the hypothesis of two extremes. On the one hand, Jeans postulated a 'super radioactive' atom, which was supposed to release energy practically independently of the physical conditions within the star. On the other hand, there was the suggestion made by Russell and Eddington that the rate of generation of energy is a function of both the temperature and pressure inside the star, and is controlled by a sort of safety-valve action wherein the star tends to expand or contract and thus correct any over-production or under-production of energy. *The theory of Russell and Eddington suffers mainly from lack of definiteness*, contested

29 Menzel, D.H., PASP **43**, 191 (1931).

30 Oppenheimer, R., Phys. Rev. **35**, 939 (1930).

Menzel,[31] who was a student of Russell. Menzel's objections were interesting. According to Russell, there are giants which produce energy and dwarfs which have exhausted their energy supply. Hence the dwarf stars should be filled with inert matter, and the only matter left for consideration was the massive, highly radioactive atoms with atomic number greater than 92. This did not appear plausible to Menzel. For these reasons, he confessed that he preferred the Jeans mass annihilation hypothesis, which would leave no ashes.

Now Menzel returned to the hypothesis he had developed with Gerasimovič. Since the temperature at the centers of stars is about constant,[32] the decrease in luminosity along the dwarf sequence had to be attributed to depletion of fuel. At this point, Menzel brought in the argument about how a giant star, which *breaks up into a binary*, would behave, and in particular the fact that its luminosity could not obey the same rule (decrease in luminosity with depletion of fuel). He concluded that:

> *[This] constitutes strong evidence against the validity of any theory that relies upon exhaustion of the transformable material to account for the decrease of stellar luminosity along the main sequence.*

Clearly, Menzel tacitly rejected Eddington's concept of the main sequence as the location of stars with different masses, rather than a mass–luminosity track.

6.7 The Masses of Nuclei

Indispensable data for all discussions on nuclear energy sources are the masses of the nuclei. The steady improvements in measuring techniques, and in particular mass spectroscopy, generated accurate data on the masses of nuclei and enabled physicists to find out how the binding energy of the elements changes from one element to the other. In particular, the general shape shown in Fig. 6.6 was discovered in the mid-1930s. The main figures to collect this data and build these curves were Weizsäcker,[33] and Bethe and Bacher.[34] The major feature of interest to us here is the fact that the curve of (minus) the binding energy has a maximum near iron. Hence any synthesis of elements up to iron releases energy and can supply energy to the star. On the other hand, energy must be invested to create nuclei heavier than

[31] Menzel, D.H., Science **65**, 432 (1927).

[32] From the run of the mass and radius along the main sequence, Eddington found that M/R changes very slowly. Next he found that, in a star composed of ideal gas, the central temperature (or the average temperature) is proportional to M/R. Thus, argued Eddington, at a critical temperature which is fixed for all stars, the energy source opens up and releases energy. The constant temperature along the main sequence is just what Menzel would claim to be the signature of an equilibrium process. A check of the HR diagram drawn by Russell in 1925 does indeed show that the main sequence corresponds to homogeneous stars with a central temperature of 32 million degrees.

[33] Weizsäcker, C.F., Zeits. f. Physik **96**, 431 (1935); Physik. Zeits. **36**, 779 (1935).

[34] Bethe, H.A., & Bacher, R.F., Rev. Mod. Phys. **8**, 83 (1936).

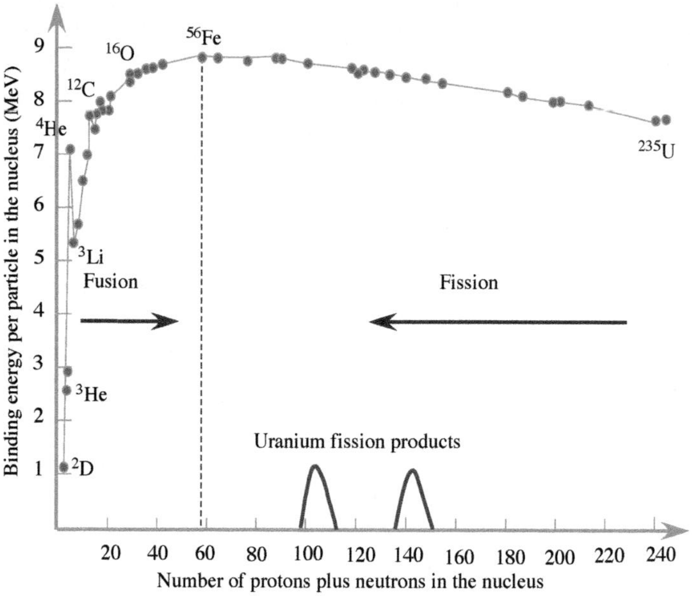

Fig. 6.6 Binding energy for different numbers of protons and neutrons in the nucleus

iron, exactly as was predicted by Eddington. Put differently, energy can be extracted by splitting (fission) of a heavy nucleus (heavier than iron) and by fusing elements lighter than iron.

6.8 The Birth of Nuclear Astrophysics

We have already considered the first attempt by Atkinson and Houtermans to calculate the rate of proton reactions in stars. We may therefore label the two long and important papers by Atkinson,[35] published in 1931, as the beginning of nuclear astrophysics. In these papers, the connection was established between nuclear reactions and stellar evolution and structure, and the first attempt was made to explain the relative abundances of elements as a consequence of element synthesis from hydrogen in stars. The three papers published by Atkinson, two in 1931 and the last one in 1936, do not actually contain calculations, and should be considered only as containing the hypothesis and scenario.

[35] Atkinson, R.d'E., Ap. J. **73**, 250 (1931); ibid. 308. In those days each issue of the Astrophysical Journal included 4–5 articles, so Atkinson's two long papers (close to 50 pages long) had to come in two consecutive issues.

At the outset, Atkinson admitted that he did not intend to derive a very detailed theory that would be able to explain, for example, the odd–even abundance ratio discovered by Russell. So Atkinson made the wise decision to leave the discussion of such details for the time being. Atkinson quoted Russell's result regarding the hydrogen in the solar interior when he stressed that the initial matter was hydrogen, and so he presumed that:

It seems very reasonable to assume that, in its initial state, any star, or indeed the entire Universe, was composed solely of hydrogen.

This was the first formulation of the generalization of what had been observed and what had been established theoretically, that hydrogen was the first element in the Universe, and the most abundant today. The situation was, as Atkinson remarked, that so much observational data had accumulated that it was no longer possible to *construct an arbitrary hypothesis without producing a contradiction*. Notwithstanding the recognition that hydrogen was the first and sole chemical element at the beginning, the difficulties in finding out just how fusion could start from a composition of pure hydrogen had led all researchers to look for alternatives. And all alternatives assumed the existence of heavy elements. But there was no explanation for how these heavy elements themselves had formed.

6.9 The Idea of Regenerative Synthesis

Neither deuterium nor the neutron were known in 1931. So, concerning the reaction $4\text{H} + 2\text{e} \rightarrow {}^4\text{He} + Q$, which had to 'jump over' the barrier arising because there was no nucleus with two nucleons, Atkinson correctly concluded that *it is almost certainly so improbable a process* that it can be ignored. The next alternative was the fusion of a proton with a helium nucleus. But neither of the possible products, namely ${}^5\text{He}_2$ or ${}^5\text{Li}_3$, seemed to exist in Nature. What it meant was that, even if the proton penetrated these nuclei and one of these other nuclei was formed, it could only live for a very short time before disintegrating back into its constituents.

But if these reactions could not take place, then a star composed purely of hydrogen and helium could not synthesize any heavier element. At this point Atkinson retreated from the assumption of a star made up of pure hydrogen and helium, and considered the possibilities for a star in which there were heavy elements. It is an observational fact that all stars do actually contain heavy elements. No star without heavy elements had ever been discovered. Could it be that the heavy elements were playing some important role? Evidently, a new idea was badly needed.

Atkinson thus invented a new concept wherein helium could nevertheless be synthesized from hydrogen. The essentials of the new idea were as follows. Protons are captured successively by light elements. In this way, heavier nuclei are built one after the other. The proton absorption process continues until the product nucleus becomes unstable and disintegrates by emitting an α particle, i.e., a helium nucleus:

The nuclei act as a sort of trap and cooking-pot combined, catching four protons and two electrons in such sequences and at such intervals as may prove practicable, fettering them by emitting as radiation most of the surplus mass brought in [...] combining its captives into an α-particle, and emitting this after a delay.

Suppose that, after several absorptions, an α particle is emitted. Then clearly a nucleus with $A-4$ and a helium nucleus will appear, and this nucleus can now absorb a further four protons, then reform the nucleus A which disintegrates. In this way a cycle is obtained, in which the nuclei A, $A+1$, $A+2$, $A+3$ are catalysts, in the sense that they absorb protons, forge them into a helium nucleus inside the nucleus, and then emit the product. The total amount of these nuclei does not change during the process, although the relative amount may change because the rate of proton absorption may be different. As a result, the amount of the slow absorber will be high and vice versa. The breakdown of the abundances of the catalysts would be a signature that such a process has taken place. For this reason Atkinson called it a regenerative process. The rationale behind the process was that, in natural radioactive decays, an α particle is emitted, rather than a proton. This is the case with heavy elements. Atkinson hypothesized correctly that it also holds for light elements.

To better appreciate the idea, let us cite Eddington[36] on this very subject:

Indeed the formation of helium is necessarily so mysterious that we distrust all predictions as to the condition required. The attention paid to temperature, so far as it concerns the cookery of the helium atom, seems to neglect the adage 'First catch your hare ...'. How the necessary materials of 4 mutually repelling protons and 2 electrons can be gathered together in one spot, baffles imagination. One cannot help thinking that this is one of the problems in which the macroscopic conception of space has ceased to be adequate, and that the material need not be at the same place (macroscopically regarded), though it is linked by a relation of proximity more fundamental than the spatial relation.

How desperate Eddington had become just a few years before quantum tunneling was discovered!

One of the main problems, as we may witness time and time again, was the attempt to explain the entire plethora of observational data without being able to separate it in some way, or identify the fact that it contained a mixture of many problems which might require different explanations.

Up to now, Atkinson had been applying Eddington's model. Now he tried to apply Milne's model, and claimed that his scenario of element synthesis agreed better with Milne's model, because of the shape of the assumed energy source. Indeed, Milne assumed a point energy generation source, while Eddington assumed a spread-out energy source. However, Atkinson did not carry out any calculations to support this claim.

The experimental data on the masses and the binding energies of nuclei did not provide Atkinson with any clues as to which element marked the end of the process, because they were not sufficiently accurate at that time. So Atkinson resorted to the recently published nuclear stability theory by Gamow.[37] Fortunately, Gamow's work was published just before Atkinson started his research. (As a matter of

[36] Eddington, A.S., *The Internal Constitution of the Stars*, p. 301.

[37] Gamow, G., Proc. Roy. Soc. **1126**, 632 (1930).

fact, Gamow, Atkinson, and Houtermans were all in contact with one another.) The theory provided good results for nuclei with atomic weights $4n$, where n is an integer less than 5. But for greater values of n, i.e., A greater than 25, the results were not very convincing. Hence Atkinson was unable to be accurate when he assumed a very high Z (around that of iron).

Worse still, since he was unable to exactly determine the 'last nucleus', Atkinson was forced to assume a very highly charged nucleus, whose reaction with the proton would be difficult and hence slow at the relatively low stellar temperatures. This in turn made the whole process extremely slow. Inevitably, Atkinson proposed an untenable scenario (an imaginary process, whose properties were the opposite of the Gamow penetration properties) and he admitted apologetically:

> *It is only with reluctance that we introduce such a hypothesis, since there is no a priori justification for it at all, as far as is known.*

It is a pity that, due to the lack of proper data, Atkinson had to retreat from his original, correct idea.

A further problem Atkinson faced was due to the general assumption that stars evolve in a completely mixed or well stirred manner. This meant that the products of the synthesis that took place in the depths of a star should be exposed on the surface. But observationally, this is not generally the case. It took some twenty years before Sweet realized that mixing between core and envelope does not take place in main sequence stars, and that nuclear energy generation takes place only in a small core, whence the products do not usually appear on the surface of stars.

Finally, there was the question as to where the elements used in the regenerative process to convert hydrogen into helium were themselves synthesized? Was the star born with them? Like everyone else at the time, Atkinson had no idea. As for the white dwarfs, Atkinson argued that: *They represent the final stage [...]. Their energy may be purely gravitational, but need not be so.* It is a pity that Atkinson added the second half of the last sentence. He was right in the first half.

The nagging problem of the giants bothered Atkinson, so he put forward the suggestion that:

> *For low-density giants some earlier source of helium must be operative. This is taken to be ^{8}Be, whose instability was already assumed by Houtermans and the writer, and has since acquired almost the status of observational fact. It must be long-lived, since it is found on the Earth and this accounts for the Hertzsprung gap and its continuation beyween the Cepheids and B stars and for the fact that ^{8}Be cannot supply helium in the main sequence.*

The hypothesis was thus that the energy supply of the giants, which at that time were thought to be an evolutionary stage before the main sequence, came from the synthesis of helium at temperatures as low as 4 million degrees.

6.10 The $^8Be_4$ Barrier

Atkinson's second paper was devoted to the problem of the giants. This time he confronted the uncertainty over whether the nucleus $^8Be_4$ is stable or not. Application of Gamow's nuclear stability theory yielded that the binding energy is very close to zero and hence practically unstable. But nobody could assume that Gamow's theory was reliable for such light nuclei. So Atkinson postulated that $^8Be_4$ was unstable, with a half-life of $10^{8.5}$ years. Everything Atkinson assumed later followed from this incorrect assumption and we shall therefore skip the discussion of the consequences.

The problem of ^{8}Be can be called 'the search for the fifth significant figure'. The basic challenge nuclear research had to meet at the beginning of the 1930s was whether ^{8}Be was stable against decay into two α particles, and the answer depended in turn on whether ^{8}Be was more or less massive than two α particles. If ^{8}Be was less massive than two α particles, then it would be stable, and the way to synthesise the heavy elements would pass by the fusion of two α particles. And vice versa, if two α particles were less massive than ^{8}Be, then the problem was to discover how Nature managed to circumvent this problem, if at all. The experimental problem was that the mass difference $m(^8\text{Be}) - 2m(\alpha)$ is less than 1% of the mass of the two α particles, and it was not easy to measure masses with an accuracy better than 1%, because ^{8}Be is not found in Nature. The straightforward way would have been to synthesize ^{8}Be in the laboratory and see what happened to it. But this was impossible at the time. So by default, the whole issue boiled down to the sign of $m(^8\text{Be}) - 2m(\alpha)$. However, as we shall see later, the sign is not the only important consideration for the synthesis to occur. The exact value is essential too. Can we already infer from the fact that ^{8}Be does not exist in Nature that it is unstable with a short lifetime, or is there some peculiar reason why it is so rare that we do not find it? The only way to answer this critical question was by producing ^{8}Be by means of nuclear reactions and observing the behavior of the product. Besides the importance to astrophysics, there was a nuclear theoretical question: what is the force between two α particles? The answer to this question has an impact on all $4n$ nuclei, where n is an integer. If the force between two α particles cannot hold them together, then how does the ^{12}C nucleus hold together? Or is the idea that ^{12}C is made of three α particles completely wrong?

As early as 1908, Lord Rayleigh[38] showed that the mineral beryl contains large amounts of helium. If helium had accumulated in this mineral as a result of atomic disintegration of beryllium, we should expect a high helium content only in older beryls. Could the helium be trapped during formation of the mineral, or was it due to some short-timescale radioactivity? The results supported the idea that the source of helium was some kind of disintegration of beryllium. If ^{8}Be was stable, but with a small binding energy, than some high-energy γ from cosmic rays could break it into two α particles, and in this way form the helium captured in the mineral. The conclusion Lord Rayleigh reached was that the marginally stable ^{8}Be had disap-

[38] Rayleigh, Nature **123**, 607 (1929).

peared from the Earth due to the action of cosmic rays. On Earth we find only the remnants of this nucleus, captured in the beryl minerals.[39]

In 1933, Bonner and Brubaker[40] investigated several reactions which form ^{8}Be, such as the reaction $^7Li_3 + {}^2D_1 \to {}^8Be_4 + {}^1n_0$, and concluded that the mass of ^{8}Be is greater by 0.3 ± 0.75 MeV than that of two α particles, which meant that it was unstable. However, the calculation based on the experiment was not sufficiently accurate (as can be seen from the error, which was greater than the result) to rule out the possibility that ^{8}Be might be less massive than two α particles and hence stable.

Later the same year, Libby (1908–1980, who won the 1950 Nobel Prize for Chemistry for the discovery of ^{14}C dating) set a minimum for the stability of natural beryllium[41] by measuring the natural radioactivity of commercial beryllium. The measurements were carried out after the publication of conflicting reports on the natural radioactivity of beryllium.[42] Libby repeated the measurements and claimed that its lifetime against decay into two α particles was longer than 4×10^{15} yrs (longer than the age of the Universe). He did not detect any unstable beryllium.

In the same year, Crane and Lauritsen[43] carried out an experiment in which lithium was bombarded by protons, and found that their results agreed well with $^7Li + {}^1H \to 2{}^4He$, suggesting that ^{8}Be was unstable. But they had a problem, because the experiment was inconclusive. They suggested therefore that the process might be more complicated than was implied by the above reaction. And they were not far from the truth!

Shortly afterwards, Oliphant, Kempton and Rutherford[44] experimented with the reaction $^9Be + {}^1H \to {}^8Be + {}^2D$, and concluded that ^{8}Be was *probably just stable*.

In 1935, Bernardini and Mando[45] carried out a different type of experiment. They bombarded ^{7}Be with γ rays. The γ knocked off a neutron which could be captured by another ^{7}Be nucleus to form ^{8}Be, and this would subsequently disintegrate (or not) into two α particles. As they did not detect any α particles coming out, they concluded that:

> *If ^{8}Be is formed in the reaction, it is stable or it has an excess mass of less than 0.56 MeV.*

The result disagreed with Crane and Lauritsen, but agreed with Bonner and Brubaker and Oliphant, Kempton, and Rutherford. ^{8}Be was stable, so Bernardini claimed.

The problem seemed to be solved when, in 1936, Oliphant[46] took all the available data (mostly measured by Aston[47]) and drew the highly instructive graph shown in

---

39 Walke, H.J., PRL **47**, 969 (1935).

40 Bonner, T.W., & Brubaker, W.M., Phys. Rev. **48**, 742 (1935).

41 Libby, W.F., PRL **45**, 513 (1933).

42 Langer & Raitt, Phys. Rev. **43**, 585 (1933), did discover the natural radioactivity of beryllium (the title of the paper was *A New Kind of Radioactivity*), while Evans, R.D., & Henderson, Phys. Rev. **44**, 59 (1933), did not.

43 Crane, H.R., & Lauritsen, C.C., Phys. Rev. **45**, 63 (1934).

44 Oliphant, M., Kempton, W., & Rutherford, E., Proc. Roy. Soc. A **150**, 241 (1935).

45 Bernardini, G., & Mando, M., PRL **48**, 468 (1935).

46 Oliphant, M., Nature, 7 March 1936, p. 396.

47 Aston, W.F., Nature **137**, 357 (1936).

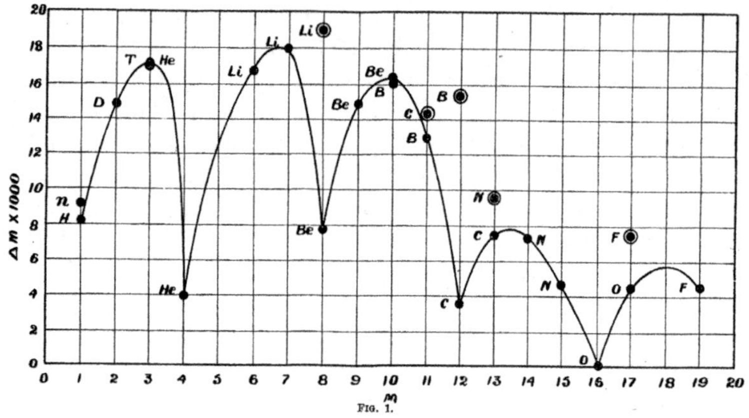

Fig. 6.7 The departure of the mass from a whole number as a function of the atomic weight, assuming $A(^{16}O) = 16.0000$, as drawn by Oliphant (1936). The curve is smooth, and we see that the α particles are the most tightly bound, and beryllium is bound. Note that the point representing ^{4}He lies exactly on the curve $\Delta m = 4$, while the point representing ^{8}Be is slightly below the line $\Delta m = 8$, implying that ^{4}Be is stable

Fig. 6.7. This shows that the binding energy possesses a quite amazing periodicity in the mass of the nuclei, and the inevitable conclusion was:

If we imagine that stable atomic species are to be built up from hydrogen by successive additions of single particles, protons or neutrons, the most strongly bound atoms are ^{4}He, ^{8}Be, ^{12}C, ^{16}O, ^{20}Ne, and so on.

This result nicely supported the idea that the α particle is probably a very strongly bound subsystem in the nucleus. Thus, according to Oliphant, ^{8}Be is a stable element. However, he gave the following numbers: $m(\alpha) = 4.0039$ and $m(^8\text{Be}) = 8.0078$, so the difference between beryllium and the two α particles was zero, to the accuracy of the numbers he provided. This actually implied that one cannot decide whether ^{8}Be is stable or not. On the other hand, if it is indeed stable, it is stable by a very small margin. So Oliphant concluded that:

The mass of ^{8}Be appears to be almost exactly equal to the sum of the masses of two α particles. The evidence is that this isotope is quite stable, as it appears in several reactions as a recoil nucleus with a kinetic energy which is high compared with the extremely small apparent binding energy. This may perhaps be regarded as evidence against the assumption that α particles exist as separate entities inside the nucleus.

As convincing as it looked, it was far from the last word on the subject. Note that there is no nucleus with $A = 5$ in Oliphant's curve.

At the same time Atkinson[48] was busy collecting all the available data and plotted the departure of the mass from a whole number (exactly as Oliphant did), as a function of atomic mass. His result is shown in Fig. 6.8. As Atkinson put it:

[48] Atkinson, d'E.R., Phys. Rev. Lett. **48**, 382 (1935).

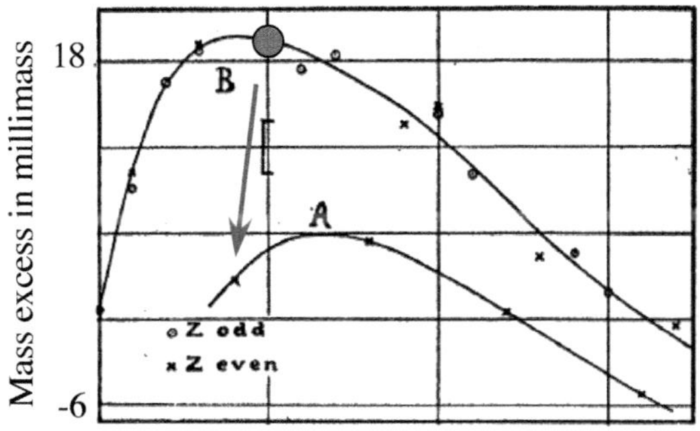

FIG. 1. Nuclear mass excess as a function of mass.

Fig. 6.8 The graph presented by Atkinson, indicating that ^{5}He is stable. Mass excess as a function of the nuclear mass. *Curve A* is for nuclei with an even number of protons and neutrons and *curve B* for all the others. The expected location of ^{5}He is marked by a *red disk*. The inferred decay is indicated by the *arrow*. The graph appeared in the PRL without the numbers for the ordinates and abscissas. These were introduced by a comment in Ap. J. **64**, 75 (1936). The *ordinates* are the mass excesses over the nearest whole number, in 'millimass' units, and the *abscissas* are atomic weights

> *It is remarkable how nearly the mass excesses of all stable class B nuclei lie on one curve; even though we have no explanation of it as yet, it would appear reasonable to expect ^{5}He to lie on the curve also; but if it did, it would certainly be violently unstable because of the great difference between the ordinates of curves A and B at this region.*

The expected location of ^{5}He was added to the figure and is marked by a full red circle. The decay mode is shown with the red arrow. After Atkinson saw Oliphant's results and arguments that ^{5}He should be stable, he added a remark in his next paper[49] saying that Oliphant's method of drawing one curve with a sharp minimum was probably the preferred way to infer stability. Yet Atkinson insisted that, given all known information, ^{5}He must be unstable. And he was right.

Bleakney et al. (1936)[50] analyzed natural beryllium in an attempt to discover traces of ^{8}Be, which was expected to exist if it were formed somehow and was stable. But no ^{8}Be traces were found at a level higher than one part in 10 000. So if ^{8}Be did not exist in Nature, it was highly likely that it was in fact unstable.

The same year, 1936, Allen[51] stated that, *although much work has been done on the disintegration of Be, it seems worthwhile to investigate*, and he went on to examine the disintegration of beryllium by protons, concluding that the mass of ^{8}Be was 8.0074 amu, just a little bit below the value of Oliphant et al. (8.0078 amu). The nucleus was therefore unstable by just 0.0003 amu! But there was not a word in the paper about the stability or instablity of the nucleus, only a discussion about some problems with the theory.

---

49 Atkinson, d'E.R., Ap. J. **84**, 75 (1936).

50 Bleakney, W., Blewett, J.P., Sherr, R., & Smoluchowski, R., Phys. Rev. **50**, 545 (1936).

51 Allen, J.S., Phys. Rev. **51**, 182 (1936).

The situation regarding ^{5}He was settled in 1937, when Williams at al.[52] carried out the experiment $^7\text{Li} + {}^2\text{D} \rightarrow 2{}^4\text{He} + \text{n}$. If ^{5}He were stable, then the outcome of the reaction would have been $^5\text{He} + {}^4\text{He}$. While in both cases α particles emerge from the reaction, the energies are completely different in the two cases. By not observing the second reaction, they provided further nuclear evidence that ^{5}He is in fact unstable.

On the other hand, the situation with ^{8}Be remained confused. In 1937, Williams, Haxby, and Shepherd[53] found that the mass of ^{8}Be was even higher, giving the value 8.0081 ± 0.00005 MeV, and thus that the nucleus was just unstable. Their value for the mass of two α particles was 8.0080. Note the extremely high accuracy claimed by the authors.

In the meantime the neutron was discovered, as will be described in the next section, and nuclear reactions with neutrons became possible. The ultimate conclusive solution to the stability conundrum of ^{8}Be was brought by two chemists, Paneth & Gläckauf,[54] who irradiated ^{9}Be with γ rays. The product was helium, and not $^8\text{Be} + \text{n}$. As a matter of fact, it was known to them from the work of Chadwick and Goldhaber,[55] published two years earlier, that the irradiation of deuterium and beryllium by γ rays produced neutrons. So Paneth & Gläckauf irradiated the beryllium, left the product to rest, and returned to measure what remained after a month. But they could not discover any trace of ^{8}Be. There were two possibilities for the product:

$$^9\text{Be} + \gamma \rightarrow {}^8\text{Be} + \text{n} \quad \text{or} \quad {}^9\text{Be} + \gamma \rightarrow 2{}^4\text{He} + \text{n}\,.$$

If it is the first reaction, i.e., the ^{8}Be lives for some time, then its lifetime is less than a month. If it is the second possibility, then ^{8}Be is unstable. Thus in both cases, the fact that no trace of ^{8}Be was found meant instability of ^{8}Be.

This result was confirmed by Kirchner and Neuert[56] by investigating the disintegration of $^{11}\text{B} + \text{H} \rightarrow {}^8\text{Be} + {}^4\text{He}$. They stated that the mass of ^{8}Be was 40–120 keV above the energy of two α particles.

But the saga did not end with Paneth and Gläckauf. Allison et al.[57] investigated $^9\text{Be} + \text{H} \rightarrow {}^8\text{Be} + \text{D}$ and found that $m(^8\text{Be}) = 8.00739$. Further, in the same volume of the Physical Review, Allison[58] gave a slightly higher mass, viz., $m(^8\text{Be}) = 8.00753$, still below the mass of two α particles, and not sufficient to change the claim regarding the stability of ^{8}Be.

During 1937, Livingston and Bethe[59] wrote an extensive review on nuclear physics, examined a set of nuclear reactions which all led to the formation of ^{8}Be (and

52 Williams, J.H., Shepherd, W.G., & Haxby, R.O., Phys. Rev. 888 (1937).

53 Williams, J.H., Haxby, R.O., & Shepherd, W.G., Phys. Rev. **52**, 1031 (1937).

54 Paneth, F.A., & Gläckauf, E., Nature **139**, 712 (1937).

55 Chadwick, J. & Goldhaber, M., Nature **135**, 65 (1935).

56 Kirchner, F., & Neuert, H., Naturwiss. **25**, 48 (1937).

57 Allison, S.K., Graves, E.R., Skaggs, L.S., & Smith, N.M., Phys. Rev. **55**, 107 (1939).

58 Allison, S.K., Phys. Rev. **55**, 624 (1939).

59 Livingston, M.S., & Bethe, H.A., Rev. Mod. Phys. **9**, 245 (1937).

subsequently its disintegration), and concluded that $m(^8\text{Be}) - 2m(^4\text{He}) = 130$ keV, whence ^{8}Be had to be unstable. Livingston and Bethe's paper was published in July, so they could not cite Paneth and Gläckauf's paper, which was published in April. But in later papers, Bethe treated Paneth and Gläckauf's paper as *the last word*.

We have based our explanation and description of the history of the instability of the ^{8}Be nucleus on the mass difference. To be more accurate, this condition is necessary, but not sufficient, because other conservation laws must be satisfied as well, such as conservation of momentum and angular momentum. We can see this simply by asking why a photon does not disintegrate into two photons, the sum of whose energies is equal to the original energy. There is no energy restriction which forbids this process. The answer lies in momentum conservation, including the spin of the photon, which is 1. If the spin of the photon is not included in the conservation relation, just the conservation of momentum and energy cannot prevent this process from happening.[60]

What is so special about the ^{8}Be nucleus? In 1937, Wheeler,[61] assuming the ^{8}Be nucleus to be stable, advanced the idea that the α particle might be a subsystem inside the nucleus, and that we should see this nucleus as composed of two particles. Note that we see the atom as composed of a nucleus, which is also composed of several particles, and the electrons. Hence one should not be surprised if the particles in the nucleus are arranged in some sort of subsystem. If indeed this were the case, it would have important ramifications elsewhere. The energy level structure of diatomic molecules was well known. Hence, if the nucleus of ^{8}Be were made of two particles like a molecule made of two atoms, it would be a trivial matter to predict the energy levels of the ^{8}Be nucleus. The evidence, however, was not sufficient to draw any general conclusion.

Two new measurements were made towards the end of the 1940s. In 1949, Hemmendinger[62] verified the $Q = 116 \pm 10$ keV observation of α particles from the reaction $^9\text{Be}(\gamma, \text{n})$. In the meantime, Tollestrup,[63] Fowler, and Lauritsen published the preliminary results in a conference. They found that the energy difference was 84.5 ± 10 keV. When Hemmendinger found out about the results of Tollestrup and his associates, he checked his data reduction and discovered[64] that he had failed to take into account the recoil of the nucleus. So he corrected his calculation and found 103 ± 10 keV. When the final paper by Tollestrup et al.[65] appeared, they gave the result $Q = 89 \pm 5$ keV, so an agreement was almost reached between the two experiments.

---

[60] In the case of a photon, the momentum is given by $p = E/c$. Hence, a photon could disintegrate into two photons if it where not for the spin. A classical particle with a non-vanishing rest mass like a proton, cannot disintegrate into two particles without violating one of the conservation laws.

[61] Wheeler, J.A., Phys. Rev. **52**, 1083 (1937).

[62] Hemmendinger, A., Phys. Rev. **75**, 1267 (1949).

[63] Tollestrup, A.V., Fowler, W.A., Lauritsen, C.C., Bull. Am. Phys. Soc. **24**, No. 2, E6 (1949).

[64] Hemmendinger, A., Phys. Rev. 1267 (1949).

[65] Tollestrup, A.V., Fowler, W.A., Lauritsen, C.C., Phys. Rev. **76**, 428 (1949).

6.11 The Discovery of the Neutron

In 1930, Bothe and Becker[66] in Germany found that, if the very energetic α particles emitted from polonium hit light elements like beryllium, boron, or lithium, an unusually penetrating radiation was produced. At first this radiation was thought to be γ radiation, although it was more penetrating than any γ rays known at the time. However, the details of the experimental results were very difficult to interpret on this basis.

A year later and on the other side of the Atlantic Ocean, Langer and Rosen[67] hypothesized about the existence of a 'neutron':

> *A combination of an electron and a proton of low energy and very small size.* The predicted mass was: *slightly smaller than that of a hydrogen, with a diameter of* 10^{-12}–10^{-13} *cm, and energy of the order of* $m_e c^2$*. The maximum is* $15 m_e c^2$*, where* m_e *is the mass of the electron.*

Langer and Rosen gave a list of experimental facts which would be easier to explain with this hypothesis. An interesting point was that they also mentioned the problem of high density matter in white dwarfs, and indicated that it would explain the features of white dwarfs in a simple way: *nor is there much danger of violating the Pauli exclusion principle by exceeding the maximum electron density for a given pressure*. However, they said nothing about the spin of the particle, and consequently they were unable to predict which statistics it would obey.

The next important contribution was reported in 1932 by Curie and Joliot[68] in Paris, who knew about the results of Bothe and Becker in Germany. Together with Webster,[69] they showed that, if this unknown radiation fell on paraffin or any other hydrogen-containing compound, it ejected protons of very high energy. This was not in itself inconsistent with the assumed γ ray nature of the new radiation, as Curie and Joliot suggested, but detailed quantitative analysis by Chadwick indicated that it would be difficult to reconcile with such an hypothesis. Later in 1932, Chadwick in England performed a series of experiments showing that the γ ray hypothesis was untenable. He suggested that the new radiation consisted of uncharged particles of approximately the mass of the proton, and he performed a series of experiments to verify his suggestion. The uncharged particles were eventually called neutrons, from the Latin root for neutral and the Greek ending -on (by imitation of electron and proton).

In his letter to Nature in 1932, Chadwick[70] reported on the experiments carried out by Bothe and Mme. Curie-Joliot and Joliot in which beryllium was bombarded by α particles emerging from polonium. As a consequence, the beryllium released a very penetrating radiation which the above authors explained as very energetic

[66] Bothe, W., & Becker, H., Z. Physik **66**, 289 (1930).

[67] Langer, R.M., & Rosen, N., Phys. Rev. **37**, 1579 (1931).

[68] Curie, I., Compt. Rendu Acad. Sci. Paris **193**, 1412 (1931); Curie, I., & Joliot, F., Compt. Rendu Acad. Sci. Paris **194**, 273 (1932).

[69] Webster, H.C., Proc. Roy. Soc. A **136**, 428 (1932).

[70] Chadwick, J., Nature, 27 February 1932, p. 312. The full paper is: Proc. Roy. Soc. A. **136**, 692 (1932).

photons. Chadwick repeated the experiment with several elements and discovered that the simplest way to explain this powerful radiation was by assuming that it was composed of particles with atomic mass 1 and charge 0, namely neutrons. It became clear that the Joliots and Bothe and Becker had misinterpreted their results, and consequently failed to discover the neutron.

Chadwick was awarded the Nobel Prize for the discovery of the neutron just three years later, in 1935. The same year, Curie and Joliot got the Nobel Prize for Chemistry for the production of artificial radioactive elements and the demonstration of the transmutation of elements. Bothe was granted the Nobel Prize in 1954 (together with Max Born) for inventing the coincidence method and for the discoveries made using this method.

The discovery of the neutron caused a revolution in nuclear physics, and for the first time made it possible to apply the new quantum theory to nuclear physics. In a way, this was the beginning of modern nuclear physics.

6.12 New Horizons. Astrophysics

In 1932, Milne[71] claimed on the basis of his analysis of the stability of stars that: *it seems to me unlikely that energy generation in a star is actually governed by a law of the type* $\varepsilon = \varepsilon_0 \rho^t T^s$ (a power law in density and temperature). This is exactly the form of power generation that results from nuclear reactions. Milne argued in favor of gravitational contraction, and in view of the new discovery of the neutron, he suggested the reaction $p^+ + e^- \rightleftharpoons n\ (+Q)$ which is completely analogous to an ionization reaction like $Ca^+ + e^- \rightleftharpoons Ca\ (+Q)$:

> *Just as cooling encourages recombination of ions and electrons*, namely the recapture of the electrons by the ions to give neutral atoms at low temperatures, *so cooling will encourage the formation of neutrons with liberation of energy.*

Milne calculated the formation of neutron to occur at about 10^{10} K, and proposed that:

> *The energy liberated maintains the star's radiation to space; and the neutrons produced are presumably removed by the formation of nuclei of higher order. The star is an unclosed system and it is precisely the star's own radiation to space which tends to depress the internal temperature and so stimulate both energy generation and element synthesis. The star acts as its own stoker.*

Two comments are warranted. It seems that Milne misunderstood Eddington's paradox, namely that the star loses energy but heats up rather than cools down, and conversion of protons to neutrons takes place when the density and temperature increase and not vice versa.

However, the most important astrophysical implication of the discovery of the neutron was that there was now a way to synthesize the very heavy elements. It was

---

[71] Milne, E.A., Zeit. f. Astrophys. **5**, 337 (1932).

already apparent to Atkinson in the late 1920s that the Coulomb barrier penetration works only for the light elements, because the temperatures in stars are not sufficiently high to allow the protons to penetrate into very heavy nuclei. The discovery of the neutron changed this, because the neutron has no Coulomb barrier. The first rather extreme demonstration of the power of the neutrons came two years later when Fermi[72] bombarded uranium with neutrons, and created a new element with atomic number 93. Thus even the heaviest of all nuclei could easily absorb neutrons. Fermi's discovery (as well as other discoveries concerning the element with atomic number 93, which is essentially the new element neptunium) was questioned and criticized by the chemist Noddack,[73] who was known for the discovery of several new elements (rhenium, for example) and abundance measurements in meteorites. Noddack claimed that Fermi had in fact fissioned the uranium, and had not produced a heavier nucleus. However, Fermi was not wrong.

6.13 New Horizons. Nuclear Physics

Many researchers in the field of nuclear physics consider the discovery of the neutron as the birth of nuclear physics. We noted above that the papers by Atkinson and Houtermans and Atkinson should be considered as the birth of nuclear astrophysics, although they predated the birth of nuclear physics.

The first pressing question brought up by the discovery of the neutron was whether the neutron was just a proton combined with an electron, as suggested by Heisenberg,[74] or whether it really was an elementary particle like the proton, as suggested by Wigner.[75] The two possibilities led to different consequences regarding the nature of the force acting between the neutron and the proton.

The idea that the neutron and the proton exert the same nuclear force led Heisenberg to a very interesting suggestion. He assumed that the neutron and the proton were in fact the same particle, the nucleon, but in two different energy states. The mass of the neutron is 939.566 MeV, while the mass of the proton is 938.272 MeV, and the difference is 1.29329 MeV. The mass of the electron is 0.51 MeV. Hence, there is enough energy in the decay of a neutron into a proton to form the mass of the electron and endow the electron with sufficient kinetic energy to separate from the attracting proton. The conversion of the neutron into a proton was, according to Heisenberg's idea, just the transition of the nucleon from a high energy state where it appeared as a neutron, to a lower state where it appeared as a proton. We also note that protons, and hence neutrons as well, and electrons all satisfy Pauli's exclusion principle. If so, whenever a neutron finds an empty energy level, it will decay into it and vice versa. A free neutron has plenty of energy levels and hence decays into a

[72] Fermi, E., Nature **133**, 898 (1934).

[73] Noddack, I., Zeit. f. Angewandte Chem. **47**, 653 (1943).

[74] Heisenberg, W., Zeits. f. Physik **77** (1932); **78**, 156 (1932).

[75] Wigner, E., Phys. Rev. **43**, 252 (1933).

proton with a half-life of 11 minutes. Most neutrons in nuclei see filled energy levels and hence cannot decay. If we could pack free neutrons into energy levels in such a way that they were all full of neutrons, the neutrons would not have free energy levels into which to decay and remain as neutrons. This is exactly what happens in neutron stars. As in a white dwarf, where all the energy levels of the electrons are occupied, all energy levels in a neutron star are filled and so the neutrons cannot decay.

In his attempt to describe the two states of the proton and the neutron, Heisenberg invented a new quantum number which he called isospin, by analogy with the spin of the proton. The spin of the proton has just two states $+1/2$ and $-1/2$. Similarly, isospin has just two states: $+1/2$ which is observed as a neutron, and $-1/2$ which is observed as a proton. It was a purely quantum idea and had no corresponding analog in classical physics, like the spin itself. Indeed, it mimics the invention of spin, in the sense that it has no classical analogue and takes just two values.

In 1933, Eckart[76] compared the nuclear theories of Wigner and Heisenberg, and found that Heisenberg's theory predicted that the nuclei ^{3}H and ^{3}He would be unstable, in contrast to observation. Thus, the neutron had to be treated as an elementary particle rather than a composite of a proton and an electron. And yet, one should remain open to the possibility of transforming a neutron into a proton and back.

In 1933, Landé[77] put forward the hypothesis that the nucleus consists of α particles, neutrons, and zero or one proton. The previous structural scheme of the nucleus had been α particles, electrons, and 0,1,2, or 3 protons. Landé succeeded in getting good agreement between the calculated masses of the nuclei and the measured ones. Thus the new idea was that the nucleus was not made only of protons and neutrons, as discussed in the first chapter, but that the protons and the neutrons grouped together to form α particles inside the nucleus. There were therefore subsystems inside the nucleus, a view supported by Rutherford. The question of how strong the α particle subsystem might be had not been settled by that time.

6.14 Deuterium Exists and Is Stable

As had been well known for a long time, the atomic mass of hydrogen, when measured by chemical methods, was found to be 1.00777 ± 0.00002, while Aston had found the value 1.00756 ± 0.00015 using the mass spectrometer. So the difference in the atomic weight was greater than the quoted error. In 1931, Menzel[78] speculated that the difference might be due to an isotope of hydrogen of mass 2, with a relative abundance of $^1\mathrm{H}/^2\mathrm{H} = 4500$, and claimed that: *It should be possible to detect such an isotope by means of spectra.*

---

[76] Eckart, C., Phys. Rev. **44**, 109 (1933).

[77] Landé, A., Phys. Rev. **43**, 620 (1933).

[78] Menzel, D.H., Phys. Rev. **37**, 1669 (1931).

Hardly a year after this speculation was announced, Urey (1893–1981m), Brickwedde, and Murphy discovered deuterium,[79] and with about the predicted abundance in water. As no stable nucleus with two nucleons had previously been known, the synthesis of the elements starting from hydrogen alone was not considered as a viable possiblity. Now new avenues for the synthesis were opened up.

Soon after the discovery of the $A=2$ nucleus, tritium, the $A=3$ nucleus, was discovered by a group of scientists, viz., Rutherford and Cockroft, Lawrence, Alvarez and Libby. The story has its own peculiarities, as Rutherford made an error of judgement and thus was not credited for the discovery of tritium. Rutherford bombarded heavy water with a beam of deuterons accelerated by Cockroft and Walton's machine. Examination of the products showed the existence of two nuclei with $A=3$, namely tritium and helium 3. Rutherford assumed that tritium was the stable isotope and helium 3 was unstable. It was Luis Alvarez (1911–1988, Nobel laureate 1968) who realized Rutherford's error, i.e., that the situation was the other way round: tritium is unstable while helium 3 is stable. Tritium is formed continuously by cosmic ray neutrons. It is used today to date underground water and wine using a method developed by Libby.

What is the nuclear state of deuterium? Bainbridge[80] measured the mass of deuterium very carefully and found that, if it was composed of two protons and one electron, then the binding energy was $2\times 938.27+0.511-1875.05=2$ MeV (twice the proton mass plus the mass of the electron, all in MeV), whereas if it was composed of *one proton and one Chadwick neutron of mass 1.0067*, the binding energy was 0.97 MeV. As the measured binding energy of the deuterium was at that time taken as 2.2 MeV (the present day value is 2.224 MeV), it seemed that the deuterium must be made of two protons and an electron.

In 1934, Murphy and Johnston[81] analyzed the atomic spectra of deuterium and reached the conclusion that the spin of the deuterium nucleus was 1 and that it obeyed Bose–Einstein statistics. Recall that the neutron, the proton, and the electron have spin 1/2. If the neutron was a proton and an electron, it should have either spin 1 or spin 0. Since neither is the case, it seems that the neutron is not a simple composition of the two (unless there is an additional third particle with spin 1/2 inside the neutron which somehow cancels the spin of the electron). On the other hand, if the neutron is not such a compound particle and there is no electron involved, the simple picture of the deuterium is of a proton and a neutron with aligned spins, whence the total spin is $1/2+1/2=1$. Could the proton and the neutron combine to form deuterium with opposite spins, giving a vanishing total spin? Nature indicates that such a deuterium nucleus does not exist. We conclude, therefore, that the force holding together the proton and the neutron inside the deuterium nucleus depends on the spin of the reacting particles. The spin of the deuteron, whether it is 1 or zero, is crucial, as will be shown later.

---

[79] Urey, H.C., Brickwedde, F.G., & Murphy, G.M., Phys. Rev. **39**, 164 (1932).

[80] Bainbridge, K.T., Phys. Rev. **42**, 1 (1932).

[81] Murphy, G.M., & Johnston, H., Phys. Rev. **46**, 95 (1934).

In 1936, Bethe and Bacher[82] reached the conclusion that:

The forces between proton and neutron can therefore depend only slightly, if at all, upon the relative spin directions of the two particles.

They refer to Heisenberg,[83] who originally assumed an interaction which was attractive for parallel spins and repulsive for opposite spins. The argument of Bethe and Bacher was based on the comparison of the binding energy of deuterium (known at that time to be about 2 MeV) with the binding energy of an α particle, which they took as 28 MeV.

The discovery of deuterium gave a new twist to the space of possibilities for proton–proton capture. The question of interest with regard to the synthesis of the elements was whether it was possible that during the collision between two protons, one proton might convert into a neutron via the weak interaction and form a deuterium? If this were possible, then the way was open to build up elements without resorting to heavy elements as catalysts. However, the interaction which converts a proton into a neutron by emitting a positron is a weak interaction.

6.15 The Weak Force. Key to Stellar Longevity

If stars extract energy from the conversion of hydrogen into helium, then the process must convert two protons into two neutrons. The process which can convert neutrons into protons and vice versa is β decay.

If β decay were a two-body decay (for example, neutron $\to$ proton + electron), then the laws of conservation of energy and momentum would require the energy of the electron to have a unique value $E(\mathrm{e}) = (m_\mathrm{n} - m_\mathrm{p} - m_\mathrm{e})c^2$. However, experiment shows a continuous spectrum of values, i.e., the electrons appear to have energies between the maximal value $E(\mathrm{e})$ and zero. The initial state of the nucleus has a well-defined energy, and so does the final state. It is therefore impossible for the emitted electron to come out with a distribution of energies and not with a unique value of energy, under the above assumptions. In radioactive α decay, for example, the system also disintegrates from a well-defined initial state into a well-defined final state, and the emitted α particle always has the same energy, viz., the difference between the initial and final energies. This is not the case in β decays. The electron has a continuous energy spectrum!

The puzzle shook the physics community, in particular after the dramatic successes of quantum theory. Why are energy and momentum not conserved in β decay? Bohr[84] suggested that energy might not be conserved in a single β decay, but only on the average, because the energy emitted by the collection of radioac-

[82] Bethe, H.A., & Bacher, R.F., Rev. Mod. Phys. **8**, 82 (1936).

[83] Heisenberg, W., Zeits. f. Physik **78**, 156 (1932); ibid. **80**, 587 (1932).

[84] Bohr, N., in *Atomic Stability and Conservation Laws*, Reale Accademia d'Italia, Rome (1932). See also, Bohr, N., Faraday Lecture, J. Chem. Soc. 349 (1932).

tive nuclei, in contrast with the energy emitted by a single nucleus, does satisfy the conservation law. Such a heresy was anathema to many physicists.

Pauli was invited to give a talk at a conference on radioactivity. He could not come, and instead sent a letter to be read before the audience.[85] In the letter, he suggested that the extra energy might be taken by an elusive particle which escaped detection. Pauli admitted that he did not feel *secure enough to publish anything about this idea*, and asked the audience to look for a possible experimental confirmation. Pauli did not carry out calculations and the idea was never published in a journal, until Fermi picked up the problem in 1934.[86] He adopted Pauli's hypothesis and named the particle the neutrino, meaning 'little neutral one', to distinguish it from the neutron, because it was expected to have a much smaller mass than the electron and no electric charge.[87] Fermi thereby created a very successful theory of β decay. The idea hovered over physics for more than twenty years. On the one hand, Fermi's theory of β decay, which assumed the existence of Pauli's particle was very successful, and on the other hand, there was no sign of a direct proof that the elusive neutrino really did exist. Pauli himself could not have imagined how elusive the particle was going to be, and his lack of confidence in his own idea arose from the (fallacious) thinking that, if the particle existed, it would have been detected long before then. Fermi's theory did explain the β decays, and gradually became adopted in physics. In 1956, Cowan, Reines et al.[88] managed to detect the neutrino and informed Pauli about the discovery of the particle he had predicted to exist 26 years earlier.[89]

It soon became clear that the neutron discovered by Chadwick two years earlier was not the particle Pauli hypothesized. Fermi adopted Heisenberg's nuclear theory that the nucleus contains protons and neutrons which can transform from one to another or change state. The idea was that inside the neutron there exists a force, the action of which is the conversion of a neutron into a proton and the emission of an electron and a neutrino. Thus, Fermi hypothesized the existence of a new force, the force that gives rise to the β decay and was later named the weak force.

The beauty of Fermi's ingenious theory was that many details of the β decay could be inferred and compared with experiment without knowing all the properties

---

[85] Pauli, W., 1930, letter to Group on Radioacitvity (Tübingen, 4 December, unpublished). See also Rappts. Septième Conseil Phys. Solvay, Bruxelles, 1993, Gautier-Villars, Paris. See also Physics Today, September 1978.

[86] Fermi, E., Zeit. Fur Phys. 161 (1934).

[87] It is easy to infer that the new particle should be neutral, because the account of existing charges showed no deficit. To find the mass of the particle was more tricky. If the particle had mass, say m_ν, then the maximum energy the electron could come out with was $(m_n - m_p - m_e - m_\nu)c^2$. But the maximum observed energy was very close to $(m_n - m_p - m_e)c^2$, only leaving room for a very small neutrino mass, and the accuracy of the measurements was insufficient to be able to decide. From statistical considerations, the most energetic electron is the rarest event.

[88] Cowan, C.L., Reines, F., Harrison, F.B., Kruse, H.W., & McGuire, A.D., Science **124**, 103 (1956).

[89] The 1995 Nobel Prize was shared by Reines and Perl. Reines' address at the Nobel Prize reception, along with the history of the discovery, can be read in Reines, F., Rev. Mod. Phys. **68**, 317 (1996).

of the weak force. It was just the space of probabilities into which each particle was ejected that determined the distribution of energy of the electron. So all possible states that the three ejected particles could take had equal probability according to Fermi. There was no preferred state. This simple assumption was sufficient to yield the basic results of β decay, and without any knowledge of the details of the weak force! However, Fermi went one step further and modeled the properties of the force on the basis of what was known about the Coulomb force. Comparison with observation allowed Fermi to set a limit on the mass of the neutrino. It was merely a question of the difference in energy between the initial state and the final state. If the most energetic electrons had energies very close to the difference in energy between the initial and final state, there was no energy/mass left for the neutrino. Fermi concluded that the best agreement with observations was obtained when the neutrino had a vanishing rest mass.

In 1937, Gamow and Teller[90] postulated an extremely important addition to Fermi's β decay theory. They realized that there were cases where the Fermi theory failed to explain the decays. Consequently, Gamow and Teller proposed an ad hoc solution to explain the discrepancy. For our purposes here, we can simplify the difference between the Fermi and the Gamow–Teller interactions as they are expressed in the reaction relevant to stars, namely $\mathrm{p}+\mathrm{p} \rightarrow {}^{2}\mathrm{D}+\mathrm{e}^{+}+\nu$. In a Fermi interaction, which converts a proton into a neutron and vice versa, the sum of all the spins of the particles does not change. In a Gamow–Teller interaction, the total spin must change by one unit. The implementation for the above reaction will be carried out later.

What Gamow and Teller actually discovered was that the weak force, which is responsible for β decay, has two different components, which behave and act differently and have different strengths. The best alternative example is the electromagnetic force, which can appear as a Coulomb force between electric charges or as a magnetic force acting on moving charges. The electric and magnetic components behave differently. Fierz[91] generalized the theory by combining the Fermi and the Gamow–Teller conditions into a unified theory of this complicated force. The strength of the two components of the force, when compared to the so-called strong force acting between the protons and the neutrons, for example, is very small, whence the name 'weak' force.

Tolman[92] was already prepared to carry Bohr's ideas about the non-conservation of energy to cosmology, and considered the possible consequences. He claimed:

> *We might assume that different atoms of a given chemical isotope have nuclei which are not really exactly alike, so that different amounts of energy actually are available.*

On the other hand, α decays show very accurate energy conservation, and this disproves such a far-reaching assumption. Tolman continued, therefore, to claim that:[93]

---

[90] Gamow, G., & Teller, E., Phys. Rev. **49**, 895 (1936).

[91] Fierz, M., Zeit. für Phys. **194**, 553 (1937).

[92] Tolman, R.C., Proc. Nat. Acad. Sci. USA **20**, 379 (1934).

[93] Pauli, W., Paper read before the Am. Phys. Soc., Pasadena, 16 June 1931. Here Pauli published his idea about the neutrino. Fermi, E., La Ricerca Scientifica, Anno IV, 143 (1930).

> *To retain the principle of conservation, we might also assume as an alternative explanation that the emission of electrons is not the sole process accompanying a β ray decomposition, but in addition that some very penetrating radiation is simultaneously emitted, which carries off the balance of energy left by the electrons and then escapes through the walls of the container without being calorimetrically detected. For this purpose neutrons of very small mass have been postulated.*

In view of the fact that positron emissions appeared to violate energy conservation as well, Tolman set out to explore the possible consequences.

Tolman reported a conversation with Bohr, who had pointed out that, if energy conservation is only statistical, i.e., in some cases energy is conserved while in others it is not conserved, but that the average energy of all cases is conserved, then the reverse process, i.e., electrons entering the nucleus and rebuilding the parent substance, must also be statistical, leading to atoms of the same element being different. The final conclusions Tolman drew were:

> *(a) The experiment should determine whether energy conservation is not valid. (b) If the experimental outcome should indicate that energy in its familiar forms can be created and destroyed by such processes as discussed above, it should be noted that the principle of conservation might perhaps still be preserved by the device of adding to the expression for energy a new term purposely so chosen as to maintain conservation.*

Tolman mentioned that this is exactly the trick one used in the general theory of relativity. However, Tolman stressed that violation of energy conservation would require modifications to the special theory of relativity (because the latter does not allow creation/destruction of matter + energy):

> *These modifications might prove of interest for the problem of relativistic cosmology.* At present, nevertheless, concluded Tolman, *there are no additional facts to support such an hypothesis.*

6.16 The Return of Atkinson

In 1936, Atkinson[94] returned and reexamined his scenario from 1931, in view of the important recent discoveries. He realized that ^{8}Be was apparently unstable, although the existence of large amounts of He in beryllium minerals was still unexplained.

The existence of the neutron and the deuterium were established by now, and Atkinson tried to contend that he *postulated (Process B) as an ad hoc assumption.* Atkinson considered the consequences that the discoveries of the neutron and deuterium had for the physics of the energy sources in stars. Clearly, reactions with neutrons do not have the problem of overcoming a Coulomb repulsion, and so can proceed at any temperature. The question was whether neutrons could be produced in sufficient amounts in stars. A check of all possible neutron production reactions revealed that they are very slow and consequently unable to produce large amounts

94 Atkinson, R.d'E, Ap. J. **84**, 73 (1936).

of neutrons. For example, the reaction $p+e^- \rightarrow n$, namely the absorption of an electron by a proton resulting in a neutron, was examined in the laboratory,[95] and the results were all negative. The only alternative left to generate neutrons was by producing plenty of deuterium, and this was possible via the reaction: $p+p \rightarrow {}^2D+e^+$, where e^+ is the positron. After the discovery of deuterium, this reaction entered the realm of the possible.

In this way, Atkinson discovered the first reaction which leads to what is known today as the pp chain, namely, the synthesis of helium out of hydrogen, starting from pure hydrogen. However, he called for this reaction to produce the deuterium, and from it the neutrons. Atkinson expected it to be easy to check this reaction in the laboratory. He could not have been more wrong in his expectations. This is almost the only nuclear reaction in stars that cannot be measured in the laboratory. As a matter of fact, Atkinson erred in the suggestion of the p + p reaction, because the Gamow–Teller selection rule was not yet known. According to the Fermi transition rule, the only one known to Atkinson, this reaction could not take place (see details and later). Atkinson realized that his previous hypothetical regenerative process, which reached elements with atomic number as high as $Z = 28$, was probably not correct, but he found comfort by stating that there was not yet enough data. And there, he was right.

The stellar stability problem, as formulated by Jeans, bothered Atkinson, and he cited Eddington's hypothesis regarding the way to overcome it. Eddington, worried about the instability that any temperature-sensitive energy source would induce, hypothesized that the actual energy release was delayed by a time factor that did not depend directly on the temperature or pressure, and in this way 'freed' the energy generation process from the choking grip of the stability condition. Ten years after Jeans' mistaken paper, Eddington himself stumbled.

When all attempts to generate neutrons by light elements had failed, Atkinson considered the possibility that neutrons might be formed by means of catalysts in a two-step regenerative process like $M+{}^1H \rightarrow N$ followed by $N+e^- \rightarrow M+n$, where N and M were two nuclei which carried the reaction, and the transformation of a proton into a neutron was to take place inside nucleus N. This was actually what von Weizsäcker and Bethe assumed later to happen, following Atkinson's 'regenerative' process. The problem was that the two processes had to occur rather fast. If either of the reactions required a few billion years, the suggested process could still operate in stars, but the rate of energy generation became meaningless.

The possibility of deuterium disintegration and recombination of a neutron and a proton was treated theoretically by Bethe and Peierls,[96] who assumed that, during the merger of the particles, the interaction resembled an electromagnetic interaction. Fermi[97] was quick to show that the theoretical results of Bethe and Peierls does not agree with experiment. According to Fermi, either the reaction could not take place,

95 Livingood, J.J., & Snell, A.H., Phys. Rev. **48**, 851 (1935).

96 Bethe, H., & Peierls, R., Proc. Roy. Soc. A **148**, 146 (1935).

97 Fermi, E., PRL **48**, 570 (1935).

or otherwise an additional interaction was needed to bring the theory into agreement with experiment, but he did not go as far as Gamow and Teller two years later.

6.17 The Last Paper on Liquid Stars

Even as late as 1932, the idea of liquid stars lingered on, and papers discussing such models for stars were still being published.[98] Many years later, astrophysicists reached the conclusion that certain layers in neutron stars and white dwarfs might behave like liquids. However, the idea of liquid stars for main sequence (or dwarf) stars was defunct by 1932, and never rekindled.

6.18 The Nuclear Dilemma: Equilibrium Versus Rate

When we have any reaction of the type

$$A + B \rightleftharpoons C + Q\,,$$

where A and B combine to form the system C releasing heat Q, the reaction can go both ways. As a rule, when the temperature is high, the reaction will go to the left (C disintegrates), and when it is low, the reaction will go to the right. By high and low temperature we mean a temperature relative to the characteristic temperature dictated by the heat released Q, namely $T_{\text{reaction}} = Q/k_{\text{B}}$, where k_{B} is the Boltzmann constant. When the reaction goes rightwards, heat is released and absorbed by the surroundings, and vice versa. In equilibrium the rate at which the reaction goes to the right is equal to the rate at which it goes to the left, and the concentrations of the species A, B, and C do not change in time unless the temperature and density change. The equations for equilibrium were described in detail by Fowler.[99] When a reaction is in equilibrium, one can calculate the concentration on both sides of the equation, without any knowledge of the rate. On the other hand, if the reaction is not in equilibrium, we have to know the rate of the reaction, and when A and B are two nuclei that combine to form a new nucleus C, this requires a theory of nuclear reactions, as well as the temperature and density.

The situation is actually a litte more complicated than this. We have to make sure that the equilibrium assumption is valid, and in order to guarantee this, we have to make sure that the reaction is fast relative to the available time. If a star lives,

---

[98] Gunn, R., Phys. Rev. **39**, 130 (1932); Narlikar, V.V., & Larmor, J., Proc. Roy. Soc. London A **144**, 28 (1934). The paper discusses the Kelvin–Poincaré problem of stellar evolution assuming liquid stars.

[99] Darwin, C.G., & Fowler, R.H., Phil. Mag. **44**, 450 (1922). The method is described in detail in Fowler, R.H., *Statistical Mechanics: The Theory of the Properties of Matter in Equilibrium*, Cambridge University Press, Cambridge (1929).

say, a million years but the reaction needs just a few years for equilibrium, then it is fine to assume equilibrium. At this time, in the early 1930s, the rates of many nuclear reactions had not yet been measured. On the other hand, quite a number of the required properties of the nuclei needed to calculate the equilibrium were not known either, although the lack of this kind of data was less detrimental to the accuracy of the calculation. In a nutshell, the assumption of equilibrium simplifies the calculation if, and this is a big if, the conditions are appropriate.

As early as 1931, Urey and Bradley[100] set about examining the extent to which the observed abundance ratios of certain isotopes agreed with the assumption that they were formed in equilibrium. They concluded that *the atomic nuclei on Earth do not represent an equilibrium mixture at any temperature*. Note the qualifier 'on Earth'. The authors mentioned that Tolman, back in 1922, had reached the same conclusion with respect to hydrogen and helium. The conclusion that equilibrium just does not work in stars was not accepted by subsequent researchers. Despite this finding, we continue to see more 'equilibrium' calculations in subsequent years.

Some little known research assuming equilibrium was carried out by Farkas and Harteck,[101] who suggested that equilibrium must have been reached in stars at the very high temperature and density of about 10^9 K and 10^5 g/cm^3, respectively, whereupon the new mixture 'somehow' froze, was removed unaltered from the star, and cooled off. The calculation ignored the fact that, at these high densities, electrons no longer obey the ideal gas laws, but the results did show the same trend as observation. Pokrowski[102] carried out a similar calculation using a somewhat different formalism that required the fitting of three constants to reach the conclusion that the equilibrium theory did not succeed in explaining the observed abundances. In view of this disagreement, he suggested the existence of the inverse process, namely the breaking of very heavy nuclei. However, no attempt was made to estimate the consequences.

The first detailed equilibrium calculation based on the new formalism developed by Darwin and Fowler[103] was carried out in 1933 by Sterne.[104] Sterne realized that mass annihilation involved some difficulties, and commented on the situation as follows:

> *There does not apear to be any direct experimental evidence to show that the 'annihilation' of matter can ever take place; and not only are there no experiments to show that particles can ever be 'annihilated', but the thermodynamic and astrophysical consequences of supposing that particles can be 'annihilated', and converted entirely into energy, are of quite an unsatisfactory nature.*

So Sterne analyzed the possibility that proton capture by light elements like lithium could provide the energy for the Sun, concluding that this could not be the correct process which supplies the energy in main sequence stars. The main argument

[100] Urey, H.C., & Bradley, C.A., Phys. Rev. **38**, 718 (1931).

[101] Farkas, E., & Harteck, P., Naturwiss. **19**, 705 (1931).

[102] Pokrowski, G.I., Phys. Zeit. **32**, 374 (1931). The paper contains an error of 10^4 in one of the constants.

[103] Fowler, R.H., *Statistical Mechanics*, Cambridge Press, London (1929).

[104] Sterne, T.E., MNRAS **93**, 736, 767, & 770 (1933).

against this possibility was his calculation of the time it would take for the light element to capture the proton. Sterne found that, at the temperatures prevailing in main sequence stars, the reaction would go too fast, and be so temperature-sensitive, that the star would not be stable. But that was according to Jeans' incorrect stability condition. Hence, he argued that the the energy source could not be driven by simple nuclear reaction, and that the release of energy should, by default, take place in equilibrium. Then, as the temperature and density changed, the elemental composition would change, and with it the energy released, while Jeans' stability condition would play no role. The advantage here was that there was no problem of rates. The only important factor was the binding energy of the nuclei. And these were relatively easily measured quantities, compared to nuclear reactions, which were much more complicated. The disadvantage was that the nuclear reactions were not in equilibrium.

Sterne considered the equilibrium reaction in which a nucleus A_M^N disintegrates into M protons and $M-N$ electrons, viz.,

$$A_M^N \rightleftharpoons (M-N)\ \text{electrons} + M\ \text{protons}\ ,$$

for all possible choices of M and N corresponding to real nuclei. Each nucleus was considered as composed of electrons and protons. The research was carried out before the neutron was discovered, but published after the discovery of the neutron. Obviously, this discovery invalidated the results. Clearly, Sterne had to know only the binding energy of the nuclei to be able to calculate the equilibrium situation.[105]

Moreover, Sterne assumed that, once a particle had penetrated into the nucleus, it would stay there and had no chance of escape. He concluded that:

> *The possibility that an important source of stellar energy could be the gradual disappearance of the element of large fraction, through transmutations which are not accompanied by the reverse processes of their manufacture, appears to be ruled out.*

The results Sterne derived are interesting, and given in Table 6.1. We see that, at low temperatures, the matter is mostly in heavy nuclei, while at high temperatures, the heavy nuclei disintegrate into hydrogen. So how did stars derive their energy à la Sterne? By gradual cooling. The stars form hot, at temperatures exceeding 4×10^9 K. Since Sterne assumed equilibrium, the question of what constituted the original matter became irrelevant. As the star cooled, the light element hydrogen converted into helium, which in turn converted into heavier nuclei until iron was formed. To carry out an accurate calculation, Sterne needed the binding energy of all nuclei (as well as some other information needed for the equilibrium formula).[106]

Sterne based his calculations on the treatment of equilibrium reactions due to Darwin and Fowler,[107] and was probably unaware of the papers by Tolman[108] and

[105] Sterne had to know the nuclear energy levels as well, but this factor could be neglected without impairing the not so accurate calculation.

[106] For accuracy, to calculate the equilibrium abundances, the statistical weight of the nucleus in equilibrium is required. The statistical weight is just the number of different ways to combine the nucleus out of its constituents.

[107] Fowler, R.H., *Statistical Mechanics*, Cambridge Press (1929).

[108] Tolman, R.C., Am. J. Chem. Soc. **44**, 1902 (1922).

Table 6.1 Composition as a function of temperature from the Sterne statistical equilibrium model (1933). All temperatures are in degrees K

Nucleus	$T = 100$	$T = 10^6$	$T = 2 \times 10^9$	$T = 3 \times 10^9$	$T = 4 \times 10^9$
^{1}H	0	0	1.3×10^{-7}	0.39	10
^{4}He	0	0	0.016	9.61	10^{-5}
^{16}O	0	0	4×10^{-19}	10^{-16}	0
^{56}Fe	10	10	9.984	10^{-28}	0

Suzuki,[109] who preceded him by about 10 years and used more primitive methods for calculating the equilibrium, but reached practically the same results and conclusions.

6.19 At Last: A Fatal Blow to Classical Mass Annihilation

In 1933, Eddington[110] returned to the problem of stellar energy and inferred that:

> *So long as free protons and free electrons are not combined in complex nuclei, protons and electrons are immune from annihilation. The reason is momentum conservation. Annihilation of a proton and electron, if it ever occurs, can happen only when they form part of a complex system which will leave a residuum to carry the recoil.*

In other words, the annihilation of a proton with an electron cannot take place without violating the law of momentum conservation. Eddington attributed this argument to Alfred Ewing (without giving any reference), but it was Hughes and Jauncey who had already clearly stated, back in 1926[111] and again in 1934,[112] that the annihilation process as envisaged by Jeans violated too many physical laws.

Any microscopic physical process, and mass annihilation is no exception, must satisfy the following conditions:

- Conservation of energy.
- Conservation of momentum.
- Conservation of charge.[113]
- The process must be reversible, i.e., it must be able to go both ways, from right to left and vice versa.
- Velocities never exceed the velocity of light, otherwise there is a violation of the special theory of relativity.

[109] Suzuki, S., Proc. Phys. Math. Soc. Japan **10**, 166 (1928); ibid. **11**, 119 (1929); ibid. **13**, 277 (1931).

[110] Eddington, A.S., *The Expanding Universe*, Macmillan (1933) pp. 78–79.

[111] Hughes, A.L., & Jauncey, G.E.M., Nature, 6 February 1926, p. 193.

[112] Hughes, A.L., & Jauncey, G.E.M., Phys. Rev. **45**, 217 (1934).

[113] No electric charge is created or destroyed. If an electron with negative charge is formed, it must be accompanied by the formation of a positron which has a positive charge.

The above laws had been confirmed in many physical experiments. Hughes and Jauncey showed that it was impossible for a proton and an electron to destroy each other and leave only a photon without violating one or more of these requirements, a result that looks trivial today. But Hughes and Jauncey complained in their paper that, in spite of the nice proof they had already provided in 1926, people continued to raise this possibility, either overlooking the solid proof that it was impossible, or else assuming that the stars must have their own laws of physics, in contrast to Eddington's postulate.

However, bad ideas die hard. Bramley, in 1934,[114] made the observation that the very energetic gamma rays observed in cosmic rays might be the result of proton annihilation *as suggested by Jeans.*[115] Bramley applied the Fermi method and Heisenberg's idea of the proton and neutron as two separate states of the same particle:

> *The basic idea is that part of the primary cosmic rays are protons which, on striking the outer atmosphere, are converted in the majority of cases into a photon and a positron. Until the proton strikes a nucleus and is converted into a photon and a positron, it loses its energy.*

Thus, even a piece of experimental evidence was suggested to back Jeans' hypothesis.

Hughes and Jauncey pointed to a suggestion by Blackett and Occhialini[116] that a high energy photon could collide with a nucleus and give rise to an electron, a positron, and a proton. The positron was discovered by Anderson,[117] but Anderson's note apparently went unnoticed, and for this reason Hughes and Jauncey refer to Blackett and Occhialini, who were actually scooped by Anderson by a few months. The discovery of the positron, although predicted by Dirac, indicated that, if a positron annihilates an electron, the 5 conditions put forward by Hughes and Jauncey are satisfied. Blackett and Occhialini did not refer explicitly to the 'positron', but wrote:

> *The tracks [in the photographic plates] must be due to a particle with a positive charge, but whose mass is much smaller than that of a proton.*

No such particle was known except for the Dirac positron. Indeed, the authors mentioned that they had consulted Dirac.

In 1934, Gamow[118] suggested the existence of a proton with negative charge, something known today as an antiproton. In this way, the proton and antiproton pair become analogous to the electron and the positron, for which Dirac had constructed his theory. The idea of annihilation as a viable process got 'theoretical support' and

---

[114] Bramely, A., PRL **46**, 438 (1934).

[115] Jeans, J., Nature **116**, 861 (1925).

[116] Blackett, P.M.S., & Occhialini, G.P.S., Nature **130**, 363 (1933); Proc. Roy. Soc. A **139**, 699 (1933). The papers are mostly cited for the new technique used, and not because they made mass annihilation plausible.

[117] Anderson, C.D., *The apparent existence of easily deflectable positives*, Science **76**, 238 (1932); ibid., *Positron confirmed as new particle of matter*, Science News Lett., 25 February 1933, p. 115; ibid., *The positive electron*, Phys. Rev. **43**, 491 (1933).

[118] Gamow, G., PRL **45**, 728 (1934).

injected new energy into the old discussion, but this time the exact process was proton–antiproton or electron–positron annihilation. The antiproton was discovered twenty one years later by Chamberlain, Segré, Wiegand, and Ypsilantis.[119]

6.20 Back to the Nuclear Barrier

The stability of ^{5}He was still in question in 1935, when Atkinson[120] collected and summarized the evidence regarding the properties of this nucleus. At that time, the expectation was that every nucleus between $A = 1$ and $A = 238$, and maybe even higher atomic weights, would be possible. Actually, there was no known reason why the periodic table should end with uranium, and not continue to ever more massive elements, and nor was there any reason why certain nuclei should not exist. On the one hand, Atkinson claimed that ^{5}He *has consistently failed to appear in reactions*, whence it had to be unstable. On the other hand, when Oliphant[121] drew the masses of the nuclei (see Fig. 6.7), he got a smooth curve, and hence there was no reason why this particular nucleus should deviate from the pattern shown by all the other nuclei. Oliphant himself did not put the point corresponding to ^{5}He on his figure, as if this nucleus simply did not exist.

After the discovery of the neutron in the nucleus, it became clear from observed β decays that, in those decays in which an electron is emitted, a neutron was converted into a proton. It was a bit tricky, however. At a certain moment, the neutron became a proton and an electron. The electron and the proton attract each other so as to prevent the decay of the neutron. However, the electron has sufficient energy to overcome the attraction of the proton and escape from the nucleus, together with the neutral neutrino. Could the process be reversed? Obviously, if the process could go in the opposite direction, then one could conceive of the reaction $\mathrm{p} + \mathrm{e}^- \rightarrow \mathrm{n}$ being possible. And if it was possible, then here were the neutrons needed to overcome the hurdle of the first step in the synthesis. Nonetheless, attempts to discover this reaction failed.[122] In fact, this reaction is theoretically possible and even takes place in stars, but under much more extreme conditions, towards the end of their life.[123]

The most important possible way to overcome the $A = 5$ barrier is the following pair of reactions. Assuming that ^{7}Be is somehow synthesized in stars, then these reactions are

$$^7\mathrm{Be} + \mathrm{e}^- \rightarrow {}^7\mathrm{Li} + \nu \quad \text{and} \quad {}^7\mathrm{Li} + \mathrm{p} \rightarrow 2\,{}^4\mathrm{He}\,. \tag{6.1}$$

119 Chamberlain, O., Segré, E., Wiegand, C., & Ypsilantis, T., Phys. Rev. **100**, 947 (1955).

120 Atkinson, R.d'E., Phys. Rev. **48**, 382 (1935).

121 Oliphant, M.L., Nature **137**, 396 (1936).

122 Livingood, J.J., & Snell, A.H., Phys. Rev. **48**, 851 (1935).

123 The correct form of the reaction is $\mathrm{p} + \mathrm{e}^- \rightarrow \mathrm{n} + \bar{\nu}$, where $\bar{\nu}$ is the antineutrino (not known to exist at that time).

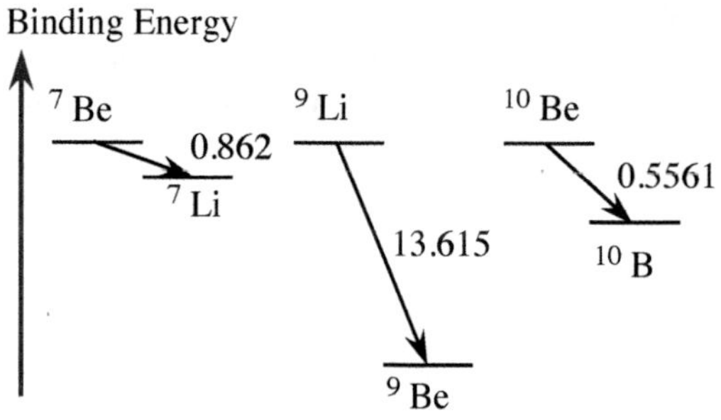

Fig. 6.9 A comparison of the ground states of nearby nuclei. The first decay is an inverse process, an electron being absorbed. The other two involve emission of a positron

In the first reaction, an electron from the sea of electrons is captured by the nucleus of ^{7}Be and converts, inside the nucleus, one of the protons into a neutron. The new neutron remains in the nucleus, which is the stable ^{7}Li. This nucleus quickly captures a proton to become a system of 4 protons and 4 neutrons. The system of 8 particles disintegrates into two helium nuclei.

The nucleus ^{7}Be does not exist on Earth (only in stars). The formation of this unstable isotope was discovered in 1938 by Roberts, Heydenburg, and Locher,[124] when they carried out the experiment $^6\text{Li} + {}^2\text{D} \rightarrow {}^7\text{Be} + \text{n}$. They performed the experiment and left the products aside (it had happened to others before!). After a while, they discovered γ radioactivity, which originated from the ^{7}Be capturing in the nucleus the closest electron from the electronic shells around it, and converting into ^{7}Li. The closest electron to the nucleus has a finite probability of crossing the nucleus. It is during such a crossing that this reaction takes place. The half-life was measured to be 43 ± 6 days. This is a very interesting result, because under the right conditions it opens up the possibility that the free electrons in the star can be absorbed by the nucleus and, once inside, convert a proton into a neutron. This process is important at high densities and in the Sun. Clearly, this explains why ^{7}Be does not exist on Earth. Beryllium in Nature has two isotopes. The first is ^{9}Be, which is stable, and the second is ^{10}Be, which is formed by cosmic radiation and has a half-life of 1.5 million years. The ratio $^{10}\text{Be}/^9\text{Be}$ is used to trace sedimentation and subduction of tectonic plates.

The explanation as to why only ^{7}Be is so special is the following. In Fig. 6.9, we plot the energy configurations of the pairs ^{7}Li–^{7}Be, ^{9}Li–^{9}Be, and ^{10}Be–^{10}B. The numbers are the differences in energy given in MeV. In the case of $A = 7$, the difference is small, and the ^{7}Be can absorb an electron from an inner shell or a free electron and convert back into ^{7}Li. The present measured half-life for electron absorption is 53.29 ± 0.07 days. In the case of $A = 9$, the difference between the pair is 13.606 MeV, and there is no electron around with such an energy. So what happens is that ^{9}Li decays into ^{9}Be. There is no level below ^{9}Be into which it can

[124] Roberts, R.B., Heydenburg, N.P., and Locher, G.L., PRL 1016 (1938).

decay. So ^{9}Be is stable. The last nucleus, ^{10}Be, decays into ^{10}B with a half-life of 1.51×10^6 yrs, if there are no sufficiently energetic electrons to prevent it.

The above decays take place in the laboratory, implying that on Earth the stable nuclei are ^{7}Li, ^{9}Be, and ^{10}B. If the elements are in a star and the surrounding density increases gradually, then the stable nuclei become, in this order, ^{10}Be, ^{7}Be, and ^{9}Li, because the decays in the laboratory become absorptions of an electron from the sea of electrons in the stars.

In 1937, Moller[125] raised the possibility that the nuclei of heavy elements here on Earth might capture an electron from the closest electronic shell and convert a proton into a neutron, thus gradually increasing the number of neutrons in the nucleus, or in short, start a process of neutronization as described and predicted by Hund. At the same time, but independently, Alvarez came up with the idea[126] that any radioactive nucleus which emits a positron might in principle capture an electron, instead of emitting a positron. In other words, the reactions could go in both directions.

6.21 Weizsäcker

Weizsäcker (1912–2007) was a nuclear physicist interested in astrophysics. From time to time he came up with various original ideas in astrophysics. In 1937, Weizsäcker published two important papers[127] about the energy source of stars.

6.22 The First Paper

This paper, like those of Atkinson, was qualitative, and no real calculations were carried out. It is basically scientific reasoning without the backing of detailed calculations. Weizsäcker started by considering reactions of charged particles. Weizsäcker rediscovered what was already known to all his predecessors, namely, that it is difficult to synthesise the heavy elements by means of charged particles, due to the high Coulomb barrier, and even the Gamow tunneling process, as applied by Atkinson and Houtermans, yields negligible abundances. Weizsäcker's solution was to look for neutron reactions. The first task was then to find sources of neutrons. Figure 6.10 shows the information about the stability of the light nuclei as assumed by Weizsäcker.

The following reactions are sources of neutrons as known to Weizsäcker. First, generate deuterium as follows:

---

[125] Moller, Chr., Phys. Rev. **51**, 84 (1937).

[126] Alvarez, L.W., PRL 133 (1937).

[127] Weizsäcker, C.F., Physik Zeitschr. **38**, 176 (1937) (paper I); ibid. **39**, 633 (1938) (paper II).

$$^{4}\mathrm{He}+{}^{1}\mathrm{H}+1.8\ \mathrm{MeV}\rightarrow{}^{5}\mathrm{Li}$$
$$^{5}\mathrm{Li}\rightarrow{}^{5}\mathrm{He}+\mathrm{e}^{+}$$
$$^{5}\mathrm{He}+{}^{1}\mathrm{H}\rightarrow{}^{4}\mathrm{He}+{}^{2}\mathrm{D}\,.$$

The cycle produces deuterium, and the helium returns to its original state. Examination of the energetics involved shows that at least 1.80 MeV of kinetic energy is needed to bring the ^{4}He + ^{1}H to the level of ^{5}Li. This is a very high energy and therefore requires a huge temperature, of the order of 5–10 $\times 10^{10}$ K, which simply does not exist in main sequence stars.

Weizsäcker assumed that the $A = 5$ nuclei were unstable but with a long decay time, sufficiently long to allow the above reactions. The lifetimes of the $A = 5$ nuclei were not specified, however. What Weizsäcker really conceived was the way a proton could convert into a neutron inside the nucleus $A = 5$. The first proton is absorbed and forms ^{5}Li. The nucleus is unstable, emits a positron and converts a proton into a neutron. Now this neutron has to be emitted. So an additional proton is absorbed and forms a deuterium, from which it is relatively easy to eject a neutron. If ^{5}He is indeed formed in this way, it is not clear why Weizsäcker did not consider the simple decay $^{5}\mathrm{He}\rightarrow{}^{4}\mathrm{He}+\mathrm{n}+1.00$ MeV, and hence avoid the assumption that the unstable ^{5}He lives long enough to collide with the abundant proton to yield ^{4}He and ^{2}D.

Once the deuterium is formed, then helium and neutrons can be formed through the following reactions:

$$\begin{aligned}
^{2}\mathrm{D}+{}^{2}\mathrm{D}&\rightarrow{}^{3}\mathrm{He}+{}^{1}\mathrm{n}\\
^{2}\mathrm{D}+{}^{2}\mathrm{D}&\rightarrow{}^{3}\mathrm{H}+{}^{1}\mathrm{H}\\
^{2}\mathrm{D}+{}^{3}\mathrm{H}&\rightarrow{}^{4}\mathrm{He}+{}^{1}\mathrm{n}\\
^{3}\mathrm{H}+{}^{3}\mathrm{H}&\rightarrow{}^{4}\mathrm{He}+2{}^{1}\mathrm{n}\,.
\end{aligned}\tag{6.2}$$

The advantage is that the first two reactions had already been observed in the laboratory and were not just hypotheses. Thus, the light elements can generate deuterium and the neutrons needed for the formation of the heavy elements, provided of course that the abundances are significant. Notwithstanding, the abundance of deuterium in ocean water is 1:100 000 relative to hydrogen. If one assumes the same abundance in stars, then there is no hope for these reactions to meet the energy requirements.

It seems that Weizsäcker missed the competition faced by deuterium from other nuclei. For example, once deuterium forms, it can very quickly undergo the reaction $^{2}\mathrm{D}+{}^{1}\mathrm{H}\rightarrow{}^{3}\mathrm{He}+\gamma$, in which the deuterium is destroyed by the much more abundant hydrogen. Hence, the vast majority of the deuterium is consumed by protons, and not by other deuteriums, so the reactions listed by Weizsäcker in (6.2) are very rare.

The difficulties in finding the energy-producing reaction led Weizsäcker to conclude in this paper that:

> *We should keep in mind that it is highly possible that the elements were formed before the stars, and now the stars change the composition only slightly.*

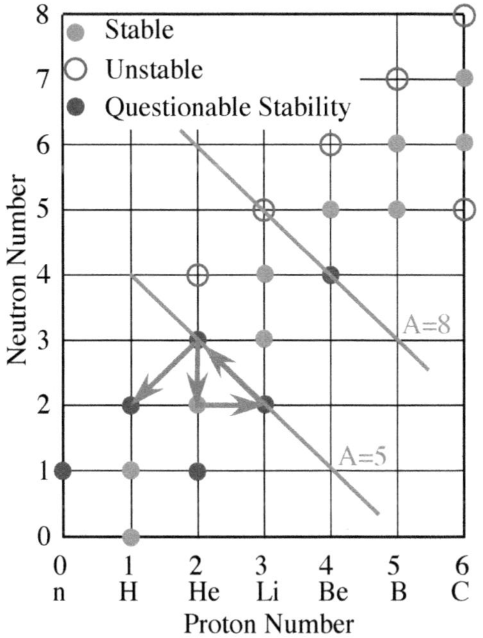

Fig. 6.10 The table of nuclei and their stability, as assumed by Weizsäcker in 1937. *Green arrows* describe the cycle which produces neutrons

The assumption was provoked by the problems with the $^{1}H + ^{1}H$ reaction, which he did not mention at all, and by the CN cycle which requires the previous existence of carbon and nitrogen. There was no discussion of the sources of energy for stars. As a matter of fact, the title of Weizsäcker's paper was *On the element transformation inside the stars*. Weizsäcker was ready to split the discussion into the energy of stars and the creation of the chemical elements,

6.23 The Second Paper

Weizsäcker's second paper was published in 1938, just a year after the first publication. Weizsäcker started by stating that this second paper (with the same title as the first) would also be qualitative and without calculations, and once more we read that:

> *We cannot exclude the possibility that the elements were created before the stars were formed, and presently the stars generate energy by creating minor changes in their composition.*

This time Weizsäcker discussed the reaction $^1H + ^1H \rightarrow ^2D + e^+$, as hypothesized by Atkinson in his third paper, and according to a footnote, Weizsäcker knew from Gamow that this set of reactions was at that time being investigated by Bethe. His assessment was: *our knowledge is insufficient to exclude this reaction.* More importantly, Weizsäcker renounced the ideas he had developed in the first paper and proposed that the chemical elements in stars were present from the beginning, whence the reactions did not need to begin with hydrogen. This led him to the rediscovery of regenerative processes, in the form of the CN cycle, and to claim that this process was the source of energy in main sequence stars.

Weizsäcker explained why his first paper was wrong:

> *(1) It is not clear that sufficient amounts of neutrons can be formed. If indeed the neutrons are formed in this way, then appreciable amounts of helium must be formed and the observations of the helium abundance in stars appear not to confirm this. (2) It seems impossible to explain the formation of large quantities of uranium and thorium by means of this process. (3) The process of neutron formation does not explain the relative abundances of the elements.*

It is not clear what large quantities of U and Th Weizsäcker was referring to. It is possible for stars to form the elements beyond Fe only via neutron capture, otherwise extremely high temperatures are needed. But such high temperatures do not exist in stars.

Another argument given by Weizsäcker was:

> *If the Sun started as pure hydrogen, its lifetime should be 30 billion years, while the present age is estimated at about 3 billion years. According to Hubble, the age of the universe is only 3 billion years. If so the Sun has burnt so far only 10% of its hydrogen. This age agrees also with the rough estimate for the time of the formation of the uranium and thorium.*

After reviewing the problems with the neutron production and synthesis hypothesis, he finally rejected this idea. So it was back to square one. What could the energy production mechanisms be? Weizsäcker enumerated the following possibilities:

- Contraction without change of composition, i.e., gravitational energy release.
- Build-up of elements from hydrogen and energy release is nuclear energy.
- Contraction resulting from conversion of a part of the matter into densely packed neutrons, as suggested by Landau and Hund some time earlier.[128]
- Mass annihilation.

Weizsäcker chose to investigate only the second case, because he did not consider the three other possibilities as likely, and he wrote:

> *Mass annihilation of an electron with a proton has a forlorn probability, since physics, to date, has discovered no cause that would be in a position to bring it about. Regarding the discovery of positrons and neutrons, it seems that, in the balancing of positive and negative charges, only the electron mass is converted into radiation energy, while the proton mass remains conserved. Complete annihilation has never been observed.*

---

[128] Landau, L., Sov. Phys. **1**, 285 (1932); Nature **141**, 333 (1938); Hund, F., Erg. Exact. Naturwiss. **15**, 189 (1936); Anderson, O., Veröff. d. Univ. Sternw. Dorpat. Gamow, G., & Teller, E., Phys. Rev. **53**, 929 (1938).

It is not clear at all why this particular process of mass annihilation, which was shown by Hughes and Jauncey to be impossible, was mentioned by Weizsäcker. Was it not already quite clear at this time that nuclear reactions represented the only way, even if the details were not known?

One of the first important experimental results was the measurement by Döpel[129] of the important nuclear reaction:

$$^{2}\mathrm{D}_1 + {}^{2}\mathrm{D}_1 \rightarrow \begin{pmatrix} ^{3}\mathrm{H}_1 + {}^{1}\mathrm{H}_1 \\ ^{3}\mathrm{He}_2 + {}^{1}\mathrm{n}_0 \end{pmatrix} . \tag{6.3}$$

The reaction was also considered by Oliphant, Harteck and Rutherford,[130] and Bonner and Brubaker.[131] This was a favorite reaction for synthesis, because it produces neutrons. But how could the deuterium be formed? Döpel suggested $^{1}\mathrm{H} + {}^{1}\mathrm{H} \rightarrow {}^{2}\mathrm{D} + \mathrm{e}^{+}$ and $^{1}\mathrm{H} + \mathrm{e}^{-} \rightarrow \mathrm{n} + \gamma$. All attempts to observe these reactions in the laboratory failed. The alternative was that the neutron would capture a proton to form deuterium, and in this way start the synthesis.

Measurements of the above reaction yielded very low rates, to the point where Döpel remarked that, at stellar temperatures, particles do not have sufficient energy to overcome the potential barrier, even with the help of the Gamow penetration factor. So if one believed that nuclear reactions did indeed take place in stars, there had to be, so hypothesized Döpel, powerful electric fields (about 10^6 volts) inside the Sun, which could accelerate the particles to sufficient energies to overcome the barrier. The suggestion by Döpel that an electric field might accelerate the particles in the core of the Sun was rejected by Weizsäcker, and quite correctly. A strong field could survive only in the outer layers.

Then Weizsäcker proposed the CN cycle, which was a concrete version of what Atkinson called regenerative processes, but could not write down the relevant reactions. The set of reactions is in fact:

$$\begin{aligned} ^{12}\mathrm{C} + {}^{1}\mathrm{H} &\rightarrow {}^{13}\mathrm{N} \\ ^{13}\mathrm{N} &\rightarrow {}^{13}\mathrm{C} + \mathrm{e}^{+} \\ ^{13}\mathrm{C} + {}^{1}\mathrm{H} &\rightarrow {}^{14}\mathrm{N} \\ ^{14}\mathrm{N} + {}^{1}\mathrm{H} &\rightarrow {}^{15}\mathrm{O} \\ ^{15}\mathrm{O} &\rightarrow {}^{15}\mathrm{N} + \mathrm{e}^{+} \\ ^{15}\mathrm{N} + {}^{1}\mathrm{H} &\rightarrow {}^{12}\mathrm{C} + {}^{4}\mathrm{He} . \end{aligned} \tag{6.4}$$

Weizsäcker wrote the reactions in this form, neglecting the emitted γ. A γ is emitted in all the proton capture reactions shown above. The two conversions of protons into neutrons take place inside a nucleus, and the extra positive charges are emitted

[129] Döpel, R., Zeit. f. Phys. **14**, 139 (1937).

[130] Oliphant, M., Harteck, P., & Rutherford, E., Proc. Roy. Soc. London **144**, 1936 (1936).

[131] Bonner, T., & Brubaker, W., Phys. Rev. **49**, 22 (1936).

as positrons. The cycle ends when no further building is possible, and the product emits an α particle. The crucial issue is why the pattern does not continue, and why $^{15}N + {}^{1}H \rightarrow {}^{16}O + \gamma$ does not occur? Or better, why it occurs only once in a thousand times.

An extensive discussion of the stability of nuclei against α decay following capture of a proton was given by Bethe and Bacher and by Livingston and Bethe.[132] It is interesting that, although the last reaction is the most important for us here, it was not discussed at all by the above researchers, even though the data for this reaction appear in the table as $^{15}N + {}^{1}H \rightarrow {}^{12}C + {}^{4}He + 4.79$ MeV. The reaction had not been measured at the time the CN cycle was proposed. As the table by Livingston and Bethe is interesting in showing the pattern, we display part of it here (see Table 6.2).

Several conclusions can be drawn from the table. Helium can already be formed with lithium, but not in a cycle. Moreover, as mentioned previously, lithium is very rare in stars. The first elements with appreciable abundances are carbon and nitrogen, and indeed these are the ones that best suit our purpose. The prediction was that ^{15}N, ^{17}O, ^{18}O, and ^{19}F eject an α particle after the absorption of a proton, and release energy (if the energy release is negative, it means that we have to invest energy to get the reaction to occur). As a matter of fact, we envisage here a unique peculiarity of the nuclear forces and the special role the α particle plays in the nucleus. Carbon 12 is composed of three α particles which hold together (while two α particles cannot hold together). When we add protons to the 3α system one after the other, a new α particle forms inside the nucleus, so that when the fourth proton enters the nucleus the formation of the new α particle is complete, and it disintegrates into the original nucleus and a helium. In summary, thanks to the unique properties of the nuclear forces which make the two-proton and two-neutron system the most strongly bound, we have a regenerative process.

The ^{15}N is made of three α particles plus two neutrons plus a proton. When the additional proton enters, it is easier for the fourth α. So we have a system of 4α with an energy of 12.110 MeV above the ground state level of ^{16}O. Due to differences in angular momentum, the decay of the 4α into the ground state of ^{16}O is complicated and largely inhibited. The decay into $3\alpha + \alpha$ releases less energy, but is favored by a factor of 1000 over the alternative. This all has to do with the particular properties of the α particle (strongly bound and spin zero) and the $4n$ nuclei like ^{16}O made of an integer number of α particles, which also has spin zero. The emitted photon carries spin 1, so the transition from the spin zero state to another spin zero state cannot take place directly (it must go through an intermediate state), and it is easier for the compound nucleus of $n\alpha + \alpha$ to form at a high energy above the ground state, then decay into an $n\alpha$ nucleus in the ground state and an α, while the extra energy is emitted as a γ. This is the 'secret' of the regenerative process.

---

[132] During 1936–37, Bethe published a trio of reviews which covered everything known in nuclear physics: Bethe, H.A., & Bacher, R.F., Stationary states of nuclei, Nuclear Physics A Rev. Mod. Phys. **8**, 82 (1936); Bethe, H.A., Nuclear dynamics, theoretical, Nuclear Physics B Rev. Mod. Phys. **9**, 69 (1937); Livingston, M.S., & Bethe, H.A., Nuclear dynamics, experimental, Nuclear Physics C Rev. Mod. Phys. **9**, 245 (1937). These three reviews became known later as the Bethe Bible.

Table 6.2 Summary of proton capture. α emission reactions as given by Livingston and Bethe in 1937

Z	Isotope	Product	Q (theory)	Q (observation)
3	^{6}Li	^{3}He	3.76	3.72
	^{7}Li	^{4}He	17.25	17.13
4	^{9}Be	^{6}Li	2.25	2.28
5	^{11}B	^{8}Be	8.60	8.60
	^{11}B	^{4}He	8.72	8.7
6	^{13}C	^{10}B	−4.15	–
7	^{14}N	^{11}C	−3.3	–
	^{15}N	^{12}C	4.79	–
8	^{16}O	^{13}N	−5.4	–
	^{17}O	^{14}N	1.3	–
	^{18}O	^{15}N	2.82	–
9	^{19}F	^{16}O	8.14	–

To solve the system of nuclear reactions, Weizsäcker assumed that the CN cycle was in thermal equilibrium, an assumption for which he needed only the masses of the nuclei. Since the energy difference between the different nuclei is of the order of 10 MeV, the temperatures required for the process to be in equilibrium are given by $T = 10\ \text{MeV}/k_{\text{B}} = 1.1 \times 10^{11}$ K. To be precise, Weizsäcker found a temperature of 2.3×10^{11} K, which is way beyond what stellar models predicted to exist in stars. Consequently, Weizsäcker realized that his CN cycle could not operate in stars. Last but not least, Weizsäcker did not calculate the energy release in the reaction.

So where could such high temperatures occur? The hypothesis put forward by Weizsäcker was that the elements were formed in explosions that took place before the stars were formed. *Great primeval aggregates of matter perhaps consisting of hydrogen* would collapse under their own gravitational attraction and thereby raise the temperature at the center to such high temperatures as to cause nuclear explosions of the stars, and in these explosions the elements might be formed. How large a mass was needed? Probably the size of the Galaxy, or even the entire Universe. Weizsäcker essentially agreed with Milne,[133] who did not accept the theory that the Universe expands (based on Hubble's observations) and proposed the steady-state theory as an alternative.

Some of the conclusions Weizsäcker reached were therefore:

- The assumption that all known chemical elements have originated and are still originating in stars existing today must be abandoned.
- The heavier elements must have been built up by neutrons whose production is necessarily coupled to helium formation. The quantitative investigation of this

[133] Milne, E.A., Zeit. für Astrophys. **6**, 1 (1933).

mechanism leads to the establishment of a lower bound for the helium abundance in a star which is incompatible with observation.

- Uranium and thorium must have been built up from rapidly decaying intermediate nuclei on the way. Spatial concentration of the energy source is insufficient to provide the rate required for the build-up process.
- The explanation of Harkin's rule about the abundances of even versus odd elements is explained by nuclear physics. The more stable nuclei, those with an even number of protons and neutrons, are more abundant than the nearby nuclei. (For example, oxygen 16, which has an even number of protons and neutrons, is more abundant than nitrogen, which has 7 protons and 7 neutrons.)
- Energy production must depend solely upon reactions of light nuclei.
- The most probable cycle is the carbon cycle.
- The predicted abundances of the light elements agree with observations.
- The most important assumption is that the elements are produced in a thermodynamic equilibrium of nuclear reactions. If so, the formation temperature was 2×10^{11} K. The fine details are determined at a temperature of 5×10^{9} K, when the temperature becomes too low for the reaction to continue. Not a word about the problem with such high temperatures.
- A very massive star could generate such temperatures, but would subsequently explode. But there were no calculations whatsoever.

The main assumption which misled Weizsäcker was that:

> *The relation between mass defect and frequency urges the assumption that, in the formation of the elements, kinetic energies of reaction partners of the order of nuclear binding energies are available. At such high energies a thermodynamic equilibrium of nuclear reaction must appear quickly. For firstly, the Coulomb repulsion, at least for the lighter nuclei, plays hardly any role; and secondly, the conversion of free protons into neutrons now becomes a very frequent process so that, for the building up and breaking down of heavier nuclei, neutrons are available in arbitrary quantities.*

The new and crucial element that would be brought to bear by Bethe was the treatment of the nuclear reactions in terms of rates, rather than assuming equilibrium. When the reactions are not in equilibrium, they are much slower and consequently the energy production is smaller. But to calculate the rates of the nuclear reactions, a knowledge of the masses is not sufficient, and an actual measurement of the reaction rate must be carried out. But these were the years when physicists were starting to measure the rates of the relevant reactions. Moreover, the theory of nuclear reactions was being developed at the time.

In 1938, Strömgren[134] returned to examine the stars on the main sequence in view of Weizsäcker's predictions regarding the helium abundance. The assumption of a constant ratio of helium to heavy elements led to curves of constant hydrogen in the HR diagram, which were very similar in nature to those previously obtained on the assumption of negligible helium content. So the observations appeared not to be sufficiently accurate to either confirm or rule out Weizsäcker's new hypothesis. The

[134] Strömgren, B., Ap. J. **87**, 520 (1938).

assumption about helium was compatible with observations of the mass–luminosity relation. But no more than this.

6.24 Gamow Again

In 1938, Gamow[135] started to play with the idea of neutron reactions, and calculated the equilibrium of a star containing thermonuclear reactions. Gamow wanted some of the reactions to release neutrons so that the reactions of higher Z could proceed at a significant rate. He argued that the only way to produce neutrons at low energy (i.e., temperatures corresponding to about 10 keV) was by preliminary formation of deuterium, which in mutual collisions would undergo the reactions (6.2) considered by Weizsäcker.

Gamow estimated that both channels had equal probability. Another possibility raised by Gamow was:

$$\begin{aligned}
{}^4\mathrm{He} + {}^4\mathrm{He} &\rightarrow {}^8\mathrm{Be} + \gamma \\
{}^8\mathrm{Be} + {}^1\mathrm{H} &\rightarrow {}^9\mathrm{B} + \gamma \\
{}^9\mathrm{B} &\rightarrow {}^9\mathrm{Be} + \beta^+ \\
{}^9\mathrm{Be} + {}^1\mathrm{H} &\rightarrow 2{}^4\mathrm{He} + {}^2\mathrm{H}
\end{aligned}$$

Again the uncertainty over the stability of ^{8}Be placed a question mark on the viability of the process. As Gamow put it:

> *Most recent determinations give, however, for the mass of this nucleus almost exactly twice the mass of an α particle, the sign of the small difference being uncertain.*

These were the only reactions that could generate deuterium and subsequently neutrons.

It is interesting to note that Gamow assumed ^{8}Be to be stable, and considered also the equilibrium reaction (provided that the binding energy is very small but the nucleus is still bound):

$$^8\mathrm{Be} \rightleftharpoons 2{}^4\mathrm{He}\,. \tag{6.5}$$

In words, he wanted the temperature to dissociate the beryllium nucleus under equilibrium conditions. For this process to be possible at the assumed temperatures at the center of stars, the binding energy must be 10–50 keV, because the temperature at which the equilibrium condition is satisfied, namely $T = E_{\mathrm{binding}}/k_{\mathrm{B}}$, must correspond to stellar temperatures. According to Gamow, the low stellar temperature implied that, if this reaction were possible, then this must be in the range of the binding energy of ^{8}Be. Whenever there are two states and the energy difference bet-

---

135 Gamow, G., Phys. Rev. **53**, 595 (1938).

ween them is ΔE, then at a temperature of about $T = \Delta E/k_B$, the states will be in equilibrium with one another.

Gamow assumed evolution at constant mass because:

The so-called 'annihilation of stellar mass' does not deserve credence in the present development of physics, chiefly because the sources of stellar energy can be sufficiently explained by well known nuclear transformations.

Gamow assumed that the movement of the star in the HR diagram is only due to changes in the molecular weight (he was right), and that the stars were fully mixed (but here he failed to recognize the error). The results were very unsatisfactory. There were many details which did not fit the observations. The most important was that his models produced an incorrect mass–luminosity relation, namely $L \sim M^5$, rather than $L \sim M^3$.

The discrepancy with observation bothered Gamow to such an extent that he concluded that:

It seems that the real behavior of stars cannot be interpreted in terms of ordinary thermonuclear sources of energy.

He then directed his attention to the possibilities for selective temperature effects. He thus hypothesized a resonance in the nuclear reactions which fixes the temperature, but he had no laboratory experiment to support this hypothesis.

6.25 The Energy Source

In 1938, Öpik[136] examined a set of reactions, all involving light elements. For example, he considered the possibility of

$$^{7}\mathrm{Li} + {}^{1}\mathrm{H} \rightarrow 2\,{}^{4}\mathrm{He}\,. \qquad (6.6)$$

The reaction is correct and valid, and the only problem is that it requires an unacceptable amount of lithium, something clearly not observed in stars or anywhere else. Next he suggested that the first reactions might be

$$\begin{aligned} &^{1}\mathrm{H} + {}^{1}\mathrm{H} \rightarrow {}^{2}\mathrm{H} + \mathrm{e}^{-} \\ &^{1}\mathrm{H} + \mathrm{e}^{-} \rightarrow \mathrm{n} - Q\,. \end{aligned} \qquad (6.7)$$

In words, pure hydrogen reacts with itself. The problem was that there was no experimental data for this reaction. The second reaction requires the energy of the electron to be more than the difference in mass between the neutron and the proton, which is about 1.2 MeV. This is a very high energy, and would occur only in very degenerate stars. As for the first reaction, Öpik assumed that the rate was equal to the rate of the reaction (6.6). This was a big mistake, because he was unaware of the

[136] Öpik, E., Pub. Obs. Univ. Tartu **30**, 37 (1938).

problem with the spin, and the fact that the reaction can only go via the Gamow–Teller transition (published two years earlier).

An interesting point is this. Öpik realized that:

> *There is no hope of ever detecting the nuclear reaction (6.7) in the laboratory. We are here confronted with a dilemma: if the reaction were detectable experimentally, the Sun would blow up from the immense energy generation; or rather, the Sun could exist only as a diffuse star.*

Consequently, Öpik raised the possibility that the interior of the Sun might actually be devoid of hydrogen, to prevent it from blowing up.

6.26 A Problem of Stability

Öpik wanted to build a genuine stellar model, and for this reason he needed an expression which described the dependence of the nuclear reactions on density and temperature. However, he knew about the stability condition Jeans had found, which ruled out any dependence of the nuclear reactions on the temperature. Recall that, as a way to escape Jeans' stability criterion, Eddington and Atkinson had hypothesized a two-step process, so that the first step depended on temperature whereas the second, during which the major part of the energy is released, was independent of temperature. As Öpik pointed out, *it appears that the danger of instability has been exaggerated, on account of imperfect analysis*, and he cited Cowling. However, there was no attempt to provide any details of the two-step process, because this hypothetical process does not exist.

Just a year prior to this, in 1937, Schwarzschild[137] had tried to construct stellar models. Facing the fact that there was no known stellar energy source, Schwarzschild chose to represent the unknown by $\varepsilon = AM^pX^q\rho^mT^n$, where ε is the energy per gram of matter, A is a numerical constant, X is the hydrogen content, and M is the mass of the star. Schwarzschild tried to use the observations by Strömgren[138] to fix the parameters in this hypothetical expression. As a result, Schwarzschild got two relations[139] between the 4 exponents, and hence found p and q as functions of m and n. Öpik did not like Schwarzschild's assumption at all, and argued against it. Öpik reached the correct conclusion that p must vanish, i.e., the rate of the nuclear reaction could not depend on the mass of the star. Next, m had to be unity, because this was a reaction between two constituents. Substituting these values into Schwarzschild's relation, Öpik found that $n = 73$ and $q = -25$, which are absolutely wild values. A negative q implies that the rate is infinite when the hydrogen content vanishes! As Öpik put it, *the result for q is absurd*. Öpik tried to play around with the relations, but whatever he did resulted in completely meaningless values. Moreover, if one took the values Schwarzschild got for his fit and calculated the ratio between

[137] Schwarzschild, M., Zeit. für Astr. **13**, 126 (1937).

[138] Strömgren, B., Zeit. für Astr. **7**, 222 (1933).

[139] The relations are $p = 2.29 + 1.34m - 0.05n$ and $q = -0.77 - 1.64m - 0.32n$.

the luminosity of Capella and that of the Sun, one found 10^{-24}. The conclusion to be drawn was therefore that a single process of energy generation could not work for all stars. It would turn out that all three, Öpik, Schwarzschild, and Strömgren had been misled by treating the dwarfs and the giants together.

6.27 Hans Bethe

Hans Bethe (1906–2005) was the high priest of nuclear physics. He published two papers almost at the same time. In these two papers Bethe reached the pinnacle of twenty years' investigations which had started with Eddington's subatomic energy hypothesis. Eddington was alive to see this extremely important hypothesis of his confirmed. However, Eddington did not publish any paper concerning Bethe's long-awaited and incredible discovery. When the Royal Astronomical Society established the Eddington Medal in 1953, it was first awarded to Georges Lemaître in 1953 and then to Hendrik van de Hulst, Horace Babcock, and James Hey, before those who had contributed to understanding the energy source of stars were recognized with the award of the medal to Robert d'Escourt Atkinson in 1960 and Hans Bethe in 1961. Bethe's achievement also won him the 1967 Nobel Prize for Physics.

6.28 The First Calculation of the pp Reaction

Critchfield was a doctoral student of Gamow. The subject of the thesis was to calculate the rate of the proton–proton reaction in stars. When Critchfield had finished the calculation, Gamow suggested that he present the calculation to Bethe, who had a unique reputation as world leader in nuclear physics and reactions. It was in 1938 that Critchfield presented his calculations to Bethe, who found them to be correct, and in the same year, Bethe and Critchfield[140] published their calculation. The authors gave credit to Weizsäcker[141] but not to Atkinson (to whom Weizsäcker himself gave credit). They wrote:

> *There seems to be a general belief that reaction $H+H-D+e^{+}+0.42$ MeV is too rare to account for any appreciable fraction of the energy production in stars, and that it can serve only to start the evolution of elements in a star which will then be carried on by other, more probable processes. It is the purpose of this paper to show that this belief is unfounded, but that reaction $p+p$ gives an energy evolution of the correct order of magnitude for the Sun.*

The paper was published while Bethe was working on the CN cycle. So he added that:

> *We do not want to imply that the pp reaction is the only important source of energy [...] the capture of protons by carbon and nitrogen will also play an important role.*

[140] Bethe, H.A., & Critchfield, C.L., Phys. Rev. **54**, 248 (1938).

[141] Weizsäcker, C.F., Physik. Zeits. **38**, 176 (1937).

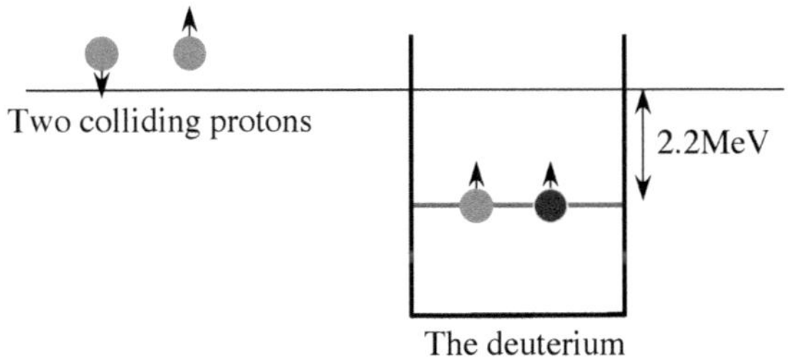

Fig. 6.11 The transition from two colliding protons to a bound deuterium requires the flip of the spin of one of the protons (*green*) and conversion to a neutron (*blue*), so as to allow the proton and neutron to attract one another and thereby form the deuterium nucleus, in which the spins of the particles are aligned. The attraction between a proton and a neutron with opposite spins is not sufficient to allow for a bound state, and hence a Fermi transition cannot lead to a bound state of deuterium

Due to the very low energy of the colliding protons in the Sun (a few keV), only states with no angular momentum (s-waves) contribute significantly. Consider it as a head-on collision, so that the angular momentum plays no role. Consequently, the total angular momentum is the sum of the spins, and only the spins control the reaction. Because of Pauli exclusion principle, the incoming protons must have opposite spins. On the other side, in the only bound state of deuterium, the spins of the neutron and the proton are aligned. Hence a spin flip must take place. The strength of the nuclear force which holds the neutron and the proton together depends on the spin of the particles. The force between the aligned proton and neutron is sufficient to give a bound state, but the interaction between two protons does not yield a bound state under any situation. The deuterium has only one bound state. Thus, the transition from two free protons to a bound deuterium nucleus requires the flip of the spin of one of the protons, and this can be done only by the Gamow–Teller modification of the Fermi beta theory.

To provide additional support to their theory of how two colliding protons can be converted into deuterium, Bethe and Critchfield cited Goldhaber,[142] who investigated a similar process (in terms of the need for a spin flip), namely, the fast decay $^6\mathrm{He} \to {}^6\mathrm{Li}$, which is a β decay with similar properties.

As Bethe and Critchfield pointed out:

The transition is therefore allowed only if the Gamow–Teller form of the β theory is used.

---

[142] The reference given is Goldhaber, Physical Review to be published. However, no such paper by Goldhaber could be found in this journal. A year later, Margenau [Phys. Rev. **55** (1939)] investigated the structure of ^{6}He, and Grönblom, from Cornell University where Bethe was then working, investigated the β decay of ^{6}He [Phys. Rev. **56**, 508 (1939)]. Bethe was aware of Grönblom's results, since he suggested the problem to him. As a matter of fact, Grönblom found that the strength of the interaction between the protons, the factor which determines the speed of the reaction, was significantly stronger than what Bethe and Critchfield had assumed, whence the rate of energy production would be greater. Bethe and Critchfield estimated their data from the reaction $^{13}\mathrm{N} \to {}^{12}\mathrm{C} + e^+ + \nu$.

At the time Bethe and Critchfield published their paper, the Gamow–Teller theory was barely two years old, and consequently had not acquired sufficient credibility, so they felt the need to point out various experimental data which confirmed it. Bethe and Critchfield commented that, if the original Fermi theory were taken instead of the Gamow–Teller theory, the reaction would still go ahead, but it would be suppressed by a factor of about 100 000, and hence:

> *In this case, then, the energy evolution in the Sun due to proton combination would be negligibly small.*

As for the full process, they calculated that it would yield about 2 ergs/g/s in the Sun, which is the average value of the energy production in the Sun and not the rate in the center, and hence they concluded that:

> *This is the energy source for stars smaller than the Sun.*

After the publication of the Physical Review paper, someone[143] suggested to the authors to amplify the symmetry considerations leading to the process. So Bethe and Critchfield added a short communication.[144] At the time, it was not clear what type of interaction was involved, and hence the comment was short and far from conclusive. However, several years later it became clear that this is a unique interaction which distinguishes between right and left, or between an image and the mirror image. This is like the difference between the right and the left hands. If you see the right hand in the mirror, you can still identify it as the right hand. Thus the basic interaction is one that distinguishes between right and left. When a physical system is invariant under a certain transformation, a conservation law must follow. If we cannot tell whether an experiment is seen in the laboratory or in the mirror, the system must conserve what is known as parity, and the conservation of parity dictates certain properties of the system. But the Gamow–Teller interaction was found not to conserve parity.

After the difficult formation of deuterium, the road lies open, and all subsequent reactions go fast. The reason for the difficulty is the weak force, and the reason for the speed of the subsequent reactions is that they are controlled by the strong force. Hence, very soon after the formation of deuterium, it will be consumed by the fast reaction $D + H \rightarrow {}^3He + \gamma$, leaving only a tiny amount of deuterium. The discovery that ${}^3He_2$ is stable was made a short while after by Alvarez and Cornog.[145] Helium has two isotopes, one extremely stable, which is regular helium in the form ${}^4He$, and the other is a light isotope ${}^3He_2$. When Bethe devised the CN cycle, it was not known that helium had this stable isotope. So he considered only the formation of deuterium. When the formation reaction is very slow but the destruction reaction is very fast, the constituent disappears quickly and only small quantities of the species are left. The concentration of deuterium in the Sun is about 10^{-17}!

143 On the basis of a footnote in the Physical Review paper, we could guess that it was Oppenheimer.

144 Bethe, H.A., & Critchfield, C.L., PRL **54**, 862 (1938).

145 Alvarez, L.W., & Cornog, R., Phys. Rev. **56**, 379 (1939).

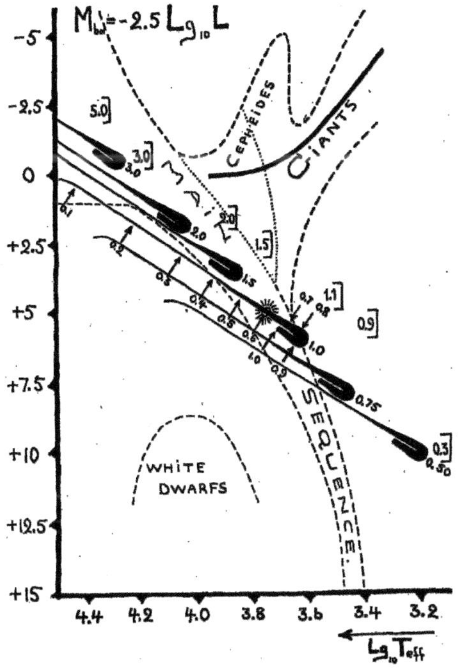

Fig. 6.12 Stellar evolution of main sequence stars which produce energy via the pp reaction, according to Gamow (1938). The *vertical axis* is the luminosity and the *horizontal axis* is the temperature. Note that the star evolves from low temperatures and low luminosities to high temperatures and high luminosities. This is not what is observed. There are no stars below the main sequence

In conclusion, the proton–proton reaction gives an energy evolution of the right order of magnitude for the Sun. But, argued Bethe and Critchfield:

> *It seems that there must be another process contributing somewhat more to the energy evolution in the Sun. This is probably the capture of protons by carbon.*

While the paper by Bethe and Critchfield was in print, Gamow[146] evaluated the effect of the newly calculated pp rate on stellar evolution. First Gamow derived the mass–luminosity law on the basis of the new result as $L \sim M^{5.5}$, which did not agree with observations. Next he calculated the evolutionary tracks, and found that they hardly differed from what he, Gamow, had calculated a year before. In short, the new expression for the proton–proton reaction rate did not have any major impact on the poor agreement between theory and observation.

On top of this, there remained the perpetual hurdle of the classical giant stars:

> *The giant stars cannot be fuelled by nuclear reactions.* If they were, argued Gamow, *they would be lying along a parallel line to the Main Sequence but with another slope. The*

[146] Gamow, G., PRL **53**, 907 (1938).

giants may be on their way to get a neutron core because they are more massive than the Chandrasekhar limiting mass.

As we shall see, the main problem with Gamow's arguments was that he assumed that the stars evolve in a completely mixed way, which is not correct. He did not know about Öpik's paper, which had been published a year earlier. Note that Öpik's paper did not appear in a professional journal, but was published as a booklet by the university observatory of Dorpat, and consequently had rather limited circulation.

6.29 Hans Bethe and the CN Cycle

Every spring a small conference sponsored by the Carnegie Institution for Science and George Washington University was held in Washington. Gamow and Teller, who were at the time affiliated with the George Washington University, usually suggested the topic of the conference, and for the 1938 meeting put forward the problem of energy production in stars. The invited participants were five astrophysicists and ten physicists. Bethe was invited, but admitted that he did not want to come to the meeting because he was working on quantum electrodynamics (which had to wait for another 10 years).

Under pressure from Teller, Bethe came and met Strömgren. The latter informed Bethe that, once he had discovered that the Sun contained mainly hydrogen, with 25% helium, the central temperature of the Sun was estimated to be only 15 million degrees, and not 40 million as calculated previously. This information was important, because it provided Bethe with the right range of temperatures over which the nuclear energy source had to operate. This low temperature signaled to Bethe that the proton–proton reaction could not be important in the Sun. Moreover, the pp reaction did not solve the problem of massive stars, which demanded a very high rate of energy generation at a modest temperature.[147] Consequently, Bethe looked to the carbon cycle to provide the answer. Almost all the reactions had by then been well measured in the laboratory. Bethe was particularly surprised by the closing reaction, namely,

$$^{15}\mathrm{N}+\mathrm{H}\rightarrow{}^{12}\mathrm{C}+{}^{4}\mathrm{He}\,, \tag{6.8}$$

because it did not yield ^{16}O, but rather disintegrated to smaller nuclei (as Atkinson had effectively predicted). Indeed, only for one part in a 1000 does it actually go to oxygen.

Upon his return from the conference, Bethe worked for about two weeks and discovered the full CN cycle.[148] So by 1938, Bethe had the full story, and even sent it for publication in the Physical Review. By good luck, Bethe got a new student

---

[147] If $L\sim M^3$ is the mass–luminosity relation for high-mass stars, the energy production per unit mass is $L/M\sim M^2$, and it increases with stellar mass.

[148] Bethe, H.A., Ann. Rev. Astron. Astrophys. **41**, 1 (2003).

at that time, Robert Marshak, who suggested that he should submit the work for a prize awarded by the New York Academy of Science for the best original paper on energy production in stars.[149] However, according to the regulations for applying for the prize, the paper had to be unpublished. So Bethe withdrew the paper from the Physical Review and submitted it to the New York Academy, whereupon he did indeed win the prize. The jackpot was 500 US$, of which Marshak got 50 as the prize broker. The rest was used as a 'donation' to the German government to secure the release of Bethe's mother's belongings, because she had finally decided to emigrate. After winning the prize, the paper was sent once again to the Physical Review and quite rightly accepted for publication, with a note that it had won the prize awarded by the New York Academy. Years later, the paper[150] also won the Nobel Prize.

It is evident that the crucial reaction is the one in which the resulting nucleus disintegrates and emits an α particle, i.e., the reaction (6.8). At the time of writing the paper, there was no measurement of this reaction, so Bethe claimed that he had estimated the rate by analogy with other reactions. However, the reaction with which it was compared was not specified. Fortunately, the exact value of the rate of this reaction is not so important, because the rate of the entire cycle depends on the slowest reaction in the cycle, and if one assumes that the slowest reaction is not this one, the results hardly change. The slowest reaction, which actually determines the rate of the cycle is

$$^{14}\mathrm{N} + {}^{1}\mathrm{H} \rightarrow {}^{15}\mathrm{O}\,, \tag{6.9}$$

which has a time scale of 5×10^7 yrs, while the reaction in question was estimated (at this stage) to have a time scale of just 2000 yrs. Hence, the fact that this reaction had not been measured presented no problem.

It is interesting to understand how and why the CN cycle ends with ^{16}O. The reaction $^{15}\mathrm{N} + \mathrm{p}$ can go either to $^{12}\mathrm{C} + \alpha$ or to $^{16}\mathrm{O} + \gamma$. In the first case this is the end of the cycle, while in the second case the build-up continues. It so happens that the ratio between the two rates is 1000:1, so that in most cases the first option wins. An examination of the structure of the ^{16}O nucleus (see Fig. 6.13) shows the following. $^{15}\mathrm{N} + \mathrm{p}$ has energy 12.1276 MeV above the ground state of ^{16}O, where a proper energy level lies very close. Once the excited ^{16}O has formed, it can only decay to the 7.1169 MeV level (because the photon takes the extra spin), which is slightly below (but sufficiently wide of) the disintegration into $^{12}\mathrm{C} + \alpha$. Recall that, in order to descend to a lower level with the emission of a γ photon, there must be a difference of one unit (because of the spin of the photon) between the spin of the initial and final levels. This process is indicated by green arrows in the figure. Once the ^{16}O nucleus is in the 7.1169 MeV level, it can descend to the ground state (red arrow), but the probability for this to happen is smaller. In nuclear astrophysics

[149] The prize was the A. Cressy Morrison Prize, awarded by the New York Academy of Sciences. This is the same prize that was awarded to Geasimovič and Menzel in 1929 for the idea of mass creation in stars.

[150] Bethe, H.A., Phys. Rev. **55**, 434 (1939).

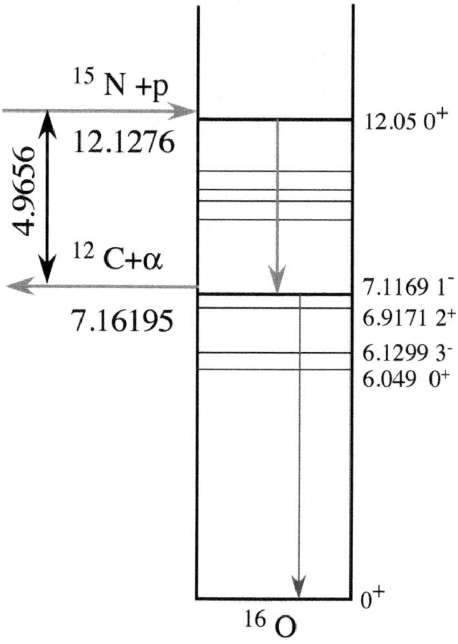

Fig. 6.13 The structure of the ^{16}O nucleus. The important levels are marked with *thick lines*. Unimportant levels are marked with *thin lines*. The energy and the spin of the levels are marked *on the right*

textbooks the reaction appears as:

$$^{15}\mathrm{N}+\mathrm{p}\to {}^{12}\mathrm{C}+\alpha+\gamma+4.9656\ \mathrm{MeV}\,,\quad {}^{15}\mathrm{N}+\mathrm{p}\to {}^{16}\mathrm{O}+2\gamma+12.1276\ \mathrm{MeV}\,. \tag{6.10}$$

A similar situation occurs also with ^{20}Ne–^{24}Mg, which forms another cycle.

In principle, the clues to the process were already in Bethe's hands in 1935,[151] when he examined the masses of the nuclei and found $m(^{12}\mathrm{C})+m(\alpha)=16.0071$, while $^{15}\mathrm{N}+m(\mathrm{p})=16.013$, which meant that the disintegration of the product nucleus ^{16}O into a carbon nucleus and a helium nucleus could take place (at least from the purely energetic point of view) with a release of about 0.06 amu in the form of energy. However, this fact went unnoticed. Similarly, in the paper by Livingston and Bethe, the possibility of the ^{20}Ne–^{24}Mg cycle, which is very similar to the CN cycle but involves heavier catalysts, is already evident. The reactions of the CN cycle are given in Table 6.3, together with the p + p reaction for comparison. The times are calculated for the conditions in the Sun. The CN cycle is faster than the pp chain, because the β decays are much faster in the first case.

[151] Bethe, H.A., *Masses of light atoms from transmutation data*, Phys. Rev. **47**, 747 (1935).

Table 6.3 Typical reaction times in the CN cycle under solar conditions, according to Bethe 1939

$H+H$	$\rightarrow$	$^{2}D+e^{+}+\nu$	1.2×10^{11} yrs
$^{12}C+H$	$\rightarrow$	^{13}N	2.6×10^{6} yrs
^{13}N	$\rightarrow$	$^{13}C+e^{+}+\nu$	870 s
$^{13}C+H$	$\rightarrow$	^{14}N	5×10^{4} yrs
$^{14}N+H$	$\rightarrow$	^{15}O	5×10^{7} yrs
^{15}O	$\rightarrow$	$^{15}N+e^{+}+\nu$	870 s
$^{15}N+H$	$\rightarrow$	$^{12}C+^{4}He$	2000 yrs

Energy generation crosses over from the pp chain to the CN cycle at a temperature of about 16×10^{7} K (see Fig. 6.14). So low mass stars with a central temperature below 16×10^{7} K derive their energy by means of the pp chain, while more massive stars implement the CN cycle.

The agreement with the main sequence was excellent. First, the total energy released at the temperatures required by Eddington's models agreed with the observed luminosity. Second, the predicted mass–luminosity relation agreed nicely with the observations. Third, the details of the model as calculated from the energy requirements (to produce the observed luminosity) agreed well with the calculated central temperature according to a stellar model which is essentially based on the radiation absorption coefficient of the matter. For example, the central temperature of the Sun was calculated to be 18.5 million degrees while integration of the Eddington model yielded 19 million K. This was a major achievement because the nuclear physics, which controls energy production, agreed with the atomic physics which controls radiative transfer in the Sun. The two completely different disciplines yielded the same result.

Still enslaved to the prevailing idea that the giants precede the main sequence stars, Bethe examined the possibility that the giants might still be breaking down lithium, beryllium, and boron, while the main sequence had already started the carbon cycle. Bethe soon reached the conclusion that:

> *It seems, however, doubtful whether the energy production in giants is due to nuclear reactions at all.*[152] *The reaction $^{7}Li+H\rightarrow2\,^{4}He$ was known to be 'improbable'.*[153]

These were among the first signs that something was wrong with the hypothesis that stellar evolution goes from the giants to the dwarfs, i.e., that there is continuous contraction of the star, or that Eddington's gaseous radiation-dominated and homogeneous model was adequate to describe the giants.

As was typical in Bethe's papers, and this paper was no exception, all possibilities were examined exhaustively. That being so, Bethe estimated that ^{8}Be would have a half-life of 10^{-13} s *if ^{8}Be is heavier than two α by 50 keV*. The instability of this nucleus posed a problem that Bethe was unable to solve. Inevitably, Bethe rejected

[152] Here Bethe cited a private communication with Gamow.

[153] Goldhaber, M. Proc. Camb. Phil. Soc. **30**, 560 (1934).

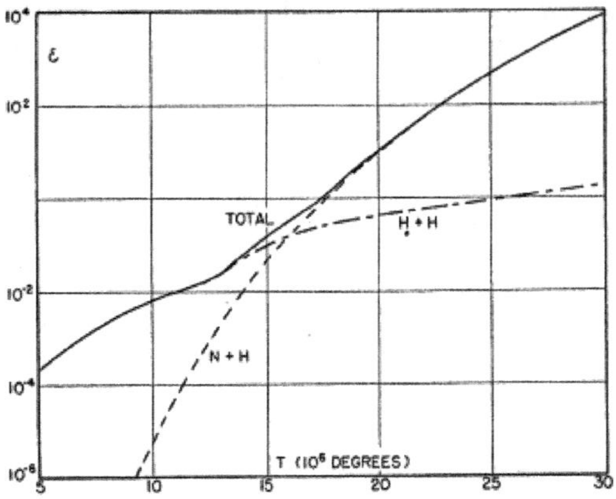

Fig. 6.14 The sum of the CN and the pp energy mechanisms for main sequence stars according to Bethe (1939). The H+H curve is the pp energy release, and the N+H curve is the CN energy release. At a temperature of about 15 million K, the dominant energy release changes from the pp chain to the CN cycle

the $3\alpha \to {}^{12}C$ reaction on account of the large Coulomb barrier (which would require temperatures in excess of what any stellar structure model could yield), which meant that it would not work in the core temperatures of main sequence stars. Bethe estimated that a temperature of 10^9 K was needed to make the triple α reaction compete with the pp chain.

Last but not least, there was the question of the stability of the stars. Bethe's first calculation for the rate of energy generation according to the CN cycle gave a rate $\varepsilon \sim T^{16}$, so the energy release appeared to be extremely sensitive to the temperature. Bethe thus retreated from the Eddington model to the Cowling model,[154] and assumed that the stars were made of a convective core and a radiative envelope, so that the temperature-sensitive energy source would be stable.[155]

6.30 Acceptance by the Community and New Puzzles

Russell[156] got the manuscript from Bethe prior to its publication, and hailed Bethe for his discovery. While the mass–luminosity relation and the main sequence stars

[154] Cowling, T.G., MNRAS **94**, 768 (1934); **96**, 42 (1935).

[155] In a convective core, the energy generated spreads quickly (by convective currents) over a large region, unlike the radiative case in which the energy is released in a small volume. Consequently, energy release in a convective region does not impair the stability of the star. The convective currents are much faster than the slow radiative diffusion of energy, so they easily overcome any attempt by the nuclear reactions to 'run away'.

[156] Russell, H.N., Proc. Am. Phil. Soc. **81**, 295 (1939). See also JRASC **33**, 287 (1939).

were finally understood from first principles, Russell pointed to the following emerging fundamental questions:

- Why do so many stars have the same amount of hydrogen?
- Where did the carbon and heavier nuclei come from?

The problems according to Russell were severe, because for some unclear reason:

> *Present evidence indicates that nearly two-thirds of the Sun is composed of such heavy atoms – for if there were fewer of them, and more helium, the molecular weight would be less, and the internal temperature too low to start the carbon chain running.*

Consequently, Russell assumed that these elements were older than the stars. Russell mentioned at this point that Weizsäcker had suggested that *at some remote time* the elements were formed in a high temperature high density explosion. Weizsäcker had sown the seeds for the Big Bang nucleosynthesis to come almost a decade later.

The victory achieved by Bethe with his nuclear scheme was considerable, since it solved a problem that was almost a century old. But there were still unsolved problems with stellar energy sources. Consider, for example, certain very luminous stars like Y Cygni. In particular, asked Russell, how had this star evolved over the past 2000 million years since the Universe had started to expand? Such a star produces energy per unit mass at a rate about 500 times higher than the Sun. This meant that the transformation of the entire supply of hydrogen into helium would last for only 200 million years:

> *Does it mean,* asked Russell, *that the massive stars have begun to shine late in the history of the Galaxy, or that after all, they have still greater stores of energy to draw on, in ways as yet unknown.*

The dilemma was obvious: did all stars have the same age, or was there some other possible energy source that remained to be discovered?

6.31 The Last Piece of Nuclear Data

Soon after the publication of his theoretical work, Bethe felt that an experimental measurement of the last reaction (6.10) was needed. This was the only reaction for which he did not have any experimental proof, although energetically it was clear that it could indeed take place. As already remarked, this was the most crucial reaction, because it was this that closed the cycle and produced helium on the one hand and returned the original nucleus on the other. So after a year's collaboration with Holloway,[157] they reported the first measurement of the reaction. The agreement with the theoretical value was satisfactory. Thus, in contrast with the pp chain, in which the fundamental reaction cannot be measured in the laboratory, all reactions in the CN cycle had been measured and confirmed directly.

[157] Holloway, M.G., & Bethe, H.A., PRL **57**, 747 (1940).

6.32 The First Full Model of the Sun

Once the rate of nuclear reactions was available, it became possible to calculate the evolution of the Sun and compare it with observations. Applying Bethe's newly discovered expression for energy generation through the CN cycle, Blanch, Lowan, Marshak, and Bethe[158] were the first to calculate such a model. To facilitate the complex computations, they assumed that (a) all the energy is produced at the center and (b) the Sun has a convective core, which is essentially the decade-old Cowling model for stars. Since they did not carry out a real calculation but assumed a model, they concluded:

> *The impossibility of estimating the amount of error introduced by this procedure made the temperatures and densities thereby obtained quite unreliable. The only check on the values was their approximate agreement with the values of the temperatures and densities given by the standard model of Eddington.*

In short, they did not compare their numerical results with observations, but only with another old model.

The results were that a hydrogen abundance of 35% by mass in the present day core of the Sun gave the best fit to Eddington's model. Consequently, they assumed the same hydrogen abundance throughout the entire Sun, i.e., a fully mixed Sun! The predicted luminosity came out to be a factor 145 higher than the observed solar luminosity. They ended the paper by stating that:

> *In view of the uncertainties enumerated above* (mainly in the radiative opacity of the Sun), *we do not regard the discrepancy of a factor of 100 between the predicted and observed luminosities of the Sun as any argument against the carbon cycle as a source of energy of main sequence stars.* And they went on to claim that: *All other reactions (except possibly the proton–proton reaction for less luminous stars of the main sequence) are excluded by much larger factors than 100. Moreover, it is almost certain that an improved calculation will remove the discrepancy completely.*

The authors were wrong. A discrepancy by a factor of 145 is too large even allowing for all the uncertainties of an astrophysical calculation. The supposed success in explaining the Sun meant that research on the proton–proton reaction was subsequently left aside. But today we know that only 6% of solar energy is derived from the CN cycle and the rest comes from the proton–proton reaction. Furthermore, the Sun does not have a convective core.

Sen and Burman[159] tried to improve the solar model of Blanch and his associates, and improved the agreement with the present day Sun. However, the values they got were still too far off the observed values to justify declaring a complete victory. It was clear that something was still missing.

---

[158] Blanch, G., Lowan, A.N., Marshak, R.E., & Bethe, H.A., Ap. J. **94**, 37 (1941).

[159] Sen, N.R., and Burman, U.R., Ap. J. **100**, 355 (1944).

6.33 The Fate of the Sun

Forgetting about the Chandrasekhar limiting mass, Gamow, the great popularizer of astrophysics, wrote in 1941[160] that *our Sun is bound to explode*. Gamow assumed that the fate of the Sun would be a supernova explosion like the 'Star of Bethlehem' and the Tycho supernova. Gamow did not appreciate, two years after Chandrasekhar had published his book, that stars with masses below the Chandrasekhar limiting mass never explode, and thus predicted in this popular article, an explosive end to the Sun.

6.34 Accretion: Revival of Helmholtz?

What happens to a star moving through space? Clearly, space is not empty but contains matter in the form of interstellar clouds. When a star moves through such a cloud, it can accrete mass, and this could therefore be a source of energy exactly like the old Helmholtz supposition. The first to calculate how a star might accrete in this way were Hoyle and Lyttleton.[161] So they developed the first formula to calculate how much mass a moving star could gain as it moved through space. Atkinson[162] examined the problem of the very bright stars of spectral class O. He argued that, if these stars live off the transmutation of hydrogen, their lifetime must be *inconveniently short*. Nuclear energy could not supply the observed luminosity for over a billion years. Indeed, recent observations and calculations confirmed that these stars had a very short lifetime, only a few million years, or about a thousand times shorter than what Atkinson assumed. However, Atkinson, not knowing the correct number, was searching for a source of energy for longer lifetimes. The conclusion Atkinson reached was that accretion could not provide the energy source for the massive O type stars for the necessary extended periods of time. (Recall Russell, who reached a similar conclusion several years later.)

Some fifty years later it would become clear that accretion powers the most energetic engines in the cosmos, namely, the massive black holes in the centers of the galaxies. But at this time it was a suggestion that was checked and proven ineffective in polluting the surface of stars. Unbelievably, by observing the composition of old stars, we should be able to observe the composition of the material out of which the old stars formed. The pollution of the surface of stars is unimportant.

[160] Gamow, G., Popular Astronomy **49**, 360 (1941).

[161] Hoyle, F., & Lyttleton, R.A., Proc. Camb. Phil. Soc. **35**, 405, 592 (1935).

[162] Atkinson, R.d'E., MNRAS **100**, 500 (1940). See also, Atkinson, R.d'E., Ap. J. **84**, 83 (1936).

Chapter 7
How the Low Mass Stars Perish

7.1 The Energy Source of White Dwarfs

The last chapter in Chandrasekhar's 1939 book[1] was dedicated to the energy source of stars. This is the only chapter which Chandrasekhar described in the introduction as different from the rest in terms of completeness, and in particular in terms of its temporary nature due to developments in the theory of stellar energy that were taking place during the writing of the book. Indeed, Chandrasekhar discussed only Weiszacker's theory of the CN cycle and did not discuss the energy source of white dwarfs.[2]

In 1939, the year Chandrasekhar published his book, Bethe and Marshak (1916–1992)[3] reviewed the theory of stellar evolution and discussed the possible energy source of white dwarfs. Adopting the theory of Chandrasekhar, they concluded that:

> *The carbon concentration is the same as in main sequence stars, and the hydrogen concentration must be* 10^{-8} *for Sirius B and* 10^{-12} *for Eridani B. Even if carbon and nitrogen were completely absent, the proton–proton reaction would still occur, and from this an upper limit for the hydrogen concentration of* 10^{-5} *for Sirius and* 10^{-4} *for Eridani B is obtained; it can easily be seen that such low hydrogen contents could supply the energy of the white dwarfs only for a very short time. The figure of* 10^{-12} *for Eridani B leads to a life of only ten years, which is manifestly wrong. This leads to the conclusion that gravitational contraction is the primary source of energy of the white dwarf stars. With this source of energy it takes at least* 10^{8} *years before a white dwarf becomes a dark object.*

Thus, Bethe and Marshak, in an almost completely forgotten and hardly ever cited paper, identified the source of energy for the white dwarfs as the Ritter–Kelvin–Helmholtz gravitational contraction mechanism. At long last, the Ritter–Kelvin–Helmholtz idea could be said to work, if not for all stars, at least for some of them.

---

[1] Chandrasekhar, S., *An Introduction to the Study of Stellar Structure*, Dover, New York (1939).

[2] At the end of the references to the chapter on energy sources, there appears a strange comment about the discovery by Joliot and Zlotowski [Jour. d. Phys. **9**, 403 (1938)] to the effect that $^{5}He_2$ exists and is stable.

[3] Bethe, H.A., & Marshak, R.E., Rep. Prog. Phys. **6**, 1 (1939).

G. Shaviv, *The Life of Stars*, DOI 10.1007/978-3-642-02088-9_7,

In 1940, Marshak[4] extended the calculations which implied that white dwarfs extract their energy from gravitational contraction. However, Marshak admitted that *the above arguments for gravitational contraction as the source of energy in white dwarfs stars are admittedly inconclusive*. Since all attempts by Marshak to reproduce the observed radius of Sirius B failed,[5] he claimed that:

> *The only conclusion we can draw, therefore, is that the observations of the radius of Sirius B are in error.*

Much like Eddington's motto: do not believe the observations if you do not understand them.

Can white dwarfs really burn nuclear fuel in a stable way? Ledoux (1914–1988) and Sauvenier-Goffin[6] noticed that white dwarfs cannot ignite nuclear fuel in a stable way. This is exactly the opposite argument to Jeans' (incorrect) stellar stability criterion. The reason is simple. The gas in white dwarfs is degenerate and consequently the pressure is insensitive to the temperature. On the other hand, nuclear reactions are extremely sensitive to the temperature. Thus, if a nuclear reaction starts in a white dwarf, it heats the gas, and as a consequence the temperature rises. But since the pressure of the gas is not sensitive to the temperature, it does not change, so the gas does not expand and it does not cool as a heated ideal gas does. The cooling of a normal gas upon expansion serves as a 'safety valve', which controls the nuclear reactions and prevents runaways. If the gas cannot expand, the temperature rises as more heat is poured in by the nuclear reactions. The higher temperature causes a still faster energy release, and the process accelerates to become precisely what we call a runaway. Such a process can cause the star to explode. So Ledoux and Sauvenier-Goffin concluded that:

> *It seems that the generation of energy in these stars must definitely be attributed to some other cause.*

In 1947, Hoyle[7] returned to examine Eddington's ideas about the origin of white dwarfs. He formulated the options in the following way:

> *If the Eddington theory is correct then the following conclusions follow: (a) The material of the Galaxy was originally composed of pure hydrogen, (b) the majority of white dwarfs are composed largely of hydrogen, and (c) The Fermi form of the β decay interaction applies to the proton–proton reaction.*

On the other hand, if Eddington's theory were wrong, Hoyle argued that:

> *(a) All white dwarfs arise from novae and supernovae. (b) Either the supply of white dwarfs is greater by a factor of order 100 than is suggested by the accepted statistics on the frequency of novae and supernovae, or the density of white dwarfs in the neighborhood of the Sun is much larger than the average density for the Galaxy as a whole. (c) The high hydrogen content of Sirius B is due to accretion of interstellar hydrogen occurring after the formation of this white dwarf.*

---

[4] Marshak, R.E., Ap. J. **92**, 321 (1940).

[5] It was the large radius of Sirius B which led Eddington to assume a high concentration of hydrogen in this white dwarf.

[6] Ledoux, P., & Sauvenier-Goffin, E., Ap. J. **111**, 611 (1950).

[7] Hoyle, F., MNRAS **107**, 253 (1947).

After a long discussion, Hoyle refrained from casting his vote and left it to the reader. It would turn out that Eddington was wrong. As more data accumulated, the implied conclusions set out by Hoyle were found to be wrong as well.

In 1945, Schatzman[8] came out with an extensive survey of nuclear reactions in white dwarfs, and reached the conclusion that the proton–proton reaction leading to the formation of deuterium could not take place in white dwarfs. Something must prevent the Gamow–Teller selection rule, which allows the conversion of a proton into a neutron, from taking place in white dwarfs. The dilemma originated from the fact that the observed white dwarfs appear to classify into DAs and DBs. The first-class contains about 80% of all WDs, while the second class includes most of the rest. The DAs show hydrogen spectral lines, and hence must have a pure hydrogen atmosphere. On the other hand, the DBs show helium lines and no hydrogen lines, and hence must have a helium atmosphere. The rest of the WDs, less than one percent of the total, have different peculiarities. So it was clear that at least the surface of the WD contained hydrogen. The situation with the white dwarfs was so confusing that, two years later, Schatzman[9] claimed that, unless some unknown process prevented nuclear reactions from taking place in white dwarfs, the origin of the white dwarfs could not be from main sequence stars.

The problem was taken up by Lee.[10] He soon reached the widely held conclusion about the impossibility of nuclear reactions in these stars: *No stable white dwarf can live on the nuclear energy produced in the interior*, stated Lee. The emphasis was on the word 'interior'. Lee mentioned the possibility of gravitational contraction, but chose to discuss in some detail the idea that nuclear reactions might take place only near the surface. Lee found reasonable numbers for the ages of white dwarfs by assuming that the outer 10^{-4} fraction of the mass of the star was composed of hydrogen. Being aware of the stability problem of the star, Lee examined the stability of the envelope of the star. He claimed that, if the envelope radiated more than the energy production due to nuclear reactions at the base of the envelope, the star would be stable. Lee did not provide any reference to this stability criterion, and indeed it is not a valid one. If anything, equilibrium requires the equality of the nuclear production with the luminosity of the star, and the stability condition applies only to the changes in this condition upon a perturbation to the temperature.[11] However, the model was found to be essentially wrong on the day of its publication. Lee simply did not know about the stability condition found by Ledoux and Sauvenier-Goffin (who examined nuclear reactions both in the core and close to the outer surface). The latter was published in the same issue of the Astrophysical Journal, just before

---

[8] Schatzman, E., Ann. d'Ap. **8**, 143 (1945).

[9] Schatzman, E., Ann. d'Ap. **10**, 93 (1947).

[10] Lee, T.D., Ap. J. **111**, 625 (1950). Lee was the 1957 Nobel Laureate in Physics for the discovery of parity violation.

[11] The derivative of the equality with respect to the temperature.

Lee's paper. We may suppose that the publication of the two papers back-to-back was not a trick played by the typesetter.[12]

The solution to the white dwarf puzzle came in 1952, when Mestel[13] discussed the origin of white dwarfs in two papers published one after the other. The basic question was the stubborn problem of the energy source in white dwarfs. Mestel began his paper by setting the history straight: he stated that Fowler[14] solved the problem of the white dwarfs when he showed that degenerate electrons held the star up. Stone and Anderson, and later Chandrasekhar, just modified the theory to include relativistic effects which lead to the Chandrasekhar limit of $5.75 M_{\odot}/\mu^2$. Most white dwarfs were not bothered by this limit, contended Mestel. This is not what we find in today's standard textbooks. Judging from the history as described previously, much of the truth was with Mestel.

Whichever way one looked at things, claimed Mestel:

> *The white dwarf cannot contain hydrogen and produce nuclear energy. On the other hand, if stars are formed from the interstellar medium, the fact that we have never observed interstellar medium devoid of hydrogen rules out this possibility. Hence, it is implausible that white dwarfs were formed directly from the interstellar medium (as suggested by Eddington in 1939).*

But many white dwarfs are observed, and the lifetime of main sequence stars with masses equal to those of the observed WDs (less than $1 M_{\odot}$) is very long. So how come so many WDs are observed? Mestel's solution was the following. Stars more massive than the Chandrasekhar limiting mass synthesize heavy elements. These heavy elements are expelled from the stars *either spasmodically or catastrophically according to the density when the instability sets in*. This was apparently the first time such an idea was expressed this way, i.e., the massive stars end their lives by losing mass (which contains the synthesized elements), reduce their mass to below the Chandrasekhar limit, and become white dwarfs devoid of hydrogen and rich in heavy elements. As observational evidence, Mestel cited the works of Baade and Minkowski on the Crab nebula.[15] It would be shown later that stars more massive than the Chandrasekhar limiting mass and less massive than $8 M_{\odot}$ do indeed lose mass and become white dwarfs as Mestel hypothesized. But to bring in the Crab nebula as evidence, when it was already known by then to be a supernova remnant, was going a step too far.

The first to carry out a detailed calculation of gravitational contraction as an energy source for white dwarfs was Mestel, and he provided an explanation. The high observed amount of hydrogen on Sirius B had misled the researchers. The hydrogen on the surface of Sirius B was probably not original, and could have been accreted, after its formation, from the interstellar medium. As Mestel put it:

---

[12] The editors of the Astrophysical Journal were Morgan, W.W. (Managing editor), Chandrasekhar, S., both from Yerkes observatory, Merril, P.W., Shapley, H., and Mayal, W.D. Only Chandrasekhar was an expert on white dwarfs, and one can guess that he had a hand in the setting of the papers.

[13] Mestel, L., MNRAS **112**, 583 (1952); ibid. **112**, 598 (1952).

[14] Fowler, R.H., MNRAS **87**, 114 (1926).

[15] Baade, W., Ap. J. **96**, 188 (1942); Minkowski, R., Ap. J. **96**, 199 (1942).

> *Many of the published treatments of the problem are misleading in that they seem to imply that there must be energy liberation within a white dwarf in order that it may shine at all. Thus, having built a heterogeneous white dwarf model in thermal equilibrium, Schatzman assumes that all white dwarfs must exist in this state [...] but in fact neither a white dwarf nor a normal star needs a nuclear energy source to make it shine. The Kelvin–Helmholtz theory for a normal star is thermodynamically perfectly sound, and predicts a luminosity fixed by the mass and central temperature of the star. There is only the question of how long the state can exist.*

In other words, Mestel, almost hundred years after the theory was published, identified the white dwarfs as the true cooling stars à la Kelvin–Helmholtz–Ritter contracting stars.

To obtain the cooling law, Mestel made one crucial and profound assumption, namely that white dwarfs contain electrons and ions. The electrons are extremely degenerate and provide the pressure to support the star against its outer layers. The ions on the other hand behave like an ideal gas, and their pressure is very small and effectively does not contribute to the total pressure. Both the electrons and the ions have energy. As a matter of fact, most of the thermal energy is with the electrons. But as the star cools, the ions release their thermal energy while the electrons cannot lose any energy as they provide the support against gravity. Moreover, as the white dwarf cools, its radius hardly changes. Since the temperature does not affect the pressure of the electrons, the cooling has no effect on the balance between the gas pressure and gravity. On the other hand, the ions, which continue to behave as an ideal gas, cool down, whence all the energy released by them appears as the luminosity of the star. The star behaves as if it contained two independent fluids. The electrons take care of the hydrostatics and the ions supply the luminosity. With this assumption, Mestel derived the result that the luminosity L varies with the central temperature T_c according to $L \sim T_c^{7/2}$. The predicted cooling time of the van Maanen white dwarf, for example, was calculated as $10^{11}/A$ yrs, where A is the mean atomic weight of the ions. The comparison with other white dwarfs agreed reasonably well.

But Sirius B, the first white dwarf to be discovered, still defied theoretical explanations, as its displayed large amounts of hydrogen. Mestel assumed that this hydrogen was not the original hydrogen from the birth of the star, but was accreted somehow as the star wandered around the Galaxy. He therefore attempted to build a model of Sirius B, and ran into the same problem that Schatzman had faced, namely, he had to assume that, for some reason, hydrogen does not burn on Sirius B. Roughly at the same time, Gamow and Critchfield[16] published a report claiming that the measured radius of Sirius B was in error, as no theory could predict it. Recall Eddington's philosophy!

Fifteen years after Mestel had published the theory of white dwarf cooling, observational proof finally came. Degenerate electrons are excellent heat conductors. Hence, the internal temperature of the white dwarf is practically constant throughout the star. The WD resembles a piece of metal at constant temperature. Suppose now that the rate of birth of WDs is constant. Then we can calculate from Mestel's theory how fast the luminosity of a WD changes, and consequently how many WDs are expected at each luminosity. Mestel predicted that the number of cooling WDs should

[16] Gamow, G. & Critchfield, C.L., *Theory of Atomic Nucleus and Nuclear Energy Source*, Oxford University Press, Oxford (1949) p. 293.

be $N = (2/7)M_b$, where M_b is the total (bolometric) magnitude of the WDs. Weidemann[17] carried out an extensive comparison with observations and demonstrated that the WDs do indeed obey Mestel's cooling law.

7.2 The Central Stars of Planetary Nebulas

Some of the most beautiful objects in the Universe are planetary nebulas. Typical examples of these magnificent objects are shown in Figs. 7.1 and 7.2. The nebulas come in various shapes and sizes but the most famous have spherical or cylindrical symmetries. The nebulas absorb the UV radiation from the central stars and emit radiation in the visible, much like a fluorescent lamp. The colors seen in the pictures are emissions typical of different ions. The fact that the colors are so clearly separated points to the fact that these nebulas are not well mixed.

Planetary nebulas have central stars that provide the UV radiant energy which the nebulas convert into visible light. It took many years to realize that the central stars are extremely hot and radiate mostly in the UV, radiation which does not penetrate the Earth's atmosphere. For example, when he described observations of the central

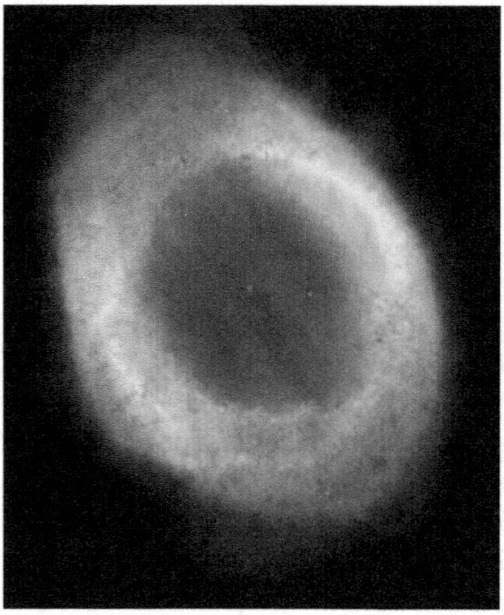

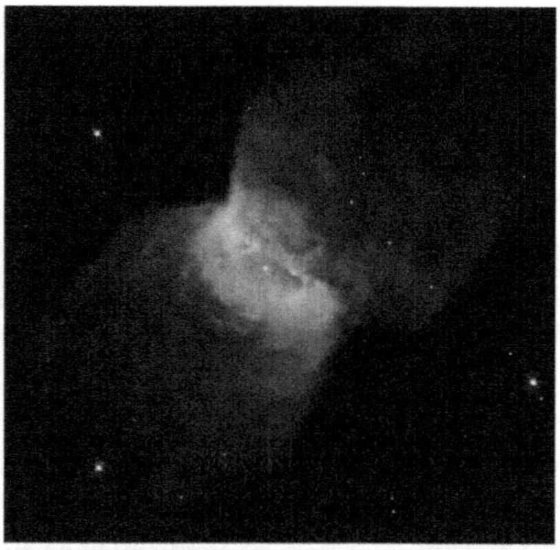

Fig. 7.1 *Left*: The famous Ring planetary nebula in the Lyra constellation, known also as M57 or NGC6720. The central star was discovered in 1800 by F. von Hahn (1742–1805). The distance is poorly determined but in excess of 2000 lyrs. Credit: NASA, Hubble, Hubble Heritage Team. *Right*: The Butterfly planetary nebula in the Monoceros constellation, known as NGC2346. The central star is a close binary. Credit: NASA, Hubble, Hubble Heritage Team

[17] Weidemann, V., Zeit. f. Astrophys. **67**, 286 (1967).

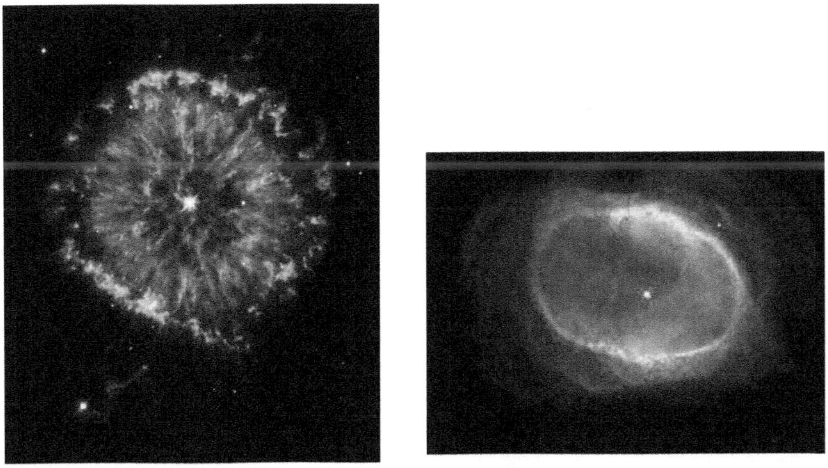

Fig. 7.2 *Left*: The planetary nebula NGC6751 in the Aquila constellation. The bright central star is conspicuous. The distance is about 6 500 lyrs. Credit: NASA, Hubble, Hubble Heritage Team. *Right*: The eight-burst planetary nebula in the Vela constellation (NGC3132). The central star is easily visible. The system is about 2000 lyrs away. Credit: NASA, Hubble, Hubble Heritage Team

stars in PNs, Keeler[18] claimed that *as in an ordinary star, the maximum of this nebula falls in or near the yellow*, which implies a surface temperature of less than 6000 K, like the surface temperature of the Sun. Keeler was very wrong, but recall that he made the claim years before quantum theory was invented.

The first planetary nebula to be discovered was M27 by Messier (see Fig. 7.3). Fifteen years later, in 1779, the French astronomer Antoine Darquier de Pellepoix (1718–1802) discovered the most famous of all planetary nebulas, the Ring nebula shown in Fig. 7.1. His discovery preceded by a couple of days the discovery of this nebula, known as M57, by Messier. As a matter of fact, the two French astronomers were following the same comet and hence the same track in the sky, so no wonder they stumbled upon the same confusing objects. Darquier described the nebula as a 'pale planet'.

It is not clear whether Darquier's description influenced Herschel or not, but the telescopes of that time were not sufficiently powerful to resolve the details we see with present day telescopes, so the nebula looked to Herschel, who discovered the planet Uranus in 1781, like a planet, i.e., a small disk of light. Consequently, Herschel called these objects planetary nebulas. Herschel's innocent choice of name dragged with it a sensational theory.

[18] Keeler, J.E., Ap. J. **10**, 193 (1899).

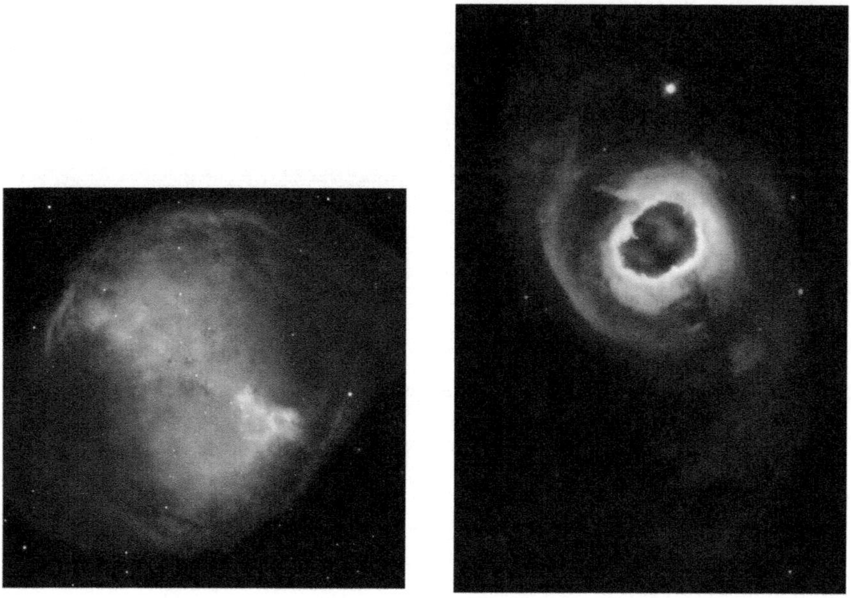

Fig. 7.3 *Left*: The planetary nebula in the Vulpecula constellation (NGC6853). The first PN to be discovered. The distance is about 1250 lyrs. Photograph by European Southern Observatory. *Right*: The Kohoutek 4-55 planetary nebula. The distance is about 4 600 lyrs. The actual size is about 1.5 lyrs across. Credit: NASA, ESA and the Hubble Heritage Team, courtesy of R. Sahai and J. Trauger (JPL)

In 1796, Laplace published his famous book about the structure of the Universe.[19] In a note which appeared towards the end of the book, Laplace proposed the nebular hypothesis: the origin of the solar system was *a large rotating nebula*. Laplace added the adjective 'rotating' even though he did not have a clue what the velocities might be. It simply suited his hypothesis. Doppler discovered his effect fifty years later, so no velocity measurements could have been made. For many years to come, the disintegration of the nebula into rings, followed by the condensation of the rings into planets, remained the fundamental theory for the formation of thc solar system. The Ring nebula was like a kind of inspiring logo for this theory, which held sway for almost 130 years.

The first measurements of the velocities of planetary nebulas were carried out by Campbell (1862–1938m) and Moore (1878–1949m) in 1916,[20] over a century later. They most likely interpreted the velocities as rotation velocities, although this is not stated explicitly, under the influence of the Laplace–Herschel nebular theory, which

---

[19] Laplace, P.S., *Exposition du Système du Monde*, Courcier, Paris, 1796. Shortly after the French Revolution, Laplace was appointed professor at the new Ecole Normale. His official duties were to give popular lectures, and the book emerged from these lectures.

[20] Campbell, W.W., & Moore, J.H., PASP **28**, 119 (1916); Lick Ob. **9**, 1 (1916).

claimed that the rings rotate. The title of the publication, namely *On the Rotation of Some Planetary Nebulae*, declared their explanation and made it appear as a well-established scientific fact. Later on, in 1918, they published the first systematic study of PNs.

While the above results were accumulating and being prepared for publication, Campbell and Moore presented the results on 10 September 1915 to the American Academy of Sciences,[21] and stated that:

> *It was shown that the planetary nebulae, those of regular form, are rapid travelers in comparison with the star, a fact which casts serious doubts upon the generally accepted hypothesis that the stars have been formed from planetary nebulae by a process of evolution.*

What they meant, and this is my interpretation, was that the measured velocities of the nebulas were significantly higher than those of the central star observed in each nebula. This was the first doubt expressed about the hypothesis that the planetary nebulas were stars in formation. To emphasise this, the title of the paper was *Radial Velocities of the Planetary and Irregular Nebulae*, which insinuated radial expansion or contraction, rather than rotation. But in the October 1915 issue of *The Scientific Monthly*, not even a month after the presentation to the academy, Campbell wrote that:

> *The hypothesis that the planetary nebulae have been the forerunners of solar systems has been considered favorably by astronomers.*

No contradictory arguments were given.

Slipher was the first to suggest that the light of the nebulas (any nebula) could be reflected light from nearby stars. Slipher's paper[22] went almost unnoticed, except by Hertzsprung and Hubble. Hertzsprung[23] was the first to put this hypothesis to the test. His subject was the Pleiades cluster (which does not constitute a planetary nebula) and the nebulosity around the stars, and found that the nebula was too faint for this hypothesis to hold. Russell[24] accepted the hypothesis, and extended it not only to reflection nebulas, but also to cases in which the light from the star excites the gas in the nebula so that it emits its own emission line. In 1922, Hubble tested this hypothesis on different kinds of nebulas, and on PNs in particular. While in non-PNs he got quite a nice correlation between the size of the nebula and the apparent luminosity of the star, he encountered problems in the case of the PNs.

Hubble's tacit assumption, as he stated it in this form in the paper, was the conservation of the radiant energy, and so he explained the reasons behind the correlation he sought:

> *If we assume that the clouds of nebulosity are illuminated by stellar light whose intensity varies inversely as the square of the distance from the stars; that each part of the nebula reflects or re-emits, without change in actinic value, all the starlight intercepted by it; and*

---

[21] Campbell, W.W., & Moore, J.H., PNAS **27**, 245 (1915).

[22] Slipher, V.M., PASP **31**, 212 (1919).

[23] Hertzsprung, E., AN **195**, 449 (1913). Hertzsprung knew about Slipher's suggestion from a Lowell Observatory Bulletin, no. 55, published in 1912.

[24] Russell, H.N., PNAS **8**, 115 (1922).

> *that the light from the stars themselves reaches us undimmed by absorption, then the square of the maximum angular extent of nebulosity a, for an exposure E, should be proportional to E times the apparent luminosity of the star I, or* $a^2/E = Const. \times I$*, or* $m + 5\log a_1 = B$*, where* a_1 *is the angular distance reduced to a uniform exposure of 60 minutes.*

The relation was between the apparent photographic luminosity and the extent of the nebula for an exposure E. Here m is the photographic magnitude.

The first attempt led Hubble to conclude that *the table indicates no trace of correlation* for PNs. However, by combining data from other observers, Hubble reached the conclusion that:

> *The general conclusion is in favor of the theory that the planetary nebulae derive their luminosity from radiation of associated stars, and the inverse-square law is at least one important factor in determining the distribution of luminosity throughout the nebula.*

Hubble noticed that the PNs have a greater UV contribution relative to non-planetary nebulas. The spectra of the central stars were mostly continuous and showed hardly any lines. Unlike Keeler, Hubble was sure that the maximum of the energy of the radiation was at wavelengths shorter than 3000 Å (and hence unobserved by ground-based telescopes), but exactly where it was situated was a matter of guesswork:

> *Thus*, concluded Hubble, *the PN receives a vast amount of energy in stellar radiation in a region of the spectrum which plays but a minor role in the determination of photographic magnitudes.*[25]

Furthermore, Hubble hypothesized that:

> *It is entirely conceivable that the energy in these continuous radiations may be absorbed by the nebulosity and, by some mechanism analogous to that of fluorescence, be re-emitted as discontinuous radiation at longer wavelengths.*

How right he was! Consider this statement in the context of the state of knowledge in atomic physics in 1922, before quantum theory had been properly established.

The first to measure the distance to several planetary nebulas was van Maanen (1884–1946m), who succeeded in 1919[26] in measuring the parallax[27] to six PNs,

[25] Photographic magnitude is the brightness of a star measured with a blue-sensitive photographic plate. This magnitude is quite inaccurate because different amounts of ultraviolet light are included, depending on whether a refractor or an aluminized reflector is used. Furthermore, the sensitivity of the plates was increased by different methods (like baking) to a point where it became difficult to calibrate them. With the introduction of CCDs (Charged Coupled Devices) this method became obsolete.

[26] Van Maanen, A., PA **27**, 410 (1919). The famous van Maanen star was discovered by van Maanen in 1917, and is a white dwarf situated inside the very faint planetary nebula NGC2440. Van Maanen's star is also famous for having a very large motion in the sky, about 2.98 arcsec per year. This particular white dwarf shows no lines of helium or hydrogen. The white dwarf is very cool, and hence should have an age of about 10 billion years. This means that it cannot be associated with the nebula. The mystery of van Maanen's star is still with us.

[27] The parallax of a celestial object is the angle at which the Earth's orbit around the Sun is seen from the object. The smaller it is, the further away the object is. The nearest star to the Sun has a parallax of 0.96 arcsec. A parallax of 1 arcsec corresponds to a distance of 1 parsec. A parallax which has been determined geometrically is called trigonometric parallax, to distinguish from methods in which the distance is measured in other ways and then converted into an angle.

and later in 1933[28] to 29 PNs. In this way, van Maanen was able to determine the physical size of the corresponding nebula. Typical distances van Maanen found were 142–403 lyrs, with diameters of 0.02–0.16 lyrs. Since these distances to the PNs lie beyond the possibility of a direct parallax measurement, the angle being much too small, various tricks were implemented. For example, van Maanen counted the number of stars equal to or brighter than the ones in question on the plate and extracted the corresponding magnitude from van Rhijn's table of average stellar densities for given galactic latitude.[29] Van Maanen applied three different methods and took the average. The fantastic result, however, appeared in the section of the paper which van Maanen described as speculation. So using his data and the method Zanstra devised to get the luminosity of the central star (see later), he found that the radius of the central star was $0.2R_{\odot}$, and since the luminosity was very high, it corresponded to a $6M_{\odot}$ star, if one used the Eddington formula. On the other hand, van Maanen showed that, if one adopted Campbell and Moore's idea of rotational velocities, one obtained a mass of $200M_{\odot}$, which according to van Maanen was *hard to accept.*

So van Maanen concluded that:

> *In any case we must expect very large densities for the central stars of the planetary nebulae, of the order of at least thousands of times that of the Sun. This fact, however, means that we are dealing with degenerate stars of the white dwarf type – a circumstance which again introduces a considerable uncertainty into the masses.*

It is truly striking just how right van Maanen turned out to be years later.

The first to show that planetary nebulas are actually expanding was Perrine. In 1930,[30] Perrine showed that there was a nice correlation between the magnitude of the velocity and the absolute (rather than apparent) size of the nebulas. He provided additional evidence to prove the claim that the planetary nebulas expand. The paper was not cited for many years, and eighteen years later Perrine repeated his arguments.[31]

Independently, but two years later, in 1931, Zanstra analyzed the conditions of the central star in PNs, and put forward the expansion hypothesis.[32] Zanstra was even able to determine the masses of several nebulas to be less than 0.01 $M_{\odot}$. Shortly after, he returned[33] to the problem and analyzed the expansion hypothesis vis-à-vis the rotation hypothesis. Zanstra assumed that the central star had a mass of 2–$4M_{\odot}$ and found that the velocities of the nebulas were all very small with respect to the velocity of escape from the stars, as if somehow the nebulas could barely manage to spread themselves out in this way. One of the crucial arguments was the fact that many spectral lines appeared doubled or very much broadened. Such an effect can

---

[28] Van Maanen, A., Ap. J. **77**, 186 (1933).

[29] Van Rhijn (1886–1960m) measured the number of stars in the Galaxy as a function of brightness and distance. The van Rhijn function in different directions in the galaxy is the result of these measurements.

[30] Perrine, C.D., AN **237**, 89 (1930).

[31] Perrrine, C.D., PA **57**, 432 (1949).

[32] Zanstra, H., Zeit. f. Astrophys. **2**, 329 (1931).

[33] Zanstra, H., MNRAS **57**, 324 (1931).

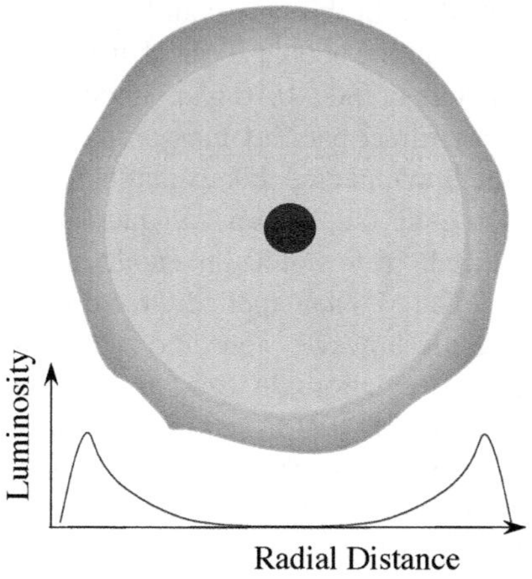

Fig. 7.4 Schematic view of a planetary nebula. The *lower curve* shows the total emission in the direction of the observer as a function of the angular distance. A spherical ring appears as a flat ring

arise due to rotation and/or expansion, and hence was not decisive in distinguishing between the two hypotheses.

On the other hand, cylindrical versus spherical symmetry is a much more critical argument, but it was not raised. As a matter of fact, in 1918, Curtis,[34] who prepared the first catalogue of PN shapes, remarked that the ring-shape hypothesis failed. His argument was statistical. If the PNs were flat and had a disk shape, we should expect to see some edge-on, and not mostly full frontal. Furthermore, the hypothesis of ellipsoidal shells of uniform thickness failed, as it did not explain the very faint central regions. This was as far as Curtis went. He classified the PNs into sphere-rings, ring shells, ellipsoidal shells, helical, and anomalous, which did not belong to any of the above. In a way, the confusing morphology just made the PNs more problematic.

Let us digress for a moment. The theory of nebulas excited by stars was developed in the years 1937 to 1945 by Menzel and Aller,[35] who worked out all the physical processes going on in gaseous nebulas. Einstein[36] developed the radiation theory for discrete states. The theory was generalized by Milne[37] to include continuous atomic states. From there, Aller and Menzel's implementation followed to yield the present day theory of nebulas.

[34] Curtis, H.D., Pub. Lick Obs. **13**, 55 (1918).

[35] Menzel, D.H., Ap. J. **85**, 330 (1937). This was the first paper. The last was Aller, L.H., & Menzel, D.H., Ap. J. **102**, 239 (1945), which was the 18th paper in the series.

[36] Einstein, A., Phys. Zeit. **18**, 121 (1917).

[37] Milne, E.A., Phil. Mag. **47**, 209 (1924); ibid. **47**, 547 (1925).

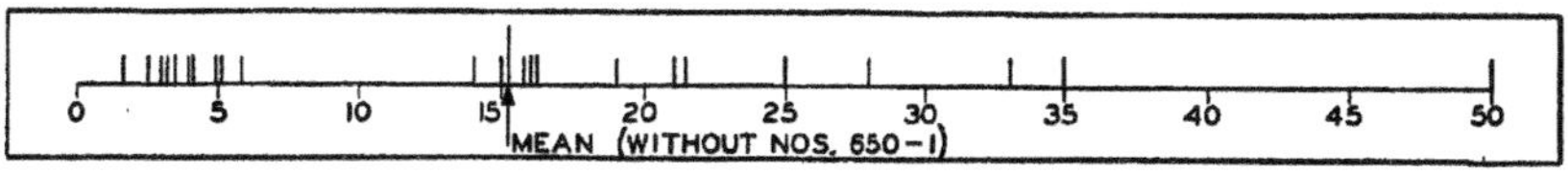

Fig. 7.5 The ages Whipple got for 29 planetary nebulas

It was only in 1938 that Whipple (1906–2004)[38] pointed out that, if Perrine and Zanstra were correct and the nebulas were expanding:

> *Quasi-ages for the best observed PNs can be determined from the linear dimension and rates of expansion by a method that is essentially the one used by Hubble in determining the age of the Crab nebula in Taurus.* Whipple also remarked that the roughly spherical symmetry of the nebulous shells, *as suggested by the ring structure, generally eliminates the need for a correction due to projection.*

Note the confusion between a flat 2D ring like a torus, and a 3D structure between two concentric spheres which appears as a ring (see Fig. 7.4). The ages Whipple obtained are shown in Fig. 7.5, from which we see that the oldest has an age of 50 000 yrs, while the mean age is about 20 000 yrs.

The beautiful planetary nebulas seen in the pictures contain a central very hot star. The hot star, with surface temperature in excess of 50 000 K, emits plenty of invisible UV radiation which is absorbed by the nebula. The nebula re-emits the absorbed radiation at many different wavelengths, and in particular in the visible. As a very significant part of the radiation of the star is in the UV, it was impossible before the era of telescopes placed on artificial satellites to measure the total luminosity of the stars directly from the ground, as UV radiation does not penetrate the Earth atmosphere.

The story of the nebulas is essentially one of energy conservation. They absorb high-energy photons and emit low-energy photons. When the idea was put forward by Slipher, it was not formulated using the idea of conservation. Slipher did not provide an equation, or a condition, and nor did he try to provide any observational evidence to show its correctness. But when one has the idea of using a conservation law to get numbers out, questions arise, e.g., whether all the photons are absorbed, whether the nebula covers the entire star, whether stellar photons escape through holes in the nebula, and so on.

In 1928, Plaskett (1865–1941)[39] suggested that the emission lines of the nebula might be created when free electrons in the nebula recombine with the ions and cascade down through the atomic levels. He stopped at this point, and did not ask how many free electrons there could be, or how come the electrons were free in the first place.

The hot UV photons, which are absorbed by the atmosphere of the Earth, are emitted by the neutral hydrogen atoms in the nebulas and excite them to higher

[38] Whipple, F.L., Harvard College Observatory Bulletin **908**, 17 (1938).

[39] Plaskett, H.H., Har. Ci. **335**, 1 (1928).

levels,[40] or even remove the electron completely and ionize the atom. The hotter the central star is, the stronger the UV radiation, and hence the stronger the resulting low energy hydrogen lines. The fundamental idea put forward by Zanstra (1894–1972m), published in a very important paper in 1931,[41] was to apply what is called today the conservation of radiant energy, and obtain the temperature of the central star. The problem Zanstra faced was that the transition probabilities between the levels, even for the simple hydrogen atom, had not yet been calculated at the time he wrote his paper, so he had to estimate them. Under these conditions, he was thus able to obtain only approximate numbers (e.g., a stellar surface temperature of over 20 000 K). The idea became a cornerstone of the nebula theory, and allowed a connection to be made between observed sizes, states, and spectra of the nebulas and properties of the central stars. The idea was crucial for subsequent discoveries. Another factor in the accuracy was the need to assume that the central star radiates like a black body, since there were no models for such hot stars. However, these were technical quibbles that could not diminish such a great idea.

In 1939, Vorontsov-Velyaminov[42] found that there exists a relation between the angular diameter of the nebula and its brightness, from which he could derive a connection between the physical diameter of the nebula and the brightness. This discovery, which was checked for nearby nebulas, made it possible to get the distance to more distant nebulas. As a matter of fact, Hubble had derived such a relation earlier. However, what Hubble did, and it was the right thing to do, was to relate the luminosity of the central star to the size of the nebula. The relation Vorontsov-Velyaminov got was similar to the one Hubble found years earlier, though the numbers were somewhat different.

By 1938, three different PN distance estimates based on different ideas and assumptions had been published: Zanstra's,[43] Vorontsov-Velyaminov's,[44] and Berman's.[45] The resulting distances differed by a factor of 2. This is a small factor in view of the crudeness of the methods, but still too large, because the luminosities derived from the distances differed by a factor of 4.

7.3 The Harman–Seaton Sequence

The Harman–Seaton sequence is an account of how all the pieces were brought together to create a beautiful tale, touching on the 'last journey' of a low mass star. The physical connection between the central star and the nebula, beside the star illuminating the nebula, was a mystery. A common idea at the beginning of

[40] The excitation is to levels ≥ 3, because the excitation to $n = 2$ leads only to scattering.

[41] Zanstra, H., Pub. Dom. Ap. Obs. **4**, 209 (1931); Zeit. f. Astrophysik **2**, 1 (1931).

[42] Vorontsov-Velyaminov, B., The Obs. **62**, 213 (1939).

[43] Zanstra, H., Zeit. f. Astrophys. **2**, 331 (1931).

[44] Vorontsov-Velyaminov, B., Poulkovo Obs. Circ. **21**, 29 (1937).

[45] Berman, L., Lick Obs. **18**, 73 (1937).

the 1930s (for example, Beals[46]) was that the central star might be an ex-nova. Gerasimovič[47] inferred from the surface temperature and the luminosity that, if the central star had a mass of $1M_{\odot}$, it must have a mean density of 5×10^8 g/cm^3, and hence *we cannot escape the conclusion* that these are collapsed stars. This comment prompted Milne[48] to suggest that nova outbursts were perhaps the transition from an extended low density star to a collapsed high density star.

In the early 1950s, Page and Greenstein[49] checked the old hypothesis that the size of the nebulas might be connected with the luminosity of the star, and claimed to find good agreement, essentially vindicating Hubble. But Wurm and Singer[50] and van de Hulst[51] criticized Page and Greenstein, claiming that what they did was nothing but *a check on an identical formula*, as van de Hulst put it, or that it was *completely unaccounted for*, as Wurm and Singer put it. The conclusion was that either not all ionizing photons are absorbed by the nebula, in which case the Zanstra idea would lead to underestimates of the temperature of the star, or there are not enough photons to ionize the entire nebula.

In 1952, Wurm and Singer suggested a way to overcome the uncertainty over the state of the nebula, and to carry out the calculation for several emission lines rather than just one, because the properties of the photons at two or more frequencies would be different. In this way, one is freed from the limiting assumption that all photons must be absorbed by the nebula. After performing the more accurate calculation, they discovered that, for many emission lines, the observed radiation was greater than the theoretical prediction, which was a new problem.

In 1956, Hattori and Yada[52] investigated the possible relation between the apparent luminosity of the central star and the angular extent of the PN. They followed a slightly modified version of Hubble's procedure and repeated his work on the PNs. In contrast to Hubble's 30 year old conclusion, they found no correlation whatsoever. A star could be bright with a small or large PN. They concluded that there might be a few small PN for which the idea was correct, but that for the majority of the nebulas, it fails.

In view of all the previous results, it is surprising to find the following description by von Weizsäcker in 1951:[53]

> *A planetary nebula would then be a special type of a giant in which the atmosphere happens to be transparent to visible light.*

It is not evident at all how von Weizsäcker reached his conclusion.

---

[46] Beals, C.S., Pub. Domin. Astrophys. Obs. IV, no. 17 (1930).

[47] Gerasimovič, B.P., Obs. **54**, 108 (1931).

[48] Milne, E.A., The Obs. **54**, 145 (1931).

[49] Page, T., & Greenstein, J.L., Ap. J. **114**, 98 (1951).

[50] Wurm, K., & Singer, O., Zeit. f. Astrophys. **30**, 153 (1952).

[51] van de Hulst, H.C., Ap. J. **115**, 331 (1952).

[52] Hattori, A., & Yada, B., PASJ **8**, 40 (1956).

[53] von Weizsäcker, C.F., Ap. J. **114**, 165 (1951).

In a beautiful, profound, and visionary analysis in 1957, Shklovskii (1916–1985[54]) put together all the elements needed to obtain the grand picture.[55] He started his review with:

> *The physical state of the planetary nebula has been investigated sufficiently well, but the main question of their origin and evolution still remains open.*

Here is Shklovskii's reasoning. Since we see how the nebulas expand to infinity, the unavoidable outcome is that with time the visibility of each PN will decrease until it fades away and can no longer be observed. If we see so many PNs today and each has a relatively short lifetime, it means that they are formed continuously at a relatively high rate. A simple calculation yields the rate at which PNs are born in the Galaxy as about one per year. As for the distance estimates to the PNs, Shklovskii claimed that *the well-known methods of Vorontsov-Velyaminov, Berman, and Camm cannot be considered as correct.* For example, the basic assumption underlying Vorontsov-Velyaminov's method was that:

> *[...] all planetary nebulas have the same luminosity, and this is not correct, because as a result of their expansion, the luminosities of the planetary nebulas decrease indefinitely.*

So Shklovskii developed a new method based on the physical mechanism responsible for the light emitted by the nebula. Shklovskii's result for the distance R of the nebula was $R = aM^{0.4}/\phi I^{0.2}$ where M is the mass of the star, ϕ the angular dimensions, and I the surface brightness. Here a is a known constant. The distances Shklovskii got for the PNs were in some cases a factor of 5 to 20 smaller. Shklovskii even ignored van Maanen's supposed parallax measurements, claiming that if van Maanen's distances were correct, the inferred mass of the central star must be over $10M_\odot$, which in his view was absurd. Shklovskii had independent estimates of the surface temperatures of the stars (made by others), and with his new distance estimates he could get the absolute luminosities of the central stars. Consequently, concluded Shklovskii, the central star must be an *overheated* white dwarf. After some tens of thousand of years, the nebula disappears and only a hot WD is observed. The nucleus, as Shklovskii called the central star, will cool and change gradually into a normal WD. Shklovskii identified the evolution process of low mass stars, and the fact that the PNs announce the gradual cooling of the star and its eventual extinction, as it sinks into the dark graveyard of the stars, the unobserved white dwarfs. PNs, declared Shklovskii, signal the last moments in the life of a star. As we observe rapid changes in the nebula, from which we learn the ages of the objects, accompanied by the observed changes in the hot star, we are actually witnessing rapid non-explosive changes in the hot star. As the distances found by Shklovskii were significantly smaller than those found by all previous methods, he also concluded that the density of PNs in the galaxy must be higher than had been predicted on the basis of previous numerical estimates. This in turn meant that they must be generated at a higher rate than previously estimated. Estimates for PN birth

---

[54] No lunar crater but Asteroid 2849 Shklovskii is named after him.

[55] Shklovskii, I.S., IAUS **3**, 83 (1957).

rates are 5–11 $\times 10^{-12}$ PN/yr,[56] while the independent birth rate of white dwarfs is 2×10^{-12} WD/yr.[57] The disagreement between the numbers is clear, and may be due to there being many unobserved WDs in binary systems.

After Shklovskii had set out the general framework, it had to be filled in with the right numbers:

> *The genetic connexion between planetary nebulae and white dwarfs suggests that some stars at a definite stage of their evolution detach a shell with zero velocity in the process of transformation into white dwarfs. This process would go for several tens of thousands of years. [...] the continuous process of PN formation is the most powerful supplier of gas to interstellar space.*

Several tens of solar masses per year return in this way from the stars to galactic space. As for the progenitor of the PN, Shklovskii hypothesized that it was, with high probability, a red giant of high luminosity.

Since Shklovskii did not know the mass of the central star, but considered that it had to be a white dwarf, he assumed that all central stars have the same mass. In this respect he traded Vorontsov-Velyaminov's assumption of constant luminosity for an assumption of constant mass. The difference between these two assumptions is that the luminosity varies by many orders of magnitude, while the mass could change only between 0.6–1.4$M_{\odot}$. As a matter of fact, luckily for Shklovskii, it turned out that the mass range was even smaller.

Vorontsov-Velyaminov did not accept his compatriot's criticism and responded,[58] but Shklovskii was ready with his reply,[59] and it did not help Vorontsov-Velyaminov much. The very poor astronomical method had to give way to the physical method devised by Shklovskii. Even Berman's method, which was based on galactic rotation, became old hat.

The basic uncertainty in Shklovskii's method was the assumption of constant mass for the central star. In 1960, Kohoutek[60] assumed that the difference between the luminosity of the star and that of the nebula was constant, and managed to improve Shklovskii's formula, so that the assumption about a fixed mass for the central star could be relaxed. In this way, Kohoutek was able to obtain the first clear correlation between the luminosity of the central star and the size of the nebula, as shown in Fig. 7.6 (left). The brighter the star, the smaller the nebula. Shklovskii's method was devised for optically thin nebulas, and Kohoutek[61] generalized it to optically thick ones.[62] In 1962, Kohoutek showed that his improved expression for the distance was actually replacing the power of I in Shklovskii's formula by 1/3 instead of 1/5, whereupon his expression yielded distances somewhere between those of Shklovskii and those of Vorontsov-Velyaminov.

---

[56] Cahn, J.H., & Wyatt, S.P., Ap. J. S. **210**, 319 (1976); Smith Jr, H., A & A **53**, 333 (1976).

[57] Weidemann, V., Ann. Rev. Ast. Astrophys. **6**, 351 (1968).

[58] Vorontsov-Velyaminov, B.A., A. J. (USSR) **33**, 809 (1956).

[59] Shklovskii, I.S., A. J. (USSR) **34**, 403 (1957).

[60] Kohoutek, L., BAICz **11**, 64 (1960).

[61] Kohoutek, L., BATCz **13**, 71 (1962).

[62] Optically thin means that radiation can get through the matter, while optically thick means that radiation cannot get through, being heavily absorbed by the matter.

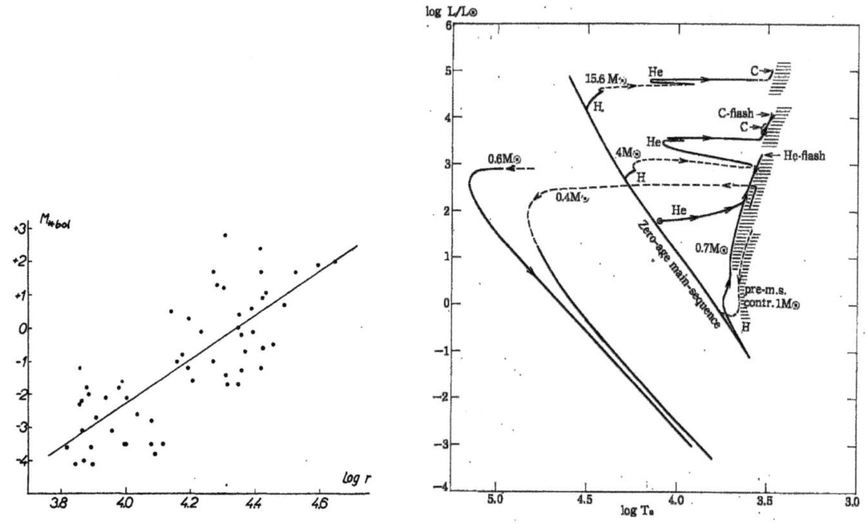

Fig. 7.6 *Left*: The log luminosity (*y* axis) versus log angular size of the nebula relation discovered by Kohoutek in 1960. Because of the particular way astronomers measure the luminosity, it increases downward. *Right*: The evolution of two low mass stars. The stars evolve to the left and then below the main sequence

Meanwhile, independently of the story of the PNs, the first calculations of the evolution of stars towards the white dwarf state were carried out in 1962 by Hayashi, Hoshi, and Sugimoto.[63] Their results are shown in Fig. 7.6 (right). The two models shown are a $0.6M_\odot$ star composed of elements heavier than helium, and a $0.4M_\odot$ population II [64] star without hydrogen or helium. The low mass star burnt its hydrogen, ignited the helium, and became a white dwarf. The interesting features of the evolution are first a contraction and heating at essentially constant luminosity, which depends only on the mass of the star, followed by a decrease in luminosity at almost constant radius. The constant luminosity is very close to Eddington's limiting luminosity, although not stated as such in the essay by Hayashi et al.

Chapter 9 of Hayashi et al. was entitled *Final Phase Toward White Dwarfs*, regarding which they declared:

> *We study in this chapter the final phase of evolution of stars, such as the remnants of the supernova or the less massive stars which have not experienced the flash phenomena, in which nuclear fuels have been completely exhausted or the temperature is too low to burn nuclear fuels.*

[63] Hayashi, C., Hoshi, R., & Sugimoto, D., Supl. Prog. Theo. Phys. (Japan) **22**, 1 (1962).

[64] Population II means the first generation of stars to form in the Galaxy, namely, the oldest stars. Population I are the young stars. The Sun belongs to population I.

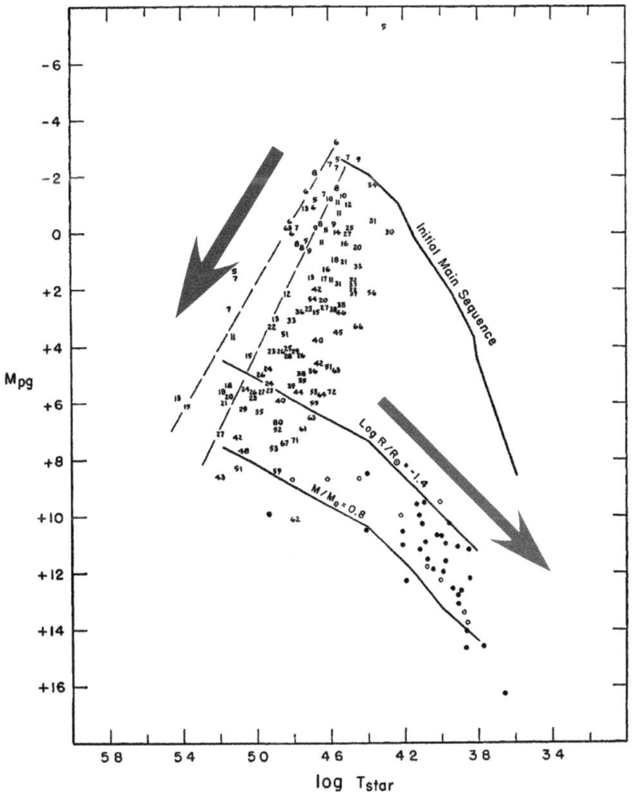

Fig. 7.7 The *blue arrow* marks the evolution of contracting and cooling stars. The *red arrow* designates the cooling of white dwarfs at practically constant radius. Based on the results of O'Dell 1963

For Hayashi et al. the only possible outcome of a supernova was a white dwarf. It is interesting to note that PNs were not mentioned in this chapter of their 1962 essay about stellar evolution.

In 1963, O'Dell[65] devised a new way to infer the temperature of the central star from the observed emission line of the nebula (a hydrogen line). The temperatures he obtained were systematically higher than those obtained by Berman. On the other hand, the mean sizes of the nebular shells were used to indicate the time scale of the changes occurring in the central star. What O'Dell discovered was that the central stars contract rapidly, and in 25 000 years (which is the average for all nebulas) shrink from $1R_\odot$ to $0.01R_\odot$, while the surface temperatures rise from 40 000 K to 150 000 K. O'Dell correctly identified what one was seeing:

> *These star + nebula systems represent the gravitational collapse phase of evolution of stars of about 1.2 solar masses as their nuclear fuel burning is completed. The stars, having*

[65] O'Dell, C.R., Ap. J. **138**, 67 (1963).

> *passed through the state with a PN, can account for a large fraction of the presently observed WDs.* He recognized that: *Many of the results of this study have been hypothesized previously by I.S. Shklovskii in 1956. Most of the qualitative features appear obvious to the evolutionary theoretician.* And predicted that: *Although many refinements of the methods used may be possible later, it is anticipated that the general features will remain the same.*

The basic results due to O'Dell are shown in Fig. 7.7, where we have added two arrows to indicate the evolution of the central star. At the beginning the central star contracts and cools. Contrary to everyday experience and unlike normal Eddington type stars, the stars contract and cool rather than heat up. The cooling and compression turn the matter into a degenerate medium, as Chandrasekhar had predicted years earlier. The evolution of the star at this phase is marked in the figure with the blue arrow. Once the star contracts to about the density of a white dwarf, the contraction continues but at a significantly slower pace. White dwarfs cool at almost a constant radius. This is reflected by the red arrow in the figure.

The final seminal paper appeared in 1964,[66] when Harman and Seaton repeated the analysis of the PNs. First, they attempted to get the best stellar temperature possible. So they followed Wurm and Singer and used the spectral lines which the latter had proven to be fully absorbed by the nebula, and hence constituted a reliable piece of information about the original spectrum of the star. In cases in which it was obvious that the nebula did not completely surround the star, they introduced a geometrical factor to correct for those stellar photons which escaped the nebula.

They classified the nebulas into two major classes:

(a) All photons at the frequency of helium absorption lines are absorbed, but only part of the photons in the hydrogen absorption lines escape.
(b) All photons are absorbed both in helium and hydrogen lines.

They admitted that the derived radii had poor accuracy, but the advantage was that they were completely independent of the rest of the data. From O'Dell, they found the connection between the radius of the nebula and the mass of the star.

With all this data, they began by producing two figures. In the first, they plotted the surface temperature as a function of the size of the nebula (see Fig. 7.8 left), while in the second, they plotted the luminosity of the central star as a function of the size of the nebula (see Fig. 7.8 right). There is no time axis. However, time is proportional to the size of the nebula. Hence the first figure shows that the temperature rises monotonically until it reaches a maximum constant value, at which it stays until the nebula fades away. At the same time, the second figure shows that the luminosity first increases and then decreases.

Having plotted these figures, Harman and Seaton placed the stars in the HR diagram of Fig. 7.9. They obtained the first well-defined demonstration of the evolution of the central stars of PNs. The nebula expands with a velocity of about 30 km/s, and it takes about 25 000 years for the nebula to disappear and the star to cool. They hypothesized that mass ejection might have continued during the phase of temperature rise and radius contraction, and attempted to explain the increase in luminosity in this way. They were wrong about this point. The rise in luminosity turned out to be

[66] Harman, R.F., & Seaton, M.J., Ap. J. **140**, 824 (1964).

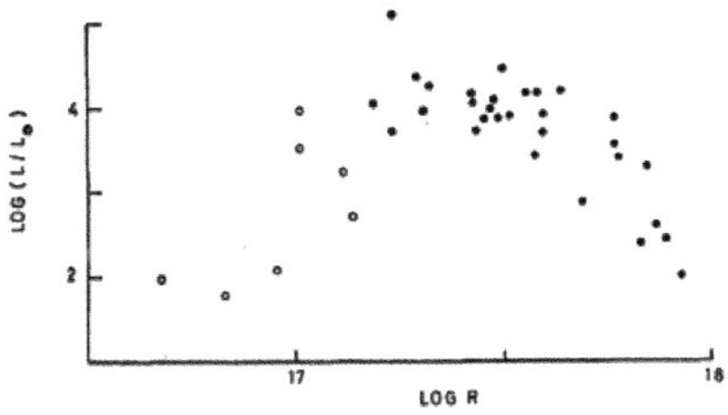

Fig. 7.8 *Left*: The star temperature in K against the radius of the nebula in cm. *Filled circles* are for class (a) and *open circles* are for class (b). *Right*: The luminosities in units of solar luminosities against the radii of the nebula. The meaning of the symbols is as before. Harman and Seaton 1964

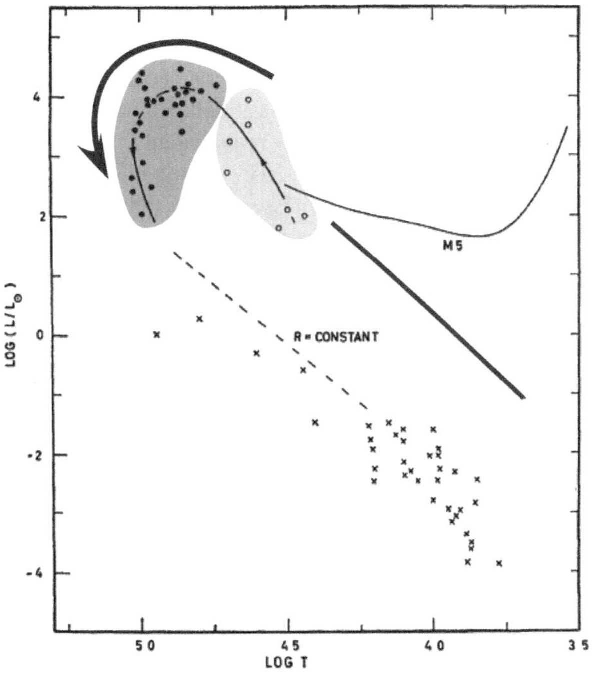

Fig. 7.9 The famous evidence for the Harman–Seaton curve. The *blue arrow* points in the direction of evolution of the star. The label MS refers to the horizontal branch (a phase in stellar evolution during which stars evolve at practically constant luminosity), and not to the main sequence. White dwarfs are marked with *crosses*. After Harman and Seaton 1964

an observational effect, and not a real one. Yet the results agreed with the theoretical results of Hayashi et al. (1962) for the evolution of a $1M_\odot$ star after the exhaustion of its nuclear fuel. The calculations by Hayashi et al. were made for 0.4 and 0.6 solar masses, and Harman and Seaton just extrapolated from the figure.

The paper ended with a sad note by Seaton:

My coauthor, Reginald Harman, was killed in a motorcycle accident while on his way to the Edinburgh meeting of the Royal Astronomical Society in September 1963.

7.4 Stellar Evolution to PN Formation

An amazing phenomenon was discovered by Paczynski[67] in 1971. Extensive calculations had shown that the luminosity of stellar models which developed cores of oxygen and carbon depended solely on the mass of the core, and not on the total stellar mass. This is like Milne's model, with a collapsed core in which the luminosity depends only on the mass of the dense core.[68] The implications are that all stars which satisfy this law must have the same fate, depending only on the mass of the core.

Stars with masses 0.8–$8M_{\odot}$ develop degenerate cores, and when the mass of the core reaches about $0.6M\odot$ (the exact mass depends on the not fully established physics of the mass loss), the envelope is removed and the degenerate core appears as the central star of a planetary nebula. The reason why the large range of stellar masses yields such a narrow range of masses for the PN is the Paczynski core mass–luminosity relation. But the physical reason for this relation is not yet understood. An estimate of the mass of the central star as a function of the progenitor's mass is shown in Fig. 7.10 (left).[69]

All stars start on the main sequence and become red giants as they consume their hydrogen in the core. Very high luminosities are typical of the red giant stage. The extremely powerful radiation exerts radiation pressure on the outer layers of the star, and gives rise to strong winds and mass loss. The processes of core collapse and envelope expansion accompanied by a continuous rise in luminosity give rise to the disintegration of the star from the outside. A competition develops between mass loss from the surface and the accelerating nuclear processes in the core. In the case of stars with original masses of less than $8M_{\odot}$, the mass loss wins, succeeding in stripping the core of its envelope before silicon ignition temperatures are reached in the core. If the core is less massive than the Chandrasekhar limit, it will not be able to continue to contract and raise the temperature, so it will contract, become degenerate, and start to cool.

The schematic track of a low mass star in the HR diagram is shown in Fig. 7.10 (right). The star starts on the main sequence and gradually converts hydrogen into

---

67 Paczynski, B., Acta Astr. **21**, 271 (1971).

68 The amazing relation was $L = 56250(M_{\text{core}} - 0.522)$ where the mass and the luminosity are given in solar units.

69 The data are from: Blöker, T., A & A **297**, 627 (1995); Iben, I., *Modern Problems of Stellar Evolution*, Ed. Wiebe (GEOS, Moscow, 1998), p. 52; Wassilidias, V., & Wood, P.R., Ap. J. **413**, 641 (1993); Weidemann, V., A & A **188**, 74 (1987); Milanova, Yu.V., & Kholtygin, A.F., Astron. Lett. **32**, 557 (2006).

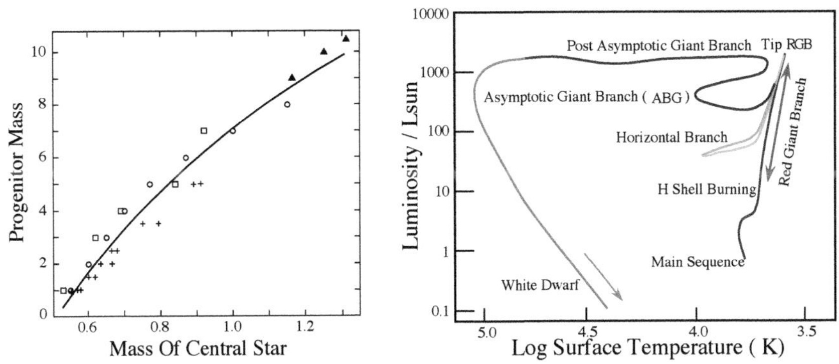

Fig. 7.10 *Left*: An estimated relation between the mass of the progenitor star and the mass of the central star. The data were collected by Milanova and Kholtygin 2006. *Right*: Schematic evolution of a low mass star from main sequence to white dwarf

helium. The initial evolution is characterized by an increase in luminosity and surface temperatures. The star burns gradually, and instead of the surface temperature increasing, it decreases due to the considerable expansion. As the hydrogen in the core is completely consumed, the burning begins in a shell.[70] The hydrogen shell moves out and the radius continues to increase. At the same time the core contracts, increasing its density and temperature. At the tip of the red giant branch, the temperature in the core reaches the helium ignition temperature, the core contraction is halted, and the star is pushed down to the horizontal branch, which is the location of helium burning.

As the helium is consumed in the core, the star starts its second climb to the tip of the red giant branch. When the helium in the core is consumed, a helium shell develops and the star has two burning shells, the old hydrogen one and the new helium one. The shells are unstable, and the star experiences a sequence of thermal pulses where the luminosity increases periodically. A typical example is shown in Fig. 7.11. The luminosity spikes at $10^9 L_\odot$ for a very short time, and in a periodic way. Various processes can take place during such a spike, for example, sporadic mixing. As the star moves to the asymptotic giant branch, it develops a strong wind, or superwind. The star loses the outer layer and reduces its mass to below the Chandrasekhar limit to become a planetary nebula. As the star is devoid of nuclear energy, it cools to become a white dwarf. The ejecta contains a bit of enhanced CNO due to the partial mixing, but most of the CNO remains in the buried white dwarf and never comes out.

[70] The burning develops into a shell because of the fact that the temperature decreases outward. Hence the rear part of the shell burns faster than the front, and in this way the shell thins down, until it becomes extremely thin and unstable. See Rakavy, G., & Shaviv, G., Ap. & S. S. **1**, 347 (1968).

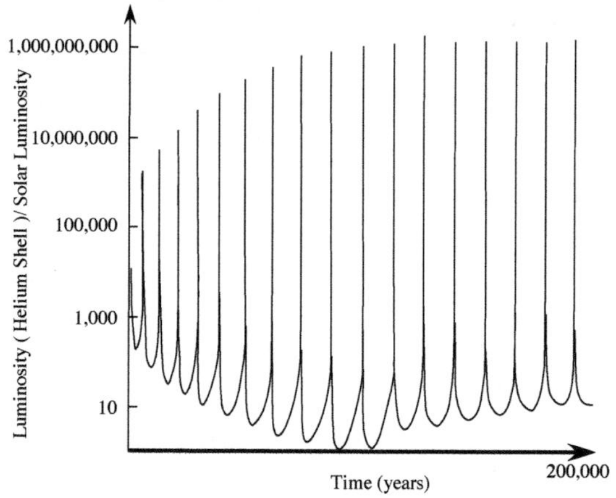

Fig. 7.11 Schematic view of the thermal pulses experienced by a low mass star on the asymptotic giant branch. The luminosity of the helium-burning shell reaches astronomical values over extremely short times. Calculation by Shaviv

7.5 Some of the Remaining Problems with the White Dwarfs

White dwarfs come in two main types: DA, which are hydrogen rich, and DB, which are helium rich. About 80% of WDs are type DA, the rest being type DB. There are WDs with peculiar compositions, but these comprise barely 1% of the total population. The DB white dwarfs appear with surface temperatures outside the range 30 000–45 000 K.

We do not know whether differences in the progenitor determine the class of the emerging WD, or whether all WDs form in the same way and something like mixing happens to change the composition of the outer layers as the WD cools.

7.6 Masses of the Central Stars

Recent results[71] obtained for the masses of the central stars in the PNs observed in the LMC are $m_{\text{central star}} = 0.65 \pm 0.07 M_{\odot}$, while the masses of these stars in PNs in the Milky Way are known with less accuracy. The reason is that the mass estimates depend on the luminosity of the star, and in order to know the luminosity, one has to measure it with an accuracy of 10% at least. But the uncertain distance gives rise to larger errors in the luminosity, and hence in the mass. However, if we observe objects in the LMC, the distance is so large that the exact location inside the LMC causes only a small error. For this reason, the distances and the masses of the central stars of the PNs are better known in the LMC than in the Milky Way. Still, the

[71] Villaver, E., Stanghellini, L., & Shaw, R., `astro-ph/0610079v1`, 2 October 2006.

Table 7.1 Modern astrometric and classical distances to selected planetary nebulas. All distances in lyrs

Name	NGC	Hipparcos	Previous estimates
Ring nebula	6720	2253	1000–4100
Helix nebula	7293	700	300–650
Dumbbell	6853	1213	850–1360

differences are minimal, indicating that the difference in heavy element abundance has no observable affect on the evolution of PNs and their central stars.[72]

Gesicki and Zijlstra[73] compared the masses of the central stars with those of WDs. The found that the mean mass of PN central stars is $0.61M_{\odot}$, while the mean mass of the field WDs is $0.58M_{\odot}$. If this difference is significant, it probably means that not all WDs undergo the PN phase. There are stars which do eject mass but somehow it is not observed (the times as discussed above disagree). So this is still a problem. The estimate is then that 30–50% of WDs may avoid the PN phase, in particular the low mass ones.

We should mention that a major uncertainty in all current calculations of stellar evolution regards the way to treat mass loss from stars, and in particular the way it depends on the general parameters of the star like luminosity, radius, and surface temperature. Only recently, the existence of a short period, lasting 100 to 300 years, of very strong winds and extensive mass loss was discovered. This so-called super-wind shows that all previous attempts to obtain highly simplified expressions for the mass loss contained a serious flaw.

7.7 Shaping the PN

The large variety of PN shapes, and in many cases cylindrical symmetry, triggered the hypothesis that at least a significant part of the central stars might be binary stars, and the shapes a consequence of the interplay between two winds blowing from the two stars. Until recently, only a handful of binary central stars were discovered.[74] Only 25 PNs have confirmed binaries. On the other hand, PN catalogues contain several thousand objects.

Could interacting winds between the binary system deform the nebula? A similar phenomenon also happens in supernovas (see Soker[75]).

[72] The abundance of the heavy elements in all stars of the LMC is 1/3 the value in our galaxy. This probably implies a different stellar history in the two neighboring galaxies.

[73] Gesicki, K., & Zijlstra, A.A., `astro-ph/0620v1`, 4 April 2007.

[74] Zijlstra, A.A., `astro-ph/0610558v1`, 18 October 2006; de Marco, O., IAU Sym. 234, Ed. Barlow & Mendez (2006).

[75] Soker, N., IAU Sym. 209, Ed. Dopita & Kwok, APS Conf. Ser. (2002).

7.8 A Few Present Day Challenges for Understanding PNs

- We still need to understand the formation of the nebulas and their shape in relation to the mass loss from the star. When does it start, how long does it last, and why does it stop? And in particular, how much hydrogen is left over on the star?
- What is the role of binaries? Does the binary help the star lose its outer layers? About 10% of the central stars are binaries, and about 10% more reveal a distant companion.
- Zanstra has found that the radiation pressure in PNs is very low. If we consider the momentum problem,[76] we find once again that the total momentum of the expanding nebula is larger by about a factor of 1000 than the radiation pressure can supply. So how come the expressions used for the mass loss depend on the luminosity of the star?
- Why are about 15–20% of the central stars hydrogen deficient? What separated them from the majority of central stars which still contain some hydrogen?
- Some PNs may have small hydrogen-deficient clumps. Consequently, one can conclude that mixing in the stars that formed the nebulas was not perfect.
- Could the dust and gas ejected by previous mass loss determine the shape, upon collision with the fast wind? Is it an environmental effect?
- There is still uncertainty regarding the mixing in the star before ejection and in the wind during the ejection. The possibility of clumps,[77] such as have recently been discovered, introduces a large error into the abundance determination (a factor of two is a plausible estimate for the error).

7.9 From the Vantage Point of Time

In an address to the American Association for the Advancement of Science, New York City, on 26 December 1916, Campbell described the two greatest problems of the Universe. The first problem will be discussed later. The second one was the *order of evolution of the stars*. As Campbell wrote:

> *While a strong case can be made out for the evolution of planetary nebulae into stars at the centers, with possible planets evolving around them, we must not conclude that all stars have been formed from planetary nebulae.*

We see in Fig. 7.12 the Cat's Eye planetary nebula with its complex expanding layers. The picture has changed since Campbell's address, from one of stellar birth to one of stellar death.

---

[76] Bujarrabal, V., Castro-Carrizo, A., Alcolea, J. & Sanchez Contreras, C., A & A **377**, 868 (2001).

[77] Alcolea, J., Neri, R., & Bujarrabal, V., `astro-ph/0701455v1`, 16 January 2007. They found an equatorial component that could be a flat disk expanding radially with a velocity proportional to the distance to the center. The kinetic age of this component is very similar to that of the two lobes. The small width of the lobe walls indicates an acceleration period of only 100–120 years. This is probably not a PN but a proto-PN, so one can see what happened at the beginning.

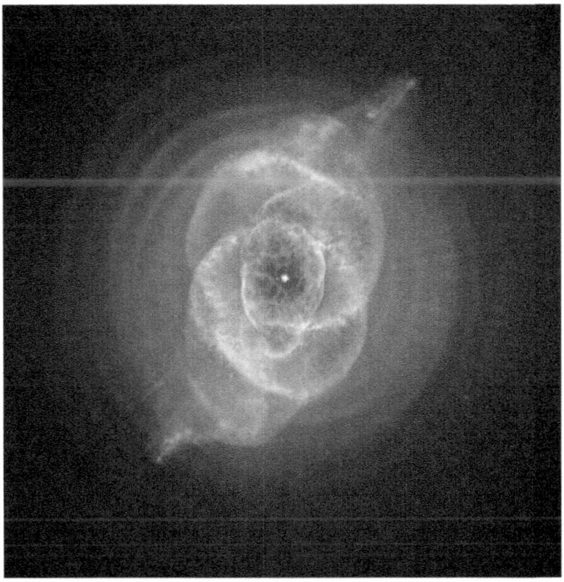

Fig. 7.12 The Cat's Eye planetary nebula. The complexity of the mass loss process and the inhomogeneities are evident. NASA, Hubble Space Telescope, Hubble Heritage Team

Chapter 8
The Life and Death of Massive Stars

8.1 New Physics: A Prelude

In 1940, Gamow and Schönberg[1] reviewed the suggestion by Baade and Zwicky[2] that stellar collapse might be triggered by an (unexplained) hypothesized formation of a large number of neutrons, and rejected it. Gamow and Schönberg pointed out correctly that, for a collapse to take place, a large amount of gravitational energy must be removed, otherwise the core would not be able to sink into a deeper gravitational potential well. Even if one calls for help from all known energy transport mechanisms, these would be unable to operate quickly enough to remove this energy from the collapsing core during the time of the collapse. The amount of energy which must be removed is to a good approximation the gravitational potential energy of the supposedly formed neutron star, and this is roughly[3] $E_{\mathrm{neut}} \sim E_{\odot}\,(R_{\odot}/R_{\mathrm{neut}})$ erg. Since the radius of a neutron star is about 10 km, the gravitational energy of a neutron star is about 7×10^4 times higher than the gravitational energy of the Sun, which is about 3×10^{50} erg. This, argued the authors, requires a new mechanism for speedy removal of the energy if an explosion is to follow. In the case of the Sun, the dynamic time scale is about 1000 seconds. So if the pressure support of the outer layers is removed for some reason, the Sun will collapse in 1000 seconds. The time scale varies inversely as the square root of the density ($\tau \propto 1/\sqrt{\rho}$). Hence, as the neutron star is about 10^{12} times denser than the Sun, it collapses in milliseconds! Gamow and Schönberg were generous and assumed that the collapse of a supernova would take a few days, since they considered the collapse time to be the time required by the signal of the collapse to propagate to the surface of the star and reach peak luminosity.[4] However, even with a time scale of

[1] Gamow, G., & Schönberg, M., Phys. Rev. **58**, 1117 (1940).

[2] Baade, W., & Zwicky, F., PNAS **20**, 259 (1934.

[3] As the initial state is highly extended, with a large radius, its gravitational energy is very small relative to the energy of the final state of a neutron star, and hence can be neglected.

[4] The discrepancy in the times is due to the fact that the progenitor is very extended and the outer layers have very low densities.

G. Shaviv, *The Life of Stars*, DOI 10.1007/978-3-642-02088-9_8,

days, the demand on the energy transport mechanisms was beyond the capabilities of the classical theory.

All energy transport mechanisms discussed so far have the property that the distance moved by the particles which carry the energy is short. In the case of radiative transfer, the distance the photon propagates inside the star is a few millimeters or less. After a few millimeters, the photon collides with an electron and generates a new photon which again moves a short distance, and so on. This process is slow, because the newly born photon can move in any direction, and not necessarily straight outward like the original photon. In the Sun it takes about 3×10^7 years until the great great ... grandchild of the original photon reaches the surface and escapes from the star. Other processes are a bit faster, but not sufficiently fast to allow an explosion. Consequently, Gamow and Schönberg realized that a new mechanism was required to allow a collapse.

Gamow and Schönberg assumed (but did not prove) that the collapse of the core leads to *rapid expansion of the outer layers and the tremendous increase of luminosity*. Stellar model calculations which showed this phenomenon were carried out only in the early 1950s. By this time the authors already knew that the luminosity of a nova flares by a factor of about 10^5, while that of a SN increases by a factor of 10^9. Further, it should be noted that no progenitor of a supernova had been observed at that time, so this was a lower limit to the energy released. As Gamow and Schönberg wrote:

> *The change of state by these stellar catastrophes strongly suggests that the process involved here is not connected with any instantaneous liberation of intra-nuclear energy due to some explosive reaction, but rather represents a rapid collapse of the entire stellar body as was first suggested by Milne.*[5]

In view of this fundamental problem, Gamow and Schönberg had a new and very imaginative idea that would revolutionize the theory of late phases of stellar evolution: removal of energy by means of neutrinos. At this time it was already well known that the neutrino shares the energy with the electron in a β decay. The neutrino escaped all detectors, and this indicated that it was hardly absorbed by matter. So if a neutrino formed in a stellar core, it would be able to escape from the star and remove energy without any problem. The basic idea was therefore that the very same particle which removes the energy in a laboratory β decay, also removes the energy in very dense and collapsing stars. The processes Gamow and Schönberg had in mind were:

$$\begin{aligned} {}^{Z}(\text{nucleus}) + \text{e} &\longrightarrow {}^{Z-1}(\text{nucleus}) + \text{neutrino}\,, \\ {}^{Z-1}(\text{nucleus}) &\longrightarrow {}^{Z}(\text{nucleus}) + \text{e} + \text{neutrino}\,, \end{aligned} \tag{8.1}$$

which are inverse β and β^- decays, respectively.[6] Despite the fact that no neutrino had yet been detected when the research was carried out, Gamow was confident that,

[5] Milne, E.A., The Obs. **54**, 145 (1931).

[6] Consider a β decay in the laboratory, viz., $(Z,A) \rightarrow (Z-1,A) + \text{e}^- + \nu$. The electron manages to get out because the space around has plenty of vacant states. As the density increases and the matter becomes degenerate, more and more states are occupied. When all states are occupied, only the

once created inside a dense star, the neutrino or antineutrino would have no problem escaping from the star. Even dense stars should be transparent to neutrinos! Hence, if energy is converted into neutrinos, fast cooling is guaranteed.

In the first paper published in Physical Review Letters, the reaction was written without distinguishing between the neutrino and the antineutrino. This was rectified in the second paper. Gamow and Schönberg described the process as *a negative energy source*. In the second paper they considered how a star could lose its pressure support due to fast cooling by neutrinos, leading to a collapse of the star. They did not distinguish between the nova and supernova phenomena, and assumed that both were triggered by the same mechanism.

However, Gamow and Schönberg noticed that the progenitor of a SN was never observed, while in some cases the nova progenitor was observed and the post-nova star was often observed. They also pointed out that:

> *The same is evidently true for the case of SNs, since the star found in the center of the Crab Nebula, and representing most probably the remainder of the galactic SN AD1054, shows the typical features of a very dense 'white dwarf'.*

In fact, this was one of the strangest stars in the sky, because it had no spectral lines and could not be classified. But for Gamow and Schönberg, it was sufficient evidence to claim that the SN leaves a white dwarf remnant. This is a bit confusing, because what they hypothesized was the collapse to a neutron star, so they should have identified the remnant of SN 1054 as a neutron star (which it is, in fact).

Schönberg called the unthinkable and fantastic process the URCA process. It was in this paper that the term URCA was invented, named after the famous Casino de URCA in the URCA[7] quarter of Rio de Janeiro, where all the punters eventually lost their money (see Fig. 8.1). Mario Schönberg (1914–1990) was a Jewish–Brazilian from São Paulo. He was awarded a J.S. Guggenheim grant which enabled him to carry out research in the USA and collaborate with Gamow and Chandrasekhar.

The process was summarized by:

$$(A,Z) + \mathrm{e}^- \rightleftharpoons (A,Z-1)\ .$$

The neutrino and antineutrino were omitted from the equilibrium because they escape from the star. This is exactly the reason why the process cannot be in strict equilibrium.[8] The process is cyclic, and the only outcome is removal of energy from the core into outer space. The calculation by Gamow and Schönberg showed that the strong cooling would induce a collapse of the star within half an hour, once

uppermost energy states remain vacant, and electrons are driven to high energy states. This process can continue only until the energy of the electron on the outside is so high that the favored energy state is when the electron moves in the opposite direction, namely $(Z-1,A) + \mathrm{e}^- \rightarrow (Z,A) + \bar{\nu}$, where $\bar{\nu}$ is an antineutrino.

[7] When I asked around, no one in Rio de Janeiro could tell me the meaning of the name URCA.

[8] If a certain participant of the reaction is not available for the back-reaction, e.g., it is removed by some process like leakage, then the reaction cannot be in equilibrium. This particular case of partial equilibrium is different because energy is absorbed from the medium. It is not a closed system.

Fig. 8.1 The URCA Casino in the URCA quarter, Rio de Janeiro circa 1930–40, when it was the center of the cultural life of Rio de Janeiro

the nucleus reached the right density to undergo inverse β-decay and absorb the electron.

The concrete examples given by Gamow and Schönberg were:

$$^{56}\mathrm{Fe} + \mathrm{e}^- \to {}^{56}\mathrm{Mn} + \text{antineutrino}\,,$$

and the inverse

$$^{56}\mathrm{Mn} \to {}^{56}\mathrm{Fe} + \mathrm{e}^- + \text{neutrino}\,.$$

They concluded that:

> *It must be emphasized that, while the neutrinos are still considered as highly hypothetical particles because of the failure of all efforts made to detect them, the phenomena of which we are making use in our considerations are supported by direct experimental evidence.*

What they were referring to was the part of the energy released in β decays that disappears in the laboratory.

Recall that this was shortly after the Oppenheimer and Volkov paper about the collapse to a neutron star. Still, Gamow and Schönberg did not make the connection between the collapsed star and what Baade and Zwicky had in mind.

8.2 The Creation of the Weak Interaction Theory

In the mid-1950s, Cowan and Reines attempted to discover the elusive neutrino. For this purpose, they needed a strong source of neutrinos. Now nuclear reactors are in fact strong sources of neutrinos, due to the many β decays of fission products. They started their experiment in Hanford, where nuclear reactors were used to produce plutonium, and soon moved to the Savannah River Plant near Augusta, Georgia, USA, where they had a better shield against cosmic rays and a large nuclear reactor. The shield was 11 meters away from the reactor and 12 meters underground. Success came in 1956,[9] when they detected neutrinos for the first time.

The radioactive decays of the fission products gave rise to a neutrino flux of about 10^{12}–10^{13} neutrinos/sec/cm^2. The researchers predicted that the cross-section would be 6×10^{-44} cm^2, and measured it to be 6.3×10^{-44} cm^2. The meaning of such a result is that the neutrino does not 'see' the real size of the nucleus, but rather a particle with an effective radius 10^7 times smaller. This explains why it is so difficult to detect the neutrino. It hardly notices the nucleus at all. This experiment, which proved Pauli's 25 year old hypothesis, won Frederick Reines the 1995 Nobel Prize. By that time Cowan (1919–1974) was dead and could not share the prize. For some obscure reason the Nobel Prize committee had to wait until the last moment, because Reines (1918–1998) passed away just three years later. As for Pauli, Reines and Cowan telegraphed him about the discovery, to which he replied with: *Thanks for the message. Everything comes to him who knows how to wait.* However, the moral was already in Pauli's famous letter to the conference on radioactivity (see p. 306) where Pauli wrote that *only those who wager can win.* Two years later Pauli passed away.

The discovery of the neutrino stimulated the theoreticians to come up with a comprehensive theory of weak interactions. Soon Feynman (1918–1988)[10] and Gell-Mann,[11,12] Sudarshan and Marshak (1916–1992),[13] Gell-Mann,[14] and Sakurai (1933–1982)[15] developed the general theory of weak interactions. The essence of the theory is that the weak interaction, which controls the β decay, behaves like a combination of an electric field denoted by V, and a magnetic field denoted by A, which gives a $V - A$ interaction when put together. The V part is essentially the Fermi interaction, while the A part is the Gamow–Teller interaction. These theoretical developments allowed the calculation of various complex processes involving weak interactions, such as neutrino interactions with matter. A detailed account would carry us too far astray, and so we shall have to leave the discussion on the

[9] Cowan, C.L., Reines, Jr., F., Harrison, F.B., Kruse, H.W., & McGuire, A.D., Science **124**, 103 (1956); Reines, F., & Cowan, C.L., Nature **178**, 446 (1956).

[10] Nobel Prize for Physics in 1965.

[11] Nobel Prize for Physics in 1969.

[12] Feynman, R.P., & Gell-Mann, M., Phys. Rev. **109**, 193 (1958).

[13] Sudarshan, E.C.G., & Marshak, R.E., Phys. Rev. **109**, 1860 (1958).

[14] Gell-Mann, M., Phys. Rev. **111**, 362 (1958).

[15] Sakurai, J.J., Nuovo Cimento **7**, 649 (1958).

theory of weak interactions and discuss only the huge impact it had on our understanding of stellar evolution.

Once the theory of weak interactions became available, there was an avalanche of suggested neutrino processes. Fowler and Hoyle reviewed all the neutrino processes which the new theory of the weak interactions would allow, including of course the URCA process. Suggestions for ν important processes came from Bethe,[16] Pontecorvo, Gandel'man and Pinaev, Chiu and Morrison, Gell-Mann, Chiu and Stabler, Chiu, Ritus, Matinyan and Tsilosani, Stothers and Chiu, Sampson, Stothers, Adams, Ruderman and Woo, Rosenberg, and Pinaev.

In 1964, Fowler and Hoyle[17] published one of the most important papers for stellar evolution. They realized that neutrino cooling is crucial in late stellar phases. This would turn out to be the dominant mode of energy loss from a star in the late phases of stellar evolution and in the collapse. The effect of the neutrinos on the evolution of the star is enormous. Up until this phase, the energy released in the core had to diffuse through the entire star to reach the surface, at which point it was radiated out into space. In contradistinction, the neutrinos escape directly from the core of the star without any interaction with the outer layers. If we could see the neutrinos, we would see the core of the star, not the relatively cool outer surface. Following Gamow's reasoning, Fowler and Hoyle realized that, unless there is a way to remove the energy from the collapsing core, there is a good chance that the energy will build up to the point that it would halt the collapse. Fowler and Hoyle examined all thirteen suggested processes and concluded that the process $e^+ + e^- \to \nu + \bar{\nu}$, named pair annihilation and suggested by Chiu and Morrison, was the most important one for massive stars.

Detailed evolution calculations carried out by Rakavy and Shaviv in 1967[18] confirmed Fowler and Hoyle's prediction that the neutrino energy losses start to dominate the evolution as soon as the temperature reaches 10^8 K, and accelerate, but do not induce the collapse.

8.3 The Evolution of Massive Stars

The theory of the evolution of massive stars is complicated, and has not yet been fully established, because several key ingredients are missing from the physical

[16] Bethe, H., Phys. Rev. **55**, 434 (1939); Pontecorvo, B., Soviet Phys. J.E.T.P. **18**, 1148 (1959); Gandel'man, G.M., & Pinaev, V.S., Soviet Phys. J.E.T.P. **10**, 764 (1960); Chiu, H.Y., & Morrison, P. PRL **5**, 573 (1960); Gell-Mann, M., PRL **6**, 70 (1961); Chiu, H.Y., & Stabler, R., Phys. Rev. **122**, 1317 (1961); Chiu, H.Y., Ann. Phys. **15**, 1 (1961); ibid. **16**, 312 (1961); Phys. Rev. **123**, 1040 (1961); Ap. J. **137**, 343 (1963); Ritus, V.I., Soviet Phys. J.E.T.P. **14**, 915 (1962); Matinyan, S.G., & Tsilosani, N.N., Soviet Phys. J.E.T.P. **14**, 1195 (1962); Stothers, R., & Chiu, H.Y., Ap. J. **135**, 963 (1962); Sampson, D.H., Ap. J. **135**, 261 (1962); Stothers, R., Ap. J. **137**, 770 (1963); Adams, J.B., Ruderman, M.A., & Woo, C.H., Phys. Rev. **129**, 1383 (1963); Rosenberg, L., Phys. Rev. **129**, 2786 (1963); Pinaev, V.S., Soviet Phys. J.E.T.P. **18**, 377 (1964).

[17] Fowler, W.A., & Hoyle, F., Ap. J. S. **9**, 201 (1964).

[18] Rakavy, G., & Shaviv, G., Ap. J. **148**, 803 (1967).

theory needed to model such stars. We saw that, along the main sequence, the luminosity increases as a high power of the mass ($L \sim M^3$). On the other hand, the maximum luminosity of a star, the Eddington luminosity, is proportional to the mass. Hence, as the mass increases, so does the luminosity, and soon the Eddington luminosity is reached at approximately $100M_\odot$. At this point the luminosity becomes so powerful as to shed mass from the surface of the star in the form of a wind. The mass loss in the wind can be prohibitive, so that a very massive star may lose over half of its mass while still burning hydrogen in the core, leading to significant consequences for its evolution.[19] Comparison between observation and theoretical values of mass loss predicted by the present theory of radiation-driven mass loss reveals a discrepancy of about a factor of two.[20]

Mixing and convective overshooting comprise another area of unsolved problems. For many years, it was assumed that the boundary between a convective and a radiative zone was sharp. All the fluid motions inside a convective region were assumed to stop abruptly at the boundary. The boundary was defined as the region where radiation could carry the entire energy flux, and no fluid motions were possible, or needed. If the fluid tried to break into the radiative region, it gave rise to forces which reacted on it and brought it to a halt. Indeed, Roxburgh[21] and Saslaw and Schwarzschild[22] argued that the diffuseness of the boundary between the region where turbulent currents carry the energy and the radiative zone where the radiative flux transfers the energy would be quite sharp and so narrow that its width could be completely neglected (see also the textbook by Schwarzschild[23]). The concept changed in 1973, when Shaviv and Salpeter[24] showed that this is not the case, and that turbulent currents penetrate extensively beyond the calculated borderline between the two regions. This implies a continuous feeding of synthesized elements into the radiative region and smoothing of compositional differences across what used to be the theoretical borderline. The treatment of Shaviv and Salpeter was rather simple and only meant to demonstrate the existence and the extent of the phenomenon. In the absence of a reliable theory of convection, all that has been done since then is to parametrize the effect using parameters chosen ad hoc. Chiosi et al.[25] compared the calculated evolution of massive stars with observations and concluded that models with extensive 'overshooting', as the phenomenon of deep penetration of convective currents into the radiative zone is called, agree better with observations than models with sharp boundaries.

---

[19] de Loore, C., de Greve, J.P., & Lamers, H.J.G.L.M., A & A **61**, 251 (1977); de Loore, C., de Greve, J.P., & Vanbeveren, D., A & A S **34**, 363 (1978); de Loore, C., IAUS **83**, 313 (1979); Chiosi, C., Nasi, E., & Sreenivasan, S.R., A & A **62**, 103 (1978); Chiosi, C., Nasi, E., Bertelli, G., A & A **74**, 62 (1979); Chiosi, C., & Maeder, A., Ann. Rev. Astron. Astrophys. **24**, 329 (1986); de Jager, C., Nieuwenhuijzen, H., & van der Hucht, K.A., A & A S **72**, 259 (1988).

[20] Leitherer, C., & Lamers, H.J.G.L., SSRv. **66**, 153 (1933).

[21] Roxburgh, I., MNRAS **130**, 315 (1965).

[22] Saslaw, W.C., & Schwarzschild, M., Ap. J. **142**, 1468 (1965).

[23] Schwarzschild, M., *Structure and Evolution of the Stars*, Princeton University Press, 1958.

[24] Shaviv, G., & Salpeter, E.E., Ap. J. **184**, 191 (1973).

[25] Chiosi, C., and 5 other authors, SSRev. **66**, 421 (1993).

As the stars expand due to their non-homogeneity and the surface cools, the convective currents become stronger until they reach a point where the pressure exerted by the convective currents (known as turbulent pressure) becomes appreciable, and even so strong as to induce mass loss (on top of the radiation pressure). Consequently, stars in this region experience even greater mass loss and can remain in this region for only a short time. The region is often called the de Jager limit.[26]

The evolution of massive stars off the main sequence takes place almost at constant luminosity. So they expand and become red giants or supergiants, keeping their high luminosity. But as the star has a large radius, the surface gravity is small, and as the luminosity does not change, one may expect a still higher mass loss. A further complication is the observation that the wind appears to be clumpy.[27] These are not direct observations of the wind, but in order to explain the structure of the spectral lines, some phenomenological model is required to describe the effect of clumps. At the present time, no theory predicts the formation of such clumps.

Due to the high central temperatures, massive stars operate via the CNO cycle, which is very sensitive to the temperature. As a consequence, the core is fully convective. The theory of convection and mixing by convective currents is problematic, even in the laboratory. The problem with stellar convection is that the range of parameters under which convection appears in the laboratory and in stars is vastly different. For example, the flow can be characterized by the Reynolds number, which is the ratio of the inertial forces to the viscous forces. The first act to keep the motion going, while the second act to stop it. The Reynolds number for winds around buildings in the Earth's atmosphere is about 10^7, while it is higher than 10^{13} for the convective currents in massive stars. This difference is very meaningful, since completely different flow regimes are established. The maximum Reynolds number in a wind tunnel is about 10^5, well below the interesting regime for stars. To this fact, one must also add stellar rotation. Massive stars are observed to be rapid rotators, at least at the surface. Owing to the lack of observations of rotational behavior inside the star and the non-existence of a credible theory to calculate the evolution of rotation, we really do not know much about this feature of movements inside the star, let alone how they evolve with time. At the present time, it is not known to what extent any of the above depends on the composition or amount of heavy elements in the star. And finally, the role of magnetic fields is far from clear. To appreciate the complexity on the one hand and the wide range of possibilities on the other, consider the present situation in the Sun, where measurements of helioseismology have provided information about the internal rotation of the Sun. No such information yet exists for other stars. Solar observations have revealed a set of surprising and unexplained phenomena concerning rotation, magnetic fields, and mixing.

All modeling of the late phases of evolution and nucleosynthesis are plagued with uncertainty due to our poor knowledge of the following two nuclear reactions: the formation of oxygen, $^{12}\mathrm{C}+\alpha \rightarrow {}^{16}\mathrm{O}+\gamma$ and the source of neutrons $^{22}\mathrm{Ne}+\alpha \rightarrow {}^{25}\mathrm{Mg}+\mathrm{n}$. Moreover, this same uncertainty affects the structure of the star in all the

[26] Chiosi, C., & Maeder, A., Ann. Rev. Astron. Astrophys. **24**, 329 (1986).

[27] Crowther, P.A., Dessart, L., Hillier, D.J., Abbott, J.B., & Fullerton, A.W., A & A **392**, 653 (2002).

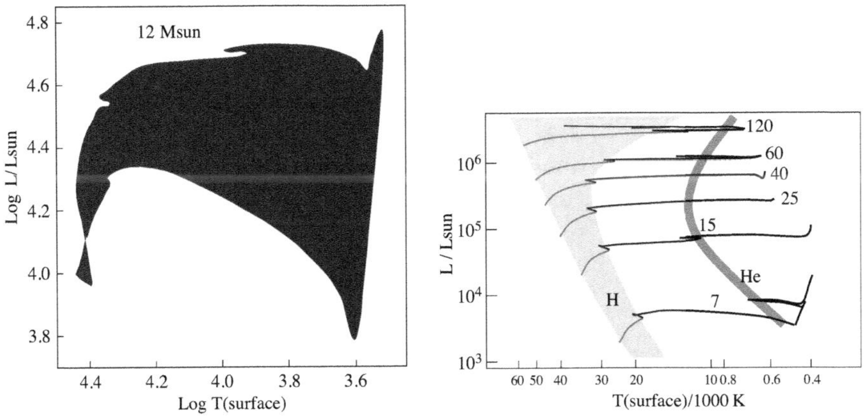

Fig. 8.2 *Left*: The effect of rotation on the evolution of a $12M_\odot$ star under various possible assumptions about the law of rotation. Based on Langer and Heger 1998. *Right*: The evolution of massive stars with low heavy element abundances. The *pink* part of the curve corresponds to hydrogen burning in the core, while *green* corresponds to helium burning. The numbers near the curves are the masses in solar masses. The star may move to the right or to the left when situated on the luminosity plateau. (Based on the Geneva grids of stellar evolution models without rotation)

subsequent phases. The effect of the uncertainty regarding rotational motions and how they are distributed inside the star can be seen in Fig. 8.2 (left), which is based on Langer and Heger,[28] who calculated the same model under different assumptions about the rotation of the star. The blue area spans the zone in the HR diagram where different evolutionary paths will pass for different assumptions about the rotation law and its behavior. In other words, depending on the rotation of the star, a $12M_\odot$ can be found anywhere in that blue region.

In view of all the above uncertainties, the following description of the evolution of a massive star is provisional. Theoretical tracks of various non-rotating massive stars in the HR diagram are shown in Fig. 8.2 (right). The evolution of a non-rotating massive star may be quite dull from the point of view of external appearances, as the star starts from the main sequence, hardly changes its luminosity, and only alters its surface temperatures. The internal structure may be quite different, but there are no simple external signs.

[28] Langer, N., & Heger, A., ASPC **131**, 76 (1998).

Table 8.1 Estimates of SN prevalence per τ years per galaxy

τ years	Estimator
612	Zwicky 1938[a]
359	Zwicky 1941[b]
105	Pakovskii 1961[c]
40	Katgert & Oort 1967[d]
316	Barbon 1968[e]
92	Rosino & Tullio 1974[f]
197	Tamman 1974[g]

[a] Zwicky, F., Ap. J. **88**, 529 (1938).
[b] Zwicky, F., Ap. J. **96**, 28 (1942).
[c] Pakovskii, Yu.P., Ast. Zh. **38**, 656 (1961).
[d] Katgert, P., & Oort, J.H., Bull. Ast. Inst. Ned. **19**, 239 (1967).
[e] Barbon, R., A. J. **83**, 1016 (1968).
[f] Rosino, L., & Tullio, G.D., *SN and SN Remnants*, Dordrecht Holland, Boston (1974) p. 19.
[g] Tamman, G.A., *SN and SN Remnants*, Dordrecht Holland, Boston (1974) p. 155.

8.4 The Explosive Climax: Properties of the 'Classical' Supernovas

SNs are observed to appear, and they exhibit a large variety of phenomena. The first to classify SNs was Minkowski in 1941[29] (see also Sect. 5.34). At his disposal were the spectra of just 14 supernovas. Nine objects were classified as Type I and five as Type II. The prototype for Type I SN was SN IC 4182,[30] which later became the model for explaining the light curve with radioactive ^{254}Cf (see Sect. 8.11). The classification was rather coarse, and contained just a short description of the spectra. Type I showed no hydrogen lines, not even faint signs of them, but Type II did show various lines of hydrogen as well as other lines belonging to several elements.

In 1964, Zwicky extended the classification,[31] realizing that Minkowski's simple classification failed to describe the richness of phenomena exhibited by SNs. Zwicky made several general statements about SNs which are in principle correct even today:

- SNs appear in all types of galaxies.
- The luminosity of SNs at maximum is comparable with that of a bright galaxy.
- There is about one SN per bright galaxy every thousand years.
- The frequency of SNs in some galaxies is significantly higher than the average. Zwicky interpreted this observation (wrongly by what we know today) by clai-

[29] Minkowski, R., PASP **53**, 224 (1941).

[30] IC stands for the Index Catalogue of nebulas, first published by Dreyer in 1895 as a supplement to the NGC.

[31] Zwicky, F., Ann. Rev. Astron. **27**, 300 (1964).

Table 8.2 Cappellaro et al. 1993 derivation of SN mean occurrence in various types of galaxy

Type of galaxy	Case I[a]	Case II[b]
E, S0	555	666
S0a, Sa	–	175
Sab, Sb	81	142
Sbc, Sc	79	81
Scd, Im	99	110
Others	244	103

[a] Sandage, A., & Tamman, G.A., *Revised Shapley–Ames Catalog of Bright Galaxies*, Carnegie Institution, Washington, 1981.
[b] de Vaucouleurs, G., de Vaucouleurs, A., and Corwin, H.G., *Second Reference Catalogue of Bright Galaxies*, Univ. Texas Press, Austin (1976).

ming that *the stellar and the material compositions of otherwise similar galaxies, may be different.*

Early estimates of SN prevalence in galaxies are shown in Table 8.1. The estimates were made when the Hubble constant was estimated at 100 km/sec/Mpc. Today the Hubble constant is estimated at 70 km/sec/Mpc, and consequently the inferred frequency of SNs is correspondingly lower.[32] However, more statistics have been accumulated in recent years, and astronomers can distinguish between the frequencies of SNs in different classes of galaxies, as well as correct the estimate for galaxies in which no supernova has been observed in recent years. For example, Table 8.2 summarizes the results of Cappellaro et al. 1993.[33]

Zwicky noticed that SNs appear preferentially in the outskirts of the galaxy, not in the core. Until 1954, Zwicky was practically the only astronomer to monitor galaxies and search for SNs. The importance of SNs was recognized in 1957 with the publication of Burbidge, Burbidge, Fowler, and Hoyle,[34] and many groups all over the world started organized searches for SNs in our own galaxy but mostly in other galaxies. The energetic Zwicky even succeeded in establishing a special committee of the International Astronomical Union to search for these objects.

While only 54 supernovas were discovered between 1885 and 1956, the extensive search yielded very encouraging results: 60 SNs were discovered just between 1956 and 1962, all of them in other galaxies. Of course, more and better telescopes were becoming available, and astronomers were soon availed of more substantial data. With the much larger sample of SNs to hand, Zwicky was able to refine the

[32] The Hubble constant enters the statistics of SNs because one has to estimate the number of galaxies per unit volume. One does not observe all galaxies, but corrects for those that are unobserved.

[33] Cappellaro, E., and 5 other authors, A & A **268**, 472 (1993).

[34] Burbidge, E.M., Burbidge, G.R., Fowler, W.A., Hoyle, F., Rev. Mod. Phys. **29**, 547 (1957).

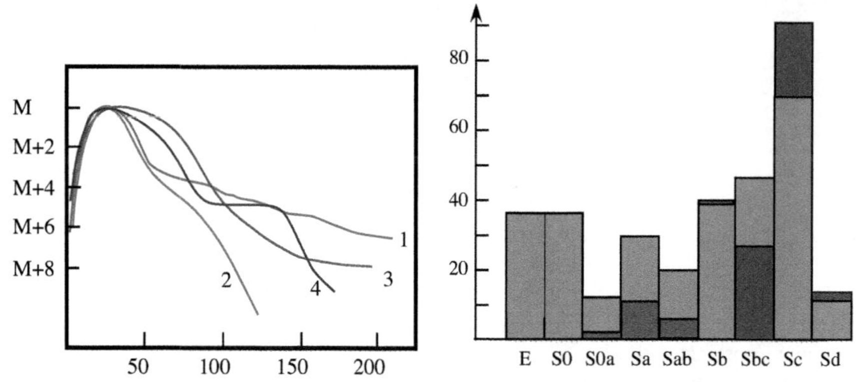

Fig. 8.3 *Left*: The light curves which characterize Zwicky's SN classes, after Zwicky 1964. The time is given in days, and the brightness relative to the peak apparent brightness. *Right*: The distribution of SN types among different types of galaxies, based on Barbon et al 1999. *Green* refers to Type I and *red* to Type II SN

classification and essentially split the Type II SN into Types II, III, IV, and V. The differences were mainly in the shape of the light curve (see Fig. 8.3 left).

Since a SN is such a violent phenomenon, one would expect no errors to occur in identifications. But this is not the case. Sometimes the SN is very strange and does not fit any type, or very far away so that it is not easy to observe, so misidentifications do indeed happen. One of the classic examples is Zwicky's class V supernova. The only member in this class was SN 1961V, and it remained the sole object of this type for many years. The SN erupted in 1961 in the Sc type galaxy NGC1058, and exhibited an extraordinarily long and erratic light curve, with maxima occurring in it from time to time. It was a matter of pure chance, but this particular object (see Fig. 8.4) was observed before the eruption took place in 1937, as a faint object showing fluctuating light. After the eruption, it remained at maximum brightness for about four months before beginning its decay, something that is highly atypical for an SN. The expansion velocities were relatively modest, just 2000 km/s, in contrast to the 5000 km/s and more exhibited by 'normal' SNs. Soon Bertola[35] and Zwicky himself noted the similarity in the spectra to that exhibited by nova after maximum light. Later, Branch and Greenstein[36] discovered that the composition was identical to that of the Sun, except that it was hydrogen-deficient. Branch and Greenstein also recognized that the spectrum was similar to that of a nova, as was suggested by Bertola.

Finally, Oke and Searle[37] concluded that: *The object perhaps should not even be considered to be a supernova.* However, some 12 years later, it was still considered

[35] Bertola, F., Contr. Asiago no. 142 (1963).

[36] Branch, D., & Greenstein, J.L., Ap. J. **167**, 89 (1971).

[37] Oke, J.B., & Searle, L., Ann. Rev. Astron. Astrophys. **12**, 315 (1974). The review was dedicated to the memory of Zwicky, who had passed away a year before.

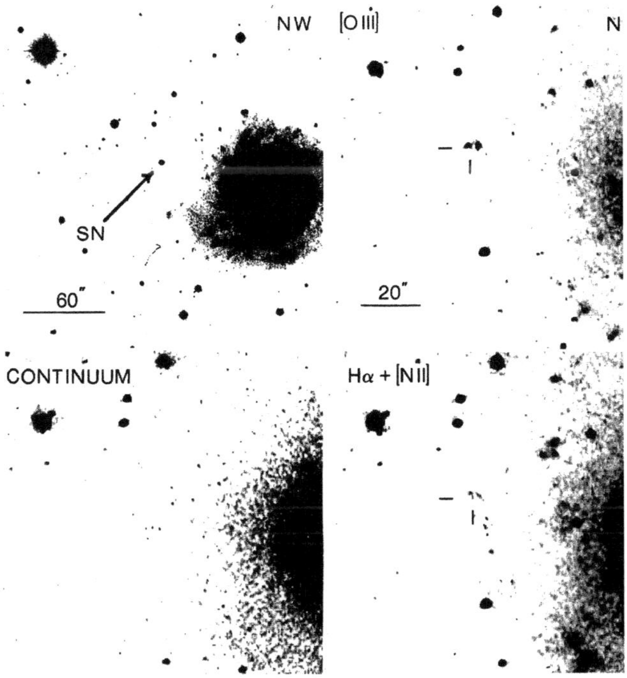

Fig. 8.4 A supernova impostor? Top left is Zwicky's original photograph of 1961V (1964). The other three images were made by Fesen in 1983, through three different filters. From Fesen 1985

as a peculiar SN (Fesen[38]). In 1988, Cowan et al.[39] discovered radio signals from possible remnants of the SN, and in 2002, Van Dyk et al.[40] seemed to recover SN 1961V in the archives of the Hubble Space Telescope. The SN was similar but not identical to a Type II SN. The peculiarity provoked the hypothesis[41] that the progenitor was a very massive star of $2000M_{\odot}$ and radius $100R_{\odot}$. The issue of whether the object is a genuine SN or a 'supernova impostor'[42] has not yet been settled.

Gradually it became clear that SNs do not separate into two distinct classes. There are significant distinctions between objects classified as the same type that are not understood. All these SNs are classified as 'peculiar' objects. For example, of Oke and Searle's list of supernovas, about 6 out of a total of 32 were listed as peculiar.

---

[38] Fesen, R.A., Ap. J. **297**, L29 (1985).

[39] Cowan, J.J., Henry, R.B.C., & Branch, D., *Supernova Remnants and the Interstellar Medium*, Proc. IAU Colloq. **101** (1987), Roger & Landecker, Cambridge University Press (1988) p. 23.

[40] Van Dyk, S.D., Fillipenko, A.V., & Li, W., Ap. J. **114**, 700 (2002).

[41] Utrobin, V.P., Ap. S. S. **98**, 115 (1984).

[42] Humphreys, R.M., *The Fate of the Most Massive Stars*, ASP Conference Series, **332** (2004) Humphreys & Stanek, Astronomical Society of the Pacific (2005) p. 93.

Table 8.3 Prevalence of SN types in various galaxies

Galaxy	Years
E, S0	666
S0a, Sa	175
Sab, Sb	143
Sbc, Sc	81
Scd, Im	110
Others	103

In 1982, Shklovskii[43] suggested a classification based on a physical principle rather than the observed morphology of the SNs. The problem with such a classification is that it is not based on proven observed physical properties, but on a suggested model for SN explosion. Since only two triggering mechanisms were considered as well established at that time, just two classes were proposed by Shklovskii:

- Type I SN is a degenerate core collapse of a star without the hydrogen layer.
- Type II SN is a core collapse of a star that still possesses its hydrogen layers.

This classification is synonymous with the statement that Type I SNs arise from low mass stars while Type II SNs arise from massive stars. In 1983, Shklovskii himself found it necessary to expand his classification. It appears today that Shklovskii's classification is too simple to describe the variety of possibilities that are actually observed.

The old idea that stars differ from one another only in the total mass has to be abandoned. Parameters which were so far considered to play a minor role, like rotation and the way different parts of a star rotate, magnetic fields, mixing on small and large scales, heavy element abundance, etc., affect the fine details of the SN to the point that all classifications have so far failed. The fact is that supernovas do not want to be put in boxes!

Recent statistics for the frequency of SNs in different types of galaxy are shown in Fig. 8.3 (right). On the basis of Barbon et al.,[44] the Type II SNs have a preference for spiral galaxies, while Type I SNs appear in all galaxies in almost equal numbers. From Table 8.3, we see that SNs in elliptical galaxies (which would be Type I SNs) are rare. On the other hand, both types of SN are relatively frequent in spiral galaxies. Independent information about the character of galaxies suggests that today new stars are generally born in spiral galaxies, while the general color of elliptical galaxies indicates the widespread presence of low-mass stars in these galaxies. This may therefore be a hint that Type I SNs occur in old low mass stars while Type II SNs occur in young massive stars.

Observations of Type I SNs as a function of distance shows that the nearby Type I SNs have a high C/O ratio, while the more distant Type I SNs have a low C/O

[43] Shklovskii, I.S., Soviet Ast. Lett. **8**, 188 (1982); ibid. **9**, 250 (1982); Nature **304**, 513 (1983).

[44] Barbon, R., Buondf, V., Cappellaro, E., & Turatto, M., A & A S **139**, 531 (1999).

Table 8.4 Spectroscopic characteristics of SN types

	Ia	Ib	Ic	II
Strong H lines	No	No	No	Yes
Strong Si lines	Yes	No	No	No
Strong He lines	No	Yes	Maybe	Early on
Location	E0–Sc	Spiral arms	Spiral arms	Spiral arms
Duration	30–40 d	30–40 days	30–40 days	9–150 days
Typical maximum visual luminosity in $10^9 L_\odot$	5.5	7.2	7.2	2–5

ratio. Apparently, the abundance of the heavy elements has an effect on the properties of the SN, but it is not clear why this should be so. A general summary of the spectroscopic properties of SNs is given in Table 8.4. Several points are worth noticing. Type I SNs are the brighter ones and contain no hydrogen at all. Furthermore, Type I SNs have a shorter peak luminosity duration. And last but not least, all peak luminosities are the same order, although the trigger is certainly different.

8.5 Supernova Theory

A supernova is actually the transition of a star from a 'normal' extended state to the state of a collapsed star. In Fig. 8.5, we plot the binding energy of a $1M_\odot$ star during the different phases of stellar evolution. Consider the star as one gigantic system which can be in several states. Stellar evolution can be characterized as a perpetual decrease in the binding energy by removal of energy from the star. The transition from the main sequence phase (like the Sun) to the giant state takes place gradually. But the transition from the giant state to the collapsed state cannot be made continuously. In the case of stars less massive than $8M_\odot$, the transition is quasi-continuous but still drastic, as the star ejects the outer layers in a continuous non-explosive mode. However, in the more massive stars, the transition is explosive. It is like a phase change from an electron-supported star to a neutron-supported star. The latent heat of the transition is the energy of the explosion.

There is no known and established state between the white dwarf and the neutron star, or between the neutron star and the black hole. There were some speculations about possible states between the neutron star and the black hole, but these have never been confirmed. Itoh,[45] Collins and Perrey,[46] Brecher and Carporaso,[47] and

[45] Itoh, N., Prog. Theor. Phys. **44**, 291 (1970).

[46] Collins, J.C., & Perrey, M.J., PRL **34**, 1353 (1975).

[47] Brecher, K., & Carporaso, G., Nature **259**, 377 (1976). The authors called their object *an obese neutron star*, but in principle it is another phase of nuclear matter.

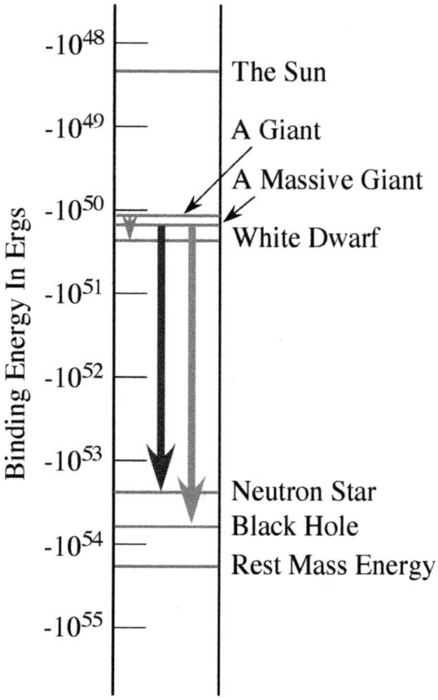

Fig. 8.5 A supernova is a transition from the giant phase to a collapsed phase. The figure depicts the binding energies of a $1M_{\odot}$ star at different phases. *Blue* and *green arrows* mark the supernova explosion with a remnant, a neutron star, or a black hole. The *small orange arrow* marks the transition to a white dwarf, or a planetary nebula. There is no known state between that of the white dwarf and the neutron star

Burrows[48] suggested, for example, a collapsed quark[49] star, which may have a mass of up to $5M_{\odot}$. But so far there has been no observational evidence that such configurations actually exist, or that stars could 'jump' into such configurations. Hence, the only possible bound states a star can 'jump' into (to borrow the quantum language) are neutron stars and black holes. The third possibility is a complete disruption of the system.

All transitions are accompanied by an energy release. These are the large arrows in Fig. 8.5. If no remnant is left, the total energy released can be significantly less.

We divide the discussion of SNs into three basic parts: the trigger, the explosion, and the light curve. While the triggers of SNs were suggested many years ago and the general belief is that they are the correct ones and probably exhaustive, the

[48] Burrows, A., PRL **44**, 1640 (1980).

[49] According to the grand unification theory, every nucleon is composed of three quarks, each with a fractional charge of $-2/3$ or $+1/3$ that of the electron. When the density is high and the mean energy of the particle in the gravitational field is correspondingly high, the energy can reach the binding energy of the nucleons themselves, and disintegrate them into their constituents to make a soup of quarks.

consequences of these triggers and what follows their action are not clear, despite many years of very intensive research. The reason is a lack of knowledge in certain essential parts of the required physics. The light curve on the other hand is quite nicely predicted by the theory, although with a non-trivial number of ad hoc assumptions. Hence, we find it appropriate to separate between what appears today as established (the trigger and probably the light curve) and what is still under intense research, without a final conclusion.

8.6 The Trigger

The idea that nuclei disintegrate at high temperatures is due to Sterne[50] who was the first to implement Fowler's *Statistical Mechanics*[51] to the problem of the statistical nuclear equilibrium in 1931. Sterne was a student of Fowler and applied his theory of statistical mechanics to physical chemistry, so it was natural for him to apply it also to stellar matter. Sterne was the first to write down the equations describing nuclear equilibrium, equations that are still in use today.

In the conclusion to the paper, he provided a table giving the composition of matter in nuclear equilibrium as a function of temperature (see Table 6.1). It shows how, at low temperatures, iron is the most stable and dominant component of matter. As the temperature reaches 3×10^9 K all the iron disintegrates, first into helium and later into hydrogen. Sterne carried out this calculation a year after the neutron was discovered. However, it was not clear at that time whether the neutron could be considered as composed of a proton and an electron. Sterne assumed this to be the case, so that he could treat the proton and the electron *as the ultimate sorts of particle out of which nuclei are made*. It is interesting to note that Sterne discussed the fact that the conversion of hydrogen into iron releases 8×10^{18} ergs per gram, and suggested this to be the energy source of stars. But while Sterne discussed the energy gain in the transformation of hydrogen into iron, he did not discuss the reverse reaction (although he did actually calculate it, as can be seen from the table), let alone the energy absorbed in the process and the consequences of this energy sink.

Six years after Sterne's publication, Öpik attempted to explain the abundances of the elements. He criticized[52] (partly correctly but mostly incorrectly) Sterne's idea of equilibrium between the elements. Öpik claimed that the neutron was more stable than hydrogen and hence that the last form of matter had to be neutrons. Öpik was wrong, because he missed the point that the neutron is stable/unstable depending on the environment. When the density is low, the neutron is unstable, while at high density the neutron is stable. On the other hand, Öpik realized that the dissociation of iron absorbs energy and reduces the adiabatic constant γ_{ad} to below 4/3. Consequently, claimed Öpik, once the dissociation starts, the star becomes unstable and

[50] Sterne, T.E., MNRAS **93**, 736 (1933).

[51] Fowler, R.H., *Statistical Mechanics*, Cambridge Press. The Macmillan Comp. (1929).

[52] Öpik, E., Publ. Tartu Obs. **30**, 3 (1938); ibid. **30**, 4 (1939).

must collapse. But the idea of stellar collapse appeared to Öpik untenable. Despite the fact that the neutron had already been known for several years, Öpik made an error in the equation for the dissociation of iron, and for this reason his numbers were not correct. He wrote $^{56}Fe_{26} \rightarrow 14^4He + 2e^-$, while the correct dissociation equation is $^{56}Fe_{26} \rightarrow 13^4He + 4n$.

By assuming that the star could be represented by Eddington's model, Öpik reached the conclusion that only stars more massive than $19M_\odot$ could reach the state of collapse, i.e., dissociation was at a magnitude sufficient to cause a collapse.[53] Öpik calculated the energy released and found that *collapse thus amounts to only a negligible decrease in radius. No real collapse thus takes place from nuclear dissociation.* In retrospect, Öpik missed the formation of a neutron star, the existence of which had already been hypothesized. It is not clear why Öpik assumed the Eddington stellar model, which was applicable only to homogeneous stars, while he wrote in the paper that the stars were not mixed and hence were non-homogeneous.

After producing some incorrect arguments, Öpik concluded that:

> *There is no escape from the conclusion that nuclear dissociation as well as pure neutron cores cannot play an appreciable role in the energy balance, stability, and structure of actual stars.*

Öpik also considered the formation of electron–positron pairs from radiation, and concluded that *it may be an important process in the core.* Öpik noticed that Sterne's calculations concerning the equilibrium of transmutations, as given in Table 6.1, were carried out for a total density of 10 g/cm^3, neglecting the radiation, and as Öpik put it:

> *It is an example of mathematical abstraction which disregards physical realities.*

Again he missed the point. Sterne gave the example to show how iron disintegrates when the conditions change. Unfortunately, he chose an absurdly low density. As regards Öpik's comment, it turns out that astrophysical reality is even more imaginative than the imagination of one of the most avant-garde astrophysicists.

Even in 1939, Tolman[54] searched for the possible mechanisms for nova (but not supernova) eruption. He carried out a mathematical stability analysis of stars, and got for the first time what is known today to be the accurate condition for the dynamic stability of stars, namely that the adiabatic constant γ_{ad} must be greater than 4/3.[55] It was the violation of this condition that would be found over twenty years

---

[53] Later Öpik corrected the number to claim that collapse could not occur for stars smaller than $200M_\odot$. Today we know that collapse takes place at much smaller masses.

[54] Tolman, R.C., Ap. J. **90**, 568 (1939).

[55] The adiabatic constant is the logarithmic change in the pressure with respect to a logarithmic change in the density under adiabatic conditions, i.e., conditions where no heat enters or leaves the system. [Mathematically, $\gamma_{\text{ad}} = (\rho/P)\partial P/\partial\rho$ at constant entropy.] If we attempt to compress the matter, that is, increase the density, the pressure rises. When a star is compressed, the gravitational force increases and compresses the star even more. However, upon compression, the pressure of the gas increases as well. If the pressure of the gas increases more than the gravitational pressure, the star is stable and vice versa. A simple calculation shows that, as the star contracts, the gravitational force increases as the density to the power 4/3. Consequently, if the increase of the pressure upon

later to be the trigger for supernovas. Since the models Tolman checked were simple Eddington-type models of ideal gases and radiation, he could not find any mechanism for getting γ_{ad} below 4/3, and consequently rejected it as a possible trigger:

> *[...] since a constant value of γ_{ad} less than 4/3 would be quite improbable for actual stellar material.*

As no mechanism was found to violate this condition, other ideas were floated. Tolman classified the ideas into three categories, claiming that a star could become a nova:

(a) as a direct result of a collision with some other astronomical body,
(b) as a consequence of some sudden alteration in the rate of internal energy generation,
(c) as a consequence of some kind of internal instability which can suddenly make a large amount of energy available.

In the same year, 1939, Whipple[56] suggested a solution along the lines of Tolman's first possibility:

> *The hypothesis that a nova may originate from a collision of two stars would long have been favored except for the theoretical rarity of such collisions.*

The estimates by Luyten[57] and Jeans[58] gave one stellar collision per galaxy per 10^7 to 10^{13} years. In view of the rarity of stellar collisions,[59] Pickering and Nolke[60] even suggested that a collision of stars with bodies of asteroidal or planetary size could be sufficient to trigger the explosion.

Whipple discovered that the SN statistics show a strong inclination to explode in or close to the spiral arms or concentrations of stars (even before Zwicky discovered it, and made a point of this fact). Supernova occurrences appeared to be spread out roughly like the total luminosity of the galaxy, with a tendency to avoid the cores

contraction is to overcome gravitation, it must increase faster than the 4/3 power of the density ($\rho^{4/3}$). Any process under equilibrium, for example $A+B+Q \rightleftharpoons C$, where Q is the binding energy of C, gives rise to $\gamma_{ad} < 4/3$, and threatens the stability of the star. The reason is simple. As the gas is compressed, the balance is shifted to the right. Q, which appears on the left, is the kinetic energy of the particles, or in other words the thermal energy or the pressure of the gas (pressure is energy per unit volume). Thus, if the balance shifts to the right, internal energy (which in this particular case is the kinetic energy of the nuclei) is converted into the binding energy of C and is eliminated from the resistance to gravity. Consequently, the star collapses. This turns out to be the trigger for supernova collapse, but not the trigger for nova explosions.

56 Whipple, F.L. PNAS **25**, 118 (1939).

57 Luyten, H.A., **85**, 73 (1923).

58 Jeans, J.H., *Astronomy and Cosmogony*, Cambridge University Press (1929) p. 319.

59 As we look at the many stars in the sky, we may wonder how frequently they collide. However, consider two baseball balls tossed anywhere in continental USA. The probability of collision between two stars is similar to the probability that the two balls collide within the USA, because the radius of the stars is so much smaller than their mean separation.

60 Pickering, W.H., & Nolke, F., in *Handbuch der Astrohysik*, Band VI, Springer-Verlag, Berlin, 1932.

of galaxies. Whipple wrote that Hubble[61] had reached the same conclusion for the nova in Andromeda. But in the abstract to the paper, Hubble said that:

> *Novae are most frequent in the nuclear region, and in a general way, the distribution follows that of the luminosity in the nebula.*

Whipple repeated the calculation of stellar collisions and found that, in the extreme case, he could predict one SN per year per galaxy.

Now came the energy argument. Whipple assumed that the energy release in an SN was up to 4×10^{49} erg. On the other hand, two stars of one solar mass moving with typical velocities in the galaxy have relative kinetic energy of about 10^{48} erg. Despite the apparent discrepancy, Whipple reached the conclusion that *the collision hypothesis deserves consideration.* Whipple was worried to what extent the kinetic energy could be converted into light which we see, and with what efficiency. Whipple could perhaps justify spending time on this hypothesis, because his estimate for the energy release in a SN was at least a factor of a 10 000 too low.

Other outlandish ideas were suggested. For example, Cernuschi (1907–1999)[62] suggested the fission of a tremendously heavy nucleus (Z of about 10 000!) which could be formed in the star and eventually give rise to a chain reaction which explodes the star. As he claimed in his Physical Review Letter:

> *It is very difficult to imagine how a supernova could result from the transformation of an ordinary star into a neutron star [...] Moreover, it is very simple to see that the process of formation of a neutron core can never produce an explosion as required to explain the appearance of a supernova.*

We can only attest that reality in astrophysics does indeed frequently go well beyond the bounds of imagination, and ask whether it is in fact so simple to show that neutron stars cannot form in an explosion? It is perhaps because it was so easy to show that no proof was provided in the letter. And that is a pity, because if a valid proof had been given, it would have saved many astrophysicists a lot of time.

Russell[63] calculated that the frequency of collisions was much too small to account for the frequency of observed nova eruptions, and in this way eliminated Tolman's first possibility. Tolman could not find a proper justification for his second mechanism (change in the rate of nuclear reactions), and so he was left with the third possibility, although he was unable to provide any real example. As a matter of fact, this possibility was also raised by Unsöld[64] when he discussed convection in the Sun. In an appendix to his paper on solar convection, and without any relation to the paper itself, Unsöld discussed a completely different subject, namely the mechanism of nova eruptions. He already realized in 1930 that $\gamma_{\rm ad}$, or in his notation $\bar{\kappa}$, must be greater than 4/3 for stability. Unsöld's idea was that an abrupt formation of the convective zone might trigger such an explosion. Unsöld even calculated

---

[61] Hubble, E., Ap. J. **69**, 103 (1929).

[62] Cernuschi, F., PRL **56**, 120 (1939). Note that Cernuschi has two PRLs with exactly the same volume and page of the journal.

[63] Russell, H.N., Dugan, R.S., & Stewart, J.Q., *Astronomy*, Ginn & Co, p. 789. The book was first published in 1926 and underwent subsequent updates over the years.

[64] Unsöld, A., Zeit. f. Astrophys. **1**, 138 (1930).

the total energy released, which he found to agree with the total energy released in nova eruptions. Tolman criticized this idea, saying that the convective zone would very quickly become stable and that the effect of γ_{ad} would then disappear. Finally, Tolman reached the idea that the fluid inside the star is not in a minimum energy state, so some sudden perturbation could drive the star 'beyond the edge', whence it would explode. However, apart from this vague idea, no details were given.

In 1948, Schatzman[65] suggested that unstable thermonuclear reactions in white dwarfs might be the basic mechanism for supernovas. Schatzman hypothesized the existence of some instability which would lead to the nuclear explosion. We already know that the extreme sensitivity of the nuclear reactions is the source of stability of stars. Here, however, the situation was different. The matter in white dwarfs is degenerate, which means that the pressure of the gas hardly depends on the temperature. Thus, the sensitivity of the nuclear reactions to temperature becomes a destabilizing mechanism rather than a stabilizing agent. The instability should give rise to mixing, and it is this mixed layer which becomes unstable. However, no details were given and there was no explanation as to why the mixed layer could suddenly become unstable while the unmixed layer would remain stable.

A year later Schatzman was more specific,[66] when he proposed for the first time that (a) the eruption of novas might be due to a nuclear explosion and (b) the outer layers could be ejected by means of a strong shock, created by the nuclear explosion, which propagates outward. No details of the nuclear reactions were given. In 1951, Schatzman went further with this idea,[67] suggesting the nuclear reaction $^3\mathrm{He} + {}^3\mathrm{He} \rightarrow {}^4\mathrm{He} + 2\mathrm{p}$ as the nova detonator. He hypothesized a detonation in the outermost layers of the surface of a white dwarf, with a subsequent shock propagating outward and blowing away the outer layers.

In 1959, several years after the pp chain had been completely worked out by Salpeter and Fowler, Gryzinski[68] noticed that (a) at low temperatures, the pp chain converts mostly hydrogen into ^{3}He, and (b) the decay time of ^{7}Be is 55 days, which agrees nicely with the light curve decay of SN IC4178 (the one which exhibits a beautiful exponential decay, discussed later when we come to consider the light curve). So Gryzinski suggested combining several ideas as follows. A very low mass star, say a fraction of a solar mass, might generate a star with large amounts of ^{3}He. The ^{3}He could then give rise to a thermonuclear explosion and generate plenty of ^{7}Be, the decay of which would power the light curve. Stars with masses less than a solar mass, according to this hypothesis, would yield Type I SNs, while more massive stars would yield Type II SNs.

There were many problems with this idea. Let us mention just one. The amount of ^{3}He in solar mass stars and more massive is very small. There is no way that such a small amount of ^{3}He could release enough energy to trigger an explosion. Furthermore, the lifetime of low mass stars is extremely long, longer than the age

[65] Schatzman, E., The Obs. **68**, 66 (1948).

[66] Schatzman, E., An. Ap. **12**, 281 (1949).

[67] Schatzman, E., An. Ap. **14**, 294 (1951).

[68] Gryzinski, M., Phys. Rev. **115**, 1087 (1959).

of the Galaxy, so these stars would only explode, if at all, in the future. The paper was published in the respectable Physical Review, without even a comment by the referee. As Gryzinski wrote at the end of his paper:

> *The suggested theory of supernovas is in many respects similar to Schatzman's theory of the explosion of nova stars.*

The only difference was that Gryzinski took Schatzman's nova theory and applied it to supernovas.

The present day standard models for the two types of SN were discovered in 1960 by Hoyle and Fowler.[69] These ideas continue to be the fundamental models for triggering SNs of both types. The first idea was to recognize the *sudden fusion of nuclear fuel as the source of the energy of SN explosion*. The basic argument was that the nuclear fuel could potentially supply the required energy to disrupt the star. However, the energy had to be supplied sufficiently fast. Stars use nuclear fuel and do not explode. The conditions should be such that ignition occurs very quickly, so that a large amount of energy is released so fast that the star cannot cope with it. Is there such a fuel? Hydrogen fusion cannot do it, because when hydrogen converts into helium, two protons must decay into neutrons, and that involves going through two β decays. Hence, the shortest time is the β decay time which is at least a few minutes long. There is no way to accelerate the explosion more than the β decay times.

Hoyle and Fowler thus converged upon the following solution: the available fuel had to be carbon–carbon or oxygen–oxygen reactions, because no β decays are involved in this fusion.[70] However, this was not sufficient. The ignition of the fuel had to be under such conditions that the energy release is unrestrained by the expansion of the star as a result of the heat poured into the volume. Mestel had already discovered in 1952[71] that the ignition of nuclear fuel is explosive when the matter is degenerate. The pressure of degenerate matter hardly depends on the temperature, and hence, when the fuel ignites and raises the temperature, the pressure does not change, so it does not cause expansion and cooling. These are just the required conditions for explosion.

Mestel essentially discovered the nova explosion mechanism, and it was this mechanism that Hoyle and Fowler rediscovered. Hence, they suggested the ignition of nuclear fuel on white dwarfs. They calculated that $0.1M_\odot$ of carbon would supply sufficient energy to generate the 10^{50} erg required to blow up the star. But as Hoyle and Fowler pointed out explicitly, the identification of the potentially explosive agent does not explain the cause or the mechanism of the explosion, and nor does it imply that the mechanism actually works. The ignition of carbon and oxygen on the surface or inside white dwarfs is still the standard model for type I SNs even

[69] Hoyle, F., & Fowler, W.A., Ap. J. **132**, 565 (1960).

[70] The fusion of hydrogen to helium differs substantially from the following steps, e.g., helium fusion to carbon, or carbon to magnesium, etc., in that hydrogen fusion requires the conversion of protons into neutrons, and this conversion requires the action of the weak force. None of the other fusion processes need the action of this force, only the action of the strong force.

[71] Mestel, L., MNRAS **112**, 598 (1952).

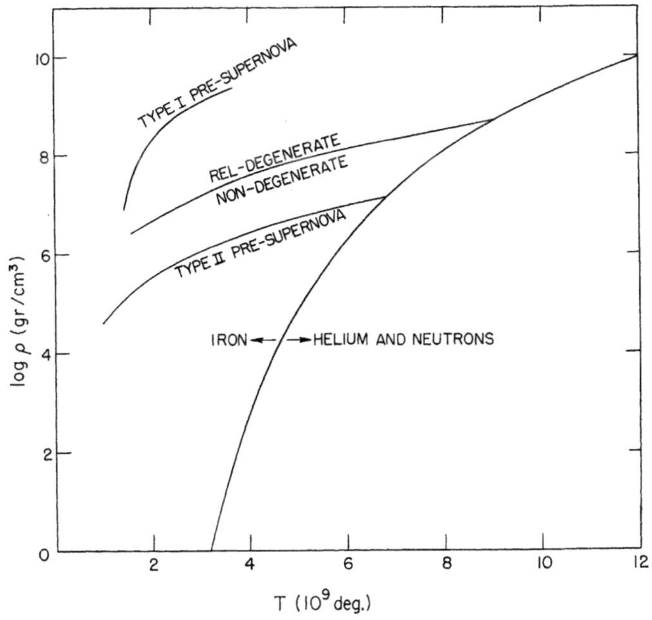

Fig. 8.6 The transition between iron and helium which absorbs 124.4 MeV per iron nucleus. Also shown is the density as a function of temperature in a Type II SN progenitor. The Type I progenitor does not reach the iron-to-helium transition. From Hoyle and Fowler 1960

today. Hoyle and Fowler also identified Type I SNs as the location where all the heavy elements are formed by successive neutron irradiation.

The massive stars do not develop a degenerate core, and Chandrasekhar's theory does not apply to them because they do not become degenerate. The massive stars evolve to higher and higher temperatures, and eventually ignite all possible nuclear fuels. Table 8.5 summarizes the nuclear fusion steps in massive stars leading to the formation of an iron core. Iron does not fuse to heavier elements, but gives in under the increasing temperature and disintegrates. This constitutes a dramatic change in the evolution of the star. The nuclear fuel cannot withstand the physical conditions needed for further fusion (on top of the fact that further fusion requires energy and does not release energy) and disintegrates. As for the trigger of Type II SNs, Hoyle and Fowler assumed that the dissociation of iron according to

$$^{56}\mathrm{Fe}+\gamma \longrightarrow 13\,^{4}\mathrm{He}+4\mathrm{n}-2.14\times 10^{18}\ \mathrm{erg/g}, \ \text{with}\ \gamma_{\mathrm{ad}}<1.2\,, \tag{8.2}$$

followed by the dissociation of helium according to

$$\gamma+{}^{4}\mathrm{He} \longrightarrow 2\mathrm{p}+2\mathrm{n}-6.82\times 10^{18}\ \mathrm{erg/g}, \ \text{with}\ \gamma_{\mathrm{ad}}<1.2\,, \tag{8.3}$$

Table 8.5 Fowler and Hoyle's estimates of SN energetics (stellar mass $30M_\odot$)

Temperature [10^9 K]	Process	Energy sources [erg/g]	Neutrino loss [erg/g]	Time interval [s]
2–3	$2\,^{16}O \rightarrow {}^{28}Si + {}^4He$	5×10^{17}	5×10^{12}	10^5
3–4	$2\,^{28}Si \rightarrow {}^{56}Ni\ (\alpha)$	2×10^{17}		
			6×10^{13}	5000
4	$^{56}Ni \rightarrow {}^{56}Fe\ (e)$	10^{17}		
4–14	$^{56}Fe \rightarrow 13\alpha + 4n$	-2×10^{18}	5×10^{15}	
	$\alpha \rightarrow 2p + 2n$	-7×10^{18}	4×10^{16}	0.3

would lead to an implosion of the core. This is the source of the core-collapse trigger, which has since been considered as the standard trigger for Type II SNs. The location in the density–temperature plane where the two types of SN take place is shown in Fig. 8.6. Recall that, at a given temperature, the more massive star has a lower density. Hence, the locus of a low mass star in the density–temperature plane is always above that of a high mass star.

Hoyle and Fowler considered a star with $T > 5\times10^9$ K, in which a nuclear equilibrium process synthesized (the updated version of Sterne's old idea) the core into iron group elements. The core is surrounded with explosive nuclear fuel at a temperature below 1.5×10^9 K. So what happens to the progenitor with this structure now? According to Hoyle and Fowler:

> *If the pressure support is withdrawn, the outer regions produce a catastrophic situation in which implosion takes place in a time of the order of free fall, or about 1 second.*

It is critical that the free fall should take place before the falling layer should have time to reorganize. Hoyle and Fowler found that the radiative and neutrino losses were insufficient to remove the heat released in the core by contraction. The heat is removed by the dissociation of iron:

$$^{56}\mathrm{Fe} \rightleftharpoons 13\alpha + 4\mathrm{n} - 124.4\ \mathrm{MeV}\,. \tag{8.4}$$

As soon as the iron disintegrates into helium and neutrons, a catastrophic implosion follows. If the star has a small mass, a degenerate core forms and then an explosion takes place. The implosion of the core of massive stars does not lead to degeneracy. On a second check, they found that a star of small mass would explode before the inner regions could implode. Fowler and Hoyle overlooked the developments which had taken place in parallel in the theory of planetary nebulas, and which predicted that the low mass stars they had in mind would actually become planetary nebulas.

For an explosion to take place, the inward motion of the imploding layers must be reversed into an outward motion. The suggested mechanism responsible for the conversion of implosion into explosion was Hoyle's old idea, namely, that the infalling matter would be braked by rotation and magnetic fields, as shown in Fig. 8.7. Hoyle and Fowler did not discuss the fate of the core and the neutron star, but re-

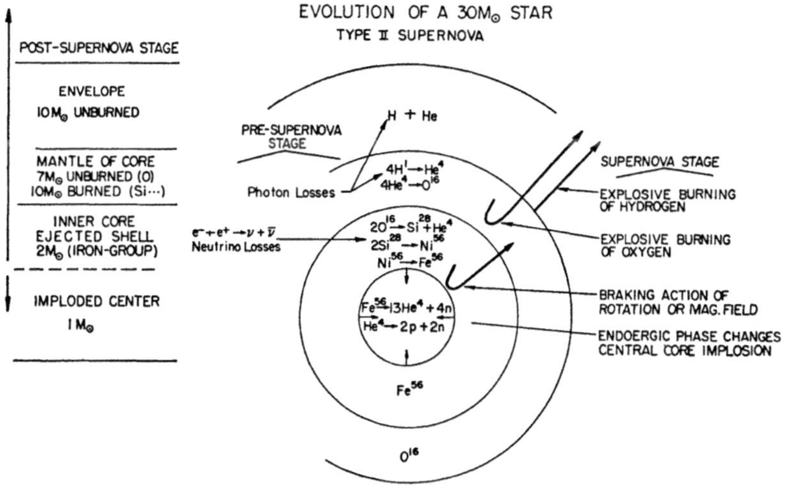

Fig. 8.7 The classical schematic description of the Fowler and Hoyle idea of core-collapse-triggered Type II SNs. In this model, the energy released in the explosion of oxygen, and braking actions of rotation and magnetic field, are supposed to convert the implosion into explosion. Fowler and Hoyle 1964

fered to previous publications by others. However, this was the first time that they mentioned a neutron star as the outcome of the implosion, and suggested a process which could lead to its formation. The possible collapse to a black hole, when the collapsing core is more massive than the maximum stable mass of a neutron star, was not discussed.

Conducting simple static calculations of nuclear equilibrium, Fowler and Hoyle got a nice fit to the relative abundance of the iron isotopes, which led them to conclude that *there must be something true in this process* (see Table 8.6). But it was not specified how these isotopes could leave the star without changing the relative abundance, or whether they all disintegrated in the collapse. Note that there are three numbers to fit and they had at their disposal two parameters, viz., $n_\mathrm{p}/n_\mathrm{n}$ (the ratio between the numbers of protons and neutrons) and the temperature. Hence the agreement with observation cannot be declared as very compelling.

Let us just say a few more words about the hypothesized details of the Type II SN. The innermost $1M_\odot$ collapses inward. All the rest must be ejected. However, the next $2M_\odot$ are lifted by means of rotation or a magnetic field.[72] The composition is a mixture of burnt and unburnt nuclear fuel, and this is ejected by the nuclear explosion. It is assumed that braking due to rotation or some other mechanism ultimately leads to mantle–envelope explosion, following core implosion caused by

[72] This would imply a non-spherical ejection, which is indeed observed today, but was not known in the 1960s. So it can be considered as a prediction that was largely ignored for almost 30 years, until the discovery of supernova 1987A (see Sect. 8.12).

Table 8.6 Iron isotope as a percentage of the total equilibrium process abundance by mass (core mass $= 20M_{\odot}$, $M_{tot} \approx 30M_{\odot}$). The *arrow* marks the choice of the parameter $\bar{Z}/\bar{N}$ by Hoyle and Fowler to explain the relative abundance of the iron isotopes

$\bar{Z}/\bar{N}$	^{54}Fe	^{56}Fe	^{57}Fe	^{58}Fe
1.000	1.7	89.1	2.9	0.0
0.950	43.4	21.9	7.2	0.0
0.900	34.0	29.6	4.7	0.04
→0.872	4.3	66.6	2.5	0.23
0.860	0.2	64.5	3.0	4.0
Solar/terrestrial values	4.2	67.2	1.6	0.25

endoergic nuclear phase changes. The explosive burning of previously unburnt oxygen is taken to be the source of energy in the explosion. The explosion results in the ejection of unburnt 'primordial' material plus products of hydrogen burning, helium burning, oxygen burning, successive capture of α particles, and nuclear equilibrium processes.

The triggering mechanisms of Type I and Type II SNs have not changed since the seminal publication by Hoyle and Fowler in 1960 and 1964, and are now considered to be well established. In 1967, Rakavy and Shaviv[73] performed stellar evolution calculations (in contrast with the static model of Hoyle and Fowler) and confirmed that the neutrino losses accelerate the evolution, but do not cause a free fall as expected by Gamow and Schönberg.

As was well known, at high temperatures and low densities, the radiation field creates spontaneous electron–positron pairs in the reaction[74]

$$\gamma + \gamma \rightleftharpoons e^- + e^+ \,. \tag{8.5}$$

This is the long-sought mass annihilation process. It takes place here as an equilibrium process that can destroy the star. The creation of such pairs takes energy from the gas and reduces its pressure. In the appendix to the 1964 paper, Fowler and Hoyle elaborated on the possible effect of electron–positron pair creation out of the radiation field. As would be expected from such a process, it causes a decrease in the adiabatic constant to below 4/3. However, the reduction is to a value of 1.317, which is only slightly below 4/3. Consequently, Fowler and Hoyle remarked that:

> *Questions concerning stability immediately arise*, but they reasoned that: *in the case considered, nuclear energy through oxygen burning prevents catastrophic collapse.*

[73] Rakavy, G., & Shaviv, G., Ap. J. **148**, 803 (1967).

[74] Fowler, W.A., & Hoyle, F., Ap. J. S. **9**, 201 (1964). Apparently, Souffrin had carried out some calculations on pair formation in 1960 [Souffrin, P., Mem. Soc. Roy. Sci. Liege Coll. **3**, 245 (1960)], but this was not known to Fowler and Hoyle, and nor was it known to Rakavy and Shaviv.

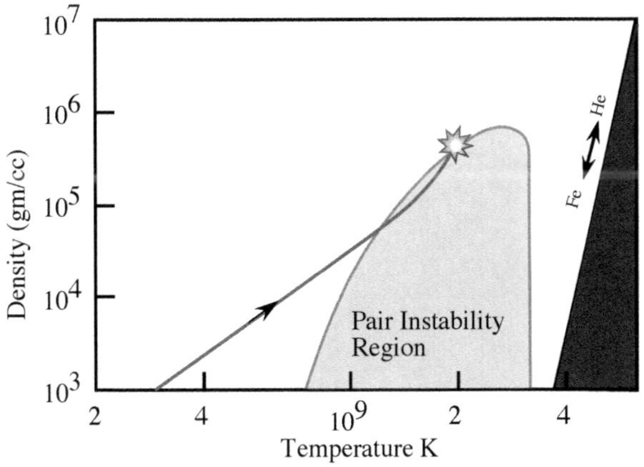

Fig. 8.8 The discovery of the collapse of an oxygen star due to formation of electron–positron pairs, from Rakavy and Shaviv 1967. The *pale green region* marks the $\gamma_{ad} < 4/3$ domain due to photons creating pairs of electrons and positrons. The *blue region* marks the $\gamma_{ad} < 4/3$ domain due to the iron-to-helium transition. The *red line ending* in a star is the track of the core of a $30M_\odot$ model which imploded when its center reached the location of the star. At the moment of collapse, a large fraction of the star lies inside the pair instability region

The first to investigate the effect of this process on the structure of massive stars were Rakavy and Shaviv,[75] who used a new numerical method, specially designed to discover stellar instabilities, to calculate the evolution of a $30M_\odot$ oxygen star. The results are shown in Fig. 8.8.[76]

The idea that such an instability could prompt a SN collapse did not gain much support until very recently, when a unique supernova eruption took place in which the progenitor was a very massive star. The figure shows the track of the core of the star in the temperature–density plane, and also the region where pair creation reduces the adiabatic constant to below 4/3. The $30M_\odot$ star collapses before reaching the iron disintegration region.

Rakavy and Shaviv[77] mapped the plane of density vs. temperature with the possible pitfalls for stars. The results are shown in Fig. 8.9 (left). Stars with masses close to the Chandrasekhar mass, but slightly below it, become unstable due to β decays. Although the paths of all stellar masses above about $1.39M_\odot$ towards extreme density and temperature conditions appear to be blocked, there was a small possibility that some stellar cores could escape and reach even more extreme conditions.

---

[75] Rakavy, G., & Shaviv, G., Ap. J. **148**, 803 (1967).

[76] The results were communicated privately to Barkat and Sack who promptly published a Physical Review Letter [PRL **18**, 379 (1967)]. However, in the text and references it was clearly stated that Rakavy and Shaviv carried out the calculation on the basis of which the Physical Review Letter had been written.

[77] Rakavy, G., & Shaviv, G., ApSS **1**, 429 (1968).

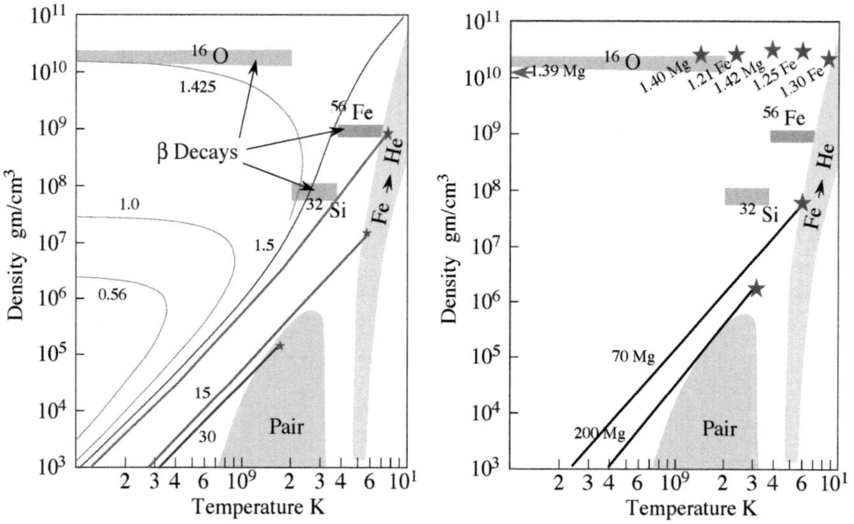

Fig. 8.9 *Left*: The instabilities encountered by stars of different mass. Based on Rakavy and Shaviv 1968. *Right*: General relativistic effects on the collapse of stars with different masses. Based on Shaviv and Kovetz 1968

At high densities effects described by the general theory of relativity become non-negligible. So Shaviv and Kovetz[78] extended the Chandrasekar theory to include general relativistic effects, and calculated the evolution of stellar models in this framework to verify that no stellar core could peacefully cross the density barrier of a few 10^{10} g/cm^3 and temperature of 6×10^9 K. The effect of general relativity is to reduce the Chandrasekahr limiting mass, and instead of a white dwarf of mass of $1.42M_\odot$ being the limit, they found that, for example, a magnesium core of $1.40M_\odot$ can reach a density slightly above $10^{10}M_\odot$ before gravity takes over and the core collapses. Similarly, general relativity defeats an iron core of $1.21M_\odot$ at about the same density. As a matter of fact, no stellar core can peacefully pass the density of a few 10^{10} g/cm^3. At these densities and temperatures the nuclei are still separated by a large distance, so that the effects of nuclear matter, which enter the game at about a density of 10^{14} g/cm^3, are not yet important.

In summary, we see that the roads whereby stellar cores may reach extreme temperature and density conditions are blocked. Either the core mass is below a revised Chandrasekhar limiting mass, in which case the star generally loses its outer layers to become a white dwarf, or the core encounters an instability which leads to a collapse of the core. There is no gradual peaceful way for a star to reach the compact state of a neutron star or a black hole.

[78] Shaviv, G., & Kovetz, A., ApSS **7**, 416 (1970).

8.7 The Explosion. A History of Unsubstantiated Proclamations

Once the collapse starts, the basic problem is how the infall of the outer layers onto the core is converted into an explosion which ejects the envelop at speeds of thousands kilometers per second. This problem is very complex, probably involves not yet known physics, and has not yet been completely solved.

It is illuminating to examine the global energetics. Consider a $10M_\odot$ star which collapses into a $1M_\odot$ neutron star and ejects the extra $9M_\odot$ at 10^4 km/s. Since the initial state is highly extended, the energy of the initial state can be neglected when compared with the binding energy of the dense neutron star. Hence the available energy is the binding energy of the neutron star or $E_{\text{neut}} = 2.4 \times 10^{53}$ erg. The kinetic energy of the ejecta is $E_{\text{kin}} = 8.5 \times 10^{51}$ erg, or about 0.036 of the total available energy. The energy radiated is likewise small. Assume that the peak luminosity of $7 \times 10^9 L_\odot$ continues for a year (and this is an exaggeration). Then the energy radiated away in photons is $E_{\text{rad}} = 8.4 \times 10^{50}$ erg, or about 0.0035 of the available energy. The rest, which amounts to 96% of the total, is radiated away by neutrinos.

The velocity of escape from a neutron star is about $0.5c$, so the expansion velocity of the ejecta is about 6% of the velocity of escape from the surface of the neutron star. The kinetic energy was calculated assuming the envelope was far away, as if it were ejected from its location prior to the collapse of the core. A particle which falls onto the surface of the neutron star and does not lose any energy will bounce and reach exactly its original position. But if the particle loses energy it will reach a lower height and never be ejected. Consequently, the neutrinos play two roles, by cooling, they allow for the collapse, and by removing the energy from the star, they reduce the available energy for lifting the envelope from the star at high speeds. In other words it is baffling that we do after all see the massive envelope expanding at these enormous speeds.

Several conclusions can be drawn. First, the fireworks we see from a supernova represent only a minute part of the energy. It is just a signal to those who use visible light to detect the external world that something dramatic has taken place. Most of the energy is removed by neutrinos. Second, a small error in the estimate of the total neutrino losses can easily eliminate the energy available for ejecting the falling envelope. Third, there is a very delicate balance between the part of the envelope which reaches the surface of the neutron star and releases gravitational energy and the rest which is still far away and can be removed more easily.

8.7.1 Type II SNs

Colgate and Johnson 1960[79] were the pioneers in numerical calculations of core-collapse supernovas. The collapse to nuclear densities created a rigid surface,[80] onto which the in-falling layers bounced to form an outgoing shock. No neutrino losses were included in the calculation and it was up to the kinetic energy of the shock to lift the in-falling layers. Colgate and Johnson postulated that, after the collapse, the energy required to eject about $1M_{\odot}$ is released. Once this energy has been released, they pursued their calculation by considering the formation of a shock wave which would then propagate into the envelope and accelerate it to observed SN velocities. As a small part of the mass reached very high energies, according to their estimate, they hypothesized that this is the way cosmic rays[81] are produced in SN.

The first full numerical calculations of the collapse of the core were carried out some 6 years later, by Arnett[82] and Colgate and White.[83] They adopted Hoyle and Fowler's 1960 trigger, but found that the collapse was much stronger than what those authors had predicted, so that more gravitational energy was released than was originally expected. Because of the problems with the bounce model, they put forward the idea that neutrinos released in the core deposit energy in the outer layers. They assumed that the neutrinos could act like radiation pressure, pushing outward and helping the shock to lift the envelope. But most of the energy was removed by neutrinos, and consequently too little energy was left for the shock. Colgate and White found that the neutrinos removed energy faster than the energy supplied by the in-falling. The collapse of the core stopped only when ultra-nuclear densities were reached and further compression of the core material became impossible.

Again, the outgoing shock was found to accelerate a small part of the mass to relativistic velocities, thus confirming the first estimates of Colgate and Johnson that SNs are the source of cosmic rays. The expansion of the shock wave cooled the matter to 5000 K before the blast of strong radiation escaped from the star. This is an interesting result, which is confirmed by observations, because it means that, despite the strong radiation of the blast, the SN appears to shine at a relatively low temperature. How could that be? The SN reaches maximum light intensity close to the moment of maximum expansion. The very large radius, more than 10^4 solar radii, implies low temperatures (despite the high luminosity). The expansion took place during the fast rise to maximum light, a phase most frequently missed. The disappointment was, as Colgate and White put it, that:

---

[79] Colgate, S.A., & Johnson, M.H., PRL **5**, 235 (1960).

[80] The repulsive part of the strong force prevents the compression, and consequently causes the nuclear matter to behave rigidly, as an incompressible fluid.

[81] Cosmic rays are energetic particles which continuously flood the Earth. The source of these particles are SNs, but the maximum energy of the particles created in SNs is much less than the maximum energy of the cosmic ray particles (about 10^{20} erg). This means that further acceleration must take place. These energies are way beyond the maximum energy of about 10^{13} erg attained in modern particle accelerators.

[82] Arnett, W.D., Canad. J. Phys. **44**, 2553 (1966); ibid. **45**, 1621 (1967.

[83] Colgate, S.A., & White, R.H., Ap. J. **143**, 626 (1966).

> *The shock-deposited internal energy is inadequate to explain the observed luminosity.* On the other hand, claimed the authors, *if* $1M_\odot$ *of radioactive material is ejected, than it can explain the peak of the light curve about a week after the explosion.*

Colgate and White did not follow the creation of the radioactive elements but assumed them to be formed in the right amounts. During the collapse, neutrons are released, and quickly absorbed by the iron to form all the elements heavier than iron, including large amounts of radioactive elements.

Arnett[84] and Wilson[85] criticized the model suggested by Colgate and White, because it did not provide sufficient energy to power the ejection. But in 1982, Bowers and Wilson returned to a modified version of the same model.[86] The revised model became known as the delayed shock model, as will be explained shortly.

In these early investigations, the neutrino interaction with the matter was not properly taken into account, due to the neglect of neutrino scattering by nuclei, which was considered unimportant. Since the strength of the interaction goes as the square of the atomic weight, it helps when the core contains heavy elements. As a consequence of the scattering, neutrinos actually stay in the core when the density reaches 10^{12} g/cm^3.

From this point on we witness how the pendulum of 'yes explosion no explosion' swings every time a new effect is discovered or the modeling improves.

A major discovery in the theory of elementary particles was made in the mid-1970s, when the weak neutral currents were identified. A proper description of the phenomenon would carry us too far astray. An important discovery for the theory of SN shocks was made by Freedman, who applied the newly confirmed Glashow–Weinberg–Salam theory[87] in 1974.[88] Freedman showed that, under the conditions of the theory of weak interactions, the resulting neutrino from the reaction $\nu + A \rightarrow \nu' + A$ goes predominantly forward like the electron in the reaction $\mathrm{e} + A \rightarrow \mathrm{e}' + A$, where the prime means the same particle, but with different energy and direction of motion. In a way, the reaction is equivalent to a forward pressure. Freedman estimated that this effect might inhibit cooling during the collapse and formation of a neutron star. Hopes rose of eventual success in getting the SN shock to eject the

---

84 Arnett, W.D., Ap. J. **153**, 341 (1968).

85 Wilson, J.R., Ap. J. **163**, 209 (1971).

86 Bowers, R.L., & Wilson, J.R., Ap. J. S. **50**, 115 (1982).

87 Weak neutral current interactions are one of the ways in which elementary particles can interact via the weak force. These interactions are mediated by the Z^0 boson and the interaction is said to be neutral because the Z^0 has no electric charge. The discovery of weak neutral currents was a significant step toward the unification of electromagnetism and the weak force into the electroweak force, and led to the discovery of the W and Z bosons. A key prediction of the Glashow–Weinberg–Salam (1926–1996) model (Nobel Prize in 1979) was the existence of weak interactions mediated by the neutral particle called Z^0.

88 Freedman, D.Z., Phys. Rev. **9**, 1389 (1974).

envelope,[89] but only for a short time. Soon it was found that improved calculations did not substantiate the hopes. As Wilson put it:[90]

> *The calculations give a very uncertain answer as to whether neutrino flows can produce a supernova explosion and a neutron star remnant.*

In 1980, Bowers and Wilson[91] claimed that:

> *The inclusion of neutrino effects may produce substantial shock damping. Current results indicate that core collapse, bounce and shock propagation does not produce an explosion when neutrino effects are included.*

The trouble was that the entire shock propagation and explosion was marginal, because a significant part of the neutrinos remained trapped in the core, leaving too few neutrinos to help the shock propagate out and lift the outer layers.

As a consequence of this disillusionment and with no new physics to come to the rescue, the original ideas of the mechanical energy in the outward moving shock were reexamined.[92] But this aspiration soon perished when it became clear that the thermal properties of the matter in the core where poorly approximated. The heat capacity of the core had been underestimated,[93] and as a consequence the core was calculated to cool less than in reality. When corrected for the higher heat capacity, the problem persisted. Recall how the theory of the structure of radiative stars was in a quandary in the 1920s, when the absorption coefficient (and the composition) were poorly known. The situation with the neutrino star is very reminiscent of that time, except that the neutrino problem may be more difficult.

These vicissitudes demonstrated how crucial the detailed nuclear physics is to the explosion mechanism. Massive effort by many investigators was invested in improving our theoretical knowledge of this problem. Recall the contributions of Eddington to stellar structure, in particular the idea that a star can be treated as a huge cavity full of radiation (and matter). Now the collapsing star creates a neutrino star. The neutrinos are trapped in the star, and like the radiation trapped in the stars, they leak out slowly. So the neutrinos leak from the core on a timescale significantly longer than the infall time of the outer layers. Like radiative stars, so neutrino stars have their neutrino sphere.

But one cannot be too careful: fine-tuning appears mandatory. The outward moving shock must be sufficiently strong to lift the outer layers and eject them. However, if the shock is too powerful, it dissociates the iron nuclei, the end product

[89] Wilson, J.R., PRL **32**, 849 (1974); Bruenn, S.W., Ann. NY Acad. Sci. **262**, 80 (1975); Schramm, D.N., & Arnett, W.D., Ap. J. **198**, 629 (1975); Wilson, J.R., Couch, R., Cochran, S., Le Blanc, J., & Barkat, Z., NYASA **262**, 54 (1975).

[90] Wilson, J.R., in *Physics and Astrophysics of Neutron Stars and Black Holes*, North-Holland (1978).

[91] Bowers, R.L. & Wilson, J.R., SSRv **27**, 537 (1980).

[92] Arnett, W.D., Ap. J. **194**, 373 (1974); ibid. **195**, 727 (1975); Barkat, Z., Rakavy, G., Reiss, Y., & Wilson, J.R., Ap. J. **196**, 633 (1975).

[93] At the extreme conditions of the core, the nuclear levels, although a few MeV above the ground level, are beginning to get populated, and beginning therefore to contribute to the heat content of the matter [Fowler, W.A., Engelbrecht, C.A., Woosley, S.E., Ap. J. **226**, 984 (1978)].

of nuclear synthesis in stars. The dissociation energy is 8.8 MeV per nucleon. The conversion of such an energy per nucleon into mass is equivalent to cooling the nucleon from a temperature of about 10^{13} K to zero! For $0.1M_{\odot}$, this energy amounts to 1.7×10^{51} erg, and hence a small increase in the mass can choke the shock, which has a total energy of about 10×10^{51} erg. The next question concerns how much matter the shock has to traverse before it reaches the surface of the iron core. This estimate depends on the approximate description of the matter and the progenitor.

Since it became a matter of such delicate fine-tuning, small changes in the physics and the numerical codes led to different results. Baron et al.,[94] who used one equation of state, got explosion in $15M_{\odot}$ models, while Wilson et al.,[95] who used another equation of state, did not. While one group[96] found an explosion in a $8.8M_{\odot}$ star, another group[97] found a dud. The controversy among nuclear physicists about the properties of nuclear matter leaked into the astrophysical community, who expressed it through their vacillations over the explosion.

However, we are not alone to waver, for the star clearly faces this dilemma, too. The collapsed core contains the 10^{53} erg released in the gravitational energy. The outer envelope continues to rain on the core. If this configuration does not lead to an explosion, but to a black hole, because the shock cannot eject the envelope, how do the heavy elements come out of the collapsed furnace? There must be a way out, because we observe it. In 1985, Bethe and Wilson[98] found the way to resuscitate and invigorate the stalled shock. Actually, Wilson let his calculation continue past the normal point where the shock usually died, and found that after the shock had stalled, neutrinos kept leaking from the neutrino star and were absorbed in the outer envelope, attempting to resuscitate the outgoing shock and the hopes of ejecting the envelope.

The hot neutrino star at the center of the collapsing star cools by neutrino leakage. Some of these neutrinos are absorbed far out and deposit enough energy to sufficiently invigorate the shock to eject the outer layer and leave behind a neutron star. As Bethe and Wilson wrote, the balance is critical. If, for example, the neutrino flux had had half the value they calculated, the shock resuscitation would not have been sufficient to eject the outer layer. Again, everything was very tightly adjusted, leaving no leeway. One particular model exploded, while others did not. Due to its time sequence, the mechanism was called the delayed explosion mechanism.

But once again, it was too soon to sound the trumpets. Careful calculations of the way the neutrinos transport the energy[99] showed that the entire effect was borderline. As a matter of fact, Arnett showed by a careful numerical experiment that, by the time the delayed mechanism had begun to appear, the accumulated error in

---

94 Baron, E., Cooperstein, J., & Kahama, S., PRL **440**, 126 (1985); Nucl. Phys. **440**, 744 (1985).

95 Wilson, J.R., Mayle, R., Woosley, S.E., & Weaver, T.A., 12th Proc. Texas Symp. Relativ. Astrophys., NY Acad. Sci., ed. by Rosen and Shaviv.

96 Hillebrandt, W., Nomoto, K., & Wolff, R.G., A & A **133**, 175 (1984).

97 Burrows, A., & Lattimer, J.M., Ap. J. L. **299**, 19 (1985).

98 Bethe, H.A., & Wilson, J.R., Ap. J. **296**, 14 (1985).

99 Arnett, W.D., IAUS **125**, 273 (1987); Hillebrandt, W., MPA Rep. No. 216 (1985).

the numerical calculation of the energy of the shock reached several times 10^{50} erg, invalidating the entire calculation.

A unique idea was suggested by Epstein in 1979,[100] namely that the non-steady-state heat-choked core would develop convection, and that the powerful convective currents could easily turn the matter over and churn the radioactive elements out. The idea was followed up by Bruenn, Buchler, and Livio,[101] and also by Colgate and Petscheck,[102] although the meaning and treatment of convection in a dynamic situation is questionable. Further calculations by the same authors[103] quenched enthusiasm, since the results were doubtful.

More recently, secondary factors, which had so far been neglected, like rotation and magnetic fields were taken into account, as suggested by Hoyle and Fowler. But the calculations[104] do not show that they can change the general picture.

And so the pendulum kept swinging between temporary success and dud. Various ideas were suggested, but when carefully tested, were found to be insufficient. In the extensive review of 1990, Bethe[105] surveyed what had been attempted up until then. The general impression is one of optimism, as if the solution to the SN conundrum might be just around the corner. In a review in 1995, Herant[106] expressed the hope that:

> *The recent progress achieved by multidimensional calculations has made investigations of supernova explosions a more attractive proposition than when one was reduced to the depressing prospect of witnessing an explosion one day, and a fizzle the next.*

The most recent review is by Woosley and Heger.[107] The authors start the review with:

> *Hans Bethe contributed in many ways to our understanding of the supernovas that happen in massive stars, but, to this day, a first-principles model of how the explosion is energized is lacking.*

The best that has been done is to assume that the explosion is like an energetic piston moving outward, whose location and speed are two free parameters, chosen so as to fit the observations as closely as possible. Thus the question of how a Type II SN takes place remains one of the most disturbing problems in modern astrophysics, and it carries along with it the problem of how the heavy elements are formed.

The situation reminds me that my friend Ed Spiegel, who contributed enormously to our understanding of the so far unsolved problem of convection, used to say that

---

[100] Epstein, R., MNRAS **18**, 305 (1979).

[101] Bruenn, S.W., Buchler, R.J., & Livio, M., Ap. J. L. **234**, 183 (1979).

[102] Colgate, S.A., & Petscheck, A.G., Dumand Symp., Honolulu, Hawaii (1980).

[103] Livio, M., Buchler, R.J., & Colgate, S.A., Ap. J. L. **238**, 139 (1980). Smarr, L., Wilson, J.R., Barton, R.T., & Bowers, R.L., Ap. J. **246**, 515 (1981).

[104] Muller, E., & Hillebrandt, W., A & A **103**, 358 (1981); Symbalisty, E.M.D., Schramm, D.N., & Wilson, J.R., Ap. J. L. **291**, 11 (1985).

[105] Bethe, H.A., Rev. Mod. Phys. **62**, 801 (1990).

[106] Herant, M., SSRv **74**, 335 (1995).

[107] Woosley, S.E., & Heger, A., Phys. Rep. **442**, 269 (2007).

he was *awarded a chaired professorship at a respectable university for not solving the convection problem.* And convection is only a part of the SN physics.

8.7.2 Type I SNs

The situation with the theory of type I supernovas is not that much better, although the consensus among investigators is more substantial. The barrier is posed by some fundamental physical questions.

The original idea of Hoyle and Fowler[108] and Schatzman,[109] i.e., the explosion of carbon and oxygen on the surface of a white dwarf, was picked up by Arnett,[110] Rose,[111] and Paczynski.[112] The current prevailing model is that of a white dwarf in a binary system which accretes matter from a close companion. When the companion star is a main sequence star, the accretion rate is slow, the matter settles onto the white dwarfs, gets compressed, becomes degenerate, ignites, and produces a nova.[113] But when the companion of the white dwarf is a giant star, the rate of mass transfer from the giant star to the white dwarf is very high, so high that the heat released in the accretion cannot be radiated away, whence it is trapped in the star and prevents a nova explosion.[114]

As the mass accumulates, the white dwarf's mass exceeds the Chandrasekhar limit, and as a consequence it collapses and ignites the oxygen and carbon in the outer layers. The violent ignition leaves no remnant since the entire white dwarf is made of carbon and oxygen. The nuclear reactions generate plenty of radioactive materials which power the light curve. The attractiveness of this model is that it requires old low mass stars, and hence should be observed in elliptical galaxies, where no massive stars are observed. Observations confirm this consequence indirectly.

Consider a simple cigar type of burning, i.e., fuel which is lit on one side. Two basic processes take place: the burning, and the heat propagating from the hot burning zone into the cold unburnt fuel. In the steady state, provided there is one, there is a balance between the two, as in a real cigar. This state is called deflagration.[115] It is possible, however, for the heat to propagate faster than the burning, heat the region in front of the burning flame, and consequently ignite the entire fuel, thereby releasing the entire energy in the fuel almost instantaneously. This process is called detonation. Clearly, the type of burning depends on the heat conduction in the fuel.

---

[108] Hoyle, F., & Fowler, W.A., 1960.

[109] Schatmann, E., in *Star Evolution*, Proc. XXVIIIth Course, *Enrico Fermi school*, Varenna 1962, Gratton, Academic Press, New York (1963), p. 389.

[110] Arnett, W.D., Nature **219**, 1344 (1968); ApSS **5**, 180 (1969).

[111] Rose, W.K., Ap. J. **155**, 491 (1969).

[112] Paczynski, B., Acta Ast. **20**, 47 (1970).

[113] Starrfield, S., *Classical Nova Explosions*, International Conference on Classical Nova Explosions, AIP Conf. Proc. **637**, ed. by Hernanz & Josž, American Institute of Physics (2002).

[114] Starrfield, S., Sparks, W.M., Truran, J.W., & Shaviv, G., STIN (1989) 9019931.

[115] From the Latin 'deflagrat', meaning to burn away.

Good heat conduction leads to detonation. The heat can be carried by conduction or by convection. The latter is difficult to handle and poses yet unsolved problems, for the theory of convection is not complete. The controversial issues among researchers relate to the nature of the ignition and its propagation, and consequently how fast the energy is released. Again one finds a high sensitivity to the details of the physics, and calculations leads unavoidably to conflicting results.

8.8 Esoteric Mechanisms?

Finzi and Wolf[116] had a very interesting and quite distinct idea regarding the mechanism of Type I SNs. The Chandrasekhar limiting mass is given by $5.75M_\odot/\mu_e$, where μ_e is the mean molecular weight per electron, appropriately averaged over the star if it is not homogeneous. Consider a white dwarf which is sufficiently massive to possess a nucleus which can undergo inverse β decay. That is, consider a nucleus (Z,A) which decays in the laboratory (on Earth) into a nucleus $(Z-1,A)$ by emitting an electron with maximum energy E_β. Assume now that the star synthesized the element $(Z-1,A)$. As long as the density is low, the nucleus $(Z-1,A)$ is stable. But as the density increases, the energy of the free electron increases as well, and approaches $E_\beta + m_e c^2$, where $m_e c^2$ is the rest mass energy of the electron. At this point an electron is absorbed by the $(Z-1,A)$ nucleus and converts it into the (Z,A) nucleus. That is, the β decay which occurred in the laboratory is reversed in the high density outside of the nucleus. In this way an electron which helped to provide the pressure to support against the gravitational pull is eliminated. Hence, it is to be expected that the corresponding limiting mass should decrease. Indeed, the change causes an increase in μ_e and a decrease in the limiting mass.[117]

Finzi and Wolf considered the two species ^{24}Mg and ^{40}Ca. If a WD of mass greater than $1.395M_\odot$ has enough ^{24}Mg, then the slow inverse β decay can take place and a collapse must follow. The reaction is $^{24}\mathrm{Mg} + \mathrm{e}^- \to {}^{24}\mathrm{Na} + \nu - 5.52$ MeV. At a density of $1.6 \times 10^9 \mu_e$ g/cm^3, the electrons have sufficient energy to enter the magnesium nucleus and convert it into sodium. A similar case was worked out with ^{40}Ca. The original mass of the progenitor can be anywhere up to $8M_\odot$. Finzi and Wolf calculated the lifetime of the white dwarf before the change in the mean molecular weight can push it towards a collapse, and found that, if the density is below 1.6×10^{10} g/cm^3, the white dwarf can live longer than the age of the Universe.[118] But as the density increases to 2×10^{10} g/cm^3, collapse follows. This is a time bomb. No such model was followed up numerically. The basic problem of how implosion turns into explosion was not tackled, nor was any estimate made of how many such objects should be expected.

---

[116] Finzi, A., & Wolf, R.A., Ap. J. **150**, 115 (1967).

[117] Recall that μ_e is defined as the number of nucleons per electron. The number of nucleons does not change in a β decay, while an electron disappears.

[118] The particular β decay Finzi and Wolf discovered is extremely slow because it is four times forbidden, so that the lifetime of the star becomes astronomical rather than just a few minutes.

8.9 Peculiar SNs. An Example

With the accumulation of more and more data, it became clear that not all SNs fall into the two distinct types. One of the most conspicuous examples is probably Cas A, the SN in the Cassiopeia constellation. When Ryle and Smith were surveying the heavens with their radio telescope in 1948, they discovered that Cas A is the strongest radio source in the sky.[119] In 1954, Baade and Minkowski[120] got the approximate location of the object in the sky, and using the 200 inch telescope were immediately able to identify the radio source with *galactic-emission nebulosity of a new type*. The location of the nebula was close, but did not coincide with the radio position supplied to them by Smith. The deviation was about 2 arc minutes (which is a lot for a telescope like the 200 inch). In this work Baade and Minkowski already realized that *there are strong indications that interstellar reddening affects the field*,[121] which in plain English means that the object is partially obscured by some opaque clouds. However, since the position of the object Baade and Minkowski discovered did not coincide very accurately with the location as given by Ryle and Smith, they carefully stated that: *The present evidence suggests strongly that no direct relation exists between optical and radio emission.*

The nebulosity discovered showed a very strange and in fact unprecedented spectrum with respect to composition and internal motions. The hydrogen lines were unusually weak. Expansion velocities of 3000 km/s were observed. More accurately, a wide range of velocities was discovered. Baade and Minkowski even reached the conclusion that:

> *There are no indications that the nebulosity as a whole is expanding. The random velocities are small relative to the expansion velocities. This is quite different from the conditions found in shells of novae and supernovae and clearly shows that the nebulosity is not a shell of this type.*

The outstanding characteristic of the filaments was the large internal difference in their velocities. On the basis of such findings Baade and Minkowski remarked that:

> *Since there is every reason to believe that the Cassiopeia source has nothing to do with a supernova, the attempt by Shklovskii*[122] *to identify the source with a new star of A.D. 369 is beside the point.*

In a paper that appeared back to back with Baade and Minkowski's contribution, in which the identification attempt was made, Minkowski and Aller[123] claimed that no obvious stellar source was visible that could excite the nebula. In other words it was not clear why the nebula was shining. They suggested that the spectral peculiarity of Cas A might be interpreted as arising from a unique excitation mechanism. In

---

[119] Ryle, K., & Smith, G., Nature **162**, 462 (1948).

[120] Baade, W., & Minkowski, R., Ap. J. **119**, 215 (1954).

[121] When light traverses dust, it is scattered. The scattering properties are such that blue light is scattered more than red. When the blue part of the light is removed, the object appears redder.

[122] Shklovskii, I.S., Astr. JUSSR **30**, 26 (1953).

[123] Minkowski, R., and Aller, L.H., Ap. J. **119**, 232 (1954).

an ordinary gaseous nebula, the energy is supplied by a hot central star. Ultraviolet quanta detach electrons from atoms. The electrons collide with ions and atoms in the gas and excite them to metastable levels, whence then cascade to lower levels with the emission of characteristic forbidden lines.[124] If the energy required to excite the electron is supplied by mechanical energy, say a shock or compression of the gas, the situation is different. So Minkowski and Aller applied the theory of Miyamamoto[125] and Chamberlain,[126] who was a student of Aller, based on the above ideas, and got a better agreement. But no suggestion was offered for the source of the mechanical energy.

Three years later, Minkowski gave an introductory lecture on the optical properties of radio sources,[127] and it was in this lecture that he made the final identification of Cas A with a supernova. Minkowski analyzed the proper motion measurements by Baade and established beyond doubt that the nebulosity corresponded to a rapidly expanding object:

> *The nebulosity consists of two different types of filaments which are so drastically different in every way that it is hard to avoid the conclusion that two masses of gas are involved in some way.*

The measured velocities were ± 5000 km/s. Minkowski claimed that only one object was known to present such velocities, and this was the Type II SN. Minkowski provided the audience with Fig. 8.10 in which the velocities and locations of the various filaments of the nebula are given. As can be seen, the nebula appears to expand from a central point which is close to the center of the radio source but does not coincide with it. Looking carefully at the figure, one sees that there are two velocity vectors (on the left-hand side of the figure), which indicate that it was probably a spherical expansion.

Not much happened for about 12 years, until van den Bergh came[128] and, following Minkowski, assumed Cas A to be a SN. Consequently, he looked for a remnant star, but could not discover any possible candidate. So van den Bergh reached the conclusion that SNs of the type observed in Cas A leave no (directly or indirectly visible) remnant.

---

[124] When the conservation laws for the transition between two levels are not satisfied, the transition is strictly speaking forbidden. In many such forbidden cases, the electron then stays in the level and does not descend to a lower level. In this case, rare events such as the emission of two photons rather then one, can take place. This is called a forbidden line, as if it violated the conservation laws, for the following reasons. In the laboratory, when the density is high, the probability that an electron will collide with the excited atom and kick the electron out of the energy level is high, and thus leaves no chance for a rare event to take place. However, in the extremely low densities of space, the probability of collision is very low, and the electron remains unperturbed for a long enough time for rare events to have a chance of occurring. In other words, the spectrum and the particular spectral lines observed depend on the density in which the excited atom is embedded.

[125] Miyamamoto, Jap. Col. Sci. **21**, 173 (1938).

[126] Chamberlain, Ap. J. **117**, 387 (1953).

[127] Minkowski, R., IAUS **4**, 107 (1957).

[128] van den Bergh, S., Nature **223**, 814 (1969).

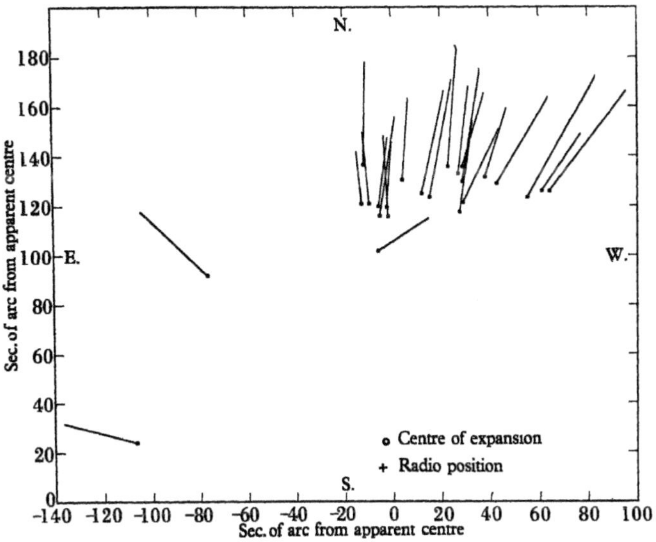

Fig. 8.10 The velocities as found by Minkowski in 1957, which convinced him that Cas A was a renmant of a supernova. From Minkowski 1957

The next twist in the story of Cas A came in 1971, when Peimpert and van den Bergh[129] discovered a high density and overabundance of nitrogen in the stationary filaments, and consequently suggested that these clouds might be circumstellar rather than interstellar in origin, ejected by the pre-supernova before the outburst. On the other hand, the fast moving knots might be part of a shell ejected by a supernova. This idea was the basis for Sgro's 1974[130] claim that what we see is a collision between the SN shock and interstellar gas. The evidence nicely fitted a story of a SN which exploded and sent the ejecta to collide with remnant gas that was either lost before or never condensed into a star. In 1976, Kamper and van den Bergh[131] measured the motion of the fast knots and estimated that the explosion took place in 1657. However, no SN event was observed around this time anywhere in the world.[132] Recall that Baade and Minkowski had already noted that there is a lot of dust between us and Cas A. So the idea was that Cas A exploded behind a screen of dust that blocked the radiation from reaching the Earth, whence it went unnoticed.[133] Attempts to relate the SN to a new star registered in Korea in the year 1592[134] do

[129] Peimpert, M., & van den Bergh, S., Ap. J. **167**, 223 (1971).

[130] Sgro, A.G., Ap. J. **197**, 621 (1975).

[131] Kamper, K., & van den Bergh, S., PASP **88**, 587 (1976).

[132] Kamper, K., & van den Bergh, S., S&T, Cassiopeia A – An Unseen Supernova, **51**, 236 (1976).

[133] As nobody saw it, was it a faint SN?

[134] Brosche, P., Bull. Var. Stars, No. 192 (1967); Chu, S. Korean Ast. Soc. **1**, 29 (1968).

not seem to agree. Although a black hole was suggested by Shklovskii,[135] Brecher and Wasserman[136] remarked that there might be no remnant at all.

Chevalier and Kirshner added to the mystery in 1977[137] when they indicated that there is very little hydrogen in the knots. Was the envelope removed from the star before the explosion that ejected the fast knots? Is it clear from the observation that complete mixing did not occur? The reason for raising these questions was their finding of abundant inhomogeneities. However, Chevalier and Kirshner suggested another solution, namely that the explosion may have been highly asymmetric.

Johnston and Yahil[138] argued that, if the ejecta of a Type II supernova does not undergo extensive mixing, then on the basis of current pre-supernova models, only a small fraction, approximately equal to or less than $0.1M_\odot$ of the mantle of a massive star, can yield abundances similar to those observed in the fast-moving knots of Cas A. This result was shown to be independent of the detailed structure of the mantle and the supernova energy. Lack of mixing in Cas A is indicated by strong upper limits on the abundance ratios Ne/O, and Fe/O. If this is confirmed by further observations, then either Cas A is not the result of a standard progenitor of mass approximately equal to or less than $25M_\odot$ disrupted by a Type II supernova, or the picture of the last stages of stellar evolution in massive stars needs substantial modifications.

Cas A is oxygen rich. Of the seven oxygen-rich SNs identified by van den Bergh, four are embedded in large hydrogen clouds where stars are still being formed. Hence the progenitor must be a relatively young massive star. Similarly, Wheeler, Harkness, and Capellaro[139] claimed that, out of 11 known Type Ib SNs, at least 5 occurred in hydrogen gas clouds. Hence, concluded van den Bergh, the similarity to Type Ib implies that the oxygen rich SN remnants came from massive stars.

8.10 Obstacles to Understanding the SN Phenomenon and the Formation of the Elements Beyond C and O

The textbook supernova does not yet exist. We will probably fail to understand the nucleosynthesis of the heavy elements beyond carbon and oxygen unless we completely solve the problems of the various types of supernovas, their progenitors, and their contribution to the interstellar medium. The problem starts from the first generation of stars.

Until recently all modeling was one-dimensional, assuming spherical symmetry. Recent SN observations teach us that secondary parameters are important, if not

[135] Shklovskii, I.S., Nature **279**, 703 (1979).

[136] Brecher, K., & Wasserman, I., Ap. J. **240**, L 105 (1980).

[137] Chevalier, R.A., & Kirshner, R.P., Ap. J. **233**, 154 (1979).

[138] Johnston, M.D., & Yahil, A., Ap. J. **285**, 587 (1984).

[139] Wheeler, J.C., Harkness, R.P., & Cappellaro, E., 13th Texas Symposium on Relativistic Astrophysics, Chicago, 1986, World Scientific, p. 402.

crucial, to the explosion mechanisms and in generating a variety of different types of supernova. The variety is greater than the standard classification implies. Secondary parameters also affect the evolution and mixing of massive stars in a critical way. As a result, the calculation is more complicated and difficult than anything yet attempted. However, the range of expected phenomena is greater. Furthermore, new physical processes are needed:

- We need to know how core collapse supernovas convert implosion into explosion.
- Some key nuclear reactions need to be better understood:
  - (a) $^{12}C + \alpha \to {}^{16}O + \gamma$, which is the bottleck for the formation of elements heavier than C/O.
  - (b) $^{22}Ne + \alpha \to {}^{25}Mg + n$, which controls secondary neutron sources.
  - (c) $^{59,60}Fe + n \to {}^{60,61}Fe$, which start the all neutron capture processes and the building of heavy elements from these seed elements.
  - (d) The rate of weak force reactions (β decays), which control the cooling of the star, and affect fast neutron capture and neutrino losses.
- A proper theory of convection is needed. The lack of a reliable and tested theory of convection pervades the entire discipline of stellar evolution.
- The history of mass loss must be better established. What is the original mass of the progenitor? The nucleosynthesis, which depends on the mass, will change accordingly.
- We need to know the values of the secondary parameters describing massive stars, like rotation and magnetic field. For example, how do these evolve with time?
- Are our theories of the neutrino correct? The role of neutrinos in the explosion appears to be very crucial. Has some special property of neutrinos not yet been discovered?
- We need to establish the properties of nuclei found along the path of the fast neutron and proton capture processes. These nuclei are unstable and not easy to study in the laboratory.

8.11 The Light Curve

The characteristic feature of Type I SNs, already discovered by Baade[140] is that, after an initial period of 50–100 days, the light curve starts to decay exponentially, with a time scale corresponding to 55 ± 1 days.

Borst (1912–2002)[141] was impressed by the following particular characteristics of Type I SNs with a time scale of 55 days:

(a) High luminosity maximum which is about 20–30 days long.

---

[140] Baade, W., Ap. J. **102**, 309 (1945).

[141] Borst, L.B., PRL **78**, 807 (1950).

Fig. 8.11 The evolution of an SN with time. As the envelope expands and thins out, we see more deeply into the hotter layers, provided the expanding envelope keeps its shape

(b) An exponential decay with a time scale of 55 days.
(c) Total energy released, as inferred from the emitted light, of 10^{49} erg.
(d) A remnant shell with no hydrogen expanding at a velocity of about 1300 km/s and radiating about 10^{36}ergs of visible light over some 900 yrs following the explosion.

Borst did not specify how he put together this list of properties, which is a mixture from several SNs.

It was the exponential decay that triggered Borst's idea that the energy might be supplied by radioactive decay, which as a rule decays exponentially. As a matter of fact, all physical processes in which the decay is proportional to the quantity itself are exponential, like the cooling of a cup of tea. The intriguing part was that the time constant of the decay was 55 days, and quite the same for all Type I SNs, with a very small variation. On the other hand, different cups of tea cool at a different exponen-

Fig. 8.12 The first thermonuclear explosion in 1952. It created the ^{254}Cf that prompted the idea of an ^{254}Cf-powered SN light curve

tial rates depending on the shape and material of the cup, and also the environment. So Borst suggested the following mechanism for the SN.

Consider a star of about $15M_{\odot}$ in which the hydrogen has been exhausted. A star without nuclear fuel must contract until a nuclear reaction starts between the helium nuclei. At a temperature of $2\text{–}3 \times 10^9$ K, the following reaction takes place: $2\alpha \rightarrow {}^7\mathrm{Be} + \mathrm{n} - 18.6$ MeV. The nucleus ^{7}Be does not exist naturally on the Earth, because it is unstable and decays. Borst had his idea about a year before Salpeter discovered how two α particles fuse to form ^{8}Be at $T = 1 \times 10^9$K, as we know today, so that there was no chance of Borst's reaction taking place. Yet what Borst had in mind was that, as this reaction removes energy from the gas, the collapse accelerates until it becomes free fall. As the density increases, the direction of the reaction reverses, the neutrons are absorbed by the beryllium, and an equilibrium is established, viz., $2\alpha \rightleftharpoons {}^7\mathrm{Be} + \mathrm{n}$. The neutrons can react with various nuclei to create still heavier nuclei. But what then? As Borst wrote:

> *After the collapse, the star becomes unstable and explodes, driving off a considerable fraction of its mass as an expanding gas cloud, probably leaving a core in a highly degenerate state. The exact mechanism of this explosion is not understood.*

To Borst it was important that the expanding envelope would contain large amounts of ^{7}Be. This nucleus is unstable and decays with a time constant of 52.9 days, which is awfully close to the decay time scale of the SN. The amount of ^{7}Be needed to power the SN was calculated to be about $0.07M_{\odot}$. Borst ended his short PRL with a promise to elaborate the details of his idea, but no further publication came out.

About six years later, Burbidge et al.[142] came up with the same basic suggestion, namely that the energy source of SNs might be radioactive decay. They realized that the three radioactive nuclei ^{7}Be, ^{89}Sr, and ^{254}Cf have half-lives close to 55 days, so the question left open was which of these nuclei might be responsible for the energy of the SN.

The authors exposed problems with the first two nuclei and reached the conclusion that production of ^{254}Cf was the best shot. Thus, the suggested appearance of this nucleus as the energy source of the SN was considered as an excellent proof that the heavy elements are formed in SNs during the explosion, through an exposure to a very intense flux of neutrons. ^{254}Cf was produced on Earth in 1952 in a thermonuclear test (the hydrogen bomb, see Fig. 8.12), when the uranium in the bomb was irradiated by a strong flux of neutrons. ^{254}Cf decays through spontaneous fission, unlike the other two candidate nuclei. The decay time is 55 days, in perfect agreement with the observed decay time of the Type I SN. The idea was that a SN is a self-exploding stellar fission bomb.

An immediate question is the following. Large quantities of many unstable nuclei are produced in the process of creating the end product ^{254}Cf. How come *the energy release in the fission of* ^{254}*Cf dominates over all other processes?* The authors could provide only a somewhat debatable answer. One would expect a sum of exponentials contributed by many unstable nuclei. What spoke in favor of their idea was the straightforward prediction that the relative abundance of the end product would be like the one found in the fission products of ^{254}Cf. The calculated amount of ^{254}Cf needed to release the 10^{47} erg estimated for a Type I SN was just $6 \times 10^{-6} M_{\odot}$, which is quite small. In their conclusion, the authors stated:

> *We wish to emphasize that the production of* ^{254}*Cf in the November 1952 thermonuclear test stands as clear evidence for the terrestrial production on a fast time-scale of the heavy elements by a neutron capture process.*

However, the bomb contained a huge amount of uranium to begin with, while in the star the process has to start from iron.

The idea of radioactive decay as the energy source of the light curve did not die, but instead changed authors. After a thorough examination, Colgate and McKee[143] suggested a more complicated radioactive decay which contained two steps:

$$^{56}\mathrm{Ni}\,(6.01\text{ days}) \longrightarrow {}^{56}\mathrm{Co}\,(77\text{ days}) \longrightarrow {}^{56}\mathrm{Fe}\,.$$

The advantage of this hypothesis is that it is well accepted that nuclei in the iron group are indeed the end product of stellar evolution. Furthermore, instead of producing a tiny amount of ^{254}Cf which releases a lot of energy per nucleus, Colgate and McKee estimated that $0.25 M_{\odot}$ of ^{56}Ni, which releases about 1/200 of the energy released by ^{254}Cf, is sufficient to power the light curve. The idea encountered difficulties because of the variability of the decay rates of supernova light curves, and

[142] Burbidge, G.R., Hoyle, F., Burbidge, E.M., Christy, R.F., & Fowler, W.A., Phys. Rev. **103**, 1145 (1956).

[143] Colgate, S.A., & McKee, C., Ap. J. **157**, 623 (1969).

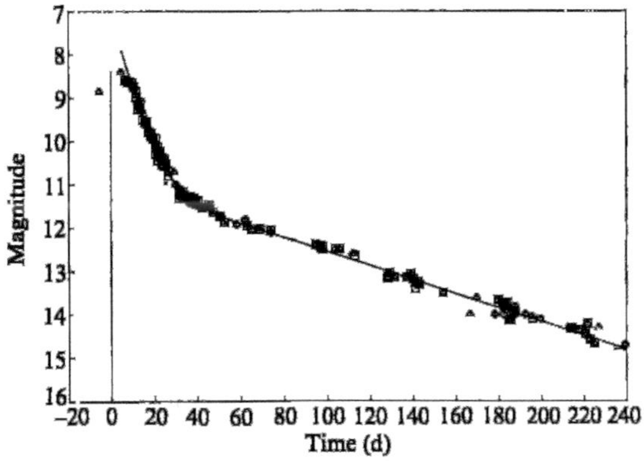

Fig. 8.13 The light curve of SN 1937c. *Points* are observations by Baade and Zwicky and Parenago and Deutsch, and the *continuous line* is the fit, using two exponentials, by Rust et al. 1976. The time runs from a hypothetical start of the explosion

consequently was not embraced by many astrophysicists. However, Leventhal and McCall[144] claimed to have overcome most of the difficulties. To confirm their model, they analyzed observations of an old SN, and as a result Rust, Leventhal, and McCall[145] were able to claim that Type I supernova light curves were well represented as a sum of two exponentials with half-lives which are approximately the half-lives of ^{56}Ni and ^{56}Co.

8.12 The Tale of a Recent Supernova. What One Good Case Can Tell Us

By the end of 2007, there were some 4000 papers published in professional journals about various fascinating aspects of supernova 1987A. Here we describe only those features relevant to the theory of stellar structure.

The estimated total rate of supernovas in our galaxy is about 1–2 per century.[146] The last SN to explode in the Milky Way was Kepler's SN in 1604. Hence, many astronomers were concerned towards the end of the twentieth century by the absence of SNs in modern times. Is the dearth of SNs in our galaxy just a matter of statistics? Observations of distant galaxies routinely reveal new SNs, but they are far away.

[144] Leventhal, M., & McCall, S.L., Nature **255**, 690 (1975.

[145] Rust, B.W., Leventhal, M., & McCall, S.L., Nature **262**, 118 (1976).

[146] van den Bergh, S., Comments Astrophys. **17**, 125 (1993). See also Ratnatunga, K.U., & van den Bergh, S., Ap. J. **343**, 713, where the above rate is given, along with an explanation as to why the estimate of 11 per century per galaxy, obtained by Bahcall and Piran [Bahcall, J.N., & Piran, T., Ap. J. L. **267**, L 77 (1983)], is off by a large factor.

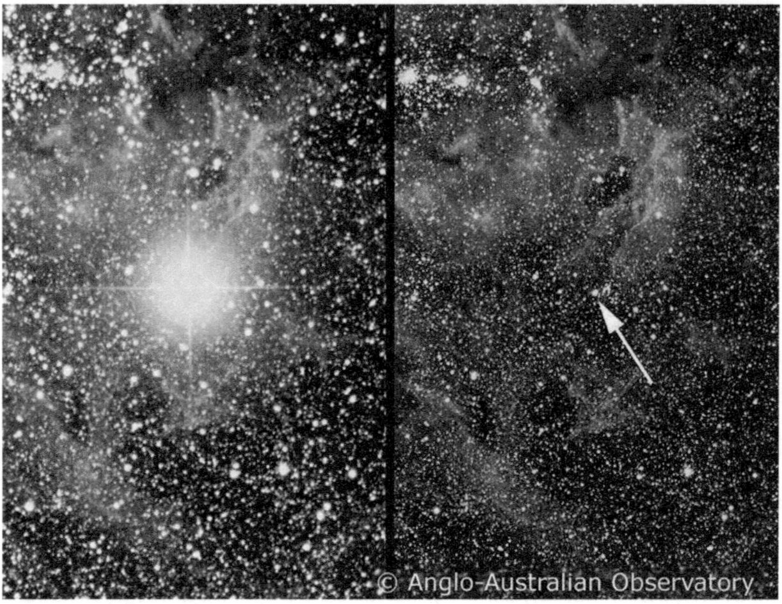

Fig. 8.14 *Right*: Tarantula nebula and the progenitor star Sk $-69°$ 202. *Left*: Peak of the explosion. By permission: Anglo Australian telescope, AAO/David Malin Images

In the mid-1960s Nicholas Sanduleak (1933–1990) investigated the Large and Small Magellanic Clouds and discovered (from the spectra of planetary nebula) that their heavy element content was only about 1/3 of the solar content. Along with this research, he prepared a catalogue of supergiant stars and their spectra, and among the catalogued stars was the star Sanduleak $-69°$ 202 (see Fig. 8.14). Sanduleak could not have imagined that 22 years after the publication of his catalogue[147] he would provide, for the first time in history, the spectrum of a star before it exploded as a supernova. The catalogue contained 1271 stars and the star in question was the 202nd on the list.[148] In the catalogue Sanduleak $-69°$ 202 appeared as an OB star of twelfth apparent magnitude.

On 23 February 1987, two astronomers located in Chile, Shelton, a graduate student from Toronto University, in the Las Campanas Observatory and Marsden in La Silla, were taking routine photographic plates of the LMC. Shelton did not realize at first what discovery was imprinted in the plates and went to look outside, where he saw the first naked-eye supernova in almost 400 years. Around 01:00 UT, the star Sanduleak $-69°$ 202 situated on the border of the Tarantula nebula was a quiet star, no different from the 10^{10} other stars of the LMC. About 10 hours later, when McNaught in Siding Spring Australia took a routine plate of the LMC, the star already had an apparent magnitude of 6, or about 250 times brighter.

---

[147] Sanduleak, N., 1970, Cerro Tololo Inter-American Obs. Contr., No. 89.

[148] The $-69°$ is the approximate declination in the celestial system of coordinates.

The morning after, around 05:00 (UT), Duhalde at Las Campanas searched the sky and discovered the SN. He notified Shelton who alerted the IAU telegram service in Cambridge, Mass.[149] Telegram 4316 was composed by Marsden, and contained a summary of the reports provided by the observers who discovered the SN.[150] This was the moment of birth of the SN 1987A saga. The SN reached an apparent magnitude of 2.9 at maximum, or about 4500 times brighter than the exploding star, which is equivalent to an intrinsic luminosity of $4.5 \times 10^8 L_\odot$. The astronomer's dream had come true. The SN is not in the Milky Way, but in the backyard, as it were, in our nearest neigbouring galaxy. Surely that was good enough.

Unfortunately, the Hubble Space Telescope (HST) was not yet in operation when the supernova exploded, since it was launched only in April 1990. The first (HST) images of SN 1987A came on 23–24 August 1990, and revealed the inner circumstellar ring in all its glory and detail, as will be discussed shortly.

8.13 Identification and Scattered Records

Regrettably, past observational records of Sk -69° 202 are limited, partly because no spectral peculiarities or variability attracted the attention of astronomers during the decades prior to its demise. Consequently, only serendipitous or survey data exist (see Fig. 8.15). As soon as the identification of the progenitor of the SN was certain, a chase began to seek out every existing piece of data. But the progenitor was such a normal and well behaved star that it had simply been of no interest to astronomers, and only scattered observations could be found.

Hardly two days after the eruption of SN 1987A, on 25 February, Shara and McLean[151] and in the same IAUC, Sanduleak himself, identified the exploding star as Sk -69° 202. Moreover, Sanduleak noticed that the image of Sk -69° 202:

> *[...] appears to be elongated northwest–southeast, suggesting a companion a magnitude or two fainter and separated by 1 arcsecond or less.*

The IAUC reported that the *duplicity* was confirmed by Lasker. Was the claim of a binary a red herring?

---

[149] IAU telegrams is a service provided by the Central Bureau for Astronomical Telegrams. Operated at the Harvard Smithsonian Center for Astrophysics (specifically under the SAO umbrella), under the auspices of Commission 6 of the International Astronomical Union (IAU). The goal is to circulate time-dependent phenomena quickly to all observatories so as to be able to monitor transient phenomena.

[150] Kunkel and Madore, Las Campanas Observatory, reported the discovery by Shelton, University of Toronto, and Duhalde from Las Campanas sent an independent report. Bateson, New Zealand, reported the independent discovery by Jones and by Moreno and Walker. McNaught from Siding Spring Observatory reported for himself and Garradd. Warner from Texas reported that Menzies from South Aftrica discovered signs typical of a Type I SN.

[151] Shara, M., & Mclean, B., IAUC No. 4318, 1987.

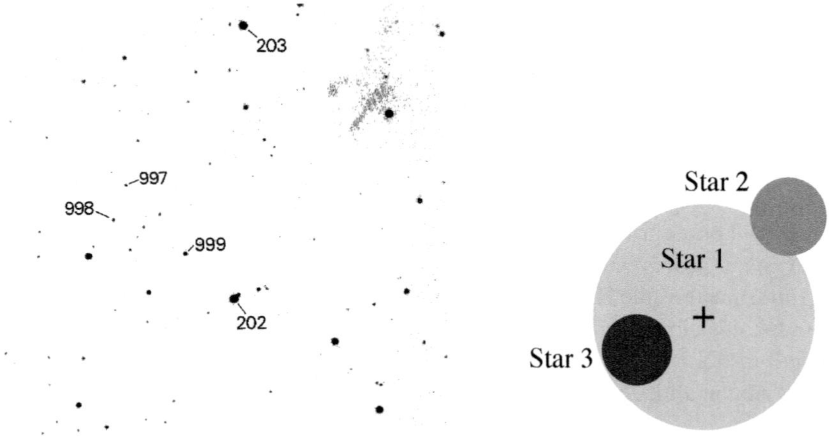

Fig. 8.15 *Left*: The 'last picture' of Sk $-69°$ 202 before the explosion, which apparently did not leave behind any compact star. Note that 202 appears elongated as if unresolved. The image of the brighter star 1 is overexposed, which explains why it is so large. After Walborn et al. 1987. *Right*: Schematic view of the progenitor field from White and Malin 1987. The separation between star 1, which exploded, and star 2 is 2.65 arcsec, which at the LMC translates to 2.18 lyrs. Hence, one would expect no interaction between the stars, even if they were supergiants

Djorgovski[152] suggested a companion to the SN, which was later found to be an unresolved nebula. Testor and Lortet[153] suggested that Sk $-69°$ 202 *was composed of at least two bright stars*. On 26 March, West et al. submitted a paper[154] in which Sk $-69°$ 202 was confirmed as the exploding star, and affirmed that none of the companions could be a progenitor. Note that West et al. used the word 'companion', rather than referring to a binary system, and yet some researchers understood it to be a binary system. They also confirmed that the progenitor was a blue star.

In May 1987, Walborn et al.[155] claimed that Sk $-69°$ 202 had two companions with different luminosities. They confirmed that it was a blue supergiant, and not a red one. On 21 July 1987, Heap and Lindler[156] claimed that:

While suggestive of a double star, we can only say that at such low count levels, the image of Sk $-69°$ 202 is consistent with either a single star or a double star.

White and Malin[157] concluded that:

The progenitor of SN 1987A was Sk $-69°$ 202 or a fainter binary companion.

---

[152] Djorgovski, S.G., IAUC No. 4376, 1987.

[153] Testor, G., & Lortet, M.-C., IAUC No. 4352, 1987.

[154] West, R.M., Lauberts, A., Jorgensen, H.E., & Schuster, H.E., A & A **177**, L1 (1987).

[155] Walborn, N.R., Lasker, B.M., Laider, V.G., & Chu, Y.-H., Ap. J. **321**, L 41 (1987).

[156] Heap, S.R., & Lindler, D.J., A & A **185**, L 10 (1987).

[157] White, G.L., & Malin, D.F., Nature **327**, 36 (1987).

The bothering feature, as stressed by Girard, van Altena, and Lopez,[158] was that the luminosities measured by the different groups were very divergent.

The final nail was hammered into the coffin of Sk $-69°$ 202 on 23 July 1987 by Gilmozzi et al.[159] Similar results were found by Sonneborn et al.[160] Moreover, this IAUC telegram contained a statement from Fransson that the behavior was consistent with a model due to Lundqvist and Fransson,[161] in which a wind from a blue star interacts with material shed by the star earlier, when it was a red giant star. This was important because at this stage it was not clear at all that the progenitor had ever passed through the red giant phase.

Once the identity of the exploding star had been established, it was possible to look at the available records. Isserstedt[162] observed the star between 1971 and 1973, got its visual luminosity, and classified it as B3I. In 1972 and 1973, Rousseau et al.[163] obtained a wavelength-limited spectrum of the star and just confirmed the classification.

In 1987, only after Sk $-69°$ 202 had exploded, Walborn et al.[164] analyzed eight CTIO 4 meter telescope plates obtained between 1974 and 1983. Sk $-69°$ 202 appeared to be normal, and no near-IR excess (which would have implied the existence of dust around the star) was detected. Furthermore, there was little or no light variability between 1974 and 1983, nor between 1970 and 1981 (Blanco et al.[165]). Plotkin and Clayton[166] examined a large number of Harvard sky patrol plates obtained between 1896 and 1954 and found no variability at the 30% level. The star did not show any obvious signs of distress. This was the 'paradox': the star appeared to the outside world as absolutely normal.

On the basis of this information, the data on the progenitor could be summarized as follows: Sk $-69°$ 202 had a luminosity of $10^5 L_\odot$, a surface temperature of 16 000 K, a mass of about $20 M_\odot$, and absolutely no peculiarities. The spectral type was B3I, which is a blue giant star.

What about the location? Was that special? Panagia et al.[167] studied neighboring stars to within a range of 90 lyrs of SN 1987A. They found that this volume of space had been forming stars in several episodes between 1 and 150 Myrs previously. These stellar youngsters were superposed on a fainter stratum of older stars, with

---

[158] Girard, T., van Altena, W.F., & Lopez, C.E., Ap. J. **96**, 58 (1988).

[159] Gilmozzi, R., Cassatella, A., Clavel, J., Fransson, C., Gonzalez, R., Gry, C., Panagia, N., Talavera, A., & Wamsteker, W., Nature **328**, 318 (1987).

[160] Sonneborn, G., Kirshner, R., Fransson, C., Cassatella, A., Wamsteker, W., Gilmozzi, R., & Panagia, N., IAUC No. 4685 (1988).

[161] Lundqvist, C., & Fransson, C., A & A **192**, 221 (1988).

[162] Isserstedt, J., A & A S **19**, 259 (1975).

[163] Rousseau, J., Martin, N., Prevot, L., Rebeirot, E., Robin, A., & Brunet, J., A & A S **31**, 243 (1978).

[164] Walborn, N.R., Lasker, M., Laidler, V.G., & Chu, Y-H., Ap. J. **32**, L 41 (1987).

[165] Blanco, V.M., Walker, A., & McCarthy, M.F., IAUC **4349**, 2 (1987).

[166] Plotkin, R.M., & Clayton, G.C., JAAVSO **32**, 89 (2004).

[167] Panagia, N., De Marchi, G., & Romaniello, M., arXiv e-print `arXiv:astro-ph/0609539`, 2006.

ages in the range $0.5–6 \times 10^9$ yrs. The dozen bright blue stars around SN 1987A are massive stars, each more massive than $6M_\odot$. With an age of about 12 Myrs, they are members of the same generation of stars that gave birth to the supernova progenitor. In short, it is a nursery of stars, in which one of the more massive ones exploded just before they had time to disperse.

8.14 Neutrinos

The energy release by the supernova was fast. During the first 10 minutes after the explosion, about 99% of the total available energy was released in the form of neutrinos. Only about $\sim 1\%$ of the energy went into the kinetic energy of the ejected material, and a trifle of only 0.01% of the energy went into the visible fireworks. Even this 'minuscule' amount of energy (about $10^{48}–10^{49}$ ergs) in the visible range turned the SN into a strong beacon which illuminated the space around and almost outshone the entire LMC galaxy. The copious neutrino losses represent a dramatic change from normal stellar evolution up to this point. Along the main sequence, stars convert nuclear energy almost exclusively into photons, and the total energy which goes into neutrinos is negligible. As the star progresses towards the ultimate collapse, the role of the neutrinos in removing energy from the star increases monotonically, until they reach complete dominance before and during the explosion. The weakest of all forces plays a dominant role in the greatest explosion in the Universe.

One of the extraordinary successes of the observation and the theory was the discovery of neutrinos from the supernova. As the neutrinos emerged from the collapsing core and not from the surface, the neutrino signal preceded the moment of maximum light.

On 23 February, at around 07:36 UT, Kamiokande II[168] recorded the arrival of 9 neutrinos within an interval of 2 seconds, followed by 3 more neutrinos 9 to 13 seconds later.[169] Simultaneously, the same event was detected by the IMB detector,[170] which counted 8 neutrinos within about 6 seconds. A third neutrino telescope, the Baksan,[171] also recorded the arrival of 5 neutrinos within 5 seconds from each other. This makes a total of 25 neutrinos detected on Earth.

---

[168] Kamiokande II is a neutrino telescope. A cylindrical tank, 15.6 meters in diameter and 16 meters deep, containing about 3000 cubic meters of water, serves as a detector. About 1000 giant photomultiplier tubes are placed on the inner walls of the tank. Kamiokande is located in the Kamioka mine in Japan, about a kilometer underground. The upgraded Super-Kamiokande contains 50 000 tons of water. The basic reaction is $\nu + e^- \rightarrow \nu + e^-$. An electron hit by the energetic neutrino recoils and emits Cerenkov radiation, which is picked up by the photomultiplier tubes. Kamiokande in Japanese means 'bite into God'.

[169] Hirata, K.S., and 23 authors, Phys. Rev. D. **38**, 448 (1988).

[170] The IMB detector is located in the Morton–Thiokol salt mine near Faiport, Ohio, at a depth of about 580 meters. It is bigger than Kamiokande II, but not as deep underground.

[171] The Baksan telescope is located in the North Caucasus Mountains of Russia, under Mount Andyrchi.

About 4 hours before the above collaborations reported the detection of neutrinos from SN 1987A, the Mont Blanc collaboration[172] reported the detection of neutrinos. This neutrino signal was not observed by the more sensitive Kamiokande neutrino telescope, a fact which caused the scientific community to doubt the detection. Schaeffer, Declais, and Jullian,[173] for example, argued that the Mont Blanc results, if true, invalidated the standard theory of core collapse and the formation of a neutron star, because of the huge energy requirements they imply (too many neutrinos detected for the sensitivity of the instrument), while the other observations confirmed the present day theory of the formation of a neutron star (which so far has not been detected). In other words, the result from the Mont Blanc team would be acceptable only if it agreed with the theory. The Mont Blanc collaboration published a detailed explanation of the results[174] in which they claimed that it *can be explained within reasonable theoretical expectations*, and refered to Hillebrandt et al. (see later), who provided a theoretical explanation (which was not accepted either). The experimental result and the theoretical hypothesis were received by the scientific community with a general skepticism.

The remarkable discovery of the SN neutrinos won for Masatoshi Koshiba the 2002 Nobel Prize for Physics. He shared half the prize with Raymond Davis (see Sect. 9.2) for *pioneering contributions to astrophysics, in particular for the detection of cosmic neutrinos*. (The second half went to Riccardo Giacconi for *pioneering contributions to astrophysics, which have led to the discovery of cosmic X-ray sources*.)

The peculiar distribution of the detected neutrinos prompted several speculations which were never confirmed. Hillebrandt et al.[175] suggested that the first pulse of neutrinos, those observed only by the Mont Blanc detector, originated from the formation of a neutron star, while the second burst resulted from the collapse to a black hole. Stella and Treves[176] claimed that it implied a close binary system with a period of ~ 0.2 s. Ogelman and Buccheri[177] analyzed the scant data and claimed to have discovered a period of 0.2024 s, which is almost a confirmation of the previous result. Saha and Chattopadhyay[178] took the 25 neutrino arrival times and claimed that they imply the existence of a neutron star which rotates with a period of 0.011 ± 0.0002 s. Harwit et al.[179] analyzed the same data and found a period of 0.00891 s. The series of period discoveries in the meager data ended when Fisher[180] analyzed the same data and concluded that no period between 0.005–0.015 s exists in the data. In view of so many different results based on the same limited data, part

172 Aglietta, M. and 24 authors, Europhys. Lett. **3**, 1315 (1987).

173 Schaeffer, R., Declais, Y., & Jullian, S., Nature **300**, 142 (1987).

174 Aglietta. M., and 24 authors, Europhys. Lett. **3** (12), 3 (1987).

175 Hillebrandt, W., Hoflich, P., Kafka, P., Muller, E., Schmidt, H.U., & Truran, J.W., A & A **180**, L 20 (1987).

176 Stella, L., & Treves, A., A & A **185**, L 5 (1987).

177 Ogelman, H., & Buccheri, R.L., A & A **180**, L 23 (1987).

178 Saha, D., & Chattopadhyay, G., Ast. Space Sci. **178**, 209 (1991).

179 Harwit, M., Biermann, P.L., Meyer, H., & Wasserman, I.M., Nature **328**, 503 (1987).

180 Fisher, H.D., A & A **185**, L 15 (1987).

of which was even considered as noise,[181] it is only safe to conclude that there was a neutrino signal and nothing beyond it. The rest is unfounded over-interpretation.[182]

All the above claims were soon refuted by Kristian et al. in 1991.[183] The question was examined once more in 2005 by Graves et al.,[184] who used the Hubble telescope to look for a stellar-type image. They could not discover any point source with luminosity higher than about $1.3L_{\odot}$. The search also failed to discover the *survivor of a possible binary system.* Only a collusion of factors could perhaps save the idea of a compact object in SN 1987A.

The most important implication of the neutrinos was that it provided support for the hydrodynamic core-collapse theory, releasing about 3×10^{53} ergs of gravitational energy mainly in the form of neutrinos of all kinds. This confirmed the old idea that it is a transition to a lower gravitational state rather than a thermonuclear explosion, for example. The collapse was conditioned by having a mechanism that could remove the energy released in the transition to the lower gravitational state, and it is the transfer of energy by neutrino emission and deposition that assists in lifting the collapsing layers, as suggested by Colgate and White[185] back in 1966, and Arnett a year later.[186] It is clear that neutrinos play a dominant role in the SN explosion, yet the details are still missing. And we must not forget the still undiscovered neutron star inside the ashes of SN 1987A, which had to be formed, according to the theory.

8.15 The Detection of Gamma Ray Line Emission

The collapse would have converted a significant fraction of the core material into radioactive isotopes with various decay times. Those with very short decay times would have disappeared while the collapsed core was still shrouded with external layers. But the radioactive isotopes with sufficiently long decay times, say days and longer, would still be alive when the expanding envelope thinned out and allowed some of the γ rays from the decays to escape through the remnant gases and hence be observed.

The end product of explosive burning of silicon is ^{56}Ni. Nickel is a quite unique nucleus. It is a doubly magic nucleus in the jargon of nuclear physics ($Z = N = 28$),[187] according to the α-model. Hence, according to all the predic-

---

[181] Arnett, W.D. & Rosner, J.L., PRL **58**, 1906 (1987).

[182] The saying at the Technion about such a situation goes like this: *A straight line can be drawn through any three points, provided the pencil is sufficiently wide.*

[183] Kristian, J. No pulsar in SN 1987A, Nature **349**, 747 (1991).

[184] Graves, G.J., and 18 authors, Ap. J. **629**, 944 (2005).

[185] Colgate, S.A., & White, R.H., Ap. J. **143**, 626 (1966).

[186] Arnett, D., Can. J. **45**, 1621 (1967).

[187] A magic number is a number of protons or neutrons that can be arranged in complete shells within the nucleus. The seven known magic numbers are 2, 8, 20, 28, 50, 82, and 126. Nuclei consisting of such a magic number of nucleons have a higher than average binding energy per nucleon.

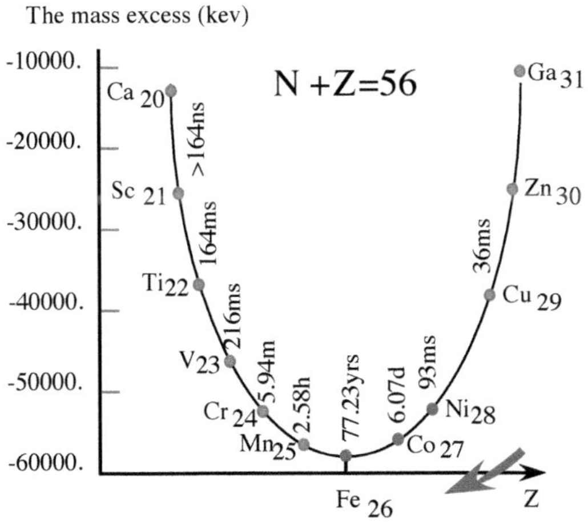

Fig. 8.16 Nuclei with mass 56 ($A+Z=56$) as a function of the mass excess. The decay times of the unstable nuclei to the daughter are shown. As one moves away from the most stable nucleus, the decay times shorten. The *red arrow* indicates the decay which powers the SN

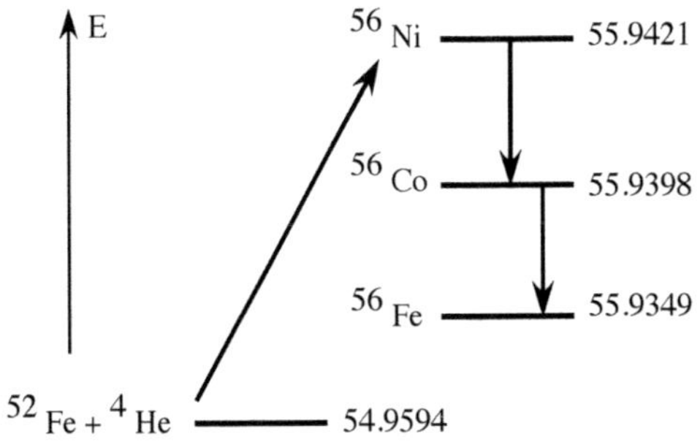

Fig. 8.17 The relative energies in the formation and decay of nickel. Formation takes place as the last step of silicon burning, while the decay to ^{56}Fe takes place after the explosion. The numbers on the right are the binding energies of the nuclei in MeV

tions, ^{56}Ni should be an unusually stable nucleus. But ^{56}Ni defies the predictions of theoretical nuclear physics. It is in fact unstable and decays. On the other hand, ^{58}Ni which has two extra neutrons, is stable, but add one more neutron to get ^{59}Ni and the result is an unstable nucleus again (with a decay time of 76 000 years).

All nuclei with 56 nucleons ($N+Z=56$) are shown in Fig. 8.16. The interplay between the nuclear forces makes ^{56}Fe the most stable nucleus with this number of nucleons. The evolution time at this stage allows for the synthesis of ^{56}Ni, which decays in 6.07 days, but not the synthesis of ^{56}Cu, which decays in 93 milliseconds. Hence, the timescale of the stellar evolution at this stage cannot be much longer then a few days. Indeed, the silicon burning takes about one day. As can be seen from the decay times, only ^{56}Ni survives until after the explosion.

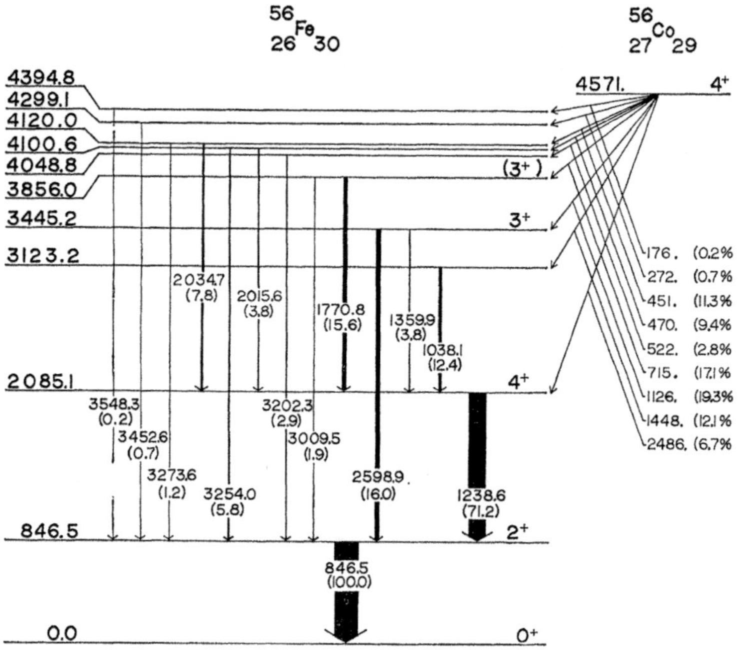

Fig. 8.18 The energy levels of ^{56}Fe into which ^{56}Co decays. From Dolan et al 1966. *Thick arrows* mark the observed γ-ray lines

If ^{56}Fe is the most stable nucleus, rather than ^{56}Ni, why is the latter formed in the first place? The temperature at which silicon burning takes place is 4–5 $\times\, 10^9$ K, and the reaction which forms the Ni is shown in Fig. 8.17. The numbers are the atomic masses. In the high temperature environment, the nuclei have large kinetic energies, and this means that the last reaction in silicon burning, i.e., $^{52}\text{Fe} + \alpha \rightarrow {}^{56}\text{Ni}$, can go through via the absorption of 5 α particles by the ^{32}Si nucleus. Due to the high temperature, the reaction goes in the direction of the Ni. Once the explosion has taken place and the temperature begins to decrease, the nickel decays into iron by two electron captures.

If Nature had followed the idea of the α model to the letter, the SN light curve would have looked completely different. The energy diagram for ^{56}Co–^{56}Fe decays is shown in Fig. 8.18 (from Dolan, McDaniel, and Wells 1966[188]). We see that cobalt decays into a large number of energy levels in the iron nucleus. The γ rays are emitted when the excited iron nucleus decays into the ground state. The strongest emission is at 846.5 keV, and all the other emissions are given relative to this transition. The width of the arrow relates to the strength of the emission. Thus the most prominent lines are $E = 846.5$, 1238.6, 1770.8, and 2598.9 keV. But note that, if

[188] Dolan, K.W., McDaniel, D.K., & Wells, D.O., Phys. Rev. **148**, 1151 (1966).

Table 8.7 Predicted detectability of γ lines in Type I SNs, from Clayton et al. 1969

γ lines (energy keV)	Time after outburst
^{56}Ni (812, 748)	10–20 days
^{56}Co (840, 1240)	15–40 days
^{55}Co (weaker lines)	
^{48}V (983, 1310)	15–30 days
^{44}Sc (1156)	15 days–50 yrs

all γ transitions in the iron nucleus were equally probable, the emitted γ rays would have been below the threshold for observation by present day satellites.

In 1969, Clayton, Colgate, and Fishman[189] predicted that the expanding remnants of a Type I SN shine by converting the radioactive decay energy of ^{56}Ni into visible photons. They predicted that ^{56}Ni would be the most abundant element resulting from silicon burning in the supernova shock conditions. They estimated that about $0.14M_{\odot}$ of ^{56}Ni would be required to power the glowing remnant. They reached this conclusion only after Colgate and McKee,[190] at the suggestion of Truran, had shown that the radioactive power delivered by the ^{56}Ni to the expanding shell while it is still opaque does maintain a sufficiently high temperature to provide a high optical luminosity when the expanding nebula becomes transparent. The predictions are given in Table 8.7. Clayton et al. predicted that the best option would be the lines from ^{56}Co.

It took 19 years to corroborate the prediction by Clayton et al. In 1988, the Solar Maximum Mission satellite[191] was fortunately still in operation (see Fig. 8.19). Among its many feats was the first observation of γ rays from SN 1987A by Matz et al.[192] Two γ lines were detected by the SMM, at 847 keV and 1238 keV. The flux during the first 38 days after the explosion was $\sim 10^{-4}$ photons/cm^{2}s. This was a tremendous success, because it directly confirmed the idea of radioactive elements powering the light curve.

It was not, however, a complete success story. In contrast with the prediction, SN 1987A is Type II, not Type I. And neither did the theory predict the γ rays from ^{56}Co to appear so soon, while the remnant was still opaque. So mixing had to be

---

[189] Clayton, D.D., Colgate, S.A., & Fishman, G.J., Ap. J. **155**, 75 (1969).

[190] Colgate, S.A., & McKee, C., Ap. J. **157**, 623 (1969). The work was announced in the 126th AAS meeting, April 1968.

[191] The Solar Maximum Mission (SMM) was launched on 14 February 1980, primarily to study the Sun during the high part of the solar cycle. The payload contained many detectors, including a gamma ray spectrometer which could also observe celestial sources. A malfunction in the satellite in January 1981 cut short the original mission. SMM was recovered by the space shuttle Challenger in April 1984 and serviced in orbit. SMM then collected data until 24 November 1989, at which time the aerodynamic forces became too great for the attitude control system to maintain accurate pointing. It subsequently served out its productive life until burning up in the Earth's atmosphere on 2 December 1989.

[192] Matz, S.M., Share, G.H., Leising, M.D., Chupp, E.L., Vestrand, W.T., Nature **331**, 416 (1988).

Fig. 8.19 The Solar Maximum Mission (SMM) satellite that observed for the first time the γ rays from radioactive cobalt synthesized in SN 1987A. Credit: NASA

Table 8.8 Radioactive sources in SN 1987A

Species					Decay constant	Contribution to light curve	Mass $[M_\odot]$
$^{56}Ni+e^-$	$\rightarrow$	$^{56}Co+\gamma$			6.077 dy	0–18 dy	0.069
		$^{56}Co+e^-$	$\rightarrow$	$^{56}Fe+\gamma$	111.3 dy	18–1100 dy	
$^{57}Ni+e^-$	$\rightarrow$	$^{57}Co+\gamma$			2.17 dy		0.003
		$^{57}Co+e^-$	$\rightarrow$	$^{57}Fe+\gamma$	390 dy	110–1800 dy	
$^{44}Ti+e^-$	$\rightarrow$	$^{44}Sc+\gamma$			87 yr	1800 dy $\rightarrow$	0.0001
		$^{44}Sc+e^-$	$\rightarrow$	$^{44}Ca+\gamma$	5.4 hr		
		^{44}Sc	$\rightarrow$	$^{44}Ca+e^+$			

invoked to allow the escape of the γ rays. This was good luck for the observers, because the radioactive decay signal weakens with time, and the satellite could not have detected the γ rays from such a distant object, had they arrived at the predicted time. So they would not have been able to verify one of the fantastic features of the not yet complete supernova story, namely, that the light curve is powered by radioactive decay, and of the right element.

The latest observation was carried out by the GSFC/Bell/Sandia collaboration[193] on days 434 and 613. In the last observation, even the line at 2598.6 keV was detected. This time, however, the detailed analysis included the shape of the line. The authors concluded that the assumption of spherical symmetry or homogeneity or both must be dropped.

[193] Tueller, J., Barthelmy, S., Gehrels, N., Teegarden, B.J., Leventhal, M., & MacCallum, C.J., Ap. J. **351**, L 41 (1990). This was a high altitude balloon flight.

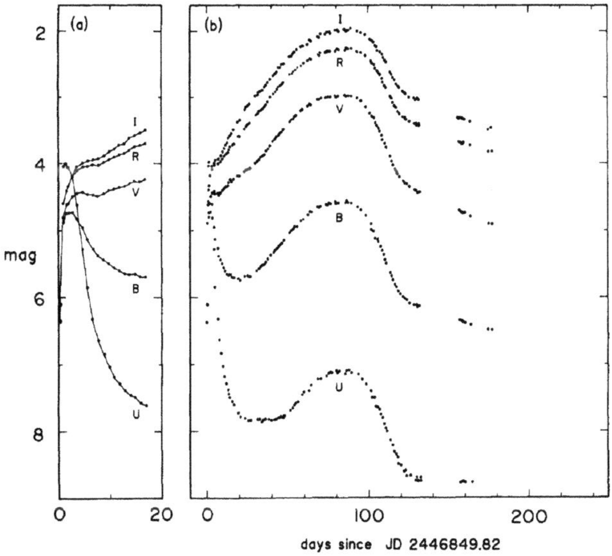

Fig. 8.20 The light curve of SN 1987A through the first 200 days, from Hamuy et al. 1987. The curves are marked according to the different filters used. The rise in the V (visible), R (red), and I (infrared), and the decline in the U (ultraviolet) and B (Blue) during this phase are so far unique to SN 1987A

8.16 The SN 1987A Light Curve

On the basis of the strong hydrogen lines in its optical spectrum, SN 1987A was classified as a Type II supernova.[194] But as it was the explosion of a blue supergiant rather than a red one, it was an atypical SN II, and hence should be Type IIpec, where the pec stands for peculiar. The unique feature was its light curve. The light curve did not reach maximum until three months after the collapse of the core had taken place (according to the neutrino signal), and at maximum it was only $\sim 10\%$ as luminous as most Type II SNs.

After a plateau of several days, the light curve decayed exponentially with a mean lifetime of 111.3 days. About 400 days after the explosion, the optical light began to drop off more rapidly, while the IR emission increased. Such an effect is known to be caused by the formation of dust. The dust obscuration was nearly independent of wavelength, indicating that the inner dust clouds were highly opaque.

The evolution of SN 1987A and the light curve was monitored by the Cerro Tololo Inter-American Observatory (CTIO) in La Serena, Chile, and the South African Astronomical Observatory, South Africa (SAAO). The time variations of the emitted light are very important in attempts to understand what powers the remnants of

[194] Parthasarathy, M., Branch, D., Baron, E., David, J., & Jeffery, D.J., Bull. Astr. Soc. India **34**, 385 (2006).

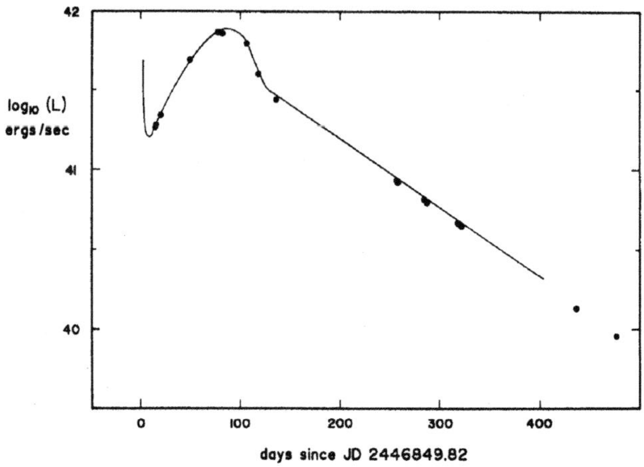

Fig. 8.21 The light curve of SN 1987A through the first 500 days, from Suntzeff et al. 1988

the supernova. The clear exponential decay of the light curve revived the old idea of radioactive powering.

About seven months after the explosion Hamuy et al.[195] published the light curve for the first 177 days. By that time the existence of radioactive cobalt had been confirmed by the detection of γ rays. However, the paper did not include any attempt to reproduce the light curve assuming radioactive cobalt as the source of energy. The authors pointed out that, according to Woosley et al.,[196] the light curve should exhibit a plateau, and only after day 25 should the radioactive ^{56}Ni power the light curve and bring it to a maximum some 50 to 150 days after the explosion, depending on the core mass and the explosion energy. The prediction was that the maximum light emission should be followed by an exponential decay, powered by ^{56}Co with a timescale of 111.3 days. On the other hand, Shigeyama et al.[197] did try to interpret the light curve, and found that, to explain the light curve after day 7, they had to assume that a buried neutron star provided a constant energy source. As a matter of fact, Shigeyama et al. claimed that:

> *Our results imply that the pulsar model is able to account for observations, in particular the plateau-like peak, better than the radioactive-decay model.*

Notwithstanding, both models had numerous fitting parameters and hence it was no surprise that both fitted the data ‘equally well’. In any case, it was impossible to rule out either of them.

---

[195] Hamuy, M., Suntzeff, N.B., Gonzalez, R., & Martin, G., A. J. **95**, 63 (1988).

[196] Woosley, S.E., Pinto, P.A., & Ensman, L., Ap. J. **324**, 466 (1988). See also Ap. J. **318**, 664 (1987).

[197] Shigeyama, T., Nomoto, K., Hashimoto, M., & Sugimoto, D., Nature **328**, 320 (1987).

It was not long before differences emerged in measurements between CTIO and SAAO. So 18 months after the eruption, the CTIO group plus Suntzeff et al.[198] discussed the differences in measurements of the decay time. The two sequences of results are given in Table 8.9 and appear quite incompatible. The difference is crucial. If you calculate the decay of the total flux according to the CTIO data, you find a time constant of 100.2 ± 0.2 days for the period 122–402 days after the explosion. On the other hand, the equivalent SAAO result was 108.5, which agreed well with the predicted 111.3 day decay for ^{56}Co only until day 265, and then started to deviate significantly. As can be seen from Fig. 8.21, the exponential behavior is evident. The controversy was over the value of the slope. The annoying fact was that the time constant was 100.5 days and not 111.3 days, which is the decay time of ^{56}Co. The quoted error in the measurements was rather small. Moreover, the SAAO result was 109.1 days for days 260–385, while later it started to deviate from this value. After attempting without success to resolve the difference, what the authors had to say about the discrepancy was this:

> *We feel that the differences between the data [...] are consistent with at least a range of e-folding times of 100–111 days.*

The results are important because deviations from the exponential law translate directly into an assumption about mixing, fast exposure of the radioactive material, or maybe a clumpy state of the matter.

Whitelock et al. of the SAAO group[199] analyzed the energy balance of SN 1987A. They showed that the initial linear decline was powered entirely by the radioactive decay of ^{56}Co, and fixed the initial mass of ^{56}Ni at $0.08M_\odot$ (because they got a good agreement with the decay time). The total energy released by the radioactive decay was estimated as 13.8×10^{48} erg, which was more than what was radiated away. Furthermore:

> *The discrepancy amounts to 6×10^{48} erg when account is taken of Woosley's (1988) theoretical prediction that, until about day 40, the radiation from SN 1987A was due to the release of energy deposited by the initial blast.*

Hence, there appeared to be an excess of energy production over radiation. It is not clear whether the effect would have disappeared if, as suggested to the authors by Woosley, mixing had been taken into account. Moreover, Whitelock et al. found that the deviation from the 111.3 day decay time increased with time. Thus the initial decay time was consistent with the time scale of decay of ^{56}Co, in particular it was consistent with the idea that the two γ lines had been detected and that the time scales agreed. But something unclear happened towards the end of this period.

On day 265, a deviation from the 111.3 day constant developed, and increased monotonically from that time on. Ignoring the explanations of Woosley (private communications to the authors), they noticed that:

> *On day 340, the γ ray and related X-ray flux from SN 1987A (Kumagai 1988, Nomoto private communication) was about 8% of the total luminosity [...] this is exactly the flux required to make up the deficit in the light curve.*

---

198 Suntzeff, N.B., Hamuy, M., Martin, G., Gomez, A., Gonzalez, R., A. J. **96**, 1864 (1988).

199 Whitelock, P.A., and 20 authors, MNRAS **234**, 5 (1988), SAAO collaboration.

Table 8.9 Decay times measured by two observatories. Letters stand for the particular filter used. Numbers correspond to days. After Suntzeff et al. 1988

Filter	CTIO (days 100–400)	SAAO (days 147–265)	SAAO (days 265–385)
U	-2714 ± 2700	-2905 ± 340	627 ± 21
B	162.7 ± 1.0	145.6 ± 0.6	160.9 ± 0.6
V	108.9 ± 0.3	112.9 ± 0.3	114.8 ± 0.2
R	130.4 ± 0.7	133.7 ± 0.4	106.8 ± 0.4
I	119.1 ± 0.9	142.6 ± 0.6	96.7 ± 0.5

The idea was simply that the decay of ^{56}Co continued to be the sole significant source of energy within the supernova in this period, but that the penetration of radioactive material into the outer layers allowed some γ rays to escape (and be detected on Earth), hence spoiling the nice energy balance.

In 1988, Salvati et al.[200] referred to an announcement by Middleditch[201] in which he claimed to have discovered a pulsar (although it was never confirmed). So Salvati et al. wrote:

> *The discovery of a pulsar inside the remnant of supernova 1987A is not in itself surprising [...]. The properties of the pulsar, however, are rather surprising, especially its extremely short period and the small value of the surface magnetic field.* Subsequently, they continued: *In the discovery observations, a weak sinusoidal modulation of the pulsar frequency is noted. If confirmed and interpreted at face value as evidence for a binary companion, the modulation would imply a companion of mass* $\sim 10^{-3} M_{\odot}$.

A year later, Bandiera et al.[202] called upon the elusive neutron star to power the nebula by its rotational energy.

In parallel, Kumagai et al.[203] showed that mixing of ^{56}Co explained the light curve and the γ rays, but not the X rays. As they stated, no mechanism for mixing was known and hence, as they pointed out quite reasonably, *the actual process of mixing is highly uncertain.* So they assumed some phenomenological mechanism and applied it to the light curve. However, most importantly, they even showed that particular mixing mechanisms or prescriptions could yield light curves with several peaks. This illustrates how sensitive the result is to the form of mixing. One can get almost everything one might want.

Kaumagai et al. returned to the problem a year later,[204] this time assuming the radioactive cobalt and a buried neutron star as energy source for the light curve. They hypothesized the formation of clumps to explain the slower decrease in X-ray luminosity, but even so, an additional X-ray source was needed. Could it be a buried

[200] Salvati, M., Pacini, F., & Bandiera, R., Nature **338**, 146 (1989).

[201] Middleditch, J., IAUC No. 4735 (1989).

[202] Bandiera, R., Pacini, F., & Salvati, M., Ap. J. **344**, 844 (1989).

[203] Kumagai, S., Shigeyama, T., Nomoto, K., Itoh, M., Nishimura, J., A & A **197**, L 7 (1988).

[204] Kumagai, S., Shigeyama, T., Nomoto, K., Itoh, M., Nishimura, J., & Tsuruta, S., Ap. J. **345**, 412 (1989).

pulsar that had so far gone undetected? They predicted that a contribution of ^{44}Ti should appear in the light curve as time progressed. However, they stated that:

> *If the reported slowdown on the decline rate of the visual luminosity (Hamuy et al. 1988) is really due to an additional energy source, the source would be either the neutron star or ^{57}Co.*

The idea of an additional radioactive cobalt nucleus, ^{57}Co, was also suggested by Suntzeff et al. 1991, along with the idea of a buried neutron star. Similar ideas were proposed by others.[205]

The need for a contribution from a neutron star did not decay with time, on the contrary. In 1991, Kumagai et al.:[206]

> *[...] calculated the theoretical light curve to derive the necessary amount of additional energy input from ^{57}Co and ^{44}Ti and the buried neutron star.*

They concluded that the ^{57}Co was less likely to be the source. The buried pulsar would be a more likely energy source.

At about the same time, Suntzeff et al. 1991[207] tried to reproduce the light curve through to day 1000. They found that they could reproduce the observations by assuming a contribution from the new radioactive element, namely ^{57}Co. So they assumed about $0.01M_{\odot}$ of ^{57}Co. On the other hand, the theory of nucleosynthesis provided upper limits on how much ^{57}Co and ^{44}Ti could be synthesized in the explosion, and these upper limits were below what was needed to explain the light curve. On that account, they claimed that either an additional energy source was needed or one of the basic assumptions in modeling the observation was in error. Not much had changed a year later,[208] save an upper limit on how much power the neutron star could provide.

In 1993, an international collaboration between France and the USA launched a ballloon with X-ray and γ-ray detectors.[209] The ballon flight was on 22 May 1989, which was day 818 for the remnant of SN 1987A. No γ lines were detected and the researchers could only place upper limits on the flux values for the γ-ray lines coming from the decay of radionuclides synthesized in this star, such as ^{56}Co, ^{57}Co, or ^{44}Ti. The null results were consistent with models incorporating the following upper limits on the masses: $0.073M_{\odot}$ of ^{56}Co, $3.1 \times 10^{-3}M_{\odot}$ of ^{57}Co, and, finally, $1.2 \times 10^{-4}M_{\odot}$ of ^{44}Ti. They saw nothing, so they only got upper limits.

Some six years later, Lundqvist et al.[210] obtained an upper limit to ^{44}Ti of $\leq 1.5 \times 10^{-4}M_{\odot}$, which was not better than the previous known limit. However, they noted the following:

---

205 Arnett, W.D., & Fu, A., Ap. J. **340**, 396 (1989); Bouchet, P., Danziger, I.J., & Lucy, L.B., AJ **102**, 1135 (1991).

206 Kumagai, S., Shigeyama, T., Nomoto, K., & Hashimoto, M., A & A **243**, L 13 (1991).

207 Suntzeff, N.B., Phillips, M.M., Depoy, D.L., Elias, J.H., Walker, A.R., AJ **102**, 1118 (1991).

208 Suntzeff, N.B., Phillips, M. M., Elias, J.H., Walker, A.R., Depoy, D.L., Ap. J. **384**, 33 (1992).

209 Chapuis, C., and 23 other authors, Ap. J. **403**, 332 (1993).

210 Lundqvist, P., and 7 authors, A & A **347**, 500 (1999).

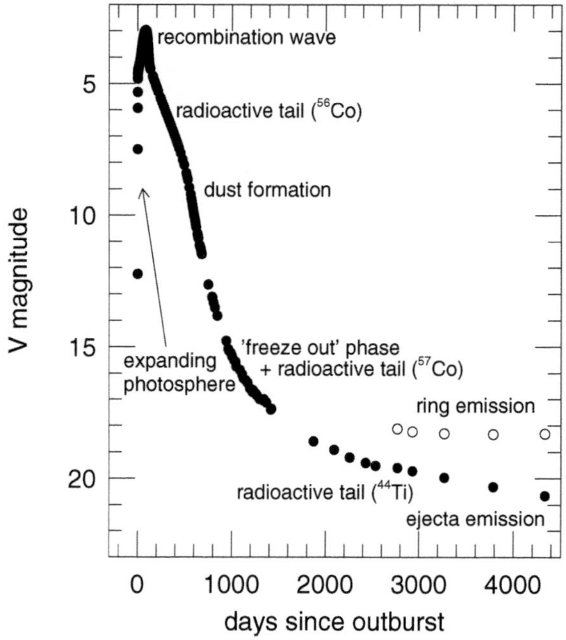

Fig. 8.22 The light curve of SN 1987A from Leibundgut and Suntzeff 2003. The V magnitude is the negative logarithm of the energy flux in the visible range. An exponential decay in time is a straight line, where the slope is one over the constant decay time

> *Models for the yield of ^{44}Ti give quite different results. This is most likely due to how the explosion is generated in the models, and how fallback onto the neutron star is treated.*

They explained that the theoretical models did not lead naturally to explosion, but were actually forced artificially to explode. Different methods were applied by different researchers, and these assumptions affected the 'predicted' amount of ^{44}Ti.

The most recent attempt to explain the light curve is due to Fransson and Kozma.[211] This is how Kozma and Fransson summarized the energetics of the light curve. After a couple of days the main energy input to the SN ejecta came from radioactive decay: first ^{56}Ni followed by ^{56}Co. Beyond 1100 days, ^{57}Co took over, while at very late epochs, beyond 2000 days, ^{44}Ti decay became dominant. There was no mention of a possible neutron star, nor the effect of the perplexing mixing.

By 2003, Leibundgut and Suntzeff[212] discussed the light curve of SN 1987A (see Fig. 8.22), but did not mention any problems. So far no remnant neutron star had been discovered.

---

[211] Fransson, C., & Kozma, C., arXiv e-print `arXiv:astro-ph/0112405` (2002).

[212] Leibundgut, B., & Suntzeff, N.B., arXiv e-print `arXiv:astro-ph/0304112`.

8.17 The Rings that Rang the Bell

It was clear that any progenitor of a supernova undergoes periods of mass loss during the red giant phase, and nobody expected Sk -69° 202 to have been any different. For this reason it was thought that sooner or later the fast-moving exploding layers of the star would catch up with the lost layers as they slowly moved away. But nobody expected the spectacle discovered shortly after the explosion of SN 1987A.

The first signs of the ejecta were discovered in June 1987, by Wamsteker et al.,[213] when they noticed the appearance of emission from certain layers moving at about 2000 km/s, and by Fransson et al.[214] In 1989, Sparks et al.[215] reported:

> *A faint arc of V band emission with a sharp inner and outer boundary centered on the supernova and of radius $\approx$ 8.3 arcsec and width $\approx$ 2.5 arcsec is clearly discernible above background.*

This was the first sign of what was about to come. No photograph was provided in the paper. The interpretation given by the authors was that the SN was illuminating the previously ejected mass, a phenomenon later called the 'light echo'. In other words, the strong beacon of the exploding supernova was illuminating the gas around the star. As time went by, more distant clumps of gas were expected to be illuminated, so the emission was expected to move away from the star.

In 1988, 10 months after the explosion, Sonneborn et al.[216] observed changes in the emissions that came from the nebula around the location where SN 1987A had exploded, and stated that:

> *The above changes are consistent with the circumstellar material having a spherical geometry, but they also indicate that recombination plays an important role in the evolution of the line fluxes.*

About a year later, Fransson et al.[217] wrote:

> *The time evolution is consistent with that expected from a fluorescent light echo by a circumstellar shell. A nebular analysis reveals a large nitrogen overabundance with N/C=7.8 $\pm$ 4 and N/O=1.6 $\pm$ 0.8. These values are respectively factors of 37 and 12 higher than the solar values, implying that the gas has undergone substantial CNO processing.*

About two years later, Crotts et al.[218] were the first to draw a ring as a description of the 'light echo', and to pose the question:

> *How did non-radial, non-spherical sheets arise so close to the SN, yet offset from it?*

This can be seen in Fig. 8.23 (left). Indeed they stated that:

---

[213] Wamsteker, W., Gilmozzi, R., Cassatela, A., & Panagia, N., IAUC No. 4410 (1987).

[214] Fransson, C., and 6 authors, Ap. J. **336**, 429 (1989).

[215] Sparks, W.B., Paresce, F., & Macchetto, D., Ap. J. **347**, L 65 (1989).

[216] Sonneborn, G., and 6 authors, IAUC No. 4685 (December 1988).

[217] Fransson, C., and 6 authors, Ap. J. **336**, 429 (1989).

[218] Crotts, A.P.S., Kunkel, W.E., & McCarthy, P.J., Ap. J. **347**, L 61 (1989).

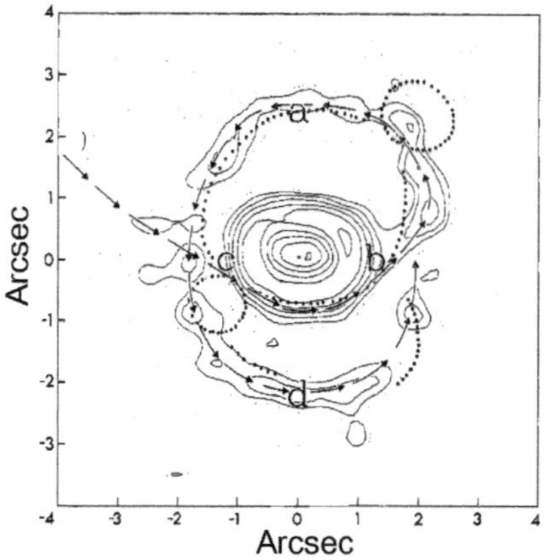

Fig. 8.23 *Left*: The first photograph by Crotts et al. 1989, in which *one complete and one partial circular loop* appear. The scale is in light years. *Right*: The first contour map obtained by Wampler et al. 1990. The scale is in arcseconds

> *They are unlike any configuration expected of the blue giant mass loss nebula, blue giant wind/red giant wind, etc.*

Many optional explanations were supplied in the conclusion, but none was worked out in any detail.

About three years later, in 1990, Wampler et al.[219] pointed out the discrepancy between the model due to Crotts et al. and any configuration expected of the blue giant mass loss, and provided the first map of the ring, shown in Fig. 8.23 (right). They concluded that the morphology of the nebulosity resembled that of a planetary nebula and that they could therefore have been formed by the mass-loss and evolution mechanisms of the SN progenitor, or so they speculated. In addition, they could not find any evidence that the 'filamentary loops' were expanding.

Eight years later, Plait et al.[220] used the Hubble telescope to photograph the the SN remnant. Their first conclusion was that the ring could not be due to reflected light (the light echo) because the ring did not change (see Fig. 8.24). The authors used the terminology 'clumpy elliptical ring'. The interpretation was that the ring was heated by the powerful flash of light from the SN, and was now cooling, whence we in fact observe the emission from the cooling gas. As far as the mechanism of formation was concerned, the similarity with planetary nebulas captivated researchers, who thus sought a similar mechanism. Another possibility raised by the authors was that the ring might be a consequence of the interaction between two stellar winds, if the progenitor was in a binary system. However, none of the hydrodynamical calculations carried out so far had resulted in such a ring.

---

[219] Wampler, E.J., and 6 authors, Ap. J. **362**, L 13 (1990).

[220] Plait, P.C., Lundqvist, P., Chevalier, R.A., & Kirshner, R.P., Ap. J. **439**, 730 (1995).

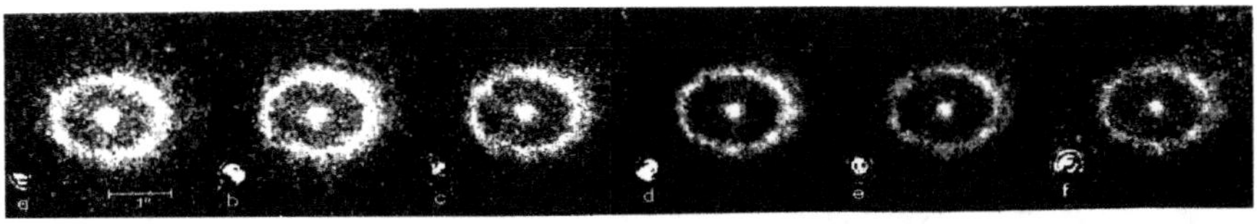

Fig. 8.24 The image of the ring in oxygen light at different times. (**a**) August 1990, (**b**) December 1991, (**c**) April 1992, (**d**) September 1992, (**e**) May 1993, and (**f**) October 1993. From Plait et al. 1995

Another explanation due to Goldstein[221] suggested that we see the light emitted from ions trapped in the magnetic field of a circular loop like a cyclotron. The requirements for the current and the magnetic field were exorbitant by all scales. There was no clue as to how such a 'cyclotron' could have formed around the progenitor and survive the explosion. But above all, it did not explain the emission lines observed, nor the particle density inside the ring which such emissions would require.

The real revolution and bewilderment came in 1995, when Burrows et al.[222] discovered a system of three rings, shown in Fig. 8.25 (right). For the first time, the authors drew a three-dimensional sketch of the rings, and provided several possible explanations, all of which they rejected. They ended by stating that:

> *We have shown that the outer nebula surrounding SN 1987A is a pair of these rings, but it is going to be very difficult to explain them without some new physical idea.*

8.18 The Progenitor and the Ring Structure

An SN is a huge lighthouse that suddenly lights up in a galaxy. This fantastic search light allows us to see what lies around the exploding star. The idea that the illumination of surrounding clouds by an SN should be a spectacular phenomenon was suggested by Oort in 1940.[223]

The first surprise was that the progenitor was a blue star, and not red.[224] However, it took time for the community to realize that Sk -69° 202 was probably never a red supergiant. Before this fact became clear, Fabian et al.,[225] Heap and Lindler,[226] Joss et al.,[227] and Testor[228] were quick to suppose that a possible fourth, red star was the progenitor.

---

[221] Goldstein, S.J., A & S S **227**, 217 (1995).

[222] Burrows, C.J., and 18 authors, Ap. J. **452**, 680 (1995).

[223] Zwicky, F., Rev. Mod. Phys. **12**, 66 (1940).

[224] See for example the calculations and predictions by Falk, S.W., & Arnett, W.D., Ap. J. S. **33**, 515 (1977).

[225] Fabian, A.C., Rees, M.J., van den Heuvel, E.P.J., & van Paradijs, J., Nature **328**, 323 (1987).

[226] Heap, S.R., & Lindler, D.J., A & A **185**, L 10 (1987).

[227] Joss, P.C., Podsiadlowski, P., Hsu, L., & Rappaport, S., Nature **331**, 237 (1988).

[228] Testor, G., A & A **190**, L 1 (1988).

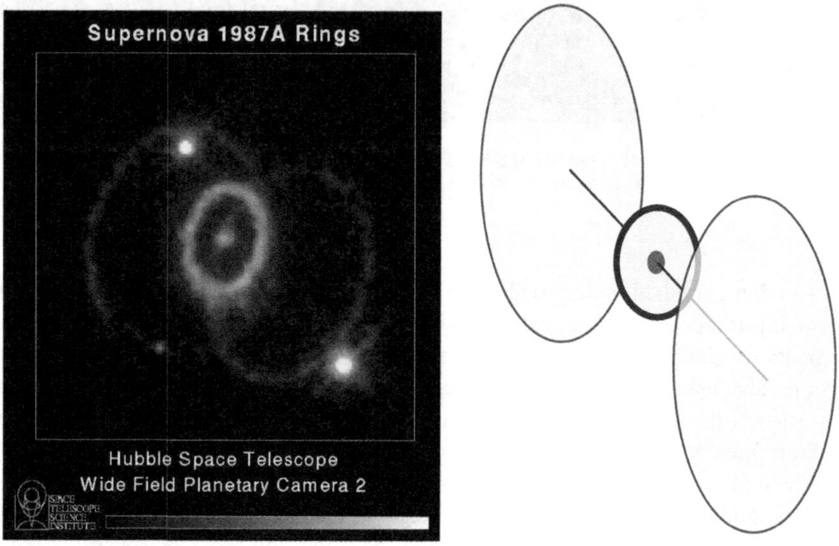

Fig. 8.25 *Left*: The double ring discovered by the HST on top of the first big ring. The two bright stars near the ring are stars in our Milky Way which happened to be in front of the SN. Credit: Hubble Space Telescope, NASA. *Right*: Schematic drawing of the spatial arrangement of the rings. After Burrows et al. 1995

When the theory-disobedient behavior of the progenitor had become quite clear, Joss and Podsiadlowski[229] revised their model by assuming that Sk-69°202 was indeed the progenitor, but that the evolution had been strongly influenced by the accretion of matter from a companion which was initially the more massive star. This nicely explained why the star was blue and not red. The explosion must have set the binary companion free to leave a solitary neutron star to be discovered. But as mentioned already, no plausible candidate for the free binary companion was found in an extensive search by the Hubble Space Telescope (HST).

There is at present no unanimously accepted model which explains all the observational facts. The physics and the model of the progenitor of SN 1987A are still a nagging problem for the theory of stellar structure and evolution, and the question of how the synthesized elements in the supernova are removed from the star and spread across interstellar space so far remains open. We summarize here the various models that have been suggested.

[229] Podsiadlowski, Ph., & Joss, P.C., Nature **338**, 401 (1989).

8.19 Single Star Evolution

Several single star models were presented in the literature. The most obvious property of the Magellanic Cloud which can affect stellar evolution is the low metal abundance. Hence, the first attempts were in this direction, that is, to see what the low heavy element abundance of the LMC could contribute. Alternatively, some extreme values, as a matter of fact, contrived ones, were assumed. The following question then arises: was Sk -69° 202 so unique as to have unusual parameters? Was it just Sk -69° 202 that was so unrepeatable?

Evolution with low metal abundance was one of the first attempts.[230] As a matter of fact, such models had been calculated several years before SN 1987A exploded,[231] with the result that the model with a lower metal abundance is indeed bluer relative to a model with solar abundance of the heavy elements. However, observations of stars in the LMC and SMC have shown the existence of red stars. As a matter of fact, Humphreys[232] concluded as follows:

> *In conclusion I want to emphasize the similarities of massive star evolution in the solar regions of our galaxy, in the large Magellanic Cloud, in M33,* and she went on to say: *the luminosity of the red giant stars in the LMC and in our galaxy is the same, and the observations show that, as the amount of heavy elements decreases, there are more red stars rather then less.*

This was contrary to what stellar models with lower amounts of heavy elements showed. It is amusing that Humphreys presented her results in a conference on *Observational Tests of the Stellar Evolution Theory*[233] with a message of universal evolution.

It appears that single star evolution always goes through the red phase, and no choice of parameters can prevent it.[234] The situation is nicely represented by the title of Barkat and Wheeler's attempt to explain the evolution:[235] *SN 1987A: Was Sanduleak -69° 202 mixed up?* But should the 'mixed up' be taken to mean 'confused'?

If there is a way to remove the hydrogen layer almost completely, then a red giant becomes blue.[236] The reason is simple. As the molecular weight becomes more uniform across the star, it becomes more compact and similar to a main sequence star, and hence bluer. We know observationally that red giants lose mass. However,

---

230 Arnett, W.D., Ap. J. **319**, 136 (1987); Hillebrandt et al., Nature **327**, 597 (1987).

231 Maeder, A., in *Observational Tests of the Stellar Evolution Theory*, 1983, ed. by Maeder & Renzini, Pub. Reidel, Dordrecht (1984) p. 299; Brunish, W.M., & Truran, J.W., Ap. J. **256**, 247 (1982); Hellings, P., & Wanbeveren, D., A & A **95**, 14 (1981).

232 Humphreys, R.M., IAUS **108**, 145 (1984).

233 International Astronomical Union Symposium No. 105, held in Geneva, Switzerland, 12–16 September 1983, ed. by A. Maeder & A. Renzini, D. Reidel Pub., Dordrecht.

234 Woosley, S.E., Heger, A., Weaver, T.A., & Langer, N., `arXiv:astro-ph/9705146v1` (1997).

235 Barkat, Z., & Wheeler, J.C., Ap. J. **342**, 940 (1989).

236 Maeder, A., Proc. ESO workshop on SN, 1987, ed. Danziger, p. 251.

in order to convert a red star into a blue one in the short available time, it must experience an extraordinary mass loss, about 10 times higher than anything that has ever been observed. The progenitor must be quite unique in adjusting the mass loss in such a way that it was blue when it exploded. The justification for this hypothesis was that SN 1987A was indeed quite unique.

Maybe there is unknown physics involved. Woosley[237] and later Langer et al.[238] and Weiss[239] demonstrated that, by adopting special restricted values for the convection, they could get models that stayed blue and did not become red. It is not clear what determines the value of the parameters in stars and why different stars may have different values for these parameters.

Saio et al.[240] just mixed the interior with the exterior artificially, which would of course lead to contraction. Playing with the amount of mixing allowed them to get an agreement with the fact that a blue star was the progenitor.

Could extremely fast rotation do the job? In the extreme case, the star would be completely mixed up. As we know, there is plenty of evidence against it. So should there be a very tricky finely tuned type of mixing that does the job?

8.20 Binary Star Evolution

About 2/3 of all stars are binaries. The binaries appear with all pairs of masses and with all kinds of separations between the component stars. We did not discuss the evolution of binary stars because hitherto it was assumed that their binary nature would not affect nucleosynthesis. SN 1987A may have convinced us to reconsider this supposition. Moreover, SN theory attempted to explain a single star SN, but most SNs should be in binary systems where the theory is only in its early stages.

Several researchers suggested that the progenitor might have been a binary star. So where is the binary? Fabian claimed that it should reappear as the photosphere of the SN disperses, but so far it has not. And then, if the binary was not dissolved in the explosion, it should appear as a massive X-ray binary, but so far it has not. The recent HST search for a companion has not discovered any candidate.

8.21 The Binary Merged

The unique phenomenon of the rings which we discussed above is best explained by assuming that there was a binary long ago, that the two stars then merged, and that the merger process expelled some mass which is responsible for the rings. There

[237] Woosley, S.E., Ap. J. **320**, 218 (1987).

[238] Langer, N., El Eid, M.F., & Baraffe, I., A & A **224**, 17 (1989).

[239] Weiss, A., Ap. J. **339**, 365 (1989).

[240] Saio, H., Nomoto, K., & Kato, M., Nature **334**, 508 (1988).

have been various attempts to apply the merger idea to calculate the shape of the rings. From the point of view of stellar evolution, the merger is like an event of massive mass accretion. The first to suggest such a coalescence were Chevalier and Soker,[241] and Hillebrandt and Meyer,[242] and later Podsiadlowski et al.[243] These papers were written before the first ring was discovered, and indeed as Chevalier and Soker wrote:

> *Our model does not clearly demonstrate a cause for the asymmetry. [...] Numerical calculations indicate that there is some tendency for the flow to become more spherical during the expansion time.*

The model by Hillebrandt and Meyer assumed that a neutron star had been detected already and attempted to explain how the evolution of a binary system would lead to such a neutron star with the period Middleditch claimed to have observed. Podsiadlowski et al. related the possible discovery of a submillisecond optical pulsar in SN 1987A,[244] an identification which was never confirmed, and before the system of rings was discovered, speaking of *the exotic post-SN binary with properties that could account for a pulsar with a period of about 2 kHz.*[245] Podsiadlowski et al. presented two possibilities, and wrote enthusiastically:

> *Both scenarios may explain all of the major observational features of this supernova event, including its most striking anomalies.*

But can the merger scenario explain the rings? Obviously this appears to be a rare event because we never had the chance to examine other SNs in such detail. The merger may lead to substantial mass loss, but naturally, in the plane of the original orbit.[246] It is not expected to lead to a ring plus two others which appear along an axis and which are not concentric! Podsiadlowski et al.[247] estimated that about 10% of all massive stars are in binaries, which would mean that something like this proportion should exhibit the phenomenon displayed by the enigmatic SN 1987A. Well, not quite. More recently, Morris and Podsiadlowski[248] argued that, due to fast rotation, mass ejection from the star would be primarily at mid-latitudes, rather than in the equatorial plane. They suggested that mid-latitude ejection might have provided the material of the outer rings, and equatorial ejection the inner ring material. None of these explanations addressed the high nitrogen abundances discovered in the ring, which implied processed material.

---

241 Chevalier, R.A., & Soker, N., Ap. J. **341**, 867 (1989).

242 Hillebrandt, W., & Meyer, F., A & A **219**, L 3 (1989).

243 Podsiadlowski, P., Joss, P.C., & Rappaport, S., A & A **227**, L 9 (1990).

244 Kristian, J., and 9 authors, Nature **338**, 234 (1989).

245 A pulsar is a neutron star which radiates in the radio, like a very fast spinning magnet. The extremely accurate periodic emission, with periods less than about 4 seconds, is the hallmark of neutron stars.

246 Livio, M., & Soker, N., Ap. J. **339**, 268 (1989).

247 Podsiadlowski, P., Morris, T.S., & Ivanova, N., *Stars with the B[e] Phenomenon*, Proc. Conf. San Francisco, ASP (2006) p. 259.

248 Morris, T., & Podsiadlowski, P., MNRAS **365**, 2 (2006).

To complicate the story even further, the progenitor of another quite strange SN, this time SN 1993J, was subsequently identified by Maund et al.[249] in images of the galaxy M81 taken before the explosion. The progenitor was found to be a non-variable red (not blue) supergiant star. Moreover, the spectrum of SN 1993J underwent a remarkable transformation from presenting the signature of a hydrogen-rich Type II supernova to exhibiting a helium-rich (hydrogen-deficient) Type Ib signature. At the position of the fading supernova, Maund et al. detected some signs of a massive star which they claimed was the binary companion to the progenitor. Thus the old theoretical prediction that the progenitor must be a red giant appeared to be correct (at least in some cases). In summary, we have not yet seen all the possibilities stellar evolution may come up with.

At this point we should mention that the entire chronicle of what kind of star the SN 1987A progenitor might have been is almost a direct re-enactment of the old problem about the evolution of Wolf–Rayet stars. These are massive stars ($20M_{\odot}$ and above), which experience huge mass loss via a powerful stellar wind. The speed of the wind can reach 2000 km/s. The stars have high surface temperatures (25 000–50 000 K) and appear blue. About 50% of these stars are members of binary systems. Peculiar abundances are observed on the surface of these stars. The controversy about the stellar evolution which leads to these stars extends over 40 years,[250] and the spectrum of proposed solutions, or scenarios, covers every conceivable line of thought, including those mentioned as scenarios for the evolution of Sk -69° 202. The additional information about the rings has complicated the problem, and so far no basically new solution has been forthcoming.

In summary, there is no consensus today among astrophysicists about the evolution of Sk -69° 202, and how it reached the point where it exploded.

8.22 A Few of the Remaining Unsolved Questions

SN 1987A has raised many questions that still remain open:

- Why was SN 1987A dimmer at maximum than most Type II SNs?
- How much mixing was there, and what were the mechanisms for it? As we have seen, the question of mixing confuses almost every analysis of the observational data. This old problem in the theory of stellar structure, which was considered to be solved in the early 1950s, has come back in force in the final stages of the evolution before and during the explosion.
- What are the composition, distribution, and relative abundances of the newly synthesized elements? Was the progenitor not spherically symmetric before the explosion, or did the observed non-spherical compositions develop during the fast expansion? The analysis of the abundances depends on the state of the mat-

[249] Maund, J.R., Smartt, S.J., Kudritzki, R.P., Podsiadlowski, P., & Gilmore, G.F., arXiv e-print arXiv:astro-ph/0401090.

[250] See Simon, N.R., & Stothers, R., Ap. J. **155**, 247 (1969).

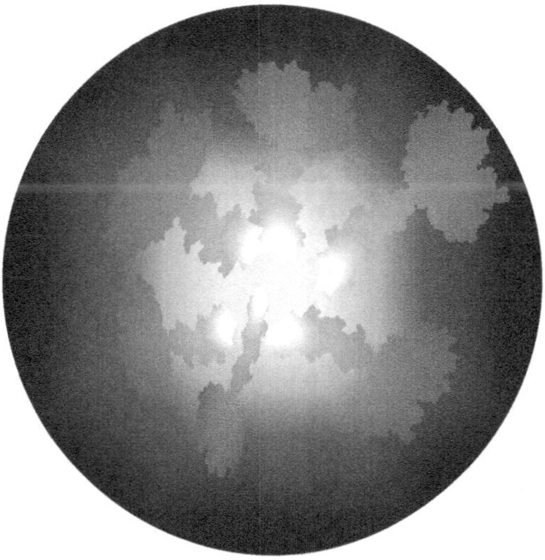

Fig. 8.26 Breakdown of nuclear burning into 'fingers' and 'mushrooms', which cause mixing. (Shaviv unpublished)

ter, i.e., whether it was mixed or homogeneous, continuous or clumpy. A lot of progress has been made in numerical simulations of the burning and explosion processes. A simple example is shown in Fig. 8.26. Inhomogeneities give rise to relatively hot and cold volumes. The hotter ones float and produce 'mushroom' structures which rise, spread, and mix the synthesized elements. Computers are beginning to be able to solve such problems, and in this way expose the patterns of this kind of mixing. But other mixing modes are not excluded. The interplay between rotation, explosion, and mixing has not yet been worked out.

- What is the nature of the compact object left behind (if such a thing is left behind) and why has it not yet been discovered? Is it a black hole or a peculiar neutron star? The present theory predicts either of the two, but not the third possibility of no stellar remnant.
- What process accounts for the triple ring system and the circumstellar matter beyond the ring?
- The initial expansion velocity was about 18 000 km/s. No Type II SN has ever had such a high expansion velocity.
- What was the evolution of the progenitor?
- The interaction between the two stars in a binary system is crucial in certain cases and changes the nature of the supernova. There are probably additional types, though rarer than the known ones, and this issue must be clarified.

The nearby SN 1987A has exposed many other unsolved problems in the theory of stellar evolution and structure.

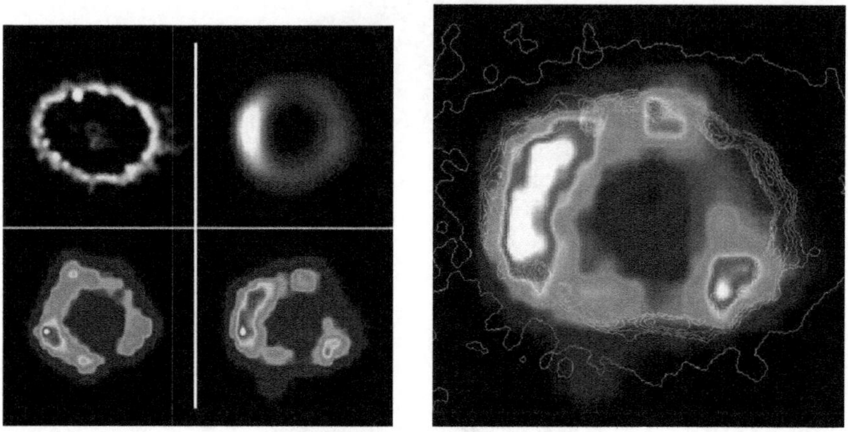

Fig. 8.27 *Left*: Four Recent Images of SN1987A. *Upper left*: Hubble Space Telescope optical image taken 2 February 2000. *Upper right*: Australian Telescope Compact Array radio image from 9 September 1999. *Lower right*: Chandra image from 17 January 2000. *Lower left*: Chandra image from 6 October 1999. (Credit: Optical: NASA/CfA/P.Challis et al; radio: MIT/ATN/Gaensler & Manchester; X-ray: NASA/PSU/D. Burrows et al.) *Right*: Chandra/Hubble composite image of SN1987A. This Chandra X-ray image of SN 1987A made in January 2000 shows an expanding shell of hot gas produced by the SN. *Colors* represent the X-ray emission intensities, with white being the brightest. The contours are from a Hubble Space Telescope optical image taken on 2 February 2000. Scale: The optical ring is 1.0×1.3 lyrs. The expanding shock encountered non-uniform blobs of gas. Blobs may have been created by non-symmetric mass loss at an earlier phase of the star, and now the fast-moving shock has caught up with it. (Credit: X-ray: NASA/PSU/D. Burrows et al.; Optical: NASA/CfA/P. Challis et al.)

8.23 The Latest Images

Figures 8.27 left and right show some of the most recent observations of the expanding rings. The X-ray emission is due to the impact of the 4 500 km/s shock wave impinging on the ring of matter. As a result, the temperature of the gas reaches a few million degrees, giving rise to X rays. The picture is the result of a 3 hour exposure time.

8.24 Life and Death Under Supernova Control

While the SN may even trigger the collapse of a protostellar gas cloud to a star, it is clear that, once a planetary system has been formed, living in the too close neighborhood of an exploding SN would be potentially disastrous to biological systems. The exploding SN could easily destroy the very sensitive life support system of the Earth by the high energy radiation it emits (UV, X rays, and γ rays). So a 'safe' neighborhood is one which is far away from massive stars. This, however, is not sufficient. Type Ia SNs emerge from very old stars. Hence being close to old stars would not guarantee our being outside the lethal zone of a SN. We may therefore need a SN at the beginning of planetary life, but we would like to stay away from SNs later. Our Solar System is far a way from the center of the Galaxy, where the density of all kinds of stars is high. But this is not the case all the time.

At present, the Solar System is revolving around the center of the Galaxy, and consequently changes its location among the stars. In particular the Solar System crosses the spiral arms.[251] The spiral arms contain gas out of which new stars, and in particular many young massive stars, form. Every time the Solar System crosses a spiral arm, the danger of a nearby star detonating as a SN increases significantly.

There are many ways in which a nearby SN could affect the Solar System. Before the Solar System was formed, a blast wave from a SN could have triggered the collapse of the cloud to form the Sun and the Solar System.[252] During this interaction, fresh material, recently synthesized by the SN, is injected into and mixed with the matter in the primeval solar nebula. Once the Solar System forms and planets like the Earth evolve to their present day constitution, a flux of high energy radiation from a nearby SN can cause severe damage to our delicate atmosphere, for example, destroying the ozone layer, and with it many of the biological systems which it protects on Earth.

During the years 1955 to 1973, Urey[253] advocated the idea that tektites, pieces of natural glass a few centimeters across, could result from collisions with comets.[254] In particular, Urey calculated the energies involved. Table 8.10 is taken from Urey's paper. It demonstrates the energetics and shows that such a comet is not a small perturbation to the Earth's climatic system. The comet is assumed to have roughly the mass of Halley's comet (which has a radius of about 10 km). A comet loses about 10^{-3} of its mass per orbit. So the danger from any given comet persists over many orbits.

One of the first discoveries of relic radioactive isotopes was in 1960 when Reynolds[255] discovered that the Richardson meteorite was much enriched in ^{129}Xe. This isotope is the decay product of of ^{129}I (half-life 16×10^6 yrs).[256] Reynolds calculated from the data that $(0.35\pm0.06)\times10^9$ yrs must have elapsed between the time of

[251] The spiral arms are like a wave propagating in the Galaxy, and rotating around the Galaxy, as does the Solar System. However, the Galaxy does not rotate like a rigid body, and different parts rotate with different rotational velocities. As a result, the Solar System moves relative to the spiral arms, crossing them about once every 145 Myrs.

[252] If the SN triggers the formation of massive stars whose lifetime is short, then these stars will live their short lifetime and explode, then trigger more star formation in a chain process. The phenomenon leads to a burst of stars, a phenomenon observed in colliding galaxies. This is another good reason why such a vulnerable system like life must stay away from SNs, once it has formed.

[253] Urey, H.C., Nature **179**, 556 (1957); ibid. **197**, 228 (1963); ibid. **242**, 32 (1973).

[254] Urey complained that he published the idea in the *Saturday Review of Literature*, but as he said: *no scientist except me, so far as I know, reads this magazine*, and for this reason he apparently published it again in Nature.

[255] Reynolds, J.H., PRL **4**, 8 (1960).

[256] The complication is that ^{129}I is produced in fission, hence present in spent nuclear fuel, high-level radioactive wastes produced by processing spent nuclear fuel, and radioactive wastes associa-

Table 8.10 Comparison of energies from Urey 1973

Solar irradiation in 1 year	3.48×10^{31} erg
Earthquake of magnitude 9.0 on the Richter scale	2×10^{25} erg
Kinetic energy of a comet with mass 10^{18} g and velocity 45 km/s	10^{31} erg

Fig. 8.28 A picture of a piece of the Allende meteorite for sale in Mexico. The size of the cube is 1 cm^3. This piece, 0.8 g in weight, is for sale for 16 $ (`www.meteoritemarket.com`)

formation of the elements and the time the meteorite crystallized (because once the meteor forms, the xenon gas remains in enclaves). The age of the Solar System is 4.6×10^9 yrs. Hence the age of the elements is close to 4.95×10^9 yrs. From the time the element was synthesized in a star, very probably in a SN, it would have taken $(0.35 \pm 0.06) \times 10^9$ yrs for it to be ejected into space, mixed with the interstellar matter, and then eventually collapse with the gas to be included in the protosolar nebula and the Solar System.

Similar ideas were expressed in 1968 by Shklovskii, who argued[257] that the Earth would have to be located within a SN remnant for cosmic radiation to produce mutations at a significant rate, and he calculated the lethal distance to be about 300 lyrs.

Two famous meteorites hit the Earth in the same year, 1969, the Allende and the Murchison[258] meteorites. The Allende meteorite was named after the town in

ted with the operation of nuclear reactors and fuel reprocessing plants. In addition, ^{129}Xe is formed in small quantities in the upper atmosphere of the Earth by cosmic rays.

257 Shklovskii, I.S., *Supernovas*, Wiley & Sons, London (1968).

258 Named after the town Murchison, Australia. It was known before 1997 that this meteor had extraterrestrial origin, because it contained several organic compounds like sarcosine (N-methyl glycine) and N-methyl alanine. However, the meteor became particularly famous after Cronin, J.R., and Pizzarello, S. [Cronin, J.R., and Pizzarello, S., Science **275**, 951 (1997)] discovered amino molecules. The intriguing fact was that the left-handed polarized molecule occurred about 7–9% more than the right-handed molecule, indicating a possible breaking of the symmetry which we know today to exist in terrestrial living systems, and already before the origin of life on Earth. The implications are far-reaching.

Table 8.11 The ratio R = Allende/Earth crust for various elements, after Wdowiak 1988

Element	R	Element	R
Na	0.145	As	0.86
Mg	5.35	Se	164
Al	0.21	Br	0.64
K	0.016	Cd	2.725
Ca	0.40	In	0.146
Sc	0.45	Sb	0.415
V	0.73	La	0.0142
Cr	29.75	Sm	0.0416
Mn	1.37	Eu	0.0538
Fe	3.81	Yb	0.103
Co	22.83	Os	165.6
Ni	134	Ir	785.0
Zn	1.565	Au	72.5
Ge	10.8		

Mexico where it fell. It is most famous for the variety of chemicals found in it. It is thought that as much as 10% of Allende is of earlier origin and is therefore older than our Solar System. About 1000 kg of the meteorite were collected. It had a fundamental impact on many questions concerning the source of the meteors, for various reasons. Firstly, it belongs to a group of relatively rare carbon-containing meteorites known as the carbonaceous chondrites, and the amount of material collected so far already exceeds the total weight of all other carbonaceous chondrites in museum collections. But in addition, it contains unique abundances.

The abundance analysis of the Allende meteorite given in Table 8.11 shows remarkable differences between the meteorite and the Earth's crust. This is evidence for various fractionation processes which operated on the Earth in such a way as to wash away the original abundances. It is also a testimony of the most primitive form of matter in the Solar System.

In 1979, Tucker[259] analyzed the effects of various cosmic phenomena on life. He calculated that SN explosions occurring within 30 lyrs of the Sun could cause a mass extinction of life on the Earth. Thus, a supernova explosion that occurred 60 Myrs ago might have been responsible for the disappearance of the dinosaurs. The galactic nucleus is found to be an inhospitable place for the evolution of life.

The picture was turned upside down in 1980, when Alvarez et al.[260] discovered an iridium anomaly in the thin clay boundary layer that commonly separates the strata of two geological periods, the layer corresponding to a date 65 Myrs ago.

[259] Tucker, W.H., *Life in the Universe*, Proc. Conf., Moffett Field, CA, 1979, MIT Press, Cambridge, MA (1981) p. 287.

[260] Alvarez, L.W., Alvarez, W., Asaro, F., & Michel, H.V., Science **208**, 1095 (1980). Luis Alvarez (1911–1988), 1968 Nobel Laureate in Physics for many contributions to nuclear physics.

Since the elements of the platinum group are very rare in the Earth's crust relative to their cosmic abundance, this rather high amount of iridium suggested that the clay might have extraterrestrial origin. The absence of ^{244}Pu in the layer led Alvarez et al. to conclude[261] that this excess of iridium by about 30, 160, and 29 (in three different locations over the globe which correspond to the same geological layer and time) was not produced by a nearby supernova. According to their calculations, it was probably the remains of a meteor of diameter 10 km hitting the Earth. The reader may consult Table 8.10 for the meaning of such an assumption. Computer simulations by O'Keefe, Ahrens, and Koschny,[262] and Roddy et al.[263] show that such an event forms a short-lived hole in the atmosphere, and induces mass loss through the hole.

So why did Alvarez et al. claim that it was not a SN? They contended that their calculation (however, no details were provided) would have placed the required SN at a distance of 0.1 lyr away. This is extremely close, and the effects would have been much more severe. They calculated the probability for such a SN to be 10^{-9}, making it highly improbable on this reckoning. A SN would produce ^{244}Pu which has a half-life of 80.5×10^6 yrs. So most of the ^{244}Pu from the origin of the Solar System would already have decayed, but they expected about 10^{-3} atoms of ^{244}Pu per atom of iridium to be left. However, no ^{244}Pu was detected. They concluded therefore that a SN could be ruled out, and that an asteroid must have struck the Earth. On the other hand, there was no explanation as to how this excess of iridium got into the asteroid. Indeed, why are there different types of asteroids? What it actually showed was that the ejecta of the SN got into a chunk of matter that later became an asteroid, and that this asteroid then wandered around the cosmos for over a billion years so that the longest-lived radioactive element produced in the SN would have sufficient time to decay. And then, by pure chance, the asteroid hit the Earth.

Van den Bergh[264] calculated that there is a probability of $\sim 6\%$ that the Sun was located within 15 lyrs of the center of a SN remnant sometime during the past 10^9 yrs. If this is the case, then as Ellis and Schramm remarked, *SN rates are certainly too low to explain all of the extinction events that occurred on the Earth.* However, these estimates could be off by a factor of a $10^{\pm 2}$ at least. For example, Ruderman[265] considered the effect of the SN on the destruction of the ozone layer, and the effect of penetrating ionizing radiation, as well as UV radiation. Ruderman calculated that the lethal radius of a SN could be as great as ~ 50 lyrs, and there is at

---

[261] Hut, P., Alvarez, W., Elder, W.P., Kauffman, E.G., Hansen, T., Keller, G., Shoemaker, E.M., Weissman, P.R., Nature **329**, 118 (1987).

[262] O'Keefe, J.D., Ahrens, T.J., & Koschny, D., LPICo **673**, 133 (1988).

[263] Roddy, D.J., Schuster, S.H., Rosenblatt, M., Grant, L.B., Hassig, P.J., & Kreyenhagen, K.N., Lunar and Planetary Inst., Global Catastrophes in Earth History: An Interdisciplinary Conference on Impacts, Volcanism, and Mass Mortality (1988) p. 158.

[264] van den Bergh, S., PASP **101**, 500 (1989); Comments on Astrophysics **17**, 125 (1993).

[265] Ruderman, M.A., Science **184**, 1079 (1974).

least one SN per few hundred Myrs at this distance from the Earth. A more detailed calculation was performed by Seward.[266]

In 1988, Wdowiak[267] pointed out that the heavy metals are toxic. Hence the resulting local density and the spread, mixing, and dilution of the metals brought in by an asteroid are very important. A meteor with the Allende composition, but with a radius of 10 km, would deposit about 40 g of nickel per square meter of the Earth's surface, and it is known that Ni is toxic to plant life at concentrations of 40 parts per million and above. Similarly, the fireball created in the impact would generate a nitric acid rainout which would cause mass extinction. Meteoric nitric acid would convert heavy metal oxides, the most quickly formed chemical species in the fireball, into highly soluble combinations of a nitric acid–heavy metal rainout that would act as a prompt and deadly destroyer of plant life.

As commented by van den Bergh:

> *The vigor of the present debate on the nature of the great extinctions lends support to the view that we are presently experiencing a shift of paradigms (Kuhn 1962), in which the view that evolution results entirely from survival of the fittest is being replaced by one in which sudden random catastrophes and survival of the fittest jointly determine the evolution of species on Earth.*

The precarious system of life on the Earth is very sensitive to drastic perturbations from outer space, the very same as those which provided the elements for the existence of biology. This also explains the large distance between stars. If the density of stars had been significantly higher, the chance of complete destruction by a nearby SN would have been much higher, and this would very likely have eliminated all possibility of life as we know it. For example, near galactic centers, the density of stars is higher, and the rate of SNs correspondingly greater.[268] Furthermore, the perturbations of a nearby star to the Oort cloud[269] of comets and asteroids would have been much more frequent, giving rise to frequent showers of comets and meteors. The Solar System has a good place in the outskirts of the Galaxy, and for good reasons.

In 1982, Wasserburg and Papanastassiou,[270] and in 2000, Goswami and Vanhala[271] argued that the existence of short-lived, and now extinct, nuclei like ^{60}Fe

[266] Seward, F.D., British Interplanetary Society, Journal (Interstellar Studies) **31**, 83 (1978).

[267] Wdowiak, T.J., LPICo **673**, 209 (1988).

[268] Most centers of galaxies contain massive black holes and are very violent. So there are many other good reasons why life could not survive near galactic centers.

[269] The Oort cloud is spherical cloud of debris which lies at roughly 50 000 AU. This is about 1/4 the distance to the nearest star. The cloud has never been observed, only hypothesized to exist in order to explain the reservoir of new comets that appear from time to time. The Oort cloud is largely composed of chunks of different ices such as water, ammonia, and methane. Perturbations by passing stars send chunks towards the Sun, and as these approach the Sun, they become comets. The cloud is a remnant from the interstellar cloud out of which the Sun formed.

[270] Wasserburg, G.J., & Papanastassiou, D.A., Essays in Nuclear Astrophysics, ed. by Barnes, Clayton, Schramm, & Fowler, Cambridge University Press, New York (1982) p. 77.

[271] Gswami, J.N., & Vanhala, H.A.T., Protostars and Planets IV, ed. by Mannings, Boss, Russell, University of Arizona Press (2000) p. 963.

(half-life 1.49 Myrs), which decayed into ^{60}Ni, provide information about the event which took place in the early history of the Solar System. However, short-lived nuclei can be synthesized either by local irradiation by energetic particles in the early phases of the formation of the Sun, or by nucleosynthesis in exploding stars. In 2000, McKeegan, Chaussidon, and Robert[272] discovered evidence for live ^{10}Be, which is produced only by energetic particle irradiation, and not in stars. This major discovery drove many to suspect that short-lived nuclei like ^{41}Ca, ^{26}Al, ^{53}Mn, and clearly ^{10}Be, are produced mainly by irradiation and not in stars. However, ^{60}Fe is not easily produced by irradiation. Hence, if it exists, it implies production in stars. The situation with ^{60}Fe is quite complicated, and there are different estimates/measurements of the upper limit.

In 1993, an Italian group led by Bignami[273] identified one of the closest ever supernova renmants. The possibility of a nearby SN was no longer a question of probability, it was real. The group identified the remnant known as Geminga,[274] a celebrated strong source of γ rays, as a SN remnant. The distance is less than 400 lyrs. The Solar System resides at the edge of a cavity of hot (10^6 K), low density (5×10^{-3} cm^{-3}), X-ray emitting gas, embedded in the interstellar medium. This void, sometimes called the Local Bubble, is thought to be less than 10^7 years old, but its origin is unknown. The next paper in the journal was by Gehrels and Chen,[275] who suggested that the Geminga SN swept the vicinity of the Solar System and drove away most of the material, leaving a rather low density bubble. The SN produced a local void where we can live in peace. From the state of the remnant, an age estimate of 340 000 yrs was derived, and a distance of 120–1200 lyrs.

Benitez et al.[276] showed that the Scorpius–Centaurus OB association, a group of young stars currently located at $\sim$ 415 lyrs from us, has generated about 20 SN explosions during the past 11 Myrs, some of them probably as close as 12 lyrs to our planet. They estimated that the deposition on Earth of ^{60}Fe atoms, produced by these explosions, can explain the recent measurements of an excess of this isotope in deep ocean crust samples. They therefore, proposed that $\sim$ 2 Myr ago, one of these SNs exploded close enough to the Earth to seriously damage the ozone layer, contributing to marine extinction.

---

[272] McKeegan, K.D., Chaussidon, M., & Robert, F., Science **289**, 1334 (2000).

[273] Bignami, G.F., Caraveo, P.A., & Mereghetti, S., Nature **361**, 704 (1993).

[274] Geminga is the name given to a neutron star in the constellation Gemini. It is the second brightest source of high-energy gamma rays in the sky. Discovered in 1972 by the SAS-2 satellite, its name is both a contraction of 'Gemini gamma-ray source' and an expression in Milanese dialect meaning 'it's not there'. In 1992, an exceptionally regular periodicity of 0.237 s in soft X-ray emission was detected by the ROSAT satellite [Halpern, J.P., & Holt, S.S., Nature **357**, 222 (1992)], indicating that Geminga is almost certainly a pulsar (a flickering neutron star), which unlike other pulsars (save the Vela γ-ray pulsar) is invisible at radio wavelengths. It has also been identified optically with an extremely dim blue star.

[275] Gehrels, N., & Chen, W., Nature **361**, 706 (1993).

[276] Benitez, N., Maiz-Apellaniz, J., Canelles, M., American Astronomical Society, 199th AAS Meeting, Bulletin of the American Astronomical Society, Vol. 33 (2001) p. 1411; PRL **88**, 081101 (2002).

In 1996, Ellis, Fields, and Schramm[277] discussed isotope anomalies as possible geological signatures of a nearby SN. They followed Alvarez et al. and discussed various anomalies like ^{10}Be discovered in ice cores. Following Ellis and Schramm,[278] it is clear that any SN within 30 lyrs would have had dramatic effects upon biology. They proposed to examine the isotopes ^{10}Be, ^{26}Al, ^{36}Cl, ^{53}Mn, ^{60}Fe, and ^{59}Ni. Combining data from the findings of the Vostok Antarctic ice cores, they pointed to the possibility that the anomalous ^{10}Be could be a consequence of a nearby SN about 300 000 yrs ago. In particular, they examined the question of whether Geminga could be the source. Actually, examining the ^{10}Be, they concluded:

> *This suggests that, if the Vostok peaks came from a SN, it was quite close and indeed may have been a near-miss.*

The distance they got for Geminga, however, had to be less than 130 lyrs. They sent the geologists to search the deep oceans, to look for long-lived isotopes produced only in SNs, and that have a sufficiently long lifetime to be traced on the Earth. It could also be a daughter of such a nucleus, if it is not produced in another way.

It took only three years for the prediction to be confirmed. In 1999, Korschinek et al.[279] reported that deep ocean ferromanganese crust was found to have an excess of ^{60}Fe radioactivity. The enhanced concentrations measured in the first two of three layers (corresponding to a time span of 0–2.8 Myr and 3.7–5.9 Myr, respectively) suggested the deposition of supernova-produced ^{60}Fe on the Earth. There was even a weak indication that the flux into the crust was higher about 5 Myr ago. But it was impossible to date the supernova accurately from those samples, because the material was distributed through several different layers of rock. As we understand things today, ^{60}Fe could only have come from SNs.

When the ^{60}Fe arrived from space, it was evenly distributed all over the Earth. But the signatures are only detectable in crust that has lain undisturbed for millions of years, such as certain parts of the Pacific Ocean floor. This particular crust sample was taken from an area a few hundred kilometers southeast of the Hawaiian Islands in 1980. It was collected by oceanographers, who were investigating the rocks as a potential source of rare mineral ores.

The authors estimated that the SN occurred between about 100 and 200 lyrs away and happened 2.8 Myrs ago, give or take 300 000 yrs. The explosion could not have been too close to the Earth, or it would have delivered enough radiation to cause mass extinctions. Conversely, if the SN was any further away, more of the ^{60}Fe would have been filtered out by the thin wisps of matter drifting between the stars.

This means that the SN would have been at the right distance to spray out a stream of cosmic rays that could have increased the cloud cover on the Earth. Korschinek calculated that there might have been 15% more cosmic rays reaching the Earth than normal for at least 100 000 yrs. This is not enough to actually kill anything, but according to their estimates, it was perhaps sufficient to change the Earth's climate.

---

277 Ellis, J., Fields, B.D., & Schramm, D.N., Ap. J. **470**, 1227 (1996).

278 Ellis, J., & Schramm, D., PNAS **92**, 235 (1995).

279 Knie, K., Korschinek, G., Faestermann, T., Wallner, C., Scholten, J., & Hillebrandt, W., PRL **83**, 18 (1999).

In 2002, Nir Shaviv[280] discovered a correlation between the passage of the Solar System through the spiral arms of the Galaxy and the periods of glaciation (see Fig. 8.29). This is a clear demonstration of the effect that not too close SN explosions can have on the Earth's climate. The relative velocity between the Solar System and the galactic system of spiral arms drives the Solar System across an arm about once every 145 Myrs. The arms are regions of intensive star formation, and in particular contain many young massive stars, which eventually explode as SNs and produce cosmic rays. The particles composing the cosmic rays diffuse out of the spiral arm into the entire volume of the Galaxy. The density of cosmic rays is therefore highest in the arms. When the cosmic ray particles penetrate the Earth's atmosphere, they cause ionization of atoms which in turn become nuclei for condensation of water vapors. The cosmic radiation produced by SNs induces more overcast skies, and consequently high reflection of solar light and subsequent cooling of the Earth, leading to periods of glaciation.[281] Fortunately, the Sun moves relative to the spi-

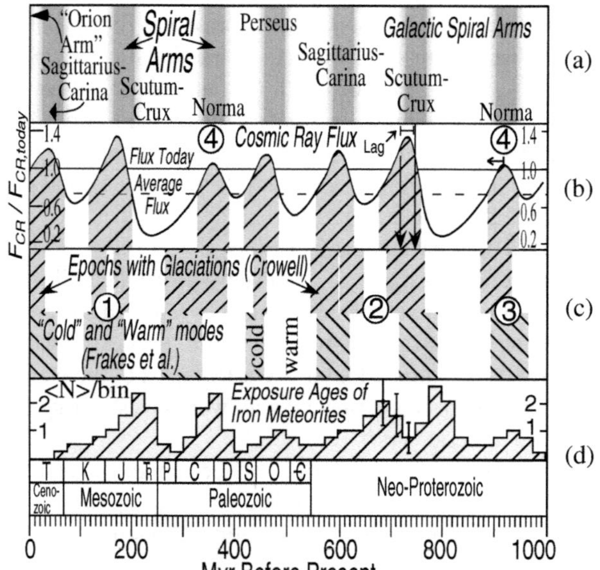

Fig. 8.29 The correlation between the Solar System passing through the spiral arms of the Galaxy and the periods of glaciation on the Earth, from N. Shaviv 2002. The *top panel* indicates the spiral arms. The *second panel* is the cosmic ray flux due to excessive SN activity in the spiral arms. The *third panel* shows glaciation epochs and the *last panel* is the exposure ages of meteorites based on the (false) assumption of steady irradiation by cosmic rays

[280] Shaviv, N.J., PRL **89**, 051102 (2002).

[281] The temperature of the Earth without atmosphere is 255 K, but with the atmosphere it is 288 K. A small change in the energy flux reaching the soil can easily cause a temperature decrease to below 273 K and give rise to glaciation.

ral arms, and hence the Earth experiences periods of higher temperatures, when far away from the spiral arms, and periods of lower temperature when the Sun and the Solar System cross the arms. The cosmic rays in this case represent the sum contribution from all SNs occurring in the spiral arms. Thus, it is not sufficient to have a Solar System and habitable planets. The system must move far away from powerful sources of ionizing radiation.

We have not covered all possible cosmic threats to a life-supporting planetary system. Additional dangers may come from gamma ray bursts (GRB),[282] which are powerful emitters of γ rays. They are probably formed by one or more SNs, in which case they would be covered by consideration of SN explosions. Here we are mainly interested in the injection of new elements from the SN synthesis furnace, and the contribution of GRBs to the enrichment of the Galaxy in heavy elements has not yet been discussed.

8.25 Recent News: Ultra-Powerful Supernovas

Several SNs detected in recent years have been found to be ultra-powerful, that is, they emit more energy than the 'standard' SN, if there is something like a standard in this game. So far we have discussed the more common SNs resulting from stars with masses in the range of 8–$30M_{\odot}$, where the upper limit remains suitably vague. As more massive stars are rarer, so are SNs with massive progenitors, but from time to time such an event does occur. The larger number of available and better telescopes, including detectors placed on satellites, not to mention the SN monitoring program, have allowed astronomers to keep a very large volume of space under constant surveillance, and thereby increase the chances of detecting even rarer events. It is their rarity that explains why the early observers did not detect these objects. The present day effect of very massive stars is small because they are so scarce. However, this was not the case in the past.

A typical example of the ultra-powerful SNs detected recently is SN 2006gy which was discovered by Smith et al.[283] in 2007. The total radiated energy was found to be $(1.2 \pm 0.2) \times 10^{51}$ erg, or about a factor of 100 more than a 'classical' SN. To explain the light curve, Smith et al. calculated that the total mass of radioactive nickel must be about $22M_{\odot}$ (which is very high) and this can result only from a very massive star which became unstable due to electron–positron pair formation. The authors produced a set of arguments leading to the conclusion that the original mass must have been well above $100M_{\odot}$. The measured expansion velocities were modest, about 4000–4500 km/s. While the temperature of the emitting region was the same as in any other SN, namely about 5000 K. The unusually high luminosity could therefore arise only if the radius of the progenitor was correspondingly greater. The low expansion velocities tell us that the progenitor was more extended than

---

[282] Dar, A., Laor, A., & Shaviv, N.J., PRL **80**, 5813 (1998).

[283] Smith, N., and 11 authors, Ap. J. **666**, 1116 (2007).

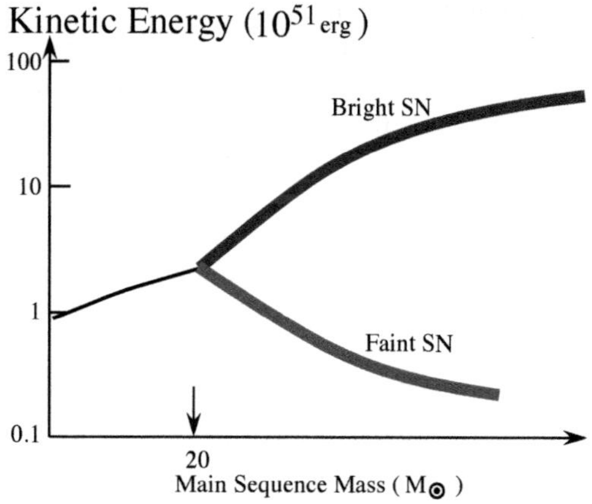

Fig. 8.30 Attempts to simulate the light curves and the total ejected mass appear to lead to two values. It is far from established whether or not there are two branches to the SN energetics as a function of mass on the main sequence

the usual SN progenitor, and it must therefore have been unusually massive. There was no direct estimate of the mass of the expanding layers. The basic mystery is how such a massive star could have been born in the host galaxy. The latter, NGC 1260, is an S0 galaxy, dominated by an old stellar population.

The term hypernova was invented to describe the particular explosion which leads to the formation of gamma ray bursters, or GRBs for short. These are extremely powerful objects that begin by radiating a burst of very powerful γ rays, and later appear as an X-ray glow. The fascinating physics of the GRBs is now under intensive investigation. However, their formation is probably accompanied by, or is the product of, an explosion of a massive star and/or massive binary, which exceeds even the canonical supernova in energy release. Hence the name hypernova.

Several particularly bright SNs have been discovered recently that are not all associated directly with GRBs. A typical example which is associated with a GRB is SN 1998bw, which was classified as peculiar Type Ic.[284] The peak luminosity was about 10 times higher than the typical value for a SN of this class. The spectral lines indicated expansion velocities of about 30 000 km/s, which imply a total kinetic energy of $\sim 2\text{–}5 \times 10^{52}$ ergs, assuming spherical symmetry. If the energy estimate is correct, then Iwamoto et al.[285] claimed that:

> *The unusual properties of these objects may require reconsideration of theories of stellar explosion mechanisms.*

[284] Sadler, E.M., Stathakis, R.A., & Boyle, B.J., Eckers, R.D., IAU Circular 6901.

[285] Iwamoto, K., Nomoto, K., Mazzali, P.A., Nakamura, T., & Maeda, K., Lecture Notes in Physics, Springer (2004) p. 243.

Fig. 8.31 SN 1994D is suspected to be an overluminous Type I SNa. Credit: High Z SN search, Hubble Space Telescope

Nomoto et al.[286] examined the energetics of a collection of SNs and realized that there may be two branches, viz., the branch of the very powerful SNs, the hypernovas, and a branch of faint or 'failed' SNs. The two branches split at about $20M_\odot$ (see Fig. 8.30). While so far undiscovered, it is plausible that the entire domain between the two branches may contain different SNs that have not yet been detected. How could that happen? Nomoto et al. explain as follows. Stars with original masses of up to $20–25M_\odot$ form neutron stars. The collapse of more massive stars is correspondingly more violent, so they would produce black holes. The power of the collapse depends on the rotation. Fast rotation can brake the collapse and give rise to a faint or failed SN. Clearly, depending on the degree of rotation, the SN finds its location between the two extremes. The large variety of SNs may imply that so-called secondary parameters, like rotation, magnetic field, and initial heavy element abundance, affect the exact evolution of the explosion and the light curve.

We have stressed the wide variety of SNs many times. A further recent example is SN 1994D, a beautiful picture of which is shown in Fig. 8.31. This SN was discovered by Treffers et al.[287] in the galaxy NGC4526. Since the distance to this galaxy is in fact quite well known, it is a relatively simple matter to check that the SN is overluminous relative to 'standard' SNs of this type. In view of such discoveries, Drenkhahn and Richtler,[288] and Richmond et al.[289] cast a question mark over the idea that all Type Ia SNs might form a homogeneous class.

---

[286] Nomoto, K., Maeda, K., Umeda, H., Ohkubo, T., Deng, J., & Mazzali, P., Proc. IAU Symp. No. 212 (2002) `arXiv:astro-ph/0209064v1` (4 September 2002).

[287] Treffers, R.R., and 6 authors, IAU Circ. 5946, (1994) p. 2.

[288] Drenkhahn, G., & Richtler, T., Annual Scientific Meeting of the Astronomische Gesellschaft at Heidelberg, September 1998, poster P95.

[289] Richmond, M.W., and 9 authors, AJ **109**, 2121 (1995).

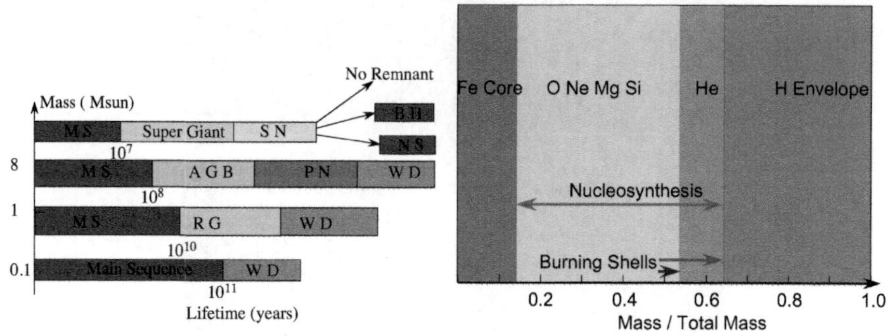

Fig. 8.32 *Left*: The phases and the final fate of different stars in the evolution of single stars. *Right*: The classical structure of a Type I SN progenitor calculated without new mixing mechanisms

8.26 Provisional Summary

The present state of the theory, and the present extensive observational as well as theoretical effort, allow only a provisional summary. The intensive observational search and theoretical research are expected to yield many new results and make many of the above theories obsolete.

The current canonical ideas about the fate of stars with various masses are shown in Fig. 8.32 (left). Stars with masses of the order of $0.1M_\odot$ have a main sequence lifetime longer than the age of the Universe, and hence are still on the main sequence, where they will evolve in due course into white dwarfs. Stars with masses less than about $0.46M_\odot$ do not reach sufficiently high central temperatures to ignite helium, and evolve into helium/hydrogen white dwarfs. The minimum mass to ignite carbon is $1M_\odot$. Stars more massive than $8M_\odot$ explode as SNs, and the outcome is a neutron star, a black hole, or no remnant at all. Since the diagram refers to single stars, Type I SNs, which the current theory claims to emerge from binary systems, are not included.

The approximate structure of the progenitor is shown in Figs. 8.32 (right) and 8.33. Note the change of scale in the structure of the core. The progenitor must include a core which collapses into a neutron star (or a black hole), a region in which nuclear reactions take place, and a pristine envelope or what is left of it after extensive mass loss.

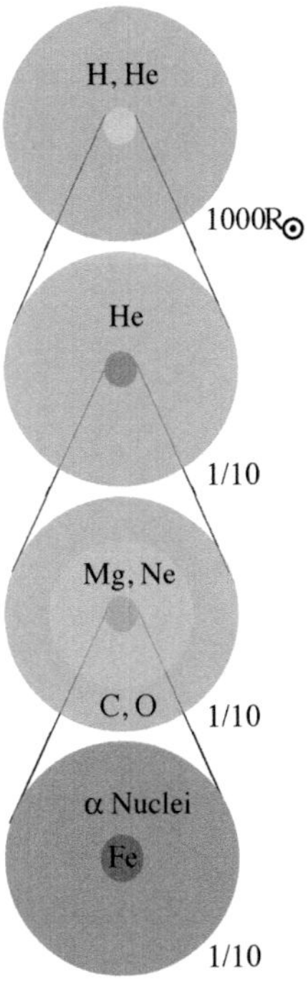

Fig. 8.33 Schematic structure and composition of a SN progenitor. The total radius is about $1000R_{\odot}$ and the scale changes by roughly a factor of 10 between the concentric spheres of different composition

Chapter 9
The Sun

9.1 How Come Life Survives on the Earth

Life on Earth depends on the Sun as an energy source. However, this is far from being sufficient. The mode in which the energy comes, or more accurately the wavelength dependence of the radiative energy distribution, is very important to the chemistry of biology.

The first calculations of the energy distribution from stars (how much energy the stars emit at each wavelength) were carried out by Biermann,[1] and by Unsöld[2] in 1933. Two years later, similar calculations were carried out by Pannekoek.[3] The three theoretical results agreed among themselves, but all were in disagreement with observation.[4]

---

[1] Biermann, L., Gött. Nachr. (1933) p. 45. Also, Veroeffentlichungen der Universitaets-Sternwarte zu Goettingen, Vol. 0003, p. 142, and Untersuchungen uber Sternatmospheren. I. Die Wellenlangenabhangigkeit des Absorptionskoeffixienten. II. Die Opazitat.

[2] Unsöld, A., Zeit. f. Astrophys. **8**, 32 (1934); Ibid. p. 225.

[3] Pannekoek, A., MNRAS **95**, 529 (1935).

[4] This fact did not prevent some disagreeable exchange between Unsöld and Pannekoek, as can be seen from Unsöld's letter to the Observatory [Obs. **58**, 247 (1935)], where he wrote that: *Exactly the same problem*, and Unsöld referred here to Pannekoek's results, *had been treated by Biermann and in more detail by the present author a year ago. Professor Pannekoek, however, does not consider it necessary even to mention these papers at all. I am afraid that this method of neglecting research done by others, which has recently become common in some quarters, will necessarily lead to severe dangers to scientific co-operation. It is only the care – and not the question of priority as such – which has induced me to clear up historical facts.* Shortly afterwards, Pannekoek wrote an apology [Obs. **58**, 328 (1935)] for overlooking the previous work. As a matter of fact, Pannekoek was so excited by the results of Greaves et al. [Greaves, Davidos, and Martin, MNRAS **94**, 488 (1934)] showing that the stars do not radiate like black bodies, that he rushed to calculate the energy distribution theoretically, and this was the reason why he overlooked the fact that it had already been done. Unfortunately, Unsöld's letter could have been written many more times since then, but it might not have yielded the same apologetic response.

G. Shaviv, *The Life of Stars*, DOI 10.1007/978-3-642-02088-9_9,

The general view in the early 1930s was that stars radiate almost like black bodies with the same effective temperature.[5] Hence, the surprise was great when astronomers started to report significant deviations from black body emission. By inverse analysis, the model calculations indicated that the Sun radiates like a gray body,[6] and the absorption coefficient decreases from 12 000 Å to 4 000 Å, and then increases once more. The evidence was clear-cut and left no doubt, so it became a major problem in astrophysics. It was not known what species could produce such absorption behavior.

For years, people suspected that the main contributors to absorption in the Sun must arise from absorption by hydrogen and the more abundant heavier species like Na, Mg, Ca, Fe, and Si.[7] However, this did not in fact provide a solution and the results by Biermann, Unsöld, and Pannekoek just proved it. At temperatures of 4500–6500 K, which is the range of temperatures in the solar photosphere, where the radiation comes from, the temperature is too low to ionize hydrogen. If this was the case, the predicted absorption had to come from the other elements mentioned above. But then the absorption should be discontinuous, and these discontinuities should be observed. But no discontinuities were observed. None of the suggested solutions for a possible mechanism was able to predict the observed behavior as a function of wavelength.

The unique H^- ion, the negative hydrogen ion, as it is called, is a proton plus two electrons. It was known to exist along with other negative ions like O^- (binding energy 2.2 eV), S^- (binding energy 2.8 eV), and F^-, Cl^-, Br^-, and I^-, which have higher binding energies. These ions were the subject of extensive research in the early 1930s.[8] The H^- ion was known to exist in the Kennelly–Heaviside layers of the Earth's atmosphere. These layers are responsible for the reflection of long wavelength radio waves back to the Earth. It was Wildt[9] who suggested in 1939 that this ion might exist in the Sun and similar stars. Very soon it became clear that, not only does it exist in the Sun, but it plays a dominant role in shaping the structure of the solar atmosphere by affecting the absorption of radiation. Since the binding energy is so small, the amount of H^- in the solar photosphere is only 1 ion per 10^7 hydrogen atoms. Yet this small amount has a crucial effect on the spectrum. Taking into account this fragile ion (with a binding energy of only 0.754 eV) results in excellent agreement between theory and observation of the Sun. Importantly for our case here, a year later, Wildt[10] showed that Unsöld's attempt to explain the discrepancy by assuming that the abundance ratio of hydrogen to metal should be

[5] If F is the radiation flux which emerges from the star, the effective temperature, which is the temperature of a black body radiating the same total flux, is given by $F/\pi = \sigma T_e^4$. The question was to what extent the energy distribution of the star matched that of a black body which emits the same total power.

[6] A body is said to be gray when its absorption does not depend on wavelength.

[7] In the range beyond the discontinuities due to Lyman and Balmer jumps in hydrogen and the corresponding jumps in the other ions.

[8] See Mayer, J.E., & Helmholtz, L., Zeit. f. Physik **75**, 19 (1932).

[9] Wildt, R., Ap. J. **89**, 89 (1939).

[10] Wildt, R., Ap. J. **90**, 611 (1939).

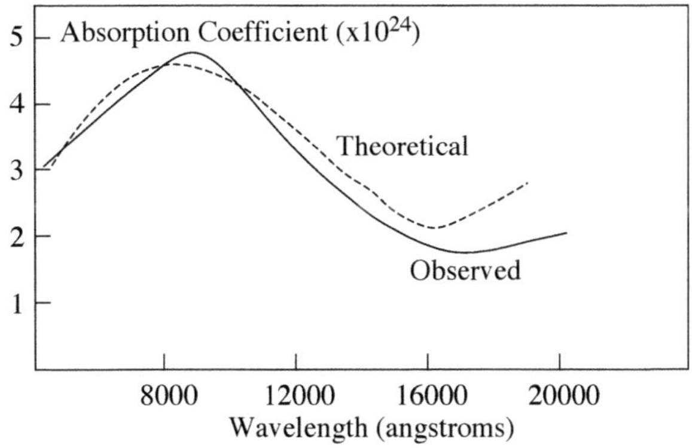

Fig. 9.1 Theoretical absorption coefficient of H^- as a function of wavelength according to Chandrasekhar and Herman 1946. The observed absorption was deduced by Chalonge and Kourganoff. The behavior is exactly as required to bring solar models into agreement with observations

50 to 1, i.e., a very high amount of metal, was not needed at all. Moreover, Wildt showed that the abundance ratio adopted by Russell and Pannekoek of 1000 atoms to 1 atom of heavy elements appeared to be compatible with observations when the absorption of the H^- ion was taken into account.

The agreement was better but not sufficient. Between the years 1944 and 1946, Chandrasekhar and Breen[11] recalculated the absorption power of the H^- ion to find a correction by a factor of 10 with respect to previous results. This time the agreement with the observations of the Sun over the range 4 000–25 000 Å was satisfactory, and abundances could be calculated with confidence. These papers by Chandrasekhar and Chandrasekhar and Breen were among the 22 most cited papers in astronomy and astrophysics in the years 1945–1955.[12] The theoretical result for the absorption as a function of wavelength is exactly as required by observation to explain the spectra of the Sun. Figure 9.1 shows the absorption coefficient as calculated by Chandrasekhar. If the Earth atmosphere had contained just 1 part in 10^7 H^- ions, then it would not have been possible to see any further away than 1 000 meters. As a matter of fact, it is the absorption by the H^- which prevents us from seeing deeper into the Sun.

Despite the fact that the formation of a negative hydrogen ion was suspected in canal ray tubes and that its stability had already been conjectured, it was not until 1930 that Bethe[13] and Hylleraas[14] carried out the first calculations which demonstrated that such an ion can exist. For many years, the only way to assess the possible existence of such an ion was theoretical. The basic problem was to calculate the

[11] Chandrasekhar, S., & Breen, H., Ap. J. **104**, 430.

[12] Brush, S.G., The Most-Cited Physical-Sciences Publications in the period 1945–1954. Current Contents **43**, 3 (1990).

[13] Bethe, H.A., Zeit. f. Physik **57**, 825 (1930).

[14] Hylleraas, E.A., Zeit. f. Physik **60**, 624 (1930).

binding energy of the ground state. The negative hydrogen atom has two electrons moving around the nucleus, and differs from a helium atom only in the mass and charge of the nucleus. The calculation of the ground state of the helium atom was at that time, and for a good many years later, one of the fundamental problems in atomic physics, and this explains the general interest in the structure of this ion, quite beyond its importance for astrophysics. In helium and negative hydrogen, one has two electrons and a nucleus. Yet there is a basic difference from a physical point of view between these two systems. In a helium atom the two electrons move with only a small interaction between them, while the relative interaction between the electrons is stronger in H^-. This fact, which makes the calculation so much more difficult, attracted the interest of physicists in the structure of this ion.[15] Fortunately, Hylleraas was able to show that the particular form of the approximation he used in his attempt to predict the state of helium resulted in an accuracy of 1 in 5000, thereby increasing the credibility of his theoretical calculations.

The result of Bethe and Hylleraas was a binding energy of 0.7 eV. The calculation for the Sun with this value[16] showed that this accuracy was sufficient to prove the existence of the ion, but not sufficient to yield a good solar model. Chandrasekhar and Breen got 0.747 eV for the binding energy and improved the accuracy significantly. The final victory of the theory came when Chalonge and Kourganoff[17] managed to derive what the behavior of the absorption coefficient should be from the observations, so that Chandrasekhar and Breen[18] could compare their theoretical calculations (see Fig. 9.1). The agreement between complex quantum mechanical calculations and what goes on in the Sun is impressive.

After some time, the question of whether H^- has any more bound states surfaced.[19] It was only in 1977 that Hill[20] proved theoretically that H^- does not have any excited bound states. Like deuterium, H^- has only a ground state. As the binding energy is so low, H^- disintegrates completely before the temperature rises to about 7 500 K. Hence, H^- plays a role only in stars with surface temperatures like that of the Sun or lower.

9.2 Victory for Stellar Structure Theory. The Solar Neutrino

As hydrogen in converted into helium, two protons must convert into two neutrons via two β decays. The process releases two neutrinos. As we know the to-

[15] The classical three-body problem is extremely difficult. Even the so-called restricted three body problem, in which the three bodies move in the same plane, is very complicated.

[16] Chandrasekhar, S., & Krogdahl, M.K., Ap. J. **98**, 295 (1943).

[17] Chalonge, D., & Kourganoff, V., AnAP **9**, 69 (1946).

[18] Chandrasekhar, S., & Breen, F.H., Ap. J. **104**, 430 (1946).

[19] Pekeris, C.L., Phys. Rev. **126**, 1470 (1962). The accurate values for the ground states of helium in particular, but also those for H^-, were of prime importance in the investigation of relativistic and quantum effects in atoms. Hence the need for very accurate results.

[20] Hill, R.N., PRL **38**, 643 (1977).

Table 9.1 Solar neutrino sources

Reaction	Neutrino energy [MeV]	Estimated flux on Earth [neutrinos/cm^2 s]
pp chain		
$p+p \longrightarrow d+e^+ +\nu$	≤ 0.429	5.59×10^{10}
$p+e^- +p \longrightarrow d+\nu$	$= 1.445$	1.42×10^8
$^7Be+e^- \longrightarrow {}^7Li+\nu$	$= 0.861$	4.82×10^{10}
$^7Be+e^- \longrightarrow {}^7Li^* +\nu$	$= 0.383$	6.57×10^5
$^8B \longrightarrow 2\alpha+e^+ +\nu$	≤ 14.06	5.15×10^6
$^3He+p \longrightarrow \alpha+e^+ +\nu$	≤ 18.773	8.04×10^3
CNO cycle		
$^{13}N \longrightarrow {}^{13}C+e^+ +\nu$	≤ 1.199	5.71×10^8
$^{15}O \longrightarrow {}^{15}N+e^+ +\nu$	≤ 1.732	5.03×10^8
$^{17}F \longrightarrow {}^{17}O+e^+ +\nu$	≤ 1.740	5.91×10^6

tal luminosity of the Sun, we know the total number of protons which are being converted into helium, and hence we immediately know the total number of neutrinos emitted by the Sun. The resulting flux of solar neutrinos on the Earth is about 6×10^{10} neutrinos/cm^{2}s. However, this is not the whole story. The different sources of neutrino release neutrinos with different energies, and our ability to detect them depends on their energy. Hence, the basic issue is the distribution they have as a function of energy.

The interaction of the neutrinos with matter depends on the energy of the neutrino. The probability of interaction, the cross-section as the physicists call it, increases as the square of the energy. Consequently, the more energetic neutrinos have a greater probability of interaction with the detector. The sources of solar neutrinos are given in Table 9.1. We divide the neutrino sources into those from the pp chain and those from the CNO cycle. All CNO neutrinos are of low energy, and therefore, as will be seen later, are not detected by some of the detectors. The pp neutrinos are unique in having one source of very energetic neutrinos, the ^8B neutrinos. Due to their higher energy, these neutrinos are the most important, although they constitute a very rare branch in the pp chain.

9.3 How the Solar Neutrino Problem Came into Being

Pauli's original idea that an additional particle is emitted during the apparently non-energy-conserving β decay, but then escapes all manner of detection, did not leave physicists with much peace of mind. Various experiments were thus suggested, with

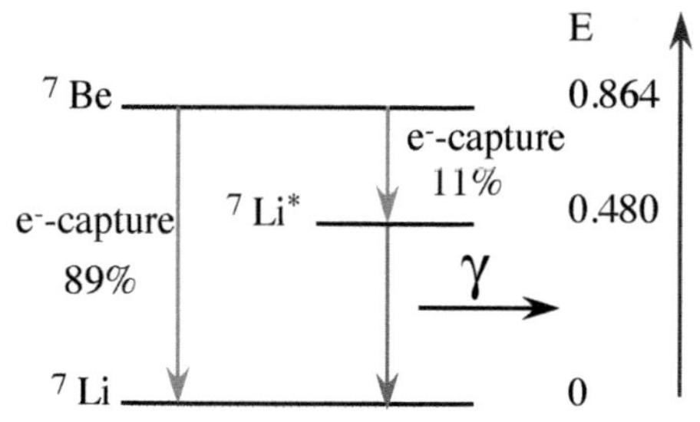

Fig. 9.2 Structure of the ^{7}Be levels and mode of decay to the stable ^{7}Li. Decay can occur directly to the stable lithium (in 90% of the cases) or through an excited state (in 10% of cases). In the latter case, a γ is emitted

a view to direct or indirect detection. But the big question was: what reaction should be chosen for the experiment?

In the late 1930s, Rumbaugh, Roberts, and Hafstad[21] were interested in the products of lithium nuclear transmutations. Lithium was the first element to be disintegrated by artificially accelerated ions, so the nature of the products was of great interest. Among the reactions these authors found was $^6\text{Li}+{}^2\text{D}\to{}^7\text{Be}+\text{n}$, i.e., they created the unstable ^{7}Be nucleus, which they identified to decay by electron capture to ^{7}Li with a half-life of 43 days. Rumbaugh et al. provided the picture shown in Fig. 9.2.

In 1942, Wang[22] suggested using the decay of ^{7}Be to demonstrate the existence of the neutrino. The particular advantages of this reaction are twofold. First, there is no electron or positron emission, so the end products are just two particles rather than three. And second, the recoil energy of the ^{7}Li amounts to 77 eV, which is relatively easy to detect.

In 1942, Allen[23] experimented with the inverse β decay of ^{7}Be. The interest in the experiment was to see whether, as result of the absorption of an electron, there would be a recoil of the nucleus. If there is a recoil, it provides a proof for the existence of the neutrino. The reactions were:

$$\begin{aligned} {}^7\text{Be}+\text{e}^-_{\text{K-capture}} &\longrightarrow {}^7\text{Li}+\nu(?)+0.87\ \text{MeV}\ (90\%)\,,\\ {}^7\text{Be}+\text{e}^-_{\text{K-capture}} &\longrightarrow {}^7\text{Li}^*+\nu(?)+0.48\ \text{MeV}\ (10\%)\,. \end{aligned} \tag{9.1}$$

The innermost electronic shell is denoted by K. Hence, K-capture means that an electron from the K shell is captured by the nucleus. Unlike the classical or even the Bohr picture of the motion of the electrons, the electrons in the K shell move in space and have a finite probability of being in the nucleus. It is then that capture can

[21] Rumbaugh, L.H., Roberts, R.B., & Hafstad, L.R., Phys. Rev. **54**, 657 (1938).

[22] Wang, K.C., PRL **61**, 97 (1942).

[23] Allen, J.S., Phys. Rev. **61**, 693 (1942).

take place. The idea of using the unstable ^{7}Be nucleus in the attempt to discover the neutrino was motivated by the fact that it is relatively light, so the expected recoil is large and easy to detect. And this is indeed what happened. Allen observed the recoil of the ^{7}Li and thereby proved the existence of the neutrino.

In 1944, Weimer et al.[24] bombarded various nuclei with protons. Among others, they attempted the reaction $^{37}\text{Cl}+\text{p} \rightarrow {}^{37}\text{Ar}+\text{n}$, and thereby created the unstable nucleus ^{37}Ar. They discovered that this nucleus decays by emitting an X ray (4.92 Å), and not by emission of a positron. At that time, the known stable isotopes of argon were $A = 36$, 38, and 40. The isotope 37 had not been shown to exist. The original detection contained the chlorine in a solid form. In the second experiment they replaced the solid Cl detector with a gas Cl detector, so that they could see the X ray, which was otherwise absorbed by the solid. The decay time of ^{37}Ar was 34.1 days, and no positron was detected. This was one of the first inverse β reactions to be discovered, and was thus of special interest, due to the fact that no positron accompanied the decay.

In 1946, while still in the West before escaping to the Soviet Union, Pontecorvo wrote a report[25] in which he discussed inverse β processes, and suggested for the first time a direct experiment to detect the elusive neutrino. Until then the best experimental proofs for the existence of the neutrino were indirect ways, mainly via Allen's recoil experiment and the experiment by Jacobsen and Kofoed-Hansen.[26] These results were known to Pontecorvo, although published two years after his report.

The underlying idea of the experiment suggested by Pontecorvo was to use a large mass of detector but to be able to isolate the few nuclei produced in the process. Pontecorvo set out several guidelines for such an experiment, the gist of which was as follows:

- The material irradiated must be cheap because large masses are needed.
- The radioactive nucleus formed should have a long decay time, at least a day, so as to allow enough time to separate it from the bulk.
- The separation between the two elements should be simple.

Then Pontecorvo gave examples. The first was:

$$\nu_\odot + {}^{37}\text{Cl} \longrightarrow \text{e}^- + {}^{37}\text{Ar}\,,$$

to be followed by the radioactive decay

$$^{37}\text{Ar} + \text{e}^-_{\text{K-capture}} \longrightarrow {}^{37}\text{Cl} + \nu_{\text{lab}}\,,$$

with a decay time of 34.1 days. In the first step, the electron is emitted from the nucleus upon the absorption of the solar neutrino $\nu_\odot$. In the second step, an electron

[24] Weimer, P.K., Kurbatov, J.D., & Pool, M.L., Phys. Rev. **66**, 209 (1944); ibid. **60**, 469 (1941).

[25] Pontecorvo, B., Chalk River Laboratories, Chalk River Ontario, 1946.

[26] Jacobsen, J.C., & Kofoed-Hansen, O., Phys. Rev. **73**, 675 (1948).

is captured from the K shell and converts the argon back to chlorine. The neutrino ν_{lab} emitted by the inverse β decay escapes from the detector.

Since the interaction of the neutrino with matter is very weak, an intense source of neutrinos is needed. Consequently, Pontecorvo suggested using the Sun as the source, although he was not especially interested in astrophysics or the structure of the Sun. Pontecorvo estimated that the flux of neutrinos from the Sun should be 10^{16}neutrinos/cm^{2}s, but that they would not be very energetic.[27] When Pontecorvo suggested the experiment, it was not yet known that the reaction $^3\mathrm{He} + {}^4\mathrm{He} \rightarrow {}^7\mathrm{Be}$, which leads to ^{8}B and the high energy neutrinos, has a large cross-section and is therefore important in the Sun.

The experiment with chlorine, as Pontecorvo envisaged it, involved irradiating a large tank of carbon tetrachloride for about a month, and extracting the resulting argon (which is a noble gas and does not interact with anything), by boiling, for example. The radioactive argon would be introduced inside a small counter to count the decay of the ^{37}Ar.

9.4 Doubts and Possible Implications

The value of the cross-section of the neutrino for interaction with matter was not well known. For this reason, the question arose as to whether the neutrino could actually escape from the Sun. In 1948, nobody had much idea about the interaction of the neutrino with matter. So some[28] reasoned that, if the interaction cross-section of the neutrino with matter had been as high as 10^{-35} cm^2, the absorption of neutrinos in the Earth would yield a heat flux of between 10 and 100 times the known rate of heat flow from the core through the surface of the Earth. So the argument was that the cross-section had to be lower than this number. It is interesting that the Earth was used as a calorimeter to set up an upper limit for the interaction of an elementary particle with matter.

A year later, Saxon[29] calculated that the number of neutrinos at the Earth would be $\sim 3.5 \times 10^{11}$ neutrinos/cm^{2}s, and was wrong about the probability of interaction by about a factor of 10^7. An upper limit could be placed on the interactions of the neutrinos with the Earth, because if the interaction had been 10^{14} times bigger, it would have caused heating of the Earth. The idea was not to detect the solar neutrino, but to find out whether these copious particles could be an important factor in the Earth's heat balance. Fortunately, they play no role in the heat balance of the Earth. Saxon claimed an upper limit for the probability of interaction which was 3.5×10^5 higher than what Pontecorvo had estimated.

[27] Pontecorvo did not specify whether such a flux of neutrinos would exist at the surface of the Sun or at the Earth, and no details of the calculation were given. The total flux at the Earth is in fact 6.4×10^{10} neutrinos/cm^{2}s.

[28] Crane, H.R., Rev. Mod. Phys. **20**, 278 (1948).

[29] Saxon, D., PRL **76**, 986 (1949).

Table 9.2 Neutrino sources on the Earth

Source	Estimated flux [neutrinos/cm^{2}s]	Comments
Sun	6×10^{10}	Mostly low energy, but a trace of high energy
Cosmic rays	0.1	Preferentially high energy
Natural Earth radioactivity	5×10^6	Low energy
Nuclear reactors	$\sim 10^{12}$	Close to a 10 GW reactor, low energy

Neutrinos are liberated in the core of the Sun and move outward through it. But the Sun is full of free electrons, and the question was to what extent the neutrinos would actually manage to escape from this massive object. Pontecorvo guesstimated the strength of the interaction between the neutrino and the electron,[30] and got a value which was a factor of about 15 higher than what we know today. This number for the interaction ensures that a negligible fraction of the neutrinos is absorbed by the Sun. However, since it was a theoretical estimate, it did not disperse all doubts.

In 1949, Pontecorvo, Kirkwood, and Hanna[31] essentially confirmed the experimental results of Weimer et al., as well as their explanation, namely that an electron from the innermost shell was absorbed by the nucleus and gave rise to the conversion of a proton into a neutron. The neutrino was not observed.

Even as late as 1952, and 20 years after Pauli's bold conjecture, the hypothesis of the neutrino was far from generally accepted. In 1952, Davis,[32] who was a physical chemist by training, repeated the ^{7}Be recoil experiment. According to him, the results of the experiment were consistent with the hypothesis of single neutrino emission in ^{7}Be.

In 1955, Davis began his long journey into the solar neutrino problem. He used tanks containing 200 and 3900 liters of carbon tetrachloride, which he irradiated outside the shield of the Brookhaven reactor in an attempt to induce the reaction $^{37}\mathrm{Cl} + \bar{\nu} \rightarrow {}^{37}\mathrm{Ar} + \mathrm{e}^-$ by means of fission-product antineutrinos released in large numbers inside the nuclear reactor. The experiments[33] served to place an upper limit of 2×10^{-42} cm^2 per atom.[34] Various experiments[35] detected very high energy

[30] Pontecorvo applied estimates and ideas due to Fierz, Bethe, and Bethe and Peierls, as well as Konopinski and Uhlenbeck, and gave no detailed reference.

[31] Pontecorvo, B., Kirkwood, D.H.W., & Hanna, G.C., PRL **5**, 982 (1949), and Kirkwood, D.H.W., Pontecorvo, B., & Hanna, G.C., Phys. Rev. **74**, 497 (1948).

[32] Davis, R., Phys. Rev. **86**, 976 (1952).

[33] Davis, R., Jr., Phys. Rev. **97**, 766 (1955).

[34] In his Nobel address, Davis commented that Luis Alvarez (who got the Nobel Prize in 1968 for his many contributions to elementary particle physics) *proposed to use the chlorine–argon reaction to detect solar neutrinos with a large tank of concentrated sodium chloride solution* [Alvarez, L., University of California Radiation Laboratory, Report UCRL 328 (1949)]. Davis replaced the compound to which the chlorine atoms were attached, for technical reasons.

[35] Firman, E.L. & Zähringer, J., Phys. Rev. **107**, 1695 (1957).

cosmic-ray-induced ^{37}Ar formation.[36] Consequently, Davis had to look for a good shield to reduce spurious ^{37}Ar formation in his detector. Measurements with the 3900 liter container, shielded from cosmic rays by 19 feet of earth, allowed him to place an upper limit on the neutrino flux from the Sun. Davis realized that the neutrinos from the pp chain have energies below the 0.816 MeV threshold of the chlorine experiment. On the other hand, the neutrinos from ^{13}N and ^{15}O have energies of 1.24 and 1.68 MeV, and hence should be observed. If the Sun produced all its energy from the CN cycle, so assumed Davis, then the flux of neutrinos at his detector should have been 6×10^{10} neutrinos/cm^{2}s. But, Davis could only place an upper limit of 1×10^{14} neutrinos/cm^{2}s. This was 1000 times too high for detecting the solar neutrinos.

In 1955, Cormack[37] improved the estimate of the number of neutrinos which reach the Earth to 7.4×10^{10} neutrinos/cm^{2}s. Cormack already had a sufficiently accurate value for the probability of absorption to conclude that such a flux has no heating effect on Earth. But, argued Cormack, if Bethe was right with his estimate for inelastic scattering in the interaction cross-section, then the heat deposited on the Earth would start to become appreciable.[38] In a later publication, the result was corrected. In fact, the energy deposited by solar neutrinos was found to be negligible relative to the energy deposited by natural radioactivity. Next it was confirmed that the Sun is practically transparent to the neutrino. By then it was already known that the cross-section is 10^{-44} cm^2 (he received a private communication from Cowan and Reines).

9.5 The Most Energetic Solar Neutrino Source Identified

In 1958, Fowler[39] attended a conference and heard how Holmgren and Johnson announced that the reaction $^3\mathrm{He} + {}^4\mathrm{He} \rightarrow {}^7\mathrm{Be}$ was significantly faster than had been previously thought. Fowler was quick to realize the implications of the new result and published a paper on the consequences of their discovery for the pp chain and for the solar neutrino flux, well before Holmgren and Johnson managed to publish their own paper. Fowler estimated that neutrinos from hot stars would give rise to a background flux on the Earth of about 2×10^{10} neutrinos/cm^2 s, and that these could, reasoned Fowler, be detected by the techniques developed by Davis. This of course implied a cosmic background of neutrinos generated by main sequence stars. Essentially, Fowler predicted the emergence of neutrino astronomy. It is hard to explain why Fowler made no mention at all of the neutrinos from the Sun in this paper. Maybe he was still thinking that most solar energy results from the CN cycle,

---

[36] By BeV protons colliding with iron for example.

[37] Cormack, A.M., PRL **95**, 580 (1954); ibid. Phys. Rev. **97**, 137 (1955).

[38] It depends on the magnetic moment of the neutron, a number not known at the time. Note that a similar story, namely the importance of the neutrino scattering relative to neutrino absorption, took place years later in the supernova story.

[39] Fowler, W.A., Ap. J. **127**, 551 (1958).

Fig. 9.3 *Light*: The scenery of the Black Hills, South Dakota, and the Homestake Gold Mine, where, at a depth of 1600 meters, Davis measured for the first time neutrinos born in the core of the Sun. *Right*: The detector in Davis' chlorine experiment

and not from the pp chain. But note the inconsistency: hot stars operate on CNO and not pp.

In 1964, Davis[40] had already obtained the first null results from a small version of the detector (two 500 gallon tanks at a depth of 766 meters underground[41]). On the basis of this experiment, in which he did not discover solar neutrinos for sure, but did test the technique, Davis estimated that he could upgrade his experiment to 100 000 gallons without *insuperable difficulties*.

The 100 000 gallon (about 520 ton) detector of tetrachloroethylene[42] was built in a rock enclosure about 1620 meters below ground in the Homestake Gold Mine (see Fig. 9.3). The deep mine location shields the detector from cosmic rays and their products.

Natural chlorine has two stable isotopes, ^{35}Cl and ^{37}Cl, and their abundances are 75.77% and 24.23%, respectively. All other chlorine isotopes are unstable. Hence, about a quarter of the molecules of C_2Cl_4 contained the right nucleus for the experiment, which had two phases. In the first, radioactive argon is allowed to build up until a steady state is reached between the buildup and the 34.1 day decay. In the second phase, the radioactive argon is flushed out of the giant tank, and its radioactivity measured.

[40] Davis, R. Jr., PRL **12**, 303 (1964).

[41] Davis' first experiment was carried out in a limestone mine at Barberton, Ohio, which belongs to the Pittsburgh Plate Glass Co.

[42] Tetrachloroethylene ($Cl_2C{=}CCl_2$), also known as PCE, is used for dry cleaning and degreasing metal surfaces.

9.6 The Neutrino Confusion

The idea that there might be two different types of neutrino emerged from the discovery that there are two components to β decay (the Fermi and the Gamow–Teller components), together with the discovery that the weak interactions do not satisfy parity conservation,[43] which is one of the most fundamental premises in physics. In 1956, Lee and Yang suggested that parity is not conserved in order to explain some kinds of weak interaction decays. The Nobel committee worked quickly in this case, and had already awarded the prize to Lee and Yang by 1957. Note that the idea of two different neutrinos was born after the existence of the neutrino had been proven indirectly, but before the existence of the first neutrino was directly demonstrated by experiment.

In 1956, Reines and Cowan[44] discovered the neutrino emitted by fission products in a nuclear reactor (see Sect. 6.15). The detector comprised 40 kilograms of $CdCl_2$ dissolved in 200 liters of water and placed about 11 meters away from the nuclear reactor and about 12 meters underground.

The puzzle over how such an 'obvious' thing as parity conservation could be violated bothered many physicists. As a way to explain how this could happen, Salam,[45] Lee and Yang,[46] and Landau[47] suggested a two-component neutrino. As Lee and Yang wrote:

The theory is possible only if parity is not conserved in interactions involving the neutrino.

However, Lee and Yang soon found out that, in such a theory, the mass of the neutrino must by zero.[48] Thus, the very strange properties of the neutrino and the weak interaction led physicists to look for creative solutions, like a two-neutrino system. It is interesting to note that such a theory had already been examined by Pauli as early as 1933,[49] although he rejected it precisely because it violated parity conservation!

In 1957, Pontecorvo wrote:[50]

If the conservation law of neutrino charge did not apply, then in principle a neutrino could convert into an antineutrino in vacuum.

---

[43] Parity conservation in physics means that the process looks the same in the laboratory and in the mirror. Parity violation means that the experiment in the mirror is not identical to the experiment in the laboratory, and the observer can tell whether he sees the experiment or its mirror image. Mathematically, conservation means that the description of the system does not change if, instead of the system of axes x, y, z, we prefer to use the system $-x, -y, -z$, or instead of the right-handed system, we prefer to use a left-handed system. We have already mentioned that the pp reaction requires the Gamow–Teller transition, which does not conserve parity.

[44] Reines, F., & Cowan, C.L., Nature **178**, 446 (1956); Cowan, C.L., et al., Science **124**, 103 (1956).

[45] Salam, A., Nuovo Cimento **5**, 299 (1957).

[46] Lee, T.D., & Yang, C.N., Phys. Rev. **105**, 1671 (1957).

[47] Landau, L., Nucl. Phys. **3**, 127 (1957).

[48] A similar result was found by Nishijima, K., PRL **108**, 907 (1957).

[49] Pauli, W., *Handbuch der Physik*, Springer, Berlin **34**, 226 (1933).

[50] Pontecorvo, B., J. Exptl. Theoret. Phys. USSR JETP **6**, 429 (1958).

Generally, physical systems have what is called CPT symmetry. This means that, given a certain elementary process, if you reverse the time (T), invert the charge (C), and look in the mirror (P), the process is invariant, i.e., it looks the same. As the weak interaction was found not to conserve parity, Pontecorvo went one step further and assumed that the charge might not be conserved either. The implications were that the neutrino could oscillate between the neutrino and antineutrino states as it propagates in vacuum.

The first time the idea of 'particle mixing' appeared in the literature was when Gell-Mann and Pais[51] tried to explain the decay of a hypothetical θ^0 particle.[52] They assumed that there exists a particle θ_1^0 and a hypothetical second neutral particle θ^0. The main distinctive feature of the two particles was that the set of decay modes that are allowed for the one are forbidden for the other. Hence, the two particles were expected to have different lifetimes. In addition, the masses of the two particles are not strictly equal. Since the propagation properties of the two particles are different, the following phenomenon arises. The particles are born in a specific combination of the two particles. As the particles propagate with different velocities (they have the same energy but different mass, hence the difference in propagation), they appear to oscillate between the two states of the particle. This particular effect was suggested to Pais and Picciono[53] by R. Server. The peculiarity of β decays led Pauli to hypothesize the existence of an elusive particle, so assuming another hypothetical particle, as Gell-Mann and Pais did, was not such a great 'chutzpa'.

In 1961, Feinberg, Gürsey, and Pais[54] explored *some possible implications of the assumption that two distinct neutrinos are involved in the weak interactions.*

In 1962, Danby et al.[55] carried out the first accelerator neutrino experiment and established the existence of two kinds of neutrinos. Once two neutrinos had been found to exist, the first to be discovered was named the electron neutrino ν_e and the new one was named the muon neutrino ν_μ, according to the associated particles in the decay. In their 1962 paper, Danby et al. gave no citation of Reines. It was almost as though the ν_e did not yet exist, although they wrote about a second kind of neutrino.[56] The confirmation of the existence of a second neutral particle, similar to the first discovered neutrino, but different from it, was very important for the theory of particle mixing.

In 1963, Bahcall et al.[57] attempted the first theoretical calculation of the neutrino flux from the Sun. However, it was Davis, in a private communication, who directed their attention to the fact that the more energetic part of the ^{7}Be neutrino flux was just above the threshold of the experiment. Most of the calculation referred to the

[51] Gell-Mann, M., & Pais, A., Phys. Rev. **97**, 1387 (1955).

[52] The symbol for this particle was later changed to K for the kaon.

[53] Pais, A., & Picciono, O., Phys. Rev. **100**, 1487 (1955).

[54] Feinberg, G., Gürsey, F., & Pais, A., PRL **7**, 208 (1961).

[55] Danby, G., Gaillard, K.J-M., Goulianos, K., Lederman, L.M., Mistry, N., Schwartz, M., & Steinberger, J., PRL **9**, 36 (1962).

[56] The title of the paper was *Observation of High-Energy Neutrino Reactions and the Existence of Two Kinds of Neutrinos.*

[57] Bahcall, J.N., Fowler, W.A., Iben, I., Jr., & Sears, R.L., Ap. J. **137**, 344 (1963).

^8B flux, which is constituted of the most energetic neutrinos coming out of the Sun. The calculations were global and came without a detailed solar model. In 1964, Bahcall[58] repeated the same kind of work but with more accurate nuclear data.

In 1964, Sears[59] used a detailed solar model for the first time, to calculate the solar neutrino flux. Sears concluded that the uncertainty in the composition of the Sun precluded an estimate of the solar neutrino flux with an accuracy better than a factor of 2. This was the first tight connection between the structure of the Sun and the neutrino flux. As Sears wrote:

> *Theoretical models of the internal structure of the Sun are no longer at the frontier of the theory of stellar structure and evolution. Since the recognition of the proton–proton chain as the major energy source, the general features of solar structure have been quite well established.*

In principle, Sears was right, but the details were, and still are, well out of our reach. And then there were surprises yet to come.

9.7 The Pioneer in Solar Neutrino Experiments

The first results from the Homestake experiment were published by Davis, Harmer, and Hoffman in 1968.[60] This time they were discussing the ^8B neutrinos. Davis et al. placed an upper limit of 2×10^6 neutrinos/cm^{2}s at the Earth.[61] In addition, they placed an upper limit of 9% on the amount of solar energy derived from the CN cycle. If the result was taken as a real detection, then the results corresponded to just 0.15–0.3 of the flux predicted by Bahcall, Bahcall, and Shaviv[62] (see also[63]). The conclusions were as follows:

- The Sun does not derive most of its energy from the CNO cycle.
- If the theory of stellar structure is correct, then the total amount of heavy elements must be less then 2% by mass.
- If the measured value of 0.019 for the ratio of the mass of the heavy elements to the mass of hydrogen is accepted, then the helium abundance is $(22 \pm 0.03)\%$.

---

[58] Bahcall, J.N., PRL **12**, 300 (1964).

[59] Sears, R.L., Ap. J. **140**, 153 (1963).

[60] Davis, R., Jr., Harmer, D.S., & Hoffman, K.C., PRL **20**, 1205 (1968).

[61] At this flux of neutrinos, Davis got about 1 neutrino count per day in the 520 tons of liquid detector. So after a month of continuous running of the experiment, they were looking for about 30 atoms floating in 520 tons of liquid. It is true that these atoms were radioactive, but filtering such a huge amount in search of 30 trace atoms was a colossal technical achievement in itself.

[62] Bahcall, J.N., Bahcall, N.A., & Shaviv, G., PRL **20**, 1209 (1968).

[63] Pochoda, P., & Reeves, H., Planetary Space Sci. **12**, 119 (1964); Bahcall, J.N., PRL **17**, 398 (1966); Ezer, D., & Cameron, A.G.W., Can. J. Phys. **43**, 1497 (1965); ibid. **44**, 593 (1966); Bahcall, J.N., Cooper, N., & Demarque, P., Ap. J. **150**, 723 (1967); Shaviv, G., Bahcall, J.N., & Fowler, W.A., Ap. J. **150**, 725 (1967); Bahcall, J.N., Bahcall, N., Fowler, W.A., & Shaviv, G., Phys. Lett. B **26**, 359 (1968).

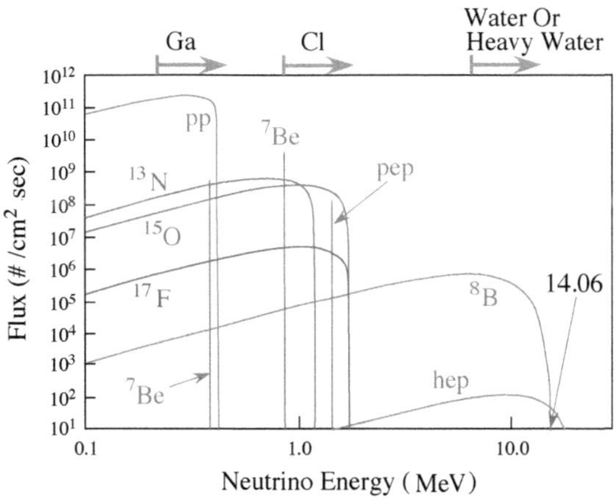

Fig. 9.4 The spectrum of neutrinos from the Sun

At that time Bahcall, Bahcall, and Shaviv thought that:

> *There is no irreconcilable discrepancy between our predictions and the experiment of Davis, Harmer, and Hoffman when the uncertainties in the various parameters that enter the calculations are taken into account.*

They were wrong. The recognition that there was a real discrepancy crept in slowly when more nuclear data and theoretical modeling of the Sun consistently reduced the uncertainty in the predictions to the point that the uncertainties in the theoretical result could no longer bridge the gap between the measured and predicted results.

Bahcall used to publish a figurative description of the solar neutrino problem every time a new measurement of the solar neutrino flux was made (see Fig. 9.5), or a new theoretical estimate published, much like a kind of weather forecast. In 1969,[64] he also introduced a new unit, the SNU, with the definition: 1 SNU = 10^{-36} captures per target particle per second.

It is amazing that, despite the fact that the solution was in principle available from the beginning (neutrino oscillations or particle mixing), it took 40 years to verify it, about 10 000 papers,[65] one Nobel Prize, and myriads of misguided suggestions. Consequently, it is impossible to cover all this flurry of papers and results, and we have to become selective. We shall therefore summarize the subject along three lines: proposed astrophysical solution, invention of elementary particle solutions, and the planning and running of new experiments with complementary properties to those of the chlorine experiment.

---

[64] Bahcall, N.J., PRL **23**, 251 (1969).

[65] A check in the Astronomical Data System gave 9882 papers with the keywords 'solar' and 'neutrino'.

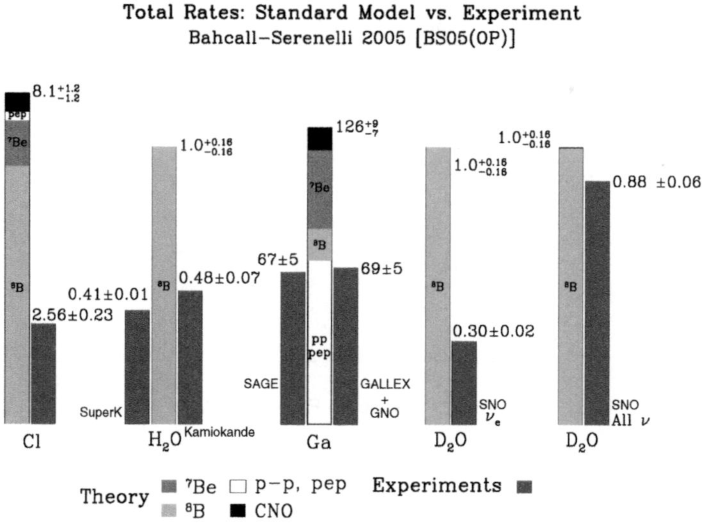

Fig. 9.5 The solar neutrino problem. Figurative summary from Bahcall and Serenelli, 2005

It just so happened that the first results were an upper limit. This is one of the vagaries of fate, for if the tank had been a factor of 2–5 bigger, the solar neutrino saga would have been completely different.

9.8 Suggested Astrophysical Solutions

The problem of the solar neutrino provided a superb opportunity to examine all the assumptions and the physics involved in the theory of the Sun and stellar structure and evolution. It is therefore interesting to see what kind of ideas were put forward, and what the solar neutrino allowed one to prove, or more accurately, to disprove.

An examination of the abundance of ^{3}He in the Sun showed that, at low temperatures, ^{3}He builds up, while at the high core temperatures, it is destroyed. Consequently, there is a peak of ^{3}He abundance somewhere out along the radius of the Sun. If for some reason ^{3}He is driven into the core, it burns quickly and releases a lot of energy,[66] and consequently reduces the mean temperature of the core, thereby reducing the amount of ^{8}B neutrinos and eliminating the problem. To drive ^{3}He into the core, you need some mixing mechanism. So the first proposed solutions were the resurrection of the old 'mixing' idea in one form or another.

[66] The reaction $^3\mathrm{He}+{}^3\mathrm{He}\rightarrow{}^4\mathrm{He}+2\mathrm{p}$ releases 12.859 MeV, and hence is the most energetic reaction after $^7\mathrm{Li}+\mathrm{p}\rightarrow 2^4\mathrm{He}$, which releases 17.347 MeV. This is a lot of energy. The $\mathrm{p}+\mathrm{p}\rightarrow{}^2\mathrm{D}+\mathrm{e}^++\nu$ releases only 1.442 MeV per reaction.

In 1972, Rood[67] proposed a temporal form of mixing. Sakurai[68] showed by numerical calculations that the effect of the Eddington–Sweet meridional flow would be to accelerate the rotation of the inner parts of the Sun. Rood argued that, in this case, the stability condition against convection must include the rotation. If so, it became possible that there would be temporal mixing of the core by means of convective currents. However, Sakurai claimed that the spinning of the core would take 10^7 to 10^8 yrs, which is much longer than the time needed to build up the ^{3}He. A similar suggestion was also proposed by Ezer and Cameron[69] in the next article after Rood's. Ezer and Cameron connected the sudden mixing of the core of the Sun with the periods of ice ages on the Earth. Kocharov[70] suggested an increase in the ^{3}He, but provided no mechanism to explain how that could happen. A year later, Prentice[71] suggested that the helium might not be homogeneous in the core of the Sun. This solution was shot down right away by Demarque, Mengel, and Sweigart.[72]

Chitre, Ezer, and Stothers[73] suggested the existence of a strong magnetic field in the core of the Sun and attributed to it the duty of reducing the gas pressure and temperature. Then, since ^{8}B production is extremely sensitive to the temperature (it varies as the fourteenth power of the temperature), they felt justified in expecting a large reduction in this neutrino flux.

Cameron looked for variations in the solar luminosity on the Kelvin–Helmholtz–Ritter timescale.[74] If such oscillations take place, it means that the luminosity which emerges from the surface is not necessarily the total energy produced by the core at that moment. We know that the Kelvin–Helmholtz–Ritter time is the calculated lifetime of the Sun if it had lived off gravitational energy. However, it is also the time it takes for the energy produced in the core of the Sun to diffuse to the surface. So if the energy production in the core is shut off somehow, it would take the Kelvin–Helmholtz–Ritter time before this was noticed on the surface. This means that, according to this suggestion, the present day luminosity does not reflect the energy produced in the core today but rather what was produced some 10^7 yrs ago. If the Sun is not in a steady state, there may be a difference between the present day energy production as reflected by the neutrino flux, and the surface luminosity which emits today the energy generated a long time ago. This would mean that the predicted neutrino flux corresponding to today's surface luminosity would be wrong.[75]

---

[67] Rood, R.T., Nature **240**, 178 (1972).

[68] Sakurai, T., Publ. Astron. Soc. Japan **24**, 153 (1972).

[69] Ezer, D., & Cameron, A.G.W., Nature **240**, 180 (1972).

[70] Kocharov, G.E., ICRC **2**, 1602 (1973).

[71] Prentice, A.J.R., MNRAS **163**, 331 (1973).

[72] Demarque, P., Mengel, F.G., & Sweigart, A.V., MNRAS **165**, 19 (1973).

[73] Chitre, S.M., Ezer, D., & Stothers, R., Astrophys. Lett. **14**, 37 (1973).

[74] Cameron, A.G.W., Rev. Geophys. Space Phys. **11**, 505 (1973).

[75] After this solution had been proposed and published in Nature, the same journal subsequently rejected a paper by Shaviv, G., who used the luminosities of a collection of solar type stars to prove that it was not after all a viable solution. The justification was that *the paper is of no interest.*

A similar solution was proposed by Shchepkin.[76] But all analyses showed that the Sun is perfectly stable, and that the total energy produced in the core is equal to the luminosity. No real calculation of the stability of the Sun was carried out by Shchepkin, and he overlooked the result by Schwarzschild and Harm,[77] who demonstrated that the Sun is perfectly stable, and that no such discrepancy exists between the energy produced in the core and the luminosity.

The standard calculation for the Sun assumes that the Sun does not rotate. The surface of the Sun actually rotates very slowly, with a period of 25–27 days (depending on the latitude). This rotation induces a centrifugal acceleration which is about 1/1000 of the gravitational acceleration on the solar surface, whence rotation is usually neglected. But the core could in principle rotate faster, and consequently, rotation in the core could be of some importance.

A fast core rotation was suggested by Demarque et al.[78] and independently by Roxburgh,[79] who claimed that:

> *The low upper limit of 1 SNU on the observed neutrino flux from the Sun obtained by Davis has proved an embarrassment to stellar physicists, and in spite of considerable intellectual gymnastics the standard solar models predict at least 6 SNU. The essential difficulty has been to produce a model with a low enough central temperature that can still produce the observed luminosity of the Sun with an age of 4.7×10^9 yrs.*

Consequently, Roxburgh supposed a fast rotating solar core.

The core of the Sun cannot be too fast a rotator, because if it were, its shape would deviate from spherical symmetry and affect the precession of the perihelion of Mercury, which is one of the best tests for the general theory of relativity. As early as 1895, long before general relativity was invented, Newcomb[80] suggested that the solution to the problem of the precession of mercury could be a 1/3600 oblateness[81] of the Sun. However, shortly after that, Poor[82] rejected the idea and gave an upper limit to the solar oblateness of less than 1.8×10^{-5}, whence it would have had a negligible effect on the precession of Mercury. A few years later, Einstein could thus carry out his calculation of the effect assuming the Sun to be a perfect sphere.

9.9 Is the Sun a Perfect Sphere?

Dicke and Brans[83] came up with a new theory of gravity which was supposed to correct Einstein's general theory of relativity. As discussed above, if the Sun is not

[76] Shchepkin, M.G., ZhETF Pis. Red. **17**, 226 (1973).

[77] Schwarzschild, M., & Harm, R., Ap. J. **184**, 5 (1973).

[78] Demarque, P., Mengel, J.G., Sweigart, A.V., Ap. J. **183**, 997 (1973).

[79] Roxburgh, I.W., Nature **248**, 209 (1974).

[80] Newcomb, S., *Fundamental Constants of Astronomy*, US GPO, Washington, D.C. (1895) p. 111.

[81] Oblateness is the difference between the polar and equatorial radii.

[82] Poor, C.L., Ann. NY Acad. Sci. **18**, 385 (1908).

[83] Brans, C., & Dicke, R.H., Phys. Rev. **124**, 925 (1961).

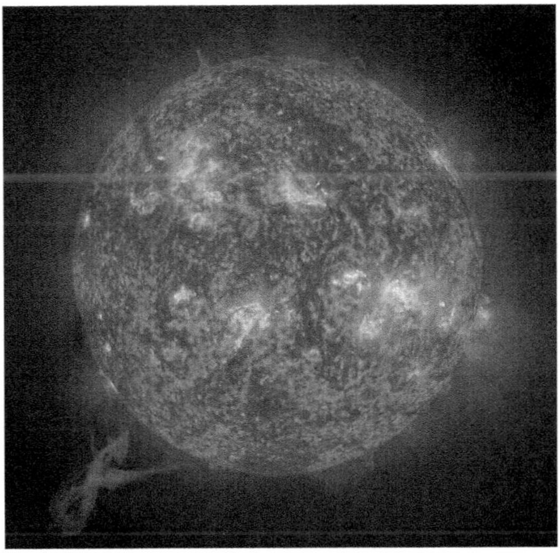

Fig. 9.6 The Sun is far from being a quiet object. The picture illustrates the difficulties in measuring the oblateness of the Sun. Credit NASA

spherical, a correction must be introduced, and this correction destroys the nice agreement between general relativity and observation. Dicke wanted his theory to step in and provide the extra precession needed in this case to restore the agreement. For this reason, he wanted to discover an oblate Sun. All he needed was an effect of about 10%, since this was sufficient to introduce a correction to Einstein's theory.

So starting in 1966, Dicke and Goldenberg[84] measured the difference in flux between the equator and the polar limb of the Sun. Their first results indicated a small oblateness. The relative difference between the polar and equatorial radii was claimed to be $(4.51 \pm 0.34) \times 10^{-5}$. Complications arose from the fact that the solar surface is never at rest, nor uniform (see Fig. 9.6). There are sunspots and eruptions. So Hill[85] repeated the measurement and found that Dicke's result was due to temporal brightening of the equator. When corrected for this effect, the result for the oblateness of the Sun was reduced to $(1/2 \pm 3) \times 10^{-6}$, thus confirming Poor's old conclusion.

Dicke did not give up, and in 1977[86] claimed that his result indicated that the Sun rotates as a solid body with a period of 12.22 ± 0.12 days (and not with the period observed on the surface). In 1979, Vasilev[87] concluded that the asphericity of the Sun could not exclude a fast core rotation with a period of 0.1–4.5 days, which is very fast indeed, and might have consequences for the possible mixing of the core.

84 Dicke, H.R., & Goldenberg, H.M., PRL **18**, 313 (1967); ApJS **241**, 131 (1974).

85 Hill, H.A., and 6 authors, PRL **33**, 1497 (1974).

86 Dicke, H.R., Ap. J. **218**, 547 (1977).

87 Vasilev, S.S., Izvestiia, Serrii Fizicheskaia **43**, 753 (1979).

Further evidence that the solar deviations from a perfect sphere can be neglected was provided by Kislik.[88]

In 1985, Dicke, Kuhn, and Libberecht[89] remeasured the oblateness and got half the value Dicke had got back in 1966. So they explained it as a time-dependent phenomenon, blaming it on periodic changes in the gravitational field of the Sun. The solar oblateness became variable in time.[90]

The issue was apparently settled when radar observations of the Sun[91] set the oblateness at $(0.66 \pm 0.9) \times 10^{-6}$. A similar value was also found by Langraf.[92]

The most recent observation by Fivian et al.[93] discovered a much larger oblateness than reported so far. However, comparison with previous measurements led the researchers to conclude that the oblateness was due to magnetic activity, and not rotation of the inner parts of the Sun.[94]

The results of helioseismology (see Sect. 9.19) show that the core is a slow rotator, and cannot have any effect on the solar neutrino problem, nor on the precession of Mercury.[95]

9.10 Is Statistical Mechanics Defective in the Sun?

Recall Eddington's basic premise, namely, that stars obey the laws of physics as we know them. Still, the annoying discrepancy between the theory and observations led astrophysicists to question some of the most basic physical assumptions.

The particles in the hot Sun are assumed to follow the Maxwell distribution of energies as a consequence of the collisions between them. In 1974, Clayton[96] made the following statement:

> *The solar neutrino discrepancy is now regarded as very serious. The general attitude is that the Sun is trying to tell us something, but no one is quite sure what. I wish in this note to add to the list of possible ad hoc explanations; namely that the scarcity of ^8B neutrinos reflects a departure of the distribution of relative kinetic energies from the Maxwellian distribution. I have not been able to calculate a cause of the indicated depletion of the high energy tail of relative energies, so this explanation must be limited to the associated effect.*

---

88 Kislik, M.D., Sov. Astron. Lett. **9**, 296 (1983).

89 Dicke, R.H., Kuhn, J.R., & Libberecht, K.G., Nature **316**, 687 (1985).

90 Dicke, R.H., Kuhn, J.R., & Libberecht, K.G., Ap. J. **311**, 1025 (1986); ibid. **318**, 451 (1987).

91 Afanas'eva, T.L., Kislik, M.D., Kolluka, Iu.F., & Tikonov, V.F., Astronomicheskii Zhurnal **67**, 1326 (1990).

92 Langraf, W., Icarus **142**, 403 (1992).

93 Fivian, M.D., Hudson, H.S., & Lin, R.P., *Variations of Solar Radius Observed with RHESSI*, Am. Geophys. Union, Fall meeting 2003, Abstract SH32A-1103.

94 Temporal variations in the oblateness were also observed by Egidi et al., Solar Physics **235**, 407 (2006).

95 For possible effects on general relativity, see Campbell, L., McDow, J.C., Moffat, J.W., Vincent, D., Nature **305**, 508 (1983).

96 Clayton, D.D., Nature **249**, 131 (1974).

In other words, the calculation of Atkinson and Houtermans in 1929 assumed that the particles in the Sun obey the Maxwell–Boltzmann distribution. But if this is not correct, then clearly all calculated reaction rates are wrong. After all, these reactions involve a very small number of particles in the high end of the distribution.

The title of Beaudet's paper[97] tells the entire story: *Desperate Models for Solar Neutrinos*. Indeed, models were being considered that would, under normal conditions, have been flatly rejected. For example, Barnothy[98] assumed that strong interactions between nuclear particles were of a gravitational nature. Due to gravitational interaction between neutrinos and nucleons, the neutrinos may transfer a momentum to nucleons and lose energy while passing through the Sun. A connection between the ice ages and solar luminosity variations[99] was even proposed as a proof that the luminosity of the Sun is not constant, despite the demonstration by Schwarzschild and Harm to the contrary.

It seems that it is difficult to kill even incorrect theories. In 1975, Vasilev[100] sounded quite desperate when he wrote the following:

> *Several hypotheses are critically examined which have been proposed to clarify the discrepancy between the theoretical and observed solar neutrino fluxes. It is noted that these hypotheses explicitly or implicitly suggest that the Sun is not a standard main-sequence star. The hypotheses attribute the discrepancy to a four-fold decrease in the neutrino flux from ^{8}B, steady intermixing of one kind or another in the Sun, sudden intermixing in the depths of the Sun, centrifugal acceleration of rapidly rotating nuclei, or the influence of the internal solar magnetic field. Each of these phenomena is shown either to contradict basic physical concepts or to be too ineffective to be the unique cause of the discrepancy.*

9.11 A Unique Mode of Formation?

In 1975, Wheeler[101] returned to the idea that the Sun was not homogeneous at the moment of formation, and had a unique evolution that maintained the inhomogeneities for a long time. Wheeler also discussed the depletion of the Maxwellian distribution (Clayton's idea), and calculated that it could reduce the ^{8}B solar neutrino flux by a factor of 1/6. This was a good result because it meant that the effect, had it existed, could do the job.

Newman and Talbot[102] put forward another version of the non-homogeneous Sun, by suggesting that it might be built by matter accretion over a long time scale, in contrast with the accepted idea that the entire mass of the Sun was assembled in one shot.

---

97 Beaudet, G., JRASC **68**, 26 (1974).

98 Barnothy, J.M., Nature **252**, 666 (1974).

99 Kuchowicz, B., ICRC **6**, 220 (1977).

100 Vasilev, S.S., Academy of Sciences, USSR, Bulletin, Physical Series **39**, No. 2, 53 (1975).

101 Wheeler, J.C., 7th Texas Symposium on Relativistic Astrophysics, Dallas, Texas, 16–20 December 1974; New York Academy of Sciences, Annals **262**, 214–218 (15 October 1975), discussion p. 218.

102 Newman, M.J., & Talbot, R.J., Jr., Nature **262**, 559 (1976).

All models assumed that main sequence stars start their evolution from a fully homogeneous state because the initial phase is completely convective. Stars condense from interstellar clouds which are transparent. At a certain moment, after a significant increase in density and temperature, the embryonic star becomes opaque, and radiation transfer through it becomes the main energy loss mechanism. At the low temperatures at which the star becomes opaque, the absorption coefficient is very large and radiation cannot easily flow through and escape from the star. Consequently, all main sequence stars, according to the present theory, were convective and fully mixed. This is called the Hayashi phase, after the discoverer.[103] In 1975, Prentice[104] suggested that, contrary to the standard theory of the Hayashi phase, the initial Sun may have been inhomogeneous. Then a special transport mechanism was hypothesized to exist.

Connected to Prentice's suggestion was the hypothesis that the heavy element abundance in the core of the Sun differs substantially from what is observed on the surface (being much lower). The core of the Sun was first to form out of an interstellar cloud that was poor in heavy elements. After the formation of the core, the Sun moved in space and encountered a cloud with the observed present day amount of heavy elements. Such contrived models[105] reflect the helplessness of astrophysicists before this enigma.

9.12 New Astrophysical Processes

Various papers suggested new transport mechanisms, able to flatten the temperature gradient and reduce the central temperature. For example, mechanical energy transport.[106] Schatzman[107] suggested mechanical energy transfer by turbulent currents. However, numerical simulations indicated that the process was not viable.

9.13 Astrophysical Summary

All astrophysical solutions had a lower limit below which they could not reduce the neutrino flux observed in the Homestake experiment. The reason is simple. The calculated neutrino flux is gauged by the total energy production, and if the latter is fixed, the only option left is some interplay between the relative importance of the different reactions. This interplay does not change the total flux of neutrinos, but it

---

[103] Hayashi, C., PASJ **13**, 450 (1961).

[104] Prentice, A.J.R., A&A **50**, 59 (1976).

[105] Ely, J.T.A., BAAS **11**, 442 (1979).

[106] Beaudet, G., Sirois, A., Tassoul, M., Fontaine, G., A&A **54**, 213 (1977).

[107] Schatzman, E., Solar neutrinos and neutrino astronomy, Proceedings of the Conference, Lead, SD, 23–25 August 1984 (A86-37626 17-93); American Institute of Physics, New York (1985) pp. 69–74.

does change the contribution of the rare ^8B neutrinos, which affect most detectors, in contrast with the pp neutrinos which escape detection by most detectors.

On the other hand, physical solutions which have something to do with the properties of the neutrinos can reduce the neutrino flux on the Earth by a substantial factor, without touching the theory of stellar structure and evolution at all. So the basic question was: whose physics needs revision, the astrophysicists', or the elementary particle physicists'? The astrophysicists won this time. The solar model was actually proven right, and with it the energy source of the stars in general, and of the Sun in particular.

9.14 Suggested Elementary Particle Solutions

By 1968, physicists had discovered several groups of particles which had the same interaction and general properties like charge and spin, but different masses. The classic example is the electron and the muon. To distinguish between the particles, a new quantum number was invented and called flavor. So the electron and the muon had different flavors (and masses). Since the electron and the muon decayed via the weak force and emitted different neutrinos, as Danby et al. had demonstrated, the corresponding neutrinos were assigned different flavors too.

In 1969, there were two basic options to explain the missing solar neutrinos:[108] either the high energy neutrinos are even rarer than calculated,[109,110,111] or some previously unknown phenomenon affects all neutrinos in their transit from the solar interior to the Earth's surface.

As early as 1969, Pontecorvo and Gribov referred to the first results of Davis et al. as a negative result, i.e., no neutrinos were detected from the Sun, and proposed[112] that the discrepancy between standard theory and the first solar neutrino experiment could be due to an inadequacy in the textbook description of particle physics, rather than in the standard solar model. Gribov and Pontecorvo suggested that neutrinos might have a dual personality, oscillating back and forth between the various facets of their existence.

Particle mixing is a purely quantum concept and phenomenon. Suppose the particle is composed of two or more oscillating ingredients and can show different

[108] Bahcall, J.N., PRL **23**, 25 (1969).

[109] Ezer, D., & Cameron, A.G.W., Astrophys. Lett. **1**, 177 (1968). The authors considered the possible effect of rotation, an effect so far ignored.

[110] Shaviv, G., & Salpeter, E.E., PRL **21**, 1602 (1968). The authors discuss the possible effect of rotation and conclude that the effect is appreciable only for unacceptable values of the rotation.

[111] Iben, I., Jr., PRL **22**, 100 (1969). The author extends the Shaviv and Salpeter treatment. In PRL **21**, 1208 (1968), Iben found that the upper limit on the solar neutrino flux set by Davis, Harmer, and Hoffman places an upper limit on the Sun's initial helium abundance that is small compared with that estimated for other galactic objects. Adopting current estimates of low-energy nuclear cross-section factors, the upper limit is essentially equal to a lower bound set by demanding that the Sun be at least 4.5×10^9 yrs old.

[112] Gribov, V., & Pontecorvo, B., Phys. Lett. **28**, 493 (1969).

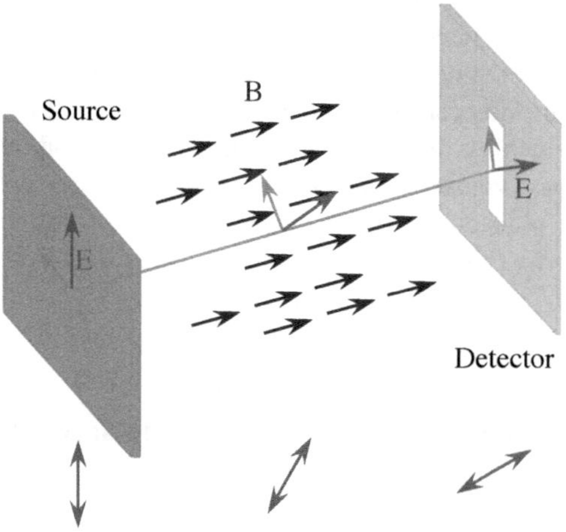

Fig. 9.7 The Faraday rotation of the polarization plane in a magnetic field. When a beam of polarized light traverses a magnetic field, the plane of polarization rotates

combinations of these ingredients to the outside world. So, depending on the moment of observation and the state of the internal mix, the 'particle' can manifest a different component. Alternatively, assume the particle has an internal property which the physicists call flavor, and suppose that the detectors are tuned to observe a definite flavor. Then if the particle oscillates between the different flavors while only one of them can be observed at any given time by a given flavor detector, the detectors will show only one type of neutrino, and consequently, a smaller than predicted total flux. Furthermore, if each flavor has a different space velocity, the result of the observation will depend on the distance between the source and the observer.

As a good example to explain the phenomenon, consider the classical phenomenon of Faraday rotation (see Fig. 9.7).[113] A plane polarized electromagnetic beam is created at the source. The electromagnetic beam is made of magnetic (orange) and electric (red) oscillating fields. Since the beam is plane polarized, it means that the electric field oscillates as marked below. As the beam propagates through a magnetic field (marked blue), the plane of polarization rotates. The longer the path through the magnetic field, the greater the rotation. Suppose the detector can only pick up a plane polarized beam in the vertical direction. Then, when the beam reaches the detector, only the vertical component of the rotated electric field can be observed. If the detector were blind to polarization, it would see no change in the intensity of the beam. Only a detector which can sense plane polarization is able to detect the change that took place in the 'structure' of the beam.

The simplest explanation of the Faraday rotation phenomenon is that the two components that make up the light propagate in the magnetic field with slightly dif-

[113] Likpin, H.J., `arXiv:hep-ph/9901399v1` (1999).

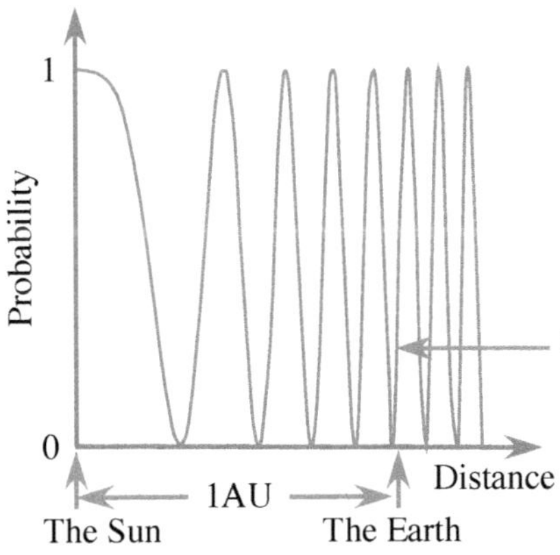

Fig. 9.8 The neutrino is born and leaves the Sun as ν_e. On the way it oscillates into another neutrino (a ν_μ or a ν_τ or both). When it has travelled the distance to the Earth, there is a probability of 0.3 that it will appear as ν_e. Note that the theory predicts the rate of oscillation to increase with distance

ferent velocities. Consequently, the mix of components which reaches the detector differs from the one which left the source.

The neutrino is born with a definite flavor and the chlorine detector detects neutrinos with this flavor. The flavor is the analog of the polarization. As the neutrino traverses the distance from the Sun to the Earth, the internal oscillations 'rotate' the flavor, and the detector 'senses' only the component of the flavor to which it is sensitive (see Fig. 9.8).

According to the suggestion of Gribov and Pontecorvo, neutrinos are produced in the Sun in a mixture of individual states. These individual states have slightly different and very small masses, rather than the zero masses attributed to them by standard particle theory. As they travel to the Earth from the Sun, the neutrinos oscillate between the easier to detect neutrino state ν_e and the more difficult to detect neutrino state ν_μ, i.e., $\nu_e \rightleftharpoons \nu_\mu$. The chlorine experiment detects only neutrinos in the easier to observe state. Neutrinos that arrive at the Earth in the state that is difficult to observe are not counted. The notation $\nu_e \rightleftharpoons \nu_\mu$ is a bit misleading. In chemical reactions in dynamic equilibrium, this notation means that, at any given pressure and temperature, the amounts of ν_e and ν_μ are fixed. However, in the neutrino case, since the two particles propagate with different velocities, the notation means that, at different distances from the source, a flavor detector will detect different amounts, and the amount oscillates with distance.

Building upon this idea, the MSW effect was discovered by Wolfenstein in 1978,[114] and by Mikheyev and Smirnov in 1985.[115] Wolfenstein discovered that, in general, if there exists an interaction through which neutrinos can change flavor (not necessarily by neutrino oscillations), this flavor change can be enhanced,

[114] Wolfenstein, L., Phys. Rev. D **17**, 2369 (1978).

[115] Mikheyev, S.P., & Smirnov, A.Yu., Sov. J. Nucl. Phys. **42**, 913 (1985).

or only be possible, if the neutrinos travel through matter. The reason is that the matter of the Sun is made solely of electrons (and nuclei), and does not contain any amounts of muons or other similar particles. The composition of the Sun is asymmetric with respect to neutrino flavors. So the assumption made is that the interaction is proportional to the number of electrons. Hence, the emitted neutrino encounters an asymmetrical material from the flavor point of view. Mikheyev and Smirnov noticed that, for specific oscillation and matter density parameters, this enhancement could even develop a resonance behavior.[116]

Neutrinos are also produced by the collisions of cosmic ray particles with other particles in the Earth's atmosphere. In 1998, the Super-Kamiokande team announced that they had observed oscillations among atmospheric neutrinos. This finding provided indirect support for the theoretical suggestion that solar neutrinos oscillate among different states.

In 1972, Bahcall et al.[117] pointed out that, if the neutrino has a mass, it could be unstable and decay on the way from the Sun to the Earth. They rejected the oscillation solution because they claimed that:

> *The oscillatory process typically leads to only a factor of 2 reduction in the terrestrially detected flux of ν_e when properly averaged over the spectrum of energies of solar neutrinos.*

This was also the opinion of Bahcall and Frautschi.[118] The observed discrepancy was $\sim 1/3$, i.e., so fewer neutrinos were detected. If the neutrino decays, it must decay into two other particles. These particles were, however, never observed.

In 1973, Trimble and Reines[119] argued that:

> *The conflict between observation and theoretical prediction of the flux of electron neutrinos from the Sun has advanced in the past year from being merely difficult to understand to being impossible to live with. We review here attempts to explore the nature of the conflict, to seek possible ways out of it, and to inquire into additional experiments that have the capability either of resolving the conflict or at least of deciding which branch of physics or astrophysics is responsible for it.*

They coined the phrase: *The Davis experiment – No SNUS is not good SNUS*. Various possibilities were examined, and like Bahcall and Frautschi, they rejected the Pontecorvo solution ($\nu_e \rightleftharpoons \nu_\mu$), which reduces the neutrinos by a factor of 2 at most. They also repeated the argument that, if the ν is unstable, the flux on the Earth could even vanish. However, this solution was rejected by an experiment carried out by Reines, Sobel, and Gurr,[120] which showed that the neutrino does not decay in at least *10^5 times the eight minutes it takes to travel from the Sun to the Earth*, and in this way:

---

[116] The interaction between the neutrino and the sea of electrons in the Sun changes the ratio of force to acceleration, as if the mass of the particle changes. Note that we are here expressing a quantum phenomenon in classical terms.

[117] Bahcall, J.N., Cabibbo, N. & Yahil, A., PRL **28**, 316 (1972).

[118] Bahcall, J.N., Frautschi, S.C., Phys. Lett. B **29**, 623 (1969).

[119] Trimble, V., & Reines, F., RMP **45**, 1 (1973).

[120] They quote the result by Reines, F., Sobel, H.W., & Gurr, H.S., PRL **32**, 180 (1974).

This mode of decay is therefore, ruled out as a possible explanation for the paucity of solar neutrinos revealed by the experiment of Davis.

In view of the final solution to the neutrino problem, it seems that this experiment contained an error.

A significant reduction could be obtained if the neutrino had a large magnetic moment,[121] so that it could be converted into an antineutrino by a large solar magnetic field.

Bagge[122] was ready to throw away the entire thirty-year-old theory of β decay when he posed the question:

Can the missing solar neutrinos be explained by a new interpretation of beta decay?

On the other hand, Chin and Stothers[123] were ready to show that Dirac's theory of mass creation is not in contradiction with the structure of the Sun.

9.15 Have All Neutrino Types been Discovered?

The only acceptable solution up until 1975 was the two-neutrino oscillation mechanism. However, this was not the end of the story. There were indications that, in the experiment $e^+ + e^- \rightarrow e^\pm + \mu^\mp +$ missing energy, where $\mu^\mp$ is a particle similar to the electron but with a mass of 206.7 times the mass of the electron, there was missing energy and momentum in such a way that an additional particle had to be assumed to exist. As Perl et al.[124] wrote in the abstract of their paper:

We have no conventional explanation for these events.

The missing particle turned out to be the τ neutrino. It was clear that this newly needed particle, the τ neutrino, was not a misidentification of a muon neutrino or an electron neutrino. Consequently, there should be a ν_e, ν_μ, and ν_τ, and, although they are similar, for example, they are all neutral, they really are different. The property which distinguishes between the neutrinos is their 'flavor'. The detectors detect flavor not mass. The mass state is a combination of flavor states, and these can oscillate. With the discovery of the third neutrino particle/state, the space of possibilities increased to include oscillations among three types of neutrinos.[125]

In 1976, Nussinov[126] examined the suggestion of Fritzsch and Minkowski[127] that neutrino mixing might account for the solar neutrino puzzle. Nussinov found that, if

[121] An elementary particle can be neutral and still possess a magnetic moment. An example is the neutron. This magnetic moment can interact with the outside world just as any magnet interacts with a magnetic field.

[122] Bagge, E., Lein. Conf. (1975) 25B.

[123] Chin, C.-W., Stothers, R., Nature **254**, 206 (1975).

[124] Perl, M.L., and 35 authors, Phys. Rev. **35**, 1489 (1975).

[125] Krauss, L., Wilczek, F., PRL **55**, 122 (1985).

[126] Nussinov, S., Phys. Lett. B **63**, 201 (1976).

[127] Fritzsch, H., & Minkowski, P., Annal. Phys. **93**, 193 (1975).

k types of neutrino are mixed, then the maximum flux reduction that can be achieved is a factor of k. Hence, if a reduction by a factor of 3 is needed, there should be just three types of neutrinos.

In 1977 and 1981,[128] Bilenky and Pontecorvo expanded the various schemes of neutrino mixing, depending of course on the number of existing neutrinos. They showed that if there were N mixing particles, the maximum reduction factor would be $N/2$, so that four particles, as Bilenky and Pontecorvo assumed, reduce the solar flux by a factor of 2, in disagreement with Nussinov's result.

Another possibility to be investigated was the idea that the rate of some critical nuclear reactions might be severely wrong. Consequently, a large campaign took place to measure all the relevant nuclear reactions. The accuracy of the various reactions improved, but no dramatic errors were discovered. Nuclear physics did not save the day.

Another set of suggested solutions was to look for strange particles which were not yet discovered, like the hypothetical elementary particle which might make up most of the matter in the Universe and which is also predicted to exist in some elementary particle theories. The possible existence of such a particle in the core of the Sun was contemplated.[129] In 1985, Krauss et al.[130] suggested cold dark matter to exist in the core of the Sun. Cold dark matter is the particle proposed as a solution to the 'dark matter problem', i.e., matter which exerts a gravitational force and whose existence is inferred from the rotation of galaxies, for example.

The hypothesis that there was a third member of the electron and muon family, as well as a third member of the ν_e and ν_μ family, flooded the literature (see for example, Pontecorvo 1971[131]). However, the tau particle was only discovered in 1978 by Perl et al. at SLAC,[132] and its decay followed the pattern of beta decay in that there was no conservation of momentum and energy. The τ particle resembles the electron (which is also called the first generation lepton) and the μ (muon which is called the second generation lepton) in many respects, but it has a different mass (in fact, 3777.5 times the mass of the electron), and consequently is called the third generation lepton. The existence of a new (third) neutrino was inferred theoretically, although it had not yet been detected directly. It was not until 2000, twenty five years after the tau lepton had been discovered, that the Fermi laboratory reported the detection of the tau neutrino. In all, 4 interactions were observed.

---

[128] Bilenky, S.M., & Pontecorvo, B., CNPPh **7**, 149B (1977); Phys. Lett. B **102**, 32 (1981).

[129] Waldrop, M.M., Science **229**, 955 (1985); Faulkner, J., Gilliland, R.L., Ap. J. **299**, 994 (1985); Kocharov, G.E., Pavlov, A.K., Cosmic ray intensity and cosmogenic isotopes. 13th Leningrad seminar on cosmophysics (1983) p. 112; Slad, L.M., Akademiia Nauk SSSR, Doklady (ISSN 0002-3264) **269**, No. 6, 1345–1349 (1983); Sur, B., & Boyd, R.N., Physical Review Letters (ISSN 0031-9007) **54**, 485–487 (4 February 1985).

[130] Krauss, L.M., Freese, K., Spergel, D.N., Press, W.H., Ap. J. **299**, 1001 (1985).

[131] Pontecorvo, B., ZhETF Pis. Red. **13**, 281 (1971), and references therein.

[132] Perl, M.L., and 15 authors, PRL **35**, 1489 (1975).

9.16 A New Generation of Neutrino Detectors

It was clear that the situation in which the Homestake experiment yielded problematic results could not be left unheeded, and new experiments were soon suggested and constructed.

9.16.1 *The Soviet–American Gallium Solar Neutrino Experiment (SAGE)*

One of the first experiments was the Soviet–American Gallium Solar Neutrino Experiment, or as it is better known, the SAGE experiment. To appreciate the magnitude of this experiment, let us note that the price of gallium is about \$ 600 per kg, and that the experiment required about 30 tons of this material. This represents about 1/6 of the world's total annual production of gallium.[133] The relevant gallium isotope (^{71}Ga) constitutes 39.892% of all the gallium. So only this gallium isotope participates in the reaction to detect neutrinos. In this respect, this is a violation of Pontecorvo's first rule: use a cheap detector. The relevant reaction is:

$$^{71}\mathrm{Ga} + \nu_{\odot} \longrightarrow {}^{71}\mathrm{Ge} + \mathrm{e}^{+} \,, \qquad \text{followed by} \quad {}^{71}\mathrm{Ge} + \mathrm{e}^{-} \longrightarrow {}^{71}\mathrm{Ga} + \nu_{\mathrm{lab}} \,,$$

where the ν_{lab} neutrino escapes from the detector. In the first step of the reaction, the gallium nucleus is in the ground state and stable. The product nucleus ^{71}Ge formed is unstable and decays in 11.43 days.

There was, however, a much more fundamental requirement for setting up such an experiment. The chlorine experiment was designed to detect an extremely rare branch of the nuclear reactions which belongs to the pp chain. If for some reason there is an error in the measured rate of the nuclear reactions, then the theoretical rate can change tremendously. But whatever happens to a rare reaction, protons must be converted into helium, and hence, even if there is an error in the set of nuclear reactions, the basic neutrinos from the conversion of protons into neutrons must be there, and these are all low energy neutrinos inaccessible to the chlorine experiment. The major advantage of the SAGE experiment is the much lower threshold, just 0.233 MeV, which means that the low energy neutrinos should in principle be detected. If these neutrinos are not detected, then it will be necessary to throw the entire theory of nuclear reactions out of the window. For this reason, despite the high price of the experiment, it was considered worthwhile.

The gallium target is kept in a liquid form (the melting point is at 29.8°C), and each measurement starts by mixing 700 μg of natural Ge carrier in each 7 ton module. After about 4 weeks, the Ge carrier and any Ge atoms produced by solar neutrinos are chemically extracted from the gallium using a complicated procedure. The difficulty of the experiment is striking: the Standard Solar Model (SSM) predicts a

[133] United States Geological Survey Mineral Resources program, 2005.

production rate of 1.2 atoms per day in 30 tons of Ga! Therefore, taking into account the one-day delay between the end of the exposure and the start of counting, as well as the chemical extraction and counting efficiencies, only about 4 atoms are expected to be detected after a 4 week exposure of 30 tons of gallium.

The best fit value for the entire 1990–1993 period was:

$$\phi^{\mathrm{SAGE}} = (72 \pm 17)\ \mathrm{SNU}\,.$$

9.16.2 GALLEX

The GALLEX detector, which is located in the Gran Sasso Underground Laboratories, Italy, detects solar neutrinos through the same reaction as SAGE, but in a 100-ton gallium chloride target solution. Counting of the resulting atoms is performed after extraction, after every 3–4 weeks of exposure.

Fig. 9.9 The GALLEX laboratory. (The GALLEX home page)

The GALLEX result for the solar neutrino flux from the data collected between May 1991 and October 1995 was:[134] $\phi^{\text{GALLEX}} = (69.7 \pm 8)$ SNU, in nice agreement with the SAGE result.

The Gallex experiment terminated in 1997, and a new experiment based on the same technology replaced it, the Gallium Neutrino Observatory (GNO). The whole existing experimental setup was modernized.

9.16.3 Super-Kamiokande

The new Super-Kamiokande (Super Kamioka Nucleon Decay Experiment) detector went into operation in April 1996. It is the follow-up of Kamiokande, which stopped taking data in February 1995. Super-Kamiokande is a 50 000 ton imaging water Cerenkov detector (Kamiokande was just 4500 tons) which detects solar neutrinos through the reactions:

$$\nu_\odot + \mathrm{e}^- \longrightarrow \nu'_\odot + \mathrm{e}^- \,.$$

The detector identifies the Cerenkov light[135] radiated by the recoiling electron in the water. The offline energy threshold for detection of solar neutrinos gradually evolved for Kamiokande from 9.3 MeV in January 1987 down to 7.0 MeV from November 1991 on. This implies that Super-Kamiokande can only detect the rare ^{8}B neutrinos. We should note here that the triggering threshold is generally lower (5 MeV in the last phase of Kamiokande), but, for reasons to do with the background, this threshold is raised at the analysis level. One major background source stems from radon contamination of the water, making water purification an essential component of detector operation.

Kamiokande was the first genuine neutrino telescope because it provided the direction from which each neutrino arrives. All other experiments just count the total number of neutrinos arriving in it, irrespective of the direction from which they came.

The observed flux[136] can be directly expressed in a neutrino flux, since only neutrinos contribute:

$$\phi^{\text{Super-Kamiokande}} = 2.51^{+0.32}_{-0.31} \times 10^6 \ \text{neutrinos/cm}^2\text{sec}\,.$$

Super-Kamiokande provides us with two more important results. No significant day–night variation of the neutrino flux is observed:

$$\phi^{\text{Super-Kamiokande}}(\text{day}) = 2.30^{+0.35}_{-0.34} \times 10^6 \ \text{neutrinos/cm}^2\text{sec}\,,$$

---

[134] GALLEX Collaboration, W. Hampel et al., Phys. Lett. B **388**, 384 (1996).

[135] When an electron moves in water at speeds greater than the speed of light in water (about 0.75 of its speed in vacuum), it emits a typical light called Cerenkov radiation.

[136] KAMIOKANDE Collaboration, Y. Fukuda et al., Phys. Rev. Lett. **77**, 1683 (1996).

Fig. 9.10 An inside view of the Super-Kamiokande experiment. By permission from the Kamioka Observatory, Institute for Cosmic Ray Research (ICRR), University of Tokyo

$$\phi^{\text{Super-Kamiokande}}(\text{night}) = 2.75^{+0.41}_{-0.40} \times 10^6 \text{ neutrinos/cm}^2\text{sec}.$$

The second point is that the observed electron energy distribution corresponds to the expected one. Last but not least, the full Kamiokande data nearly cover a full solar cycle, so as to enable a search for a possible correlation between sunspot activity and neutrino flux. No correlation could be found. The implication is that the nuclear reactions which are the source of the neutrinos are not affected by solar magnetic activity.

9.17 Sudbury Neutrino Observatory (SNO)

The Sudbury Neutrino Observatory is another neutrino telescope because, along with the detection of the neutrino, it also identifies the direction from which it came. The experiment is a major expansion of the Kamiokande experiment idea. Moreover, along with the ability to detect the ν_e, it is able to detect ν_μ and ν_τ. Hence, if the neutrino oscillates and the ν_e hides in the form of other neutrinos, this experiment can detect it. The size of the experiment is about equivalent to a ten-storey high building, placed 2 kilometers underground in INCO's Creighton Mine near Sudbury, Ontario.

The SNO detector consists of 1000 tons of ultra-pure heavy water[137], which in turn is surrounded by ultra-pure ordinary water in a giant 22-meter diameter by 34-meter high cavity. The heavy water container is surrounded with a 17-meter diameter sphere containing 9456 light sensors or photomultiplier tubes, which detect tiny flashes of light emitted as neutrinos are scattered in the heavy water. The flashes are recorded and analyzed to extract information about the neutrinos causing them. At a detection rate on the order of 10 per day, many days of operation are required to provide sufficient data for a complete analysis. But here there is no need to extract a few atoms out of an astronomical number of atoms, because each neutrino is seen as it reacts with the heavy water. This is a major advantage.

The result[138] for neutrinos undetected by previous experiments was:

$$\phi(\nu_{\mu\tau}) = (3.69 \pm 1.13) \times 10^6 \text{ neutrinos/cm}^2\text{sec},$$

while the total solar neutrino flux was:

$$\phi(\text{all } \nu) = (5.44 \pm 0.99) \times 10^6 \text{ neutrinos/cm}^2\text{sec},$$

which agrees nicely with the prediction of the standard solar model. The confirmation that ν_e released in nuclear reactions taking place in the core of the Sun transform into neutrinos of another type is very important for a full understanding of the Universe at the most microscopic level. This transformation of neutrino types is not allowed in the Standard Model of elementary particles.

[137] The cost of heavy water is about US$ 1100 per kg. However, the heavy water is on loan from the Canadian Energy Commission.

[138] SNO collaboration, Phys. Rev. Lett. A **89**, 1301A (2002).

Fig. 9.11 An outside view of the SNO experiment. Note the size of the people. (The SNO home page)

9.18 Summary of Solar Neutrino Experiments

Table 9.3 summarizes the present day results from the neutrino experiments. The first four solar neutrino experiments observe a deficit of solar neutrinos compared to the predictions of the standard solar model. More significantly, any two of the three classes of experiments indicate that the largest suppression is in the middle of the spectrum (the ^{7}Be line and the lower energy part of the ^{8}B neutrino energy distribution). Astrophysical and nuclear physical explanations generally predict that the largest suppression should be of the ^{8}B neutrinos (since they are made after and from ^{7}Be), and that the energy distribution of the ^{8}B neutrinos is not significantly distorted. On the other hand, the SNO result agrees with the predicted total

Table 9.3 The results of the solar neutrino experiments

Experiment	Measured flux in 10^6 neutrinos/cm^2 s	Ratio (observed)/(predicted)	Threshold MeV	Active
Homestake	2.56 ± 0.32 SNU	0.273 ± 0.021	0.814	1965–1995
SAGE	75 ± 10	0.526 ± 0.089	0.233	1990–2006
GALLEX	78 ± 11	0.509 ± 0.089	0.233	1991–1997
Super-Kamiokande	2.35 ± 0.1	0.379 ± 0.034	5.5	1996–
GNO	65.8 ± 14	0.429 ± 0.11	0.233	1998–
SNO	$\phi(\nu_{\mu\tau}) = 3.69 \pm 1.13$ $\phi(\nu_{^8\mathrm{B}}) = 5.44 \pm 0.99$	0.83 ± 0.15	6.75	1999–

^{8}B neutrino flux. Thus, unless most of the experiments are wrong or exhibit very large statistical fluctuations, a solution based on neutrino properties, such as MSW matter-enhanced conversions or vacuum oscillations is favored. The next generation of neutrino experiments should be able to confirm or falsify these ideas, essentially free of astrophysical uncertainties. On the other hand, with or without new neutrino properties, the solar neutrinos are and will be an important probe of the solar core, complementing data provided by helioseismology.

In summary, the solar neutrino experiments have taught us the following lessons:

Astrophysical

- The theory of stellar structure and evolution is confirmed. No new quantitative astrophysical information about the structure of the Sun has been obtained. From this point of view, it is a great victory for theoretical astrophysics, whose fundamental premises have been confirmed.
- The set of nuclear reactions taking place inside the Sun has been correctly identified. At most 3% of the solar energy comes from CNO.

Elementary Particle Physics

- The neutrinos appear to have masses, and neutrino oscillation takes place. The resolution of the solar neutrino problem is the MSW effect combined with oscillations between 2 or 3 neutrinos.
- About 1/3 of the ^{8}B neutrinos survive as ν_e on their way from the core of the Sun to the detector on Earth. The remaining 2/3 convert into ν_μ or ν_τ, or ν_{sterile} (depending on the theory).[139]

[139] A sterile neutrino is a hypothetical neutrino that does not interact via any of the fundamental interactions of the Standard Model of elementary particles except gravity. The irony is that the sterile neutrino plays a similar role to neutrinos in β-decay, but in neutrino reactions. If the sterile neutrino exists, then cosmological data limits its mass to < 0.23 eV.

Open Questions

- Do neutrinos violate the combined charge and parity symmetry?
- Have we detected all existing neutrinos?
- Are there 'sterile' neutrinos which are formed and never interact, like a kind of 'hole' in the mass-energy conservation law?
- Do neutrinos have unexpected or exotic properties?
- What can neutrinos tell us about new physics beyond the Standard Model?

The answers to the above questions lie at higher energies, and at earlier times, closer to the Big Bang.

9.19 Helioseismology: Independent Confirmation of Stellar Structure

Alongside the saga of the solar neutrino experiments, another new technology has been developed, the technology of helioseismology. The suspicion that the solar surface is not at rest was raised by Plaskett[140] as early as 1954. Plaskett actually detected the fact that bright regions in the Sun oscillate vertically with an amplitude of about 0.5 km/s. A year later, Hart[141] discovered the same phenomenon, and argued that the total energy in the motions on the surface of the Sun is not negligible relative to the radiative flux which flows from the Sun into space. A simple calculation shows that density of radiative energy is about 7.4 erg/cm^3, while the density of kinetic energy is 15.4 erg/cm^3, almost twice as much. Somehow gravitational energy is converted into kinetic energy and drives the oscillations.

The confirmation and establishment that the solar surface oscillates came in 1962 when Leighton, Noyes, and Simon[142] developed special techniques and instruments to investigate the velocities on the solar surface. The latter is not quiet, but moves all the time. While investigating the velocities, they discovered that patches of the solar surface oscillate with a period of about 5 minutes. At roughly the same time, Evan and Michard[143] discovered that a large number of points in the Sun oscillate vertically with a period of 260 ± 30 s, which is close to the period discovered by Leighton et al.

Leighton et al. established finally that there are regions over the surface of the Sun which oscillate with a period of 296 s or about 5 minutes. The effect was small. As calculated by Leighton et al., an element on the surface of the Sun moved hardly 20 km up and down, not even 3 parts in 100 000 of the radius of the Sun. The typical acceleration is not more than 5–6% of the solar gravitational acceleration.

---

140 Plaskett, H.H., MNRAS **114**, 251 (1954).

141 Hart, A.B., MNRAS **116**, 38 (1956).

142 Leighton, R.B., Noyes, R.W., & Simon, G.W., Ap. J. **135**, 474 (1962).

143 Evan, J.W., & Michard, R., IAU XI Gen. Assem. Berkeley, Calif. 1961.

All the above observers attributed the oscillations to the convective zone in the Sun. But research by Ulrich in 1970,[144] Leibacher and Stein in 1971,[145] and also by Deubner in 1975[146] and Claverie et al. in 1979,[147] indicated that this is not the correct explanation. The emerging idea was that the oscillations were due to the addition (superposition) of many resonant modes of global solar oscillation. In other words, the entire Sun oscillates in many ways which combine to yield the observed 5 minute oscillations.

In 1977, Gough[148] and Ulrich and Rhodes[149] showed how one can use the oscillation to infer the internal structure of the Sun. Dramatic progress has been made since these first steps were taken, to create the field of research known today as helioseismology. Currently, the observed oscillations are used in the same way as seismologists use earthquakes to understand the innards of the Earth, but here to investigate the detailed structure of the Sun. It should be said that helioseismology has provided information about the interior of the Sun to the point that we know its internal structure much better than we know the internal structure of the Earth. What Eddington said in 1920 in his presidential address (see Sect. 4.27) has really come true today.

What is important for our discussion here is that helioseismology has confirmed the fundamentals of stellar evolution and structure, while calling upon additional small effects, like diffusion. Independently of the solar neutrino experiments, this constitutes a confirmation of the concepts applied to understand the solar interior and evolution. The two independent disciplines, viz., the solar neutrino investigations and helioseismology, have converged to the same picture of the solar interior, in complete accord with our understanding of stellar evolution and structure.

9.20 Some Reflections

Neutrino-related topics have won a long list of Nobel Prizes:

- In 1957, the prize was awarded jointly to Chen Ning Yang and Tsung-Dao Lee *for their penetrating investigation of the so-called parity laws which has led to important discoveries regarding the elementary particles.*
- In 1979, the prize was divided equally between Sheldon L. Glashow, Abdus Salam, and Steven Weinberg *for their contributions to the theory of the unified weak and electromagnetic interaction between elementary particles, including inter alia the prediction of the weak neutral current.*

[144] Ulrich, R.K., Ap. J. **162**, 993 (1970).

[145] Leibacher, J., & Stein, R.F., Astrophys. Lett. **7**, 191 (1971).

[146] Deubner, F.L., A&A **44**, 371 (1975).

[147] Claverie, A., et al., Nature **282**, 691 (1979).

[148] Gough, D.O., Proc. IAU Coll. No. 36, eds. Bonnet & Delache (1977) p. 3.

[149] Ulrich, R.K., & Rhodes, E.J., Ap. J. **218**, 521 (1977).

- In 1980, the prize was divided equally between James W. Cronin and Va L. Fitch *for the discovery of violations of fundamental symmetry principles in the decay of neutral K-mesons*. This discovery was relevant to the theory of neutrino oscillations.
- In 1988, the prize was awarded jointly to Leon M. Lederman, Melvin Schwartz, and Jack Steinberger *for the neutrino beam method and the demonstration of the doublet structure of the leptons through the discovery of the muon neutrino*. The Nobel Prize was awarded for the discovery of the second type of neutrino before the discovery of the first neutrino was awarded the prize.
- In 1995, the prize was awarded for pioneering experimental contributions to lepton physics, with one half to Martin L. Perl *for the discovery of the tau lepton*, and and the other half to Frederick Reines *for the detection of the neutrino*. Reines and Cowan (1919–1974) discovered the first neutrino (the ν_e), but the prize was awarded after Cowan's death. Perl's discovery implied the existence of a third neutrino, but 20 years of extensive effort were needed for its experimental confirmation. Pontecorvo, who contributed so much to neutrino physics and ideas, was still alive.
- In 1999, the prize was awarded jointly to Gerardus 'T Hooft and Martinus J.G. Veltman *for elucidating the quantum structure of electroweak interactions in physics*.
- In 2002, the prize was awarded with one half jointly to Raymond Davis Jr. and Masatoshi Koshiba *for pioneering contributions to astrophysics, in particular for the detection of cosmic neutrinos*, and the other half to Riccardo Giacconi *for pioneering contributions to astrophysics, which have led to the discovery of cosmic X-ray sources*.

The Nobel Prize for the solar neutrino experiments was awarded after the death of Pontecorvo (1913–1993). It seems that the Nobel committee could not decide whether the theoretical explanations of the neutrino paucity in Davis' experiment were correct and worth a Nobel recognition.

Some names involved in neutrino physics got the Nobel Prize for other discoveries:

- In 1938, Enrico Fermi was awarded the prize *for his demonstrations of the existence of new radioactive elements produced by neutron irradiation, and for his related discovery of nuclear reactions brought about by slow neutrons*, but not for his theory of β-decay.
- In 1945, Wolfgang Pauli got the prize *for the discovery of the exclusion principle, also called the Pauli principle*, and not for his 'unofficial' suggestion of the existence of a new particle which later became the various different types of neutrino.

Fermi's discoveries came after Pauli's, but he got the prize before him. No doubt, both Fermi and Pauli could have been attributed more than one Nobel Prize.

9.21 Further Implications

The problem of the solar neutrino turned from an astrophysical problem into a problem of elementary particle physics with far-reaching consequences, some of which have returned to astrophysics. The neutrino oscillation demonstrated for the first time that the solar neutrino problem has implications for the existence of 'new physics', i.e., physics that goes beyond the Standard Model of elementary particles.[150] New phenomena are expected somewhere around energies of 10^{11}–10^{15} GeV, a scale which is well below the Planck scale of 10^{19} GeV.

Neutrino oscillations provide a possible opportunity to search for charge and parity violation, which may eventually explain why our Universe, which was born completely symmetrical between matter and antimatter, contains only matter today. Studies are under way on 'neutrino factories', to provide neutrino beams that will allow a search for CP-violating effects in the oscillations of neutrinos and antineutrinos.

9.22 Closing the Circle: Neutrino Geology

After the discovery of the neutrinos from the core of the Sun, the next challenge is to observe the neutrinos emitted by the β decays in the interior of the Earth. Will it be possible to repeat the solar experience with the Earth? Will we be able to use the neutrinos emitted by the β decays?

The basic decays in the Earth are as follows:

$$^{238}\mathrm{U} \longrightarrow {}^{206}\mathrm{Pb} + 8\,{}^{4}\mathrm{He} + 6\mathrm{e}^- + 6\bar{\nu} + 51.7\ \mathrm{MeV} \quad (0.95\ \mathrm{erg/g\,s})\,,$$

$$^{232}\mathrm{Th} \longrightarrow {}^{208}\mathrm{Pb} + 6\,{}^{4}\mathrm{He} + 4\mathrm{e}^- + 4\bar{\nu} + 42.8\ \mathrm{MeV} \quad (0.27\ \mathrm{erg/g\,s})\,,$$

$$^{40}\mathrm{K} + \mathrm{e}^- \longrightarrow \begin{cases} {}^{40}\mathrm{Ar} + \nu + 1.513\ \mathrm{MeV}\ (11\%) & (0.36\ \mathrm{erg/g\,s}) \\ {}^{40}\mathrm{Ca} + \bar{\nu} + \mathrm{e}^- + 1.321\ \mathrm{MeV}\ (89\%) & (3.3\ \mathrm{erg/g\,s})\,. \end{cases}$$

The detection of geo-neutrinos, as they are called, will allow a direct and global measurement of the actual abundances of uranium, thorium, and potassium, and provide important information for discriminating between different models for heat production and, more generally, for the formation and evolution of the Earth.

So we are flooded with neutrinos from the Sun and antineutrinos from the Earth. Is there really any chance of detecting them? In a recent paper, Fiorentini et al.[151] assessed the extent to which present day detectors would be able to detect the anti-

150 Gonzales-Garcia, M.C., & Nir, Y., RMP **75**, 345 (2003).

151 Fiorentini, G., Lissia, M., Mantovani, F., & Vannucci, R., Nuc. Phys. B **145**, 170 (2005), A brief review of geo-neutrinos.

neutrinos from the Earth. The predictions are very positive, and we may expect the beginning of a new era in which the exact distribution and quantities of the various radioactive elements will be mapped. The new discoveries are sure to open up new avenues in our understanding of our planet.

The particle which helped us confirm the energy source of the Sun and provide an explanation for the age of the Sun will in the near future help us confirm the radioactive heat source in the Earth, and explain why all previous calculations of the age of the Earth, beginning with Fourier, have failed.

Index

G. Shaviv, *The Life of Stars*, DOI 10.1007/978-3-642-02088-9_BM2,

GPSR Compliance

The European Union's (EU) General Product Safety Regulation (GPSR) is a set of rules that requires consumer products to be safe and our obligations to ensure this.

If you have any concerns about our products, you can contact us on ProductSafety@springernature.com

In case Publisher is established outside the EU, the EU authorized representative is:

Springer Nature Customer Service Center GmbH
Europaplatz 3
69115 Heidelberg, Germany

Batch number: 10682187

Printed by Printforce, the Netherlands